Visual Detection of Motion

This book is dedicated to the memory of Werner Reichardt. His masterful combination of psychophysics, neurophysiology and mathematical modelling provided one of the cornerstones for contemporary research on visual detection of motion.

Visual Detection of Motion

edited by

Andrew T. Smith
University of London

and

Robert J. Snowden
University of Wales

ACADEMIC PRESS
Harcourt Brace & Company, Publishers
London · San Diego · New York
Boston · Sydney · Tokyo · Toronto

ACADEMIC PRESS LIMITED
24–28 Oval Road
LONDON NW1 7DX

U.S. Edition Published by
ACADEMIC PRESS INC.
San Diego, CA 92101

This book is printed on acid-free paper

A catalogue record for this book is available from the British Library

ISBN 0–12–651660–X

Typeset by J&L Composition, Ltd, Filey, North Yorkshire
Printed in Great Britain by The University Printing House, Cambridge

Contents

Contributors

Thomas D. Albright   The Salk Institute for Biological Studies, La Jolla, CA 92037, USA

Franklin R. Amthor   Department of Psychology and Neurobiology Research Center, University of Alabama at Birmingham, Birmingham, AL 35294, USA

Julie R. Brannan   Department of Visual Science, State University of New York, State College of Optometry, 100 E. 24th Street, New York, NY 10010, USA

Myron L. Braunstein   School of Social Sciences, University of California, Irvine, Irvine, CA 92717, USA

Bruce Cumming   Department of Physiology, University of Oxford, Parks Road, Oxford OX1 3PT, UK

Norberto M. Grzywacz   Smith-Kettlewell Eye Research Institute, 2232 Webster Street, San Francisco, CA 94115, USA

Julie M. Harris   Smith-Kettlewell Eye Research Institute, 2232 Webster Street, San Francisco, CA 94115, USA

Laurence R. Harris   Department of Psychology, York University, 4700 Keele Street, Toronto, Ontario M3J 1P3, Canada

Michael G. Harris   Cognitive Science Research Centre, School of Psychology, University of Birmingham, Birmingham B15 2TT, UK

Richard J. Krauzlis   Laboratory of Sensorimotor Research, National Institutes of Health, Building 49, Room 2A–50, Bethesda, MD 20892, USA

Nikos K. Logothetis Division of Neuroscience, Baylor College of Medicine, Texas Medical Center, Houston, TX 77030, USA

George Mather Department of Experimental Psychology, University of Sussex, Brighton BW1 9QG, UK

Suzanne P. McKee Smith-Kettlewell Eye Research Institute, 2232 Webster Street, San Francisco, CA 94115, USA

Andrew T. Smith Department of Psychology, Royal Holloway, University of London, Egham, Surrey TW20 0EX, UK

Robert J. Snowden School of Psychology, University of Wales College of Cardiff, PO Box 901, Cardiff CF1 3YG, UK

Gene R. Stoner The Salk Institute for Biological Studies, La Jolla, CA 92037, USA

Scott N. J. Watamaniuk Smith-Kettlewell Eye Research Institute, 2232 Webster Street, San Francisco, CA 94115, USA

Douglas W. Williams Rockefeller University, Neuroscience Institute, 1230 York Avenue, New York, NY 10021, USA

Hugh R. Wilson Visual Sciences Center, University of Chicago, 939 E. 57th Street, Chicago, IL 60637, USA

Preface

Visual detection of motion refers to the way in which the brain detects and uses information arising from the motion of the images that are cast on the retina of the eye as we go about our daily business. The last 15–20 years have seen an explosion of activity in the area of vision research in general and visual detection of motion in particular. The explosion has encompassed several of the traditionally-defined disciplines, including psychology, physiology and computing, and has fostered the development of links among these disciplines of a depth and degree of success that occurs only rarely in science. There are, perhaps, two reasons why relatively large numbers of scientists have recently chosen to work in this field. The first is simply the realization that vision, though usually experienced as effortless, is one of the most complex, as well as important, tasks tackled by the brain, occupying about a third of the cerebral cortex. The second and perhaps more fundamental reason is that many students of the brain recognize that vision, along with the other sense systems, offers us the prospect of establishing general principles of information processing in the brain. The hope is that the principles so established may prove to be applicable to other aspects of brain function that are at present somewhat intangible. Within the area of vision research, the study of motion perception has developed particularly rapidly.

Our aim in compiling this book has been to provide a concise but informative summary of our current understanding of the field of visual detection of motion. We chose to solicit contributions to the book, rather than writing it entirely ourselves, because we believe that the deepest insights into, and the most accurate and complete accounts of, any given area can be provided by those who have devoted years of their lives to research in that area. We have worked hard to achieve some uniformity of style, while recognizing that this aim is rarely if ever fully achieved in an edited volume, so as to provide a volume which will function as a textbook for the newcomer to the field, as well as a reference work for established researchers to dip into according to their interests.

The brief given to the contributors was to provide a summary of our knowledge of one particular specified aspect of motion detection, emphasizing the complementarity of the approaches offered by the different disciplines. We allowed them to articulate their own views, theories and empirical findings if they so wished, but not at the expense of providing a balanced overview. We are deeply grateful to our contributors who, by and large, heeded the brief we gave them, furnished their contributions promptly and responded to our suggestions for changes to their drafts. We are pleased with the resulting book, and hope that it will prove useful to others.

Part 1
Introduction

1

Motion Detection: An Overview

Andrew T. Smith[1] and Robert J. Snowden[2]

[1]University of London and [2]University of Wales College of Cardiff, UK

1 WHY STUDY THE VISUAL DETECTION OF MOTION?

The prospect of participating in a high-speed Formula One or Indy motor race is not one which all of us would relish. The car must be negotiated at high speed through a tortuous circuit, not alone but in the company of numerous other cars whose movements are only partially predictable, separated from disaster only by judgements of visual motion of exquisite accuracy and lightning speed. The fact that some gifted members of our species can perform such feats pays high tribute to the abilities of our brains to calculate and use motion information. Predicting the future position of an object from information about the current movement of the object is clearly a most useful skill – crossing a road in busy traffic, or even pouring a drink, would be a major problem without such a skill. Indeed, in rare cases where a person loses the ability to see motion as a result of damage to the brain (Zihl *et al.*, 1983), these tasks take on a difficulty which those of us with intact motion-detection systems find hard to appreciate.

Many animals have evolved so that their appearance is similar to that of the world in which they dwell. Thus a moth resting on the bark of a tree may be indistinguishable from other parts of the bark to the inspection of possible predators – it is camouflaged. This situation changes drastically if the moth moves. Effective camouflage is now broken and the moth can readily be detected by a predator. A tiger hiding motionless in tall grass faces the inverse problem if the wind blows so that the grass moves. That master of camouflage, the chameleon, has incorporated this knowledge into its behavioural repertoire. Members of some species will only make movements towards their prey when the wind blows and their own motion is disguised by that of the surrounding foliage. We see then that the relative motion between different areas of an image can be used to segment the visual world into figure and ground relationships. Indeed our ability to define and distinguish forms using only motion information can be surprisingly accurate and indeed comparable to our performance using luminance information (Regan, 1989).

VISUAL DETECTION OF MOTION
ISBN 0–12–651660–X

The uses of information derived from motion do not stop here. Rogers and Graham (1979) devised a display in which the movements of the elements on a screen were yoked to the subject's head movements. When the elements were moved as if the subject was looking at a sheet of corrugated iron (of course the screen itself was flat) the subjects reported a vivid sensation of depth. The kinetic depth effect (Wallach & O'Connell, 1953), in which the three-dimensional structure of an object can be perceived in a moving two-dimensional projection (but not in its stationary image) is further evidence for the importance of motion in providing us with depth information. There are other uses of motion information, just as valuable as those outlined above, such as moving our eyes to keep an image in the region of the visual field where acuity is greatest, or sensing our position in space, and others to which we have not referred. Some of them are eloquently summarized by Nakayama (1985) and many are presented in the succeeding chapters of this book in a level of detail befitting their importance.

The study of motion has produced a rich phenomenology. Surely, few who have witnessed it fail to be impressed by the motion aftereffect, also known as the waterfall illusion (the illusory perception of motion in the opposite direction to one which has just been observed). Addams (1834), upon observing the rocks moving upwards after having been staring at the Falls of Foyers at Loch Ness, was driven to remark 'I saw the rocky surface as if in motion upwards, and with an apparent velocity equal to that of the descending water, which the moment before had prepared my eyes to behold this singular deception.' We are equally impressed when the moon appears to rush past the clouds instead of vice versa (the induced effect – Duncker, 1929, 1938), or we are fooled into thinking our journey has begun when in fact it is the train on the adjacent track which has begun to pull out of the station. The entertainment trade has begun to realize the importance and power of correctly stimulating our motion processing systems. Large-screen movies provide more compelling sensations of movement by stimulating the motion detectors in peripheral parts of our field of view, which would otherwise be providing contradictory signals. Although less well known, the auto-kinetic effect (the apparent wandering of a stationary light on a blank background (e.g. Gregory & Zangwill, 1963) is also most impressive and has even been used by social psychologists as a means of testing how easily a person can be influenced by the judgement of others (Sherif, 1937). Phenomena like these cannot help but nag at the minds of scientists. Why should our perceptual system think that it is the moon that is moving when we know from learning that it must be the clouds? How can it be that large rocks can suddenly appear to defy gravity?

The visual detection of motion is now one of the flagships of both psychology and neuroscience. Phenomena of the type described in the previous paragraph have been subjected to a tri-service bombardment from psychophysical, neurophysiological and computational studies in an attempt to lay bare the processes by which the changing pattern of illumination upon our retinae leads to effective behaviour. We have now been able to stimulate a tiny cluster of cells in a certain region of the brain and so change the perceptual behaviour of a primate (Salzman *et al.*, 1990). Isolating aspects of perception in this way is no mean feat and owes much to the approach of combining the resources of the various disciplines. This in turn allows us to ask fundamental questions such as: how do cells interact to produce perceptual phenomena? How can previous experience alter a perception? What happens if there is damage to the motion areas of the brain? What happens if this part of the brain is starved of its inputs? Clearly these questions are ones whose answers have major implications that extend way beyond those for motion perception. The advanced state of this discipline ensures that its voice is likely to be most influential in the coming years.

Finally, the world in which we live is, for good or bad, slowly being replaced by a simulated world. For entertainment we have no need to go to a play or to a motor race: we can bring these events into our simulated theatre or racing circuit with considerably more convenience. In the cockpit of the fighter aircraft, the danger posed by laser weapons may soon mean that the windows will be boarded up and all visual information transmitted into the helmet of the pilot. Motion information is critical to our everyday behaviour and may mean the difference between life and death in some situations. Understanding the motion processing system will allow us to design displays and simulations to maximize both performance and enjoyment. For some of us, the realistic simulation of a few laps in an Indy car travelling at 150 miles/hour is indeed a prospect to be relished.

2   A BRIEF HISTORY OF RESEARCH ON VISUAL DETECTION OF MOTION

To our minds, the starting point for research in motion perception is provided by the work of Exner (1888). He demonstrated that motion could be perceived from two stationary images (sparks of electricity in his case) presented in quick succession – a fact exploited by television and 'movies'. This had been known for some time previously (see Boring, 1942). However, Exner's great insight was that this perception of movement could be elicited from two sparks that were so close together in space that they could not be distinguished. Under these conditions it seems impossible that the observer could infer (consciously or unconsciously) motion from a knowledge of position and time. It therefore follows that motion perception must be a sensation in its own right, not one derived from a sense of position and time. Exner appreciated the importance of this point, and suggested that there must be elements within the brains of animals that are sensitive to the properties of image motion (direction and velocity). He also hinted at a specialized area within the brain dedicated to motion processing. His prediction pre-dates our confirmation of these notions by 50 to 100 years! What is more, Exner put forward a theory of how a cell could become directionally selective. Figure 1 reproduces his model. The points a–f represent inputs from retinal elements. S, E, It and If represent summating points which eventually project to C the 'organ of consciousness' as well as to the eye muscles. The model is very simple. The time taken for a signal to propagate to any given summator cell is proportional to the distance from the input element to the summator. Imagine a stimulus that moves across the retina such that it stimulates a, b, c successively. The signals arriving at S or at If would be spread out in time by the same amount of time that it takes the stimulus to move from a to b to c, etc. The signal arriving at It would be even more spread out because a is nearer than c to It. However, at summating element E it is possible (depending on the speed of motion) that the delay in the signal propagation exactly matches the time it takes for the stimulus to pass from element to element. In this case

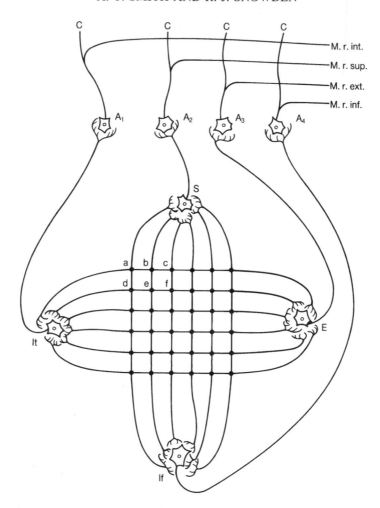

Figure 1   Exner's (1888) model of directionally selective movement detectors. The points a–f represents inputs from retinal elements. S, E, It and If represent summating points which eventually project to C, the cortex or 'organ of consciousness' as well as to the eye muscles. M. r. ext., etc. are the nuclei of the exterior eye muscles (only four are represented: externus, internus, superior and inferior). See text for further details.

all signals arrive together at E, providing a strong signal for motion in this direction (a to c). The idea of comparing the signals from two or more retinal points with an appropriate delay is still fundamental in much of our thinking about the initial extraction of motion information.

Exner's demonstrations of apparent motion led to a large body of research on the topic, mainly cataloguing the effects of duration, stimulus separation, luminance dependency, and other factors, culminating in statements or laws that governed the phenomena (e.g. Korte, 1915). However, much of this work was done in a theoretical vacuum and little further insight emerged as to how or why these effects occur. The simple idea that movement was seen despite the images presented being stationary was, however, a major influence in starting the Gestalt school of psychology (e.g. Wertheimer, 1912).

While there was considerable interest in the phenomenon of apparent motion and in measuring the absolute sensitivity to motion, there was surprisingly little interest in understanding speed discrimination and perception. This was rectified by Brown (1931) who measured perceived velocity for a variety of patterns. He was able to show that perceived velocity was influenced by a considerable number of factors such as the size of the visual field in which the movement occurred, whether the movement occurred upon a patterned or blank background, and whereabouts on the retina the stimulus was placed. His reasoning represents a significant paradigm shift for the study of motion perception. Until this point the sensory organs were regarded as mechanisms for faithfully measuring the state of the real world. The only interesting phenomena were those that did not correspond to the 'real world' – the dominant paradigm was the study of phenomena such as apparent motion, induced motion, the waterfall effect and the autokinetic effect. The job of the sensory researcher seemed to be to explain these effects as 'illusory' deviations from the correct perception (that of the 'real world'). Brown's results seemed incompatible with such an approach. Each experimental manipulation gave a different phenomenal velocity, so it was impossible to think of one as a 'correct' percept and the others as illusions that needed to be explained. His work made it possible to shift the emphasis to the factors that influence velocity and to the perhaps more important question of how any velocity might be calculated and represented in the brain. But what techniques could give insights into such a question?

Recording from single cells or fibres really commenced with the work of Adrian and colleagues (Adrian, 1928) and soon led to the idea of a 'receptive field' – the area of the skin innervated by a single fibre. This offered an analogy and a technique by which vision might be explored. The challenge was soon taken up (Hartline, 1938) and results began to emerge concerning the early processes of vision – the notions of receptive fields, lateral inhibition and adaptation all became well established firstly in lower species (particularly the horseshoe crab) and then later in mammals (Kuffler, 1953). Could there be cells whose receptive fields were specialized for the detection of motion? The answer to this was a resounding 'yes'. Barlow (1953) showed that some cells in the frog's retina respond to spots moving through their receptive fields. A little later, more detailed experimentation and theorizing were conducted by Werner Riechardt and colleagues in insects (Hassenstein & Reichardt, 1956; Reichardt, 1961). Reichardt's work stands as an exemplary model of sensory research. From a series of experiments, commencing with behavioural studies of the turning responses of a beetle as its world was systematically moved, and including much research on how responses from two ommatidia interact in the brain of the fly, these researchers suggested a model of motion detection. The model involves multiplying the inputs from two spatially separated units with a small delay introduced into one of the inputs, producing a directionally biased response in the receiving unit. This stage is followed by subtracting the responses of units with opposite directional biases to produce a unit that is directionally selective. This model has been put to the most rigorous tests over the passing years, has been refined and applied to other species (e.g. van Santen & Sperling, 1985), and still continues to be the starting point for much research on motion detection, whether physiological, computational or psychophysical. Similarly the fly still plays a significant role in the study of the visual detection of motion (e.g. Egelhaaf & Borst, 1993).

It was not too long before reports appeared of directionally selective cells in the mammalian retina (Barlow & Hill, 1963a). This was followed by a study which attempted to distinguish between possible models of how a cell achieves directional selectivity (Barlow & Levick, 1965). These authors utilized one of the psychophysicist's favourite tools, apparent motion, and recorded the responses of cells to each individual flash in the apparent movement sequence. They found that the response to the first flash was much the same for sequences in either direction. The response to the second flash, however, was much smaller in one direction but remained the same as that to the first flash in the other direction. This suggested to them that the first flash elicits an inhibitory response which travels in only one direction, thus vetoing responses from flashes which might be presented in this 'null' direction, whereas the responses to the flashes in the 'preferred' direction are not vetoed. Barlow and Hill (1963b) discovered that if one of these same units is stimulated for some time by a stimulus in its preferred direction its response declines. At the cessation of movement the response of the cell falls below its spontaneous rate. They pointed out that any cell which might be comparing the response of this cell with another of the opposite preferred direction would now be signalling the opposite direction of motion and suggested that this could explain the waterfall illusion. This tradition of attempting to relate physiological reactions to perceptual events is still a strong theme in sensory research (e.g. Newsome *et al.*, 1989).

Whilst these tremendous advances came from the study of individual fibres and neurones, psychophysical studies floundered somewhat. One major problem seemed to be what to measure. Sensitivity to motion had a long history dating back to the landmark study of Aubert (1886) yet researchers still could not agree on such basic concepts as whether sensitivity should be considered in terms of a distance moved, velocity, or some luminance/velocity trade-off (see review by Graham, 1966). Efforts elsewhere were somewhat more rewarding. Gibson (1950) pointed out that the rather simplistic stimuli and viewing conditions under which most work on motion perception, and indeed visual perception in general, was conducted does not allow the visual system to exhibit many of the behaviours of which it is capable. Gibson had become interested in how pilots land planes and was convinced that static cues to depth and orientation must be very limited, but that there was ample information in how the scene changed as the aircraft moved forward. He termed this information the 'optic flow' and argued that if this could be sensed directly it could be of great value in guiding our movements and behaviour. Direct perception has come under considerable criticism over the years (e.g. Ullman, 1980); however, the influence of Gibson's ideas remains a driving force still.

In the 1960s, two major advances were made that laid the foundation for much of the work that has since been conducted. The first was the extension of measurements of the response properties of individual sensory neurones to the mammalian cerebral cortex. Hubel and Wiesel (1959, 1962, 1968) discovered that neurones of the visual cortex (as of sub-cortical structures in the visual pathway) of cats and monkeys have highly specific response properties. They found that the typical cortical visual neurone appears to spend much if its waking life inactive. Only when a stimulus of the right orientation, size, possibly colour or depth and commonly direction of motion appears in the right location in the visual field does the neurone spring into action. As regards motion, Hubel and Wiesel discovered that many neurones in the primary visual cortex are direction-selective. Earlier work had shown that the direction-selective cells found in beetles and flies, and in the retina of frogs and rabbits, could not be found in the retina of higher mammals such as the cat. The discovery of such cells in the cortex made it possible to extend to higher mammals the notion of motion selectivity as a fundamental sensory property. The fact that neurones have small, localized receptive fields places a fundamental constraint on theories of vision: processing, including that of motion, must, at least in the first instance, be performed locally. This simple but fundamental fact has shaped the course of motion research. Indeed, the work of Hubel and Wiesel has influenced almost all the physiological studies of visual neurones, together with a sizeable proportion of the psychophysical studies, that have been carried out in the three decades since it was conducted. The focus of attention has now moved away from the primary visual cortex towards more anterior cortical regions, but still numerous studies appear annually that are concerned with documenting the response specificities of visual neurones. Motion specificity has always been, and remains, an important aspect of such work.

As well as stimulating physiological investigation, Hubel and Wiesel's work also gave impetus to a school of thought concerning pattern recognition (machine as well as biological) which said that visual analysis should proceed by breaking down the image into specific 'features' such as lines, blobs and corners. Lettvin *et al.* (1959) in their celebrated paper 'What the frog's eye tells the frog's brain' reported that the frog retina contained fibres responding only to the sharply defined boundaries between objects, and fibres responding to dimming illumination 'such as might be caused by the shadow of a bird of prey'. These neurones were often referred to as 'bug-detectors', implying a simple and direct link between single neurones and function. Following the discovery by Hubel and Wiesel of what became known as 'edge detectors' in the mammalian visual cortex, Barlow (1972) put forward the more general notion of 'feature detectors' – the idea that the firing of any given cell indicates the presence of some particular feature in the visual world. The discovery of neurones which apparently encoded specific, simple features to the exclusion of all else seemed to lend plausibility to this approach. The feature-detector approach has now largely fallen out of favour, having been replaced by the notion that any given piece of perceptual information is distributed in an ensemble or network of neurones, but can nonetheless be seen as an important stage in the development of the subject.

The second major advance of the 1960s was the application to vision, initially by Campbell and Robson (1968), of Fourier analysis. Two separate ideas can be discerned within this work. The first is that the visual image can, in principle, be described as the sum of some set of sinusoidal variations in luminance across the image. Similarly, the luminance of any part of the image will vary over time if the image moves, and these temporal variations can be described as the sum of a set of sinusoidal changes in luminance over time. Thus Fourier analysis in the luminance domain can provide a complete description of both the spatial structure of an image and image motion. The second idea is that this is how human vision actually proceeds. A heated debate raged on the latter point throughout the late 1960s and much of the 1970s. Its conclusion was a compromise: the visual system does not compute a strict, global Fourier transform of the image, but it does appear to do something rather similar on a local, patchwise basis. The current view is that visual analysis commences with filtering operations that are rather like local Fourier analysis, but then continues with processes that are quite different. In the case of motion detection, temporal filtering is the first stage, but is not the whole story. The elucidation of this state of affairs consumed most of the psychophysical effort that was invested in visual psychophysics during the 1970s.

By the late 1970s, vision research had reached a temporary stalemate. The limits of the applicability of Fourier theory to vision were beginning to be realized but the direction that should be taken next was not obvious. The important physiological findings of the 1960s had been exploited to the full in terms of derivative psychophysical work. Yet our knowledge of the fundamentals of how vision is accomplished did not appear to have advanced greatly as a result. Such excitement as there was in the field of vision largely surrounded the continuing neurophysiological studies of the response selectivity of individual visual neurones. During the 1970s the primate cortex began to replace the more evolutionarily distant cat as the primary object of investigation, and the discovery by Zeki (1974, 1977, 1978) and others of functional specialization in extrastriate visual cortex, including the 'motion area' MT (V5), was clearly a major advance. Yet, in the psychophysics community in particular, and indeed in some sections of the neurophysiology community, there was a growing awareness that the approaches then current did not look poised to reveal just how the information conveyed by individual neurones is combined to yield vision. The solution was not obvious, and progress slowed noticeably. The 1978, 1979 and 1980 volumes of the journal *Vision Research* were thinner than any others before or since.

The breakthrough came in the early 1980s with the advent of the computational approach, championed by the late David Marr. The fundamental tenet of Marr's approach is that it is necessary to know what a system does before you can explain how it does it. Although this is manifestly sensible, even obvious, once stated, the vision community had been behaving as if it did not appreciate it. Physiologists felt it sufficient merely to describe the response properties of cells, while psychologists either looked for direct correlates of physiological results or else in other ways broke down the problem into sub-problems so specific that their place in the jig-saw puzzle was often obscure. Marr (1982) advocated an approach in which clear, precise and detailed descriptions of the processes carried out by the brain are formulated. Most importantly he provided, in principle at least, a strategy for solving a problem that had defeated psychologists (and been ignored by physiologists): how to marry 'top-down' approaches concerned with high-level perceptual phenomena with 'bottom-up' reductionist approaches. Marr distinguished three levels of explanation: computational theory, in which the task accomplished is described; algorithm, in which a detailed means of accomplishment is specified in computational terms; and implementation, in which the means of implementing the algorithm in the brain (or other system) is specified. Because the algorithm bears a clear relationship both to the computational theory and to the implementation, it provides a means of moving between and linking the two.

Algorithms proliferated, and continue to do so. In motion research, as in other aspects of vision, a new impetus had been created. In the 1980s and early 1990s numerous computational models of motion detection were published (e.g. van Santen & Sperling, 1985; Adelson & Bergen, 1985) detailing methods of detecting local image motion and, more recently, ways of integrating such signals across directions and spatial scales and across space, to give global as well as local motion signals and to account for transparent motion.

Yet neither of the traditional approaches, neurophysiology and psychophysics, has been made redundant by the advent of the computational approach. On the contrary, psychophysics is, and probably always will be, necessary in order to test the plausibility of computational models. The question of whether a particular algorithm is efficient and effective is quite separate from whether it is the algorithm chosen by evolution to implement in the visual system. Sometimes the most computationally elegant and efficient algorithm is not the easiest to implement in neural hardware; sometimes evolution may simply not have arrived at a very efficient solution. As regards physiology, having (by psychophysical testing of the predictions of models) established the algorithm which best characterizes the behaviour of the visual system, it is necessary to address the question of implementation. This can only be achieved by reference to the physiological response properties of neurones, in tandem with various anatomical and histological techniques. Thus, the future is assured for neurophysiological work also.

The current approach to the study of visual detection of motion therefore involves a multi-disciplinary approach consisting of computational modelling, psychophysical testing of the predictions of models and physiological studies conducted in the context of the implementation of models. The current mood is one of optimism and it is easy to envisage this cocktail of methods surviving productively well into the 21st century. Logically, it can be argued that computational modelling is primary in that models must be developed before they can be tested for biological plausibility and their implementation considered. Our view, however, is that modelling should, to some extent, be driven by known psychophysical and physiological constraints. This view is purely pragmatic. Those engaged in computational modelling often face choices as to how to proceed. If they have and use knowledge of the empirical facts that emerge from study of the visual system, then they are likely to generate more plausible models more quickly than if they do not.

Physiologists and psychophysicists need not (and do not) feel confined to testing the plausibility of computational models. A striking example of the progress that can be made outside this framework is the discovery in the 1980s, using a combination of anatomical and physiological techniques, of two parallel 'processing streams' within the primate cerebral cortex (Maunsell & van Essen, 1983). One of these streams seemed to be concerned primarily with object recognition, the other with spatial location and motion, although recently it has become apparent that the distinction may be less sharp, both in anatomical and in functional terms, than originally conceived. Such knowledge of the functional anatomy of the brain has provided neither tests of nor constraints on computational models, yet it is clearly of great interest and value.

Much of the foregoing applies to vision research as a whole rather than to research on the visual detection of motion in particular. The history of research on motion is, in fact, that it was somewhat overshadowed by research on spatial vision until relatively recently. Traditionally, spatial vision (leading to the recognition of objects in static scenes) has been seen as the major challenge and motion was seen as a less challenging and less important adjunct. In the last 15 years, however, this state of affairs has changed dramatically. One reason is the realization that all vision is based on an initial filtering process and that this filtering is applied in both spatial and temporal domains. In this sense, the two domains are of equal importance. Another reason, perhaps, is a growing awareness of a long-established fact: that the retinal image is in almost constant motion, if only due to movements of the eyes and head. In this sense consideration of the processing of static images is inappropriate. For whatever reasons, the study of visual detection of motion is now burgeoning.

3 OVERVIEW OF THE BOOK

The starting point for all aspects of the detection and encoding of visual motion is the detection of motion of the image across the retina, an operation which is performed locally, in parallel for all points in the image. All subsequent computations (of object motion, three-dimensional trajectories, self-motion, etc.) are based on these local, retinocentric measurements. Part 2 of this book (Principles of Local Motion Detection) is concerned with the way in which local motion detection is performed in the mammalian visual system. Firstly, Grzywacz, J. Harris and Amthor provide a general account of the computational approach to local motion detection, moving on to the specifics of what type of computation best characterizes biological motion detection systems. Next, Snowden provides a survey of our knowledge of the response properties of motion-sensitive neurones, focusing on the various areas of visual cortex in primates. Finally, McKee and Watamaniuk review psychophysical studies concerning the ability of human observers to detect and discriminate motion. The reader with limited knowledge of the subject who seeks an introduction to research in motion is advised to start with whichever one of the three chapters in this section is closest to his/her interests and experience.

Part 3 (Inputs to Local Motion Detectors) concerns the same topic as Part 2, local motion detection, but the emphasis is on the characteristics of the image that enable motion detection. Motion is typically specified by moving luminance boundaries, but this is not necessarily the case. Clearly the motion system must be tailored to the detection of whatever type of image motion prevails in natural images. Firstly, Mather considers the nature of the luminance signals that need to be detected, the possible mechanisms of detection and the psychophysical evidence bearing on them. Then Smith considers image motion defined by characteristics other than luminance, the mechanisms needed to detect such motion and the psychophysical evidence for the existence of such mechanisms. Lastly, Logothetis reviews physiological evidence concerning the sensitivity of the motion detection system to the various types of image motion discussed in the two preceding chapters.

The integration, or combination, of local motion signals is considered in Part 4. One important problem of integration to be solved is that the local motion signals arising from the initial motion detection process appear to encode vectors along a single axis of motion, rather than the true directions of motion of images. Wilson discusses computational models of the integration of vectors in different directions to give the true direction of motion of the image. He also reviews some psychophysical evidence bearing on the validity of the model he favours. Stoner and Albright address the same topic but from a more empirical standpoint, reviewing psychophysical and physiological studies relevant to this issue. Finally, Williams and Brannan consider a different problem of integration: the integration of local motion signals at different locations to give rigid global motion of larger areas of the image. Also covered are other types of interaction between motion signals at different spatial locations, such as induced motion.

Part 5 (Higher-order Interpretation of Motion Signals) is devoted to subsequent stages of the processing of motion information. M. G. Harris considers the optic flow (the pattern of motion signals across the retina) that arises as we move through the environment. He then considers the psychophysical evidence for detection mechanisms that are specific to particular flow patterns. Cumming considers possible strategies, and the psychophysical evidence for or against them, for the interpretation of two-dimensional image motion in terms of three-dimensional trajectories. Finally, Braunstein considers strategies for the recovery of spatial structure from the motion gradients in an image, together with the psychophysical literature concerning human ability to detect structure from motion.

Part 6 (Motion Detection and Eye Movements) examines the issues arising from the fact that image motion can result either from the movement of objects or (more commonly) from movements of the eye, head or body. The means by which the various sources of image motion are dissociated in order to give a stable view of the world despite movements of the eye and head are discussed by L. R. Harris. Lastly, Krauzlis reviews physiological and anatomical knowledge of the neurones and pathways that are used to feed visual information to the eye movement system so as to enable eye movements of a type that facilitate efficient and stable perception.

REFERENCES

Addams, R. (1834). An account of a peculiar optical phaenomenon seen after having looked at a moving body etc. *Lond. Edin. Phil. Mag. J. Sci.*, 3rd series, **5**, 373–374.

Adelson, E. H. & Bergen, J. R. (1985). Spatiotemporal energy models for the perception of motion. *J. Opt. Soc. Am. A*, **2**, 284–299.

Adrian, E. D. (1928). *The Basis of Sensation*. Christophers, London.

Aubert, H. (1886). Die Bewegungsempfindung. *Pflugers Arch. Ges. Physiol.*, **39**, 347–370.

Barlow, H. B. (1953). Summation and inhibition in the frog's retina. *J. Physiol.*, **119**, 69–88.

Barlow, H. B. (1972). Single units and sensation: a neuron doctrine for perceptual psychology? *Perception*, **1**, 371–394.

Barlow, H. B. & Hill, R. M. (1963a). Selective sensitivity to direction of motion in the ganglion cells of the rabbit's retina. *Science*, **139**, 412–414.

Barlow, H. B. & Hill, R. M. (1963b). Evidence for a physiological explanation of the waterfall phenomenon and figural after-effects. *Nature*, **200**, 1345–1347.

Barlow, H. B. & Levick, W. R. (1965). The mechanism of directionally selective units in rabbit's retina. *J. Physiol. (Lond.)*, **178**, 477–504.

Boring, J. F. (1942). *Sensation and Perception in the History of Experimental Psychology*. Appleton Century, New York.

Brown, J. F. (1931). The visual perception of velocity. *Psychol. Forsch.*, **14**, 199–232.

Campbell, F. W. & Robson, J. G. (1968). Application of Fourier analysis to the visibility of gratings. *J. Physiol.*, **197**, 551–566.

Duncker, K. (1929/1938). Induced motion. In W. D. Ellis (Ed.) *A source of Gestalt Psychology*. Routledge and Kegan Paul, London.

Egelhaaf, M. & Borst, A. (1993). A look into the cockpit of the fly: visual orientation, algorithms, and identified neurons. *J. Neurosci.*, **13**, 4563–4574.

Exner, S. (1888). Einige beobachtungen uber bewegungsnachbilder. *Centr. Physiol.*, **1**, 135–140.

Gibson, J. J. (1950). *The Perception of the Visual World*. Houghton Mifflin, Boston.

Graham, C. (1966). *Vision and Visual Perception*. Wiley, New York.

Gregory, R. L. & Zangwill, O. L. (1963). The origin of the autokinetic effect. *Q. J. Exp. Psychol.*, **15**, 252–257.

Hartline, H. K. (1938). The response of single optic nerve fibres of the vertebrate eye to illumination of the retina. *Am. J. Physiol.*, **121**, 400–415.

Hassenstein, B. & Reichardt, W. (1956). Systemtheoretische analyse der zeit-, reihenfolgen- und vorzeichenauswertung bei der bewegungspwezeption des rüssel-kafers *Chlorophanus*. *Z. Naturf*, **11b**, 513–524.

Hubel, D. H. & Wiesel, T. N. (1959). Receptive fields of single neurons in the cat's striate cortex. *J. Physiol.*, **148**, 574–591.

Hubel, D. H. & Wiesel, T. N. (1962). Receptive fields, binocular interaction and functional architecture in the cat's striate cortex. *J. Physiol.*, **160**, 106–154.

Hubel, D. H. & Wiesel, T. N. (1968). Receptive fields and functional architecture of monkey striate cortex. *J. Physiol. (Lond.)*, **195**, 215–243.

Johansson, G. (1964). Perception of motion and changing form. *Scand. J. Psychol.*, **5**, 181–208.

Korte, A. (1915). Kinematoskopische Untersuchunden. *Z. Psychol.*, **72**, 193–296.

Kuffler, S. W. (1953). Discharge patterns and functional organization of mammalian retina. *J. Neurophysiol.*, **16**, 37–68.

Lettvin, J. Y., Maturana, R. R., McCulloch, W. S. & Pitts, W. H. (1959). What the frog's eye tells the frog's brain. *Proc. Inst. Radio Eng.*, **47**, 1940–1951.

Marr, D. (1982). *Vision*. WH Freeman, San Francisco.

Maunsell, J. H. R. & van Essen, D. (1983). The connections of the middle temporal area (MT) and their relationship to a cortical hierarchy in the macaque monkey. *J. Neurosci.*, **3**, 2563–2586.

Nakayama, K. (1985). Biological image motion processing: A review. *Vision Res.*, **25**, 625–660.

Newsome, W. T., Britten, K. H. & Movshon, J. A. (1989). Neuronal correlates of a perceptual decision. *Nature*, **341**, 52–54.

Regan, D. (1989). Orientation discrimination for objects defined by relative motion and objects defined by luminance contrast. *Vision Res.*, **29**, 1389–1400.

Reichardt, W. (1961). Autocorrelation, a principle for the evaluation of sensory information by the central nervous system. In W. A. Rosenblith (Ed.) *Sensory Communication*. Wiley, New York.

Rogers, B. J. & Graham, M. (1979). Motion parallax as an independent cue for depth perception. *Perception*, **8**, 125–134.

Salzman, C. D., Britten, K. H. & Newsome, W. T. (1990). Cortical microstimulation influences perceptual judgements of motion direction. *Nature*, **346**, 174–177.

Sherif, M. (1937). An experimental approach to the study of attitudes. *Sociometry*, **1**, 90–98.

Ullman, S. (1980). Against direct perception. *The Behavioural and Brain Sciences*, **3** (whole issue).

van Santen, J. P. H. & Sperling, G. (1985). Elaborated Reichardt detectors. *J. Opt. Soc. Am. A*, **2**, 300–321.

Wallach, H. & O'Connell, D. N. (1953). The kinetic depth effect. *J. Exp. Psychol.*, **45**, 205–217.

Wertheimer, M. (1912). Experimentelle studien über das Sehen von Bewegung. *Z. Psychol.*, **61**, 161–165.

Zeki, S. M. (1974). Functional organization of a visual area in the posterior bank of the superior temporal sulcus of the rhesus monkey. *J. Physiol. (Lond.)*, **236**, 549–573.

Zeki, S. M. (1977). Colour coding in the superior temporal sulcus of the rhesus monkey visual cortex. *Proc. R. Soc. Lond. B*, **197**, 195–223.

Zeki, S. M. (1978). Uniformity and diversity of structure and function in the monkey prestriate visual cortex. *J. Physiol.*, **277**, 273–290.

Zihl, J., Cramon, D. V. & Mai, N. (1983). Selective disturbance of movement vision after bilateral brain damage. *Brain*, **106**, 313–340.

Part 2
Principles of Local Motion Detection

2

Computational and Neural Constraints for the Measurement of Local Visual Motion

Norberto M. Grzywacz[1], Julie M. Harris[1] and
Franklin R. Amthor[2]

[1]*The Smith-Kettlewell Eye Research Institute, San Francisco and* [2]*University of Alabama at Birmingham*

1 PHILOSOPHY

A good starting point for the study of a particular brain function is to ask what benefits this function may bestow upon the animal. Such a question will not only motivate the inquiry, but might also place computational constraints on it. In the case of visual perception, for instance, the question might help to focus our attention on image measurements that are relevant to the animal. However, the ideal measurements to perform from a computational perspective might not coincide with the ideal measurements to perform from the brain's perspective. The brain uses computational elements, for example, neurons, synapses, and ionic channels, that might not be well suited to implement these 'ideal computations', but might instead be more suitable for other computations that have sufficient survival value to the animal. Hence, an effective inquiry into a brain function must be constrained both from computational and neural perspectives. This chapter will take this approach to explore the measurement of local visual motion in the brain. We will start the chapter (section 2) by asking the 'why' and 'what' questions, that is, we will start with the computational perspective. We ask *why* the visual system would benefit from having motion information and *what* information is available to the brain. Then, sections 3 and 4 will deal with the two main answers to the 'what' question, namely, speed and direction of visual motion. Finally, section 5 will deal with the 'how' question, by discussing the biophysical details of how direction of visual motion is measured in the retinas of turtles and rabbits.

VISUAL DETECTION OF MOTION
ISBN 0–12–651660–X

2   COMPUTATIONAL PERSPECTIVE

This section will begin by considering the computations performed by the visual system that benefit from measurements of visual motion. Then, we will discuss the fundamental variables of visual motion. The question of whether the visual system must measure all these variables with equal emphasis will be addressed in section 2.3. Finally, section 2.4 will deal with how local these measurements must be.

2.1   Why measure visual motion?

Visual-motion cues play a role in various functions of biological visual systems (Nakayama, 1985, gives a comprehensive review). For instance, when an animal is moving (egomotion), motion cues could help it avoid obstacles along the way (Gibson, 1950). In the same context, motion measurements play a role in helping the animal to find its direction of heading, that is, helping it navigate (Warren & Hannon, 1988; Hildreth, 1992). Besides aiding egomotion, visual motion measurements may also be used to determine whether an object coming towards the animal will hit it (see Chapter 12 by Cumming). Such information is particularly useful when attempting to escape from one's predators. In turn, if an animal is a predator or is trying to approach a mate, then motion measurements can be useful in the interception process (Lee, 1976; Reichardt & Poggio, 1976). (Of course, interception is also used in functions of more socio-biological relevance such as hitting a baseball – or perhaps we should say: in a batsman's defcnse of the wicket.)

   All of the examples above involve analyzing a scene that is being modulated by essentially translatory motion of individual objects. However, other examples illustrate the usefulness of measurements of motion that involve rotations, multiple objects, or simply trying to bring the main portions of the image into focus. For instance, the motion of an object can be used to determine its three-dimensional structure (see Chapter 13 by Braunstein). This process could be useful in determining the shape of a head when it is shaking or when viewing a car turning at an intersection. When there are multiple objects moving in the scene, differences in measured visual motion in different portions of the scene may help the animal to parse the scene for further analysis (Nakayama & Loomis, 1974). This image parsing or segregation is useful when an animal needs to penetrate another animal's camouflage (Helmholtz, 1910/ 1962; Regan & Beverley, 1984). One way in which measurements of visual motion are useful for bringing the essential portions of the scene into focus is by telling the eyes where to move and how fast to move (saccades and smooth pursuit; see Chapter 15 by Krauzlis) to bring a feature of interest into the fovea (for reviews see Sparks and Mays, 1990, and the book edited by Miles and Wallman, 1993). Another way is by deblurring (Burr *et al.*, 1986), that is, by compensating for the same type of blur that occurs when one does not set the photographic camera's shutter to a sufficiently fast speed.

2.2   Fundamental variables of visual motion

In physics, the most fundamental variables of motion are velocity (the derivative of position over time) and acceleration (the derivative of velocity over time).

Consequently, it would seem valuable for an animal's visual system to determine these variables accurately. Unfortunately, the only data available to animals about motion in the external world come from the projections onto the eyes' retinas, which by being essentially two-dimensional, cannot encode directly the three-dimensionality of the world (Wheatstone, 1838; Marr, 1982).

Therefore, in visual perception, a more fundamental variable of motion might be the 'optic flow' (see Chapter 11 by M. Harris). This flow is a spatial distribution of velocity vectors that is close to the distribution obtained by projecting the three-dimensional motion vectors of a moving scene onto the retina. The usefulness of this vector field quantity has been discussed by both psychologists (for example, Gibson, 1950; Regan, 1986) and computer scientists (Koenderink & van Doorn, 1976; Longuet-Higgins & Prazdny, 1980). To mention one example of the use of optic flow, when one moves forward in a straight line, the optic flow is that of an expansion (or contraction if translating away from the scene), whose focus signals the direction of heading (see Regan et al., 1986). A proposed definition of optic flow is in terms of the 'image constraint equation' $dE/dt = 0$ where E is the image's brightness (Fennema & Thompson, 1979; Horn & Schunck, 1981). This equation, which assumes that brightness varies slowly over time, defines an optic flow $\vec{v}$ from brightness as

$$\nabla E \cdot \vec{v} + \frac{\partial E}{\partial t} = 0. \tag{1}$$

(Because this is one equation and there are two variables, the horizontal and vertical components of $\vec{v}$, another constraint must be included so that it is possible to solve for $\vec{v}$. Typically, a smoothing assumption is made such that large discontinuities are not allowed (Horn & Schunck, 1981).

However, a handicap of considering optic flow as a fundamental variable of visual motion, is that definitions different from that expressed in equation 1 are possible (Grzywacz & Poggio, 1990; Verri et al., 1989). The definition is not unique, since image motion can be influenced by parameters unrelated to motion, such as the reflectance properties of the viewed objects. (For example, a specularity on an object – a mirror-like reflection of a nearby light source – often moves in a different direction to the object itself; Grzywacz and Poggio, 1990.) The non-uniqueness of local motion measurement is emphasized by the so-called 'correspondence problem' (Julesz, 1971; Ullman, 1979b). This problem is to find correspondences between image features at one instant in time and (hopefully the same) features at the next instant in time.

Nevertheless, despite its non-uniqueness, optic flow has been useful in emphasizing the importance of measuring image velocity vectors, and in particular their direction and speed components. (Optic flow equations also demonstrate the mathematical similarities between certain types of visual motion such as translatory egomotion and translatory external motion.) In addition, some definitions of optic flow require measuring image acceleration vectors (Verri et al., 1989) and thus one must consider this variable too. The image velocity and acceleration variables must be measured locally as required by their derivative definition. In section 2.4 below, we will consider the implications of this requirement of locality.

2.3   Are all variables of visual motion equally important?

There are several reasons why visual systems may not need to measure visual motion too precisely. One is that the portion of the visual system devoted to the measurement of motion does not work in isolation. Visual motion is just one piece in the puzzle of the reconstruction of the external world (Marr, 1982). Furthermore, since visual systems have limited resources, it might not be to their advantage to try to compute all the possibly relevant variables. It might be better to compute certain variables with precision while ignoring other variables altogether, than to try to compute everything poorly (Ratliff, 1965).

One example of how the motion system is aided by another visual module is in the computation of structure from motion (Wallach & O'Connell, 1953; Ullman, 1979a, 1984; see Chapter 13 by Braunstein). On its own, this computation essentially requires that the viewed object rotates about an axis contained by the fronto-parallel plane to obtain the object's three-dimensional structure from visual motion. Even if this very specific motion occurs, the structure-from-motion system may often be unable to tell the direction of rotation because of an ambiguity in the velocity field when perspective cues are weak (Howard, 1961). The addition of stereopsis in the recovery of three-dimensional shape disambiguates the direction of rotation (Nawrot & Blake, 1991) and may allow a full recovery of the object's shape and distance (Richards, 1985; Tittle & Braunstein, 1991). Another example of how the motion system is helped by stereopsis may occur in egomotion. Although neither motion (Longuet-Higgins, 1984) nor stereo (Foley, 1980; Johnston, 1991) information can give, on their own, the absolute distance of an obstacle, the combination of egomotion and stereo can, in principle, yield this distance.

As mentioned above, due to the limited resources of the visual system, it might not be to its advantage to measure all the variables of visual motion with equal accuracy. One variable that the visual system 'should seriously consider ignoring' is image acceleration. To begin with, acceleration is the derivative of velocity over time or the second derivative of position over time. Hence, because any method or algorithm that measures derivatives practically is unstable in the sense that small changes in the initial data do not produce correspondingly small changes in the final results (Burden & Faires, 1989), and because the second derivative is a derivative of a derivative, the measurement of acceleration as a second-order derivative should be highly susceptible to noise. This problem becomes even more serious when one notices that psychophysically determined measurements of local speed are themselves very noisy (Vaina et al., 1990; Bravo & Watamaniuk, 1991; Watamaniuk & Duchon, 1992). To add to these problems, in none of the computations mentioned in section 2.1 was acceleration needed. For instance, to measure contracting or expanding optic flows and thus determine direction of heading, one only needs image velocities (e.g. Rieger & Lawton, 1985; Hildreth, 1992). Similarly, the putative computation of divergence, curl and shear from optic flows could be performed from global estimates based on local velocity measurements (Werkhoven & Koenderink, 1991; Sekuler, 1992) and not necessarily from their derivatives as has been suggested (Koenderink & van Doorn, 1976; Longuet-Higgins & Prazdny, 1980). (To be fair, a possible exception to the irrelevance of acceleration might occur when a baseball outfielder has to catch a fly ball.) Therefore, if acceleration is difficult to compute and is not used for relevant computations by visual systems, then why should it be measured? Psychophysical data strongly suggest that it is not (e.g. Gottsdanker, 1956; Schmerler, 1976).

We also argue that visual systems might be better off concentrating on measuring the direction of motion rather than speed. Firstly, it is difficult to measure the latter, since it is the magnitude of the derivative of position over time and thus requires precise spatio-temporal information. In contrast, to measure direction, one only requires two relatively imprecise position measurements (see section 4). Moreover, one can perform many important visual motion computations with directional measurements alone. For instance, it is often theoretically possible to determine the direction of one's heading during egomotion without using speed information. When 'ego-translating' one can find the optic flow's focus of expansion, and thus heading, with only the directional measurements (Figure 1A; Koenderink, 1986; Regan, 1986). Another example of visual-motion computations without speed occurs in structure-from-motion. Human patients who have lost their ability to measure speed due to a brain stroke can still recover structure from motion (Vaina et al., 1990). This ability might be due to an independent recovery of the axis of rotation (Hoffman & Bennett, 1985, 1986). Another function that can be computed with directional information alone is image segregation. A sharp change of direction of motion at boundaries is theoretically and experimentally sufficient for the perceptual localization of the

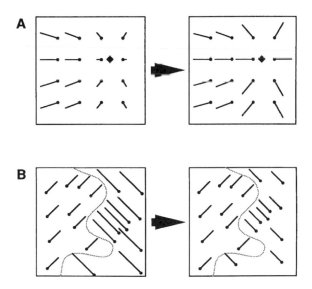

Figure 1 Examples of visual information from directional measurements. The left panels illustrate optic flows elicited by an ego-translation towards the black diamond (A) and two objects sliding past each other (B). The local velocity vectors in the flow begin at the small black circles, and the vectors' direction and speed are represented by the orientation and length of the associated lines respectively. The right panels are obtained from the left panels by setting the speeds to a constant value. In A, to determine the direction of heading, one must measure the location of the focus of expansion, which can be performed with directional measurements alone (right panel). In B, to find the discontinuity edge (curvy textured line), it is sufficient to detect discontinuities in the directional field.

boundary (Figure 1B – Nakayama & Loomis, 1974; Baker & Braddick, 1982). Finally, directional signals could, in principle, be sufficient to mediate smooth-pursuit eye movements (Komatsu & Wurtz, 1988). These signals could encode the direction of retinal slips and thus tell the eye in what direction to accelerate to minimize slip. Following these arguments, it may not be surprising that direction is the first motion-related variable neurally encoded in visual pathways (Barlow *et al.*, 1964; Hubel & Wiesel, 1962) and that its overall measurement is quite precise (to within 1° discriminability – see Chapter 4 by McKee and Watamaniuk).

Unlike acceleration, speed signals should probably not be ignored by the visual system, since they can be useful. A recent model (Hildreth, 1992) uses both direction and speed signals to account for the 1–2° precision in the estimation of heading direction (Warren & Hannon, 1988). Furthermore, models of structure from motion based on full optic-flow information (Ullman, 1979a, 1984) suggest that this computation would benefit from speed information. Psychophysical findings indicate that one can segregate images based only on gradients of speed signals (Hildreth, 1983; Sekuler, 1990). Both smooth pursuit and saccades could be more precise if speed signals were used in their computations (Komatsu & Wurtz, 1988; Sparks & Mays, 1990) and the visual system must somehow use speed signals to achieve effective deblurring (Burr *et al.*, 1986; Watamaniuk, 1992). Consequently, to take advantage of this useful information, visual systems compute speed locally (McKee *et al.*, 1986; Vaina *et al.*, 1990; Bravo & Watamaniuk, 1991). However, this computation is relatively less precise than that of the direction of motion.

2.4 How local is local?

The preceding section suggested that visual systems should devote energy to the measurement of local direction and speed of motion. We now address the implications of the 'locality' requirement. Spatial locality first emerges in the context of the computation of motion, because velocity is the derivative of position over time. By definition, this derivative requires the use of positions that are infinitesimally close. However, in practice, any physical system would have to measure this derivative by using positions that are separated by a finite amount. At a minimum, these positions must be separated by the spatial resolution of the system. But sometimes, it may make sense to measure the derivatives with positions that are separated by a distance that is significantly larger than the system's resolution. This is because the relative precision of the measurement will increase as the distance increases. For instance, if one measures 1 mm with a ruler that has 1 mm resolution, then the error may be 100%, but if one measures 30 cm with the same ruler, then the error is about 0.3%. In other words, for the derivative of position over time to be known accurately, information from a large area must be used and so the location at which the derivative is measured becomes uncertain. Conversely, if only very local information is used the location is known precisely but there will be uncertainty in the value of the calculated derivative.

Hence, there is a tradeoff between locality and precision of visual motion measurement that is reminiscent of the Heisenberg uncertainty principle in quantum mechanics, where the multiplication of errors in the measurement of position and momentum has a lower bound. This principle for measurement of visual motion will be addressed more formally in section 3.

Given this uncertainty principle in the computation of visual motion, the question arises of how much locality to give up. Visual systems cannot give up locality too much, because if they did so, they would lose their ability to locate discontinuities based on gradients of motion signals (see Figure 1B) since a more 'global' measurement would smooth (or average) out any motion discontinuities. Loss of locality would also impair the ability to detect foci of expansion (Figure 1A) and thus impair computation of direction of heading in egomotion. Perhaps most important of all, except at singularities and discontinuities, local measurements allow the visual system to decompose arbitrarily complex motions into a collection of local translations. This allows for a relatively simple interpretation of global motions by later stages of the visual pathway (e.g. Hildreth, 1992; Yuille & Grzywacz, 1988).

One way to avoid losing locality and at the same time retain precision is by using multiple scales of measurement. By this method, the visual system would make fine-scale relatively imprecise measurements of visual motion in parallel with large-scale relatively precise measurements. This could be achieved by using multiple mechanisms, each tuned to a different spatial scale. The visual system does this by using motion-selective receptive fields of various sizes (Dow, 1974). However, exactly how the visual system 'chooses' what scales of processing to use under different circumstances is an open question.

2.5 Summary of computational perspective

Consideration of the types of computation that local visual-motion measurements may contribute to, and the ease of measurement of each fundamental visual-motion variable, suggested focusing our inquiry into direction and speed separately. Direction may be a more fundamental variable than speed and thus we will pay more attention to the measurement of the former. In both cases, we will consider multi-scale measurements to address the trade-off between locality requirements and precision of measurement.

3 FORAY INTO THE MEASUREMENT OF LOCAL SPEED

In the preceding section, we argued that speed is a less fundamental visual-motion variable than direction. The argument was based on direction being neurally encoded in the visual pathway before speed, on the ease of computing direction as compared to speed, and on the large number of useful computations that one can perform based on direction measurements alone. Sections 4 and 5 will discuss the measurement of direction at length. Although direction may be a more fundamental variable than speed, we decided to consider local speed and discuss models for its measurement first. We did this because, while this section will focus only on computational issues, the sections on the measurement of direction will underscore the necessity to use a multilevel approach combining the computational, behavioral and neural perspectives.

Humans can measure speed quite precisely if provided with a relatively long trajectory of motion (McKee & Welch, 1985; and see Chapter 4 by McKee and

Watamaniuk). Under these conditions, the errors made in discriminating speed can be as low as 5%. This high precision seems to be achieved by integrating relatively imprecise local-speed signals over time, since local signals are available (McKee *et al.*, 1986; Vaina *et al.*, 1990; Bravo & Watamaniuk, 1991). The relative errors in the measurement of local speed have been estimated to be between 30 and 100% (Vaina *et al.*, 1990; Bravo & Watamaniuk, 1991).

Perhaps the simplest method for measuring local visual speed is by its derivative definition: find the positions of an image feature in two discrete instances in time and then compute the ratio between the positional distance and the temporal delay. This is essentially the approach proposed by Ullman in his Minimal Mapping theory (Ullman, 1979b). He proposed that the main problem facing the visual system to measure motion is to solve the correspondence problem (section 2.2). He suggested that the features correspond so as to minimize the total distance traveled. After correspondence is established, then distance and delay, and thus the velocity of a feature, can be calculated. Serious challenges to Ullman's emphasis of the correspondence problem, and thus his method to measure visual speed, have been raised by motion psychophysicists (Adelson & Bergen, 1985). They pointed out that his theory was not immediately consistent with known neural processes underlying motion perception. In addition, they argued that the correspondence problem is essentially nonexistent when one deals with real neural receptive fields, even if the stimulus motion is discontinuous (apparent motion). Finally, another difficulty with Ullman's Minimal Mapping theory is that it does not lend itself to a multiple scale analysis.

Models for the measurement of local speed based on known properties of neurons that are selective for the direction of motion do not begin with the intuitive definition of velocity based on derivatives, but are more suitable for multi-scale analysis (Adelson & Bergen, 1985; Watson & Ahumada, 1985; Heeger, 1987; Grzywacz & Yuille, 1990, 1991). These models begin by looking at the responses of a population of directionally selective (DS) neurons of different receptive-field sizes. These are neurons that can discriminate the direction of motion through the strength of their response to stimuli moving in different directions (the mechanisms underlying directional selectivity will be addressed in detail in sections 4 and 5). It is known that striate cortex neurons with larger and larger receptive-field sizes are tuned to lower and lower spatial frequencies when stimulated with sinusoidal gratings (Hochstein & Shapley, 1976; Maffei & Fiorentini, 1977; Andrews & Pollen, 1979). Moreover, different neurons tend to be tuned to different temporal frequencies of the gratings. The DS-cell-based models for local speed measurement consider a three-dimensional space which has two axes corresponding to the optimal horizontal and vertical spatial frequencies that drive a given DS cell, and a third axis corresponding to the optimal temporal frequency that drives the cell. (We denote the optimal temporal frequency by Ω_t, optimal spatial frequency by Ω_r, and DS cell's preferred direction by $\vec{u}$. Therefore, the optimal horizontal and vertical spatial frequencies are the projections of $\Omega_r\vec{u}$ onto the horizontal and vertical frequency axes respectively.) A point in this space corresponds to a DS cell. What is interesting about this space is that when a visual translation of velocity $\vec{v}$ covers the receptive fields of these cells, the optimal responses fall in the plane

$$\Omega_r\vec{u} \cdot \vec{v} + \Omega_t = 0. \tag{2}$$

Consequently, to measure local speed in DS-cell-based models, all one needs to do is to detect the slant of this plane relative to the temporal frequency axis (Figure 2; local direction can be obtained from the tilt of the plane relative to the spatial frequency axis). Several schemes to detect this plane have been proposed in the literature (Watson & Ahumada, 1985; Heeger, 1987; Grzywacz & Yuille, 1990, 1991).

To understand the origin of this plane, consider the spatio-temporal Fourier transform of a general translating image $I = F(\vec{r} - \vec{v}t)$, where F is the spatial profile of the image, $\vec{r}$ is spatial position, and t is time. The transform is

$$\tilde{I}(\vec{\omega}_r, \omega_t) = 2\pi\tilde{F}(\vec{\omega}_r)\delta(\vec{\omega}_r \cdot \vec{v} + \omega_t), \tag{3}$$

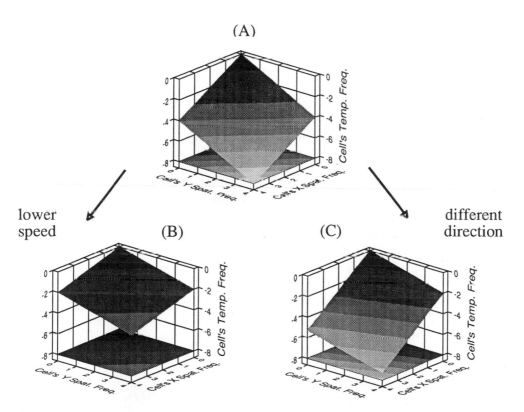

Figure 2 Measurement of velocity in the space of neuron parameters. Each triplet of abcissas and ordinate in this plot corresponds to a neuron in a population of directionally selective neurons. The values of the abcissas and ordinate are the neuron's optimal spatial frequency in the horizontal and vertical directions, and the optimal temporal frequency. If the neurons are stimulated with a translation, then the shaded slanted plane is where the maximal responses lie. The slant of the plane from the floor (also shaded) is equal to the speed of motion. The tilt of the plane relative to the spatial-frequency axes is equal to the direction of motion. The greater slant of (A) as compared to (B) indicates that in (A) the translation is faster than in (B). While the translations in (A) and (B) have identical directions of motion (45° between vertical and horizontal), the motion in (C) has much more motion in the horizontal than vertical direction (making an angle of about 20° with the horizontal axis).

where the tilde represents a Fourier transform, $\vec{\omega}_r$ and ω_t are the spatial- and temporal-frequency variables of the transform, and δ is the Dirac delta function (an infinitely sharp impulse at $x = 0$, with unit integral). Equation 3 is a multiplication of a term that depends only on $\vec{\omega}_r$ and a term that is (infinitely) sharply tuned on a plane that is very similar to the plane expressed in equation 2. (The relationship between this plane and the Fourier transform of a general translation was first pointed out by Fahle and Poggio, 1981, and then exploited by Watson and Ahumada in a model of motion detection, 1985.)

The difference between the planes expressed in equations 2 and 3 is that while the variables in equation 3 belong to the stimulus's Fourier transform, the variables in equation 2 are properties of the cells themselves. Hence, in one sense, to implement a DS-cell-based model of local speed measurement, the cells must perform a sort of spatio-temporal Fourier transform of the image. A true Fourier transform integrates the image (after multiplying it with sines and cosines) over an infinite spatio-temporal range. Because cells have finite receptive fields and temporal windows, they cannot, strictly speaking, perform a true Fourier transform of the image. They must somehow perform a finite approximation of a Fourier transform.

If one think of cells as implementing a filter, as is often done, then it turns out that there is a tradeoff between the size of their receptive fields and how sharply tuned the cell can be to spatial frequency (Hochstein & Shapley, 1976; Maffei & Fiorentini, 1977; Andrews & Pollen, 1979). In other words, there is a tradeoff between how well a cell can localize a stimulus and how precise its Fourier transform of the image is. This tradeoff is the root of the uncertainty principle for localization and velocity measurements described in section 2.4. If a cell can localize a stimulus well (if it has a small receptive field), then the cell will add a lot of thickness to the optimal plane in equation 2. This thickness will result in uncertainty about the velocity estimate, since finding velocity is essentially the detection of that plane.

Is there a receptive-field or filter profile for which this tradeoff problem is minimal? It is possible to show that under a certain minimization metric, the optimal profile is the Gabor function (Gabor, 1946; Daugman, 1985). A spatio-temporal representation of this function (Grzywacz & Yuille, 1990, 1991) is

$$G(\vec{r}, t) = \frac{1}{(2\pi)^{3/2}(\sigma_r)^2(\sigma_t)} \; \exp\left(-\frac{|\vec{r}|^2}{2\sigma_r^2}\right) \exp(-i\Omega_r \vec{u} \cdot \vec{r})$$
$$\exp\left(-\frac{t^2}{2\sigma_t^2}\right) \exp(-i\Omega_t t), \tag{4}$$

where $\sigma_r > 0$ and $\sigma_t > 0$ are scalar parameters.

Grzywacz and Yuille (1990) argued that besides being optimal from the uncertainty perspective, the Gabor function is also optimal for motion measurements, since it is the only function that will not create maxima outside the plane described by equation 2. Also, the spatial portion of this function approximates reasonably well the spatial profiles of receptive fields in striate cortex (Jones & Palmer, 1987). This function's temporal portion does not do such a good job, since it is non-causal, that is, it is non-zero at negative times. However, Grzywacz and Yuille (1990, 1991) argued that the non-causality problem is not serious if the brain's computation of velocity is slightly delayed (< 100 ms) relative to the image's motion. Alternatively, a non-Gabor filter might be used, removing the non-causality but allowing strong responses off the plane sometimes.

4   MEASUREMENT OF LOCAL DIRECTION OF MOTION

The most basic mechanism used by the visual system to measure direction of motion is directional selectivity. A visual neuron is said to be directionally selective (DS) if back and forth motions symmetric about the middle of its receptive field elicit different responses (non-symmetric motions may cause artifactual directional selectivity – see Barlow et al., 1964). For the axis along which the ratio between the responses to back and forth motions is maximal, we call the direction eliciting the largest response the preferred direction and the opposite direction the null or non-preferred direction. Such DS neurons are abundant in the retinas of insects and small-brained vertebrates (Maturana et al., 1960; Barlow & Levick, 1965; Hausen, 1981), and in the visual cortex of large-brained vertebrates (Tolhurst & Movshon, 1975; Holub & Morton-Gibson, 1981).

It is unlikely that individual DS neurons compute the direction of motion on their own. The amplitudes of their responses confound the direction of motion with other visual parameters such as contrast. To disambiguate the responses, a comparison over a population of neurons is probably performed. For example, consider a pair of neurons with opposite preferred directions. Even when the contrast varies, the response of the neuron whose preferred direction is in the direction of visual motion will be larger than the response of the neuron with opposite preferred direction. Therefore, the brain can conclude that the direction of motion is not that of the latter neuron. A computation over more than two neurons (such as that proposed in the plane calculation described by equation 2) can trim the candidate directions significantly, thus constituting a precise estimate of local direction.

In the rest of this section we will not dwell further on how the output of several DS cells may be combined to measure direction. Rather, we will focus on how the brain achieves directional selectivity. The discussion will begin with theoretical preliminaries derived from studies on insects. Then, sections 4.2 and 4.3 will deal with the two main requirements raised by the theoretical preliminaries for directional selectivity, namely, spatial asymmetries and nonlinearities.

4.1   Theoretical preliminaries

The theoretical work of Reichardt, Poggio, and collaborators (Reichardt, 1961; Poggio & Reichardt, 1976) emphasized two requirements that any model of directional selectivity must fulfill: The first is a spatially-asymmetric mechanism. The second is a nonlinear mechanism to yield two different single-number responses for preferred- and null-direction responses. (Single-number responses refer to quantities like total number of spikes, as opposed to time course of the spikes. Poggio and Reichardt emphasized the necessity of computing a single number from each response to allow a decision on the direction of motion. They proved that if there are no nonlinearities,

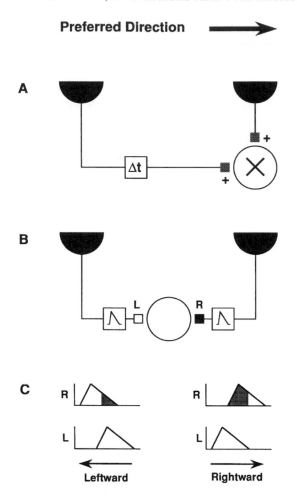

Figure 3 The Reichardt model for directional selectivity and a variant. A. This illustration of the Reichardt model has a slow input from the left (Δt box) and a fast input from the center. The inputs are excitatory (textured squares) and the nonlinear interaction is multiplicative. For rightward, but not leftward motion, the sluggishness of the left pathway is compensated by the light reaching it before reaching the center pathway. Hence, only for rightward motion, the signals arrive at the multiplication site roughly simultaneously. The black semi-circles indicate that each input integrates information over a finite area of the image. This integration helps aliasing by filtering out high spatial frequencies. B. This variant of the Reichardt model illustrates that a difference in the inputs' time courses (indicated by the traces inside the boxes) is not necessary for directional selectivity. It is sufficient that the inputs have a different role (indicated by the black (R) and white (L) squares) in the nonlinear interaction. C. We illustrate one example of how different roles in the nonlinear interaction could lead to direction selectivity. For a rightward motion (right of the figure), the L input precedes the R input (and vice versa for leftward motion – shown on the left of the figure). Suppose that the L input gates the R input (L input allows R input signal to be transmitted further) when the L input is sufficiently high. Then, the responses would be proportional to the textured area of the R input curves and therefore different for leftward and rightward motion.

then the computed numbers are identical for all directions of motion. Large portions of the system computing direction selectivity could be linear (Watson & Ahumada, 1985), but at some stage, a nonlinearity is required, as for instance, the detection of amplitude or a decision rule.)

Figure 3A illustrates the simplest model proposed by Reichardt and colleagues (Reichardt, 1969) for insects' directional selectivity, in which inputs from two spatial locations converge onto an interaction site where these inputs are multiplied (the nonlinearity) and integrated. The inputs are spatially asymmetric, since there is only a slow signal coming from the left and a fast signal coming from the center. For rightward, but not leftward motion, the sluggishness of the left pathway is compensated for by the stimulus arriving at it first. Consequently, for rightward motion, if the speed is appropriate, both signals reach the multiplication site at roughly the same time, yielding a positive multiplication. In contrast, for leftward motion, the multiplication will yield a value close to zero.

The multiplication in Figure 3A is one of many quadratic-nonlinearity models supported by insect data. Poggio and Reichardt (1973) investigated the predictions common to all quadratic models of directional selectivity. For that purpose, they used the Volterra-series formulation. The output under such a formulation for a smooth, time-invariant, nonlinear interaction between the responses to stimuli in spatial locations a (z_a) and b (z_b) is

$$y(t) = h_{0,0} + \sum_{m=1}^{\infty} \sum_{j=0}^{m} h_{j,m-j} *^m z_a^{(j)} z_b^{(m-j)} \tag{5}$$

where $*^m$ is the mth order convolution and where $h_{j,m-j}$ are the mth order kernels of the interaction. An mth order kernel describes the nonlinear interaction between the responses to stimuli at m different instants in time. A quadratic nonlinearity is one for which if $m \geq 3$, $h_{j,m-j} = 0$. This means that a quadratic nonlinearity describes multiplicative interactions between responses to stimuli at *pairs* of times. Multiplication and squaring are particular examples of general quadratic interactions.

Two predictions of models with only quadratic nonlinearities are frequency doubling and superposition of nonlinearities. The former is the appearance in the Fourier spectrum of the response to moving sinusoidal gratings of energy at a frequency of twice the fundamental, but not at frequencies higher than that. Superposition of nonlinearities is the property in which the nonlinear average response to a grating composed of two sinusoidal gratings of different frequencies, but whose ratio is a rational number, is equal to the sum of the responses to the individual gratings. These two properties of quadratic nonlinearities can serve as a tool to test whether a system computes directional selectivity with this type of nonlinearity (they will be used in section 4.3.1 for this purpose).

4.2 Spatial asymmetries

As mentioned in the preceding section, spatial asymmetries are one of the fundamental requirements for models of directional selectivity. Put in other words, this requirement simply says that if a cell responds better to a motion coming from the left than to a motion coming from the right, then something must be different between left and right.

The simplest version of the Reichardt model illustrated in Figure 3A uses arguably the simplest form of spatial asymmetry possible. Inputs from only two spatial locations arrive at the site of interaction. (An input from a single location could not produce an asymmetry, of course.) To make the inputs different, thus creating an asymmetry, Reichardt and colleagues attributed different temporal properties to each input. However, in principle, there are other ways to differentiate the inputs and thus create an asymmetry (Figure 3B). For instance, if one input cannot cause a response on its own, but its activity enables (gates) a response proportional to the other input, then if the inputs have a fast rise time and slow decay, directional selectivity could be achieved (Figure 3C). Alternatively, one input could be inhibitory and the other excitatory, and their interaction nonlinear. Hence, although spatial asymmetries are necessary for directional selectivity, temporal differences between the inputs are not. This is an important point, since all models for directional selectivity presented in the literature have used temporal differences between inputs.

The spatial scale relevant for motion computations in a Reichardt model is given by the distance between the model's two inputs. Therefore, to achieve a multi-scale measurement of local direction based on this model, a visual system must compute the direction of motion several times in parallel (possibly using several different neurons), each time with the inputs separated by a different amount. An example of such a computation has been described in the literature (Reichardt et al., 1988). Another relevant spatial scale illustrated in Figure 3A comes from the possible spatial low-pass filtering performed by each of the two inputs. The importance of this filtering was emphasized by van Santen and Sperling (1985) to get rid of spatial aliasing not observed in human perception of motion. To understand aliasing, note that in the absence of filtering, moving sinewave gratings of higher and higher spatial frequencies would stimulate the inputs with arbitrary phase difference, thus making the model report incorrect directions of motion (Figure 4). If, however, low-pass filtering removes signals induced by frequencies higher than one over the distance between the inputs, then the possible phase differences are small and the correct direction can be reported. Another issue emphasized by van Santen and Sperling, and also by Reichardt and colleagues, is motion opponency, that is, the comparison of cells of opposite preferred directions for the computation of absolute local direction (see also the introduction to section 4).

Another important class of models, called Motion-Energy models (Adelson & Bergen, 1985), uses a distributed spatial asymmetry (as opposed to the two-input spatial asymmetry in the Reichardt model). This distributed spatial asymmetry arises from different locations in the receptive field having slightly different time courses of response (Figure 5A). Imagine that the region of the receptive field first reached by a stimulus moving in the preferred direction has a slow response. The next region reached by the stimulus has a slightly faster response. In fact, the later a region is reached, the faster its response. Consequently, if the stimulus speed is appropriate, the time courses of all the responses will overlap to yield a total response of large amplitude. For a motion in the opposite direction, the responses from the different regions will not overlap and thus the amplitude of the total response will be small. A model for such

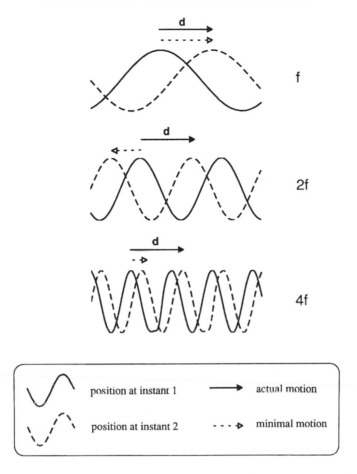

Figure 4   Aliasing of high spatial-frequency components. Imagine an image jumping to the right a distance d (indicated by the solid arrows). Let us concentrate on three of the image's Fourier components at frequencies f, 2f, and 4f. The solid curves show the position of these components before the jump and the dashed curves show the position after the jump. The open arrow shows the smallest displacement that would bring the Fourier components to the position of the dashed line. Hence, if the visual system bases its computations on this minimal motion, then only when the spatial frequency is sufficiently low will the actual and the computed motions coincide.

space-and-time co-dependence uses a receptive-field profile based on Gabor functions (Grzywacz & Yuille, 1990, 1991) such as that described in equation 4.

According to Motion-Energy models, directional selectivity is interpreted as orientation in space–time. This can be seen by plotting the time course of responses of different positions of the receptive field on space–time axes as schematically illustrated in Figure 5B.

Figure 5A illustrates the most important spatial scale in Motion-Energy models. It is the largest distance in the distributed spatial asymmetry. In terms of equation 4, this scale corresponds roughly to σ_r, the size of the receptive field. However, another spatial scale that is often discussed in conjunction with Motion-Energy models is one

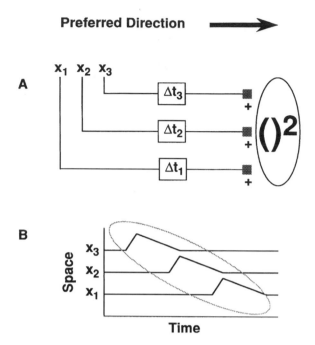

Figure 5 The Motion-Energy model for directional selectivity. A. This illustration of the model has several slow inputs from the left (at x_1, x_2 and x_3) and no inputs from the right and center. It is required that the time course for the inputs are such that $\Delta t_1 > \Delta t_2 > \Delta t_3$. The nonlinear interaction between inputs is a squaring operation after they are summed. For rightward, but not leftward motion, the decreasing sluggishness of the Δt_1, Δt_2 and Δt_3 pathways allows the responses from these pathways to overlap such that the total response has high amplitude. B. The Motion-Energy model can be viewed as a receptive-field profile tilted in space and time (dotted ellipse), that is, different spatial positions (for example x_1, x_2, x_3) will elicit responses with different time courses.

over the optimal spatial frequency that stimulates a DS cell. In terms of equation 4, this scale corresponds roughly to $1/\Omega_r$. It turns out that these two scales are essentially proportional in real DS cells (Hochstein & Shapley, 1976; Maffei & Fiorentini, 1977; Andrews & Pollen, 1979) and thus can be interchanged. Consequently, to achieve a multi-scale measurement of local direction based on Motion-Energy models, visual systems must compute the direction of motion several times in parallel, each time with a different range of spatial asymmetries or similarly, with different optimal spatial frequencies.

The above discussion of the Reichardt and Motion-Energy models does not include many of the details emphasized by the authors of the models. For instance, Adelson and Bergen (1985) postulated that pairs of space–time oriented receptive fields with different zones of excitation and inhibition (quadrature pairs) would be necessary to make the responses independent of the phase of the Fourier components of the stimuli. Our goal in this chapter was not to discuss specific details but simply to enquire into the origins of directional selectivity. We refer the reader to Chapter 5 by Mather for more details on models of early motion detectors.

4.3 Nonlinearities

As mentioned in section 4.1, the presence of a nonlinearity is one of the fundamental requirements for models of directional selectivity. Poggio and Reichardt (1973) proved mathematically that if the system is linear, then even though the time courses of the responses to motions in different directions might be different, the system cannot decide in what direction the motion is going. This section will be divided into two subsections, the first dealing with quadratic nonlinearities and the second with non-quadratic nonlinearities.

4.3.1 Quadratic nonlinearities

The simplest type of nonlinearity is a quadratic one. This type of nonlinearity implies that the behavior of the DS system depends only on the multiplications of pairs of input values. In other words, the quadratic interaction between inputs at three or more instants in time can be understood by decomposing it into pairwise interactions. The simplicity of quadratic nonlinearities has its origin in it being the lowest order nonlinear term in a Taylor-series approximation (or, more generally, in the Volterra series – equation 5).

The Reichardt model uses multiplication as its quadratic nonlinearity. The engineering motivation for using multiplication is to make the analogy between this model and the idea of statistical correlation complete. Hence, the Reichardt model can be thought of as computing the degree of correlation between the image in two different spatial positions in different instants in time. In contrast, the Motion-Energy model does not attempt to make a full analogy with statistical correlation. Instead of multiplication, many versions of the Motion-Energy model use a squaring operation, also a quadratic nonlinearity. Squaring polls the extent to which the signals passing through the various lateral asymmetric paths (Figure 5A) agree with one another. Multiplication and squaring emphasize different aspects of the signals. While multiplication is intended to make an analogy with the cross-correlation of two inputs (whether the input signals have the same waveform), squaring is intended to emphasize the temporal overlap of signals from various inputs (whether input signals arrive together).

There are some important similarities between the behaviors of the Reichardt and the Motion-Energy models, which arise because of their common use of quadratic nonlinearities. One similarity is that these models only respond correctly to the so-called Fourier motions. These motions are such that the Fourier components of the image move coherently. (Humans can also respond to motions in which this condition is not met, prompting the postulate of additional nonlinearities – Chubb & Sperling, 1988). Besides responding correctly only to Fourier motions, another similarity between the Reichardt model (in the extended version studied by van Santen and Sperling, 1985) and the Motion-Energy model is that on average, they yield identical responses to motion (Adelson & Bergen, 1985; van Santen & Sperling, 1985). (However, the details of the time courses of the various components of these models are different. Emerson et al. (1992) used these differences to argue from physiological data that the brain implements the Motion-Energy model.) The theoretical foundation for why the Extended Reichardt and Motion-Energy models share similar properties was laid down by Poggio and Reichardt (1976). They proved that a quadratic system with n inputs is equivalent to the sum of $n(n-1)/2$ two-input quadratic systems, which are all possible combinations of the n channels, two by two. Therefore, since the Motion-Energy model is a quadratic combination of multiple spatial inputs (Figure 5), it can be represented as multiple quadratic interactions of each pair of these inputs. Similarly, the Extended Reichardt model has multiple spatial inputs in the sense that each signal arriving at the quadratic interaction site is pooled from several spatial locations by the model's early spatial filtering (Figure 3). Consequently, the response of the Extended Reichardt model can also be analyzed as pairwise quadratic interactions of multiple spatial inputs.

From the computational perspective, there is a lot to be said in favor of using quadratic nonlinearities in models of directional selectivity. Besides being the simplest type of nonlinearity, quadratic interactions are optimal in terms of the 'resolution limit'. The resolving power of a DS system is determined by the spatial separation between the inputs to the system, here denoted by $\Delta\phi$. In agreement with Shannon's Sampling theorem, a periodic array of equidistant inputs can resolve uniquely the direction of movement of a periodic grating only if $\lambda \geq 2\Delta\phi$, where λ is the spatial wavelength of the grating. The resolution limit $\lambda = 2\Delta\phi$ is obtained by quadratic interactions between neighboring inputs (Poggio & Reichardt, 1976). Nonlinearities of higher order can (but will not necessarily) deteriorate the resolution limit set by the sampling theorem.

The experimental data are not conclusive about whether visual systems use quadratic nonlinearities for directional selectivity. van Santen and Sperling (1985), and Adelson and Bergen (1985) described human perceptual evidence in favor of the Reichardt and Motion-Energy models. However, all the discussed evidence hinged on the type of spatial asymmetry used by the models and their spatial scale, rather than on their nonlinearities. For instance, one widely documented piece of evidence is from the Missing-Fundamental paradigm (Adelson & Bergen, 1985; Georgeson & Shackleton, 1989). Here, a squarewave grating first jumps to the right a distance equal to $\frac{1}{4}$ of its cycle (90°) and observers see it move to the right. Next, the grating's Fourier fundamental is removed before the remaining grating makes the same jump. Now subjects see a motion to the left. This result is explained by saying that by removing the fundamental from the squarewave grating, the lowest Fourier component left has three times the spatial frequency of the fundamental. Hence, this component jumps $\frac{3}{4}$ of its cycle to the right (270°) or equivalently, $\frac{1}{4}$ of its cycle to the left (90°). Because both the Reichardt and Motion-Energy models can use motion detecting units of appropriate spatial scale to detect the Fourier component of highest amplitude (the third harmonic in this case), these models can account for the outcome of the Missing-Fundamental paradigm. However, in both cases, the explanation does not depend on the type of nonlinearity, but rather on the spatial scale of the asymmetry and on the stimulus.

Can psychophysical data argue in favor of the nonlinearity of human directional selectivity being quadratic? The task seems hard, since even if the relevant interactions are quadratic, other postulated nonlinearities such as half-wave rectification (Albrecht & De Valois, 1981) or inhibitory normalization (Heeger, 1993) might make the visual system's response appear non-quadratic. Conversely, even if the relevant interactions are higher order than quadratic, noise in the system would tend to get rid of the higher order nonlinearities, making the system appear more quadratic than it is. Nevertheless, Adelson and Bergen (1985), who analyzed the Motion-Energy model's psychophysical predictions with a quadratic nonlinearity for the sake of simplicity, argued that interaction at the output of the model (Figure 5) should probably not be quadratic. Their argument was that a quadratic nonlinearity would make the dynamic range of DS cells small. For example, a threefold increase in input would cause an almost tenfold increase in output, quickly forcing the cell's response out of its dynamic range.

Behavioral and physiological data in flies, and physiological data in cat striate cortex suggest that quadratic nonlinearities are used in these systems' DS cells. For example, in the fly, the optomotor reflex (turning in the direction of a visual motion) is independent of the relative phase of the Fourier components of the moving stimulus (phase invariance), as predicted by quadratic models (reviewed by Poggio & Reichardt, 1976). The evidence for quadratic nonlinearity in cat cortex comes from measurement of Wiener kernels (which resemble the Volterra kernels described above) by Emerson et al. (1992). In their data, kernels of order higher than second were found to be negligible.

In contrast to results in the fly and in cat cortex, experiments with the rabbit retina showed that the nonlinearity mediating directional selectivity is probably not quadratic (Grzywacz et al., 1990). In particular, the retina's DS cells failed the frequency-doubling and superposition-of-nonlinearity tests described in section 4.1. The authors of the rabbit retina study cautioned that their results do not imply that there is a difference in the type of nonlinearity used in the directional selectivity of rabbits, flies, and cats, since the tests of quadratic nonlinearity applied to the latter two species might not have been sufficiently strong. For instance, a previous theoretical study showed that some non-quadratic nonlinearities do not violate phase invariance grossly (Grzywacz & Koch, 1987), which thus might have appeared to hold in the fly experiments, even if it did not. Furthermore, the amplitude of kernels falls fast with the kernels' order (as do the coefficients of high order terms in a Taylor expansion), and thus the kernels with order higher than second might be difficult to detect.

4.3.2   Non-quadratic nonlinearities

From the preceding discussion, one may conclude that quadratic nonlinearities do not apply to DS cells in the rabbit retina, and might not even apply to DS cells in fly and in cat cortex. Why might visual systems not always use a nonlinearity that is the simplest and the optimal from a sampling-resolution perspective? One possible answer (mentioned above) is that quadratic nonlinearities are not favorable in view of a neuron's limited dynamic range (Adelson & Bergen, 1985). Another answer is that it is hard to implement quadratic nonlinearities with the biophysical processes available to neurons. The simplest way to implement this nonlinearity would be to restrict the inputs to the DS cells to small values, thus making the contribution of high order nonlinearities negligible (as in a Taylor-series approximation, when high order terms are neglected). However, Grzywacz and Koch (1987) analyzed the main biophysical mechanisms thought to be involved in directional selectivity, using criteria like frequency doubling and superposition of nonlinearities, and concluded that a small-signal approximation to a quadratic nonlinearity is not practical. They showed that the signal would have to be so small that it would be buried in the noise. We want to stress, however, that the problem of using quadratic nonlinearities is not a fundamental one, but arises from implementational difficulties.

If not quadratic, then what sorts of nonlinearities may be involved in directional selectivity? For the rabbit retina, the seminal work of Barlow and Levick (1965) suggested the involvement of a spatially-asymmetric nonlinear inhibitory process, which they called veto inhibition. Figure 6 illustrates their model in which the spatial asymmetry is due to inputs only coming from the right (inhibitory and slow) and center (excitatory and fast). The model generates directional selectivity, because when the motion comes from the right, but not the left, the sluggishness of the inhibitory pathway makes the signals from inhibitory and excitatory pathways arrive at about the same time. Hence, the excitatory signal is vetoed by the inhibitory signal. Recent experiments have shown that the underlying inhibitory nonlinearity is not all-or-none (as would be implied by the term 'veto'), but rather, it divides the excitatory response by a factor that is monotonically related to the stimulus contrast (Amthor & Grzywacz, 1991, 1993). Therefore, the nonlinearity involved in the rabbit retina directional selectivity is a division-like inhibition (Figure 6).

Striate-cortex investigators are still debating whether a nonlinear inhibitory mechanism also underlies cortical directional selectivity. Reid *et al.* (1991) indicated that, for non-preferred directions, an inhibitory mechanism might suppress responses. They suggested that this mechanism might be nonlinear inhibition similar to that which Barlow and Levick proposed for retinal directional selectivity. Reid *et al.*'s interpretation for cortical inhibition shares its asymmetry with the mechanism suggested by Douglas and Martin (1992) in their model of cortical directional selectivity. However, the nonlinearity underlying Douglas and Martin's 'functional microcircuit' for cortex is positive feedback mediated by intracortical excitation, which is prevented from being 'ignited' by relatively weak, and essentially linear inhibition. In contrast, Heeger (1993) reinterpreted Reid *et al.*'s data by suggesting that the non-preferred-

Preferred Direction ⟶

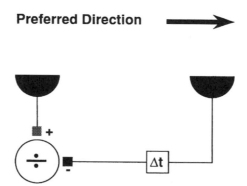

Figure 6 The Barlow and Levick model for retinal directional selectivity. This illustration of the model has a slow input from the right and a fast input from the center. The center input is excitatory and the slow input causes a nonlinear inhibition (black box). For leftward, but not rightward motion, the sluggishness of the right pathway is compensated by the light reaching it before reaching the center pathway. Therefore, the signals arrive at the interaction site roughly simultaneously, and the inhibition vetoes the excitation. Recent data indicated that the inhibition is division-like ($\div$), rather than subtraction-like or all-or-none veto.

direction suppression might be due to a cortical inhibitory network devoted to normalizing neural responses.

If the mechanism mediating cortical directional selectivity is a division-like inhibition as in the retina, then this mechanism might be an implementation of a computational model of directional selectivity called the Gradient model (Marr & Ullman, 1981; Harris, 1986). This model computes image velocity, and thus direction of motion as the ratio between temporal and spatial gradients, as suggested by equation 1. Because the model requires a division, a division-like inhibition could implement it (Poggio, 1983; Harris, 1986). (Marr and Ullman (1981) showed that one can also implement the Gradient model with an 'And' operation instead of a division-like nonlinearity.)

Finally, another nonlinearity that has been frequently mentioned in the literature in the context of directional selectivity is threshold (Grzywacz & Koch, 1987). It could replace the multiplication or the squaring in Reichardt and Motion-Energy models respectively. In this case, during preferred, but not null-direction motions, the signals arriving at the interaction site would temporally coincide and their sum would cross threshold. The nice feature of this model is that above threshold, the responses could behave linearly as reported in early studies of cortical directional selectivity (Albrecht & De Valois, 1981; Reid et al., 1987). However, an exclusive threshold mechanism for directional selectivity would have the disadvantage of preventing responses at low stimulus contrasts. At these contrasts, regardless of whether the motion is in the preferred or null direction, the signals would be too small to cross threshold.

5 BIOPHYSICS OF RETINAL DIRECTIONAL SELECTIVITY

In section 4.3.1, we provided two reasons for why, despite being the simplest nonlinearity and optimal from a sampling-resolution perspective, a quadratic mechanism might not be used by biological visual systems computing directional selectivity. The first reason was that a quadratic nonlinearity will often cause the output to be beyond the dynamic range of neurons. The second reason was that it is difficult to implement a quadratic nonlinearity with the biophysical elements available to neurons. These reasons emphasize that the mechanisms used in a particular computation, in this case directional selectivity, are not just constrained by the computation itself but also by neural limitations. Consequently, we feel that to understand what is involved in the computation of local motion, it is important to consider an example of a computation of local motion that is being worked out in as much biophysical detail as possible. The example we chose is that of retinal directional selectivity. We begin with a section describing the experimental background. This section will emphasize the importance of a nonlinear, GABAergic, inhibitory mechanism. The next section presents the first detailed biophysical model proposed in the literature, the shunting-inhibition model. Section 5.3 argues that there are serious experimental challenges to this model, and to the Barlow and Levick model. This argument is followed by a section proposing an alternative biophysical model for retinal directional selectivity.

5.1 Experimental preliminaries

In their classical investigation of the On–Off DS ganglion cells of the rabbit retina, Barlow and Levick (1965) showed the importance, for directional selectivity, of inhibition within the excitatory receptive field center. In their two-slit apparent-motion experiments, they demonstrated that turning on and off two spatially separated slits at some appropriate delay, in a manner similar to that which yields apparent motion illusions in humans, elicited direction-associated, asymmetric responses from the On–Off DS ganglion cells. When the two-slit apparent motion corresponded to the null direction, the number of spikes elicited by the second slit was much smaller than that elicited by the second slit when it was presented alone. This reduction occurred even though the response to the first slit itself was excitatory. The inhibitory effect of the first slit on the second was not due to saturation, could be observed virtually everywhere within the excitatory receptive field center, and was effective for interslit distances that were much less than the width of the excitatory center of the receptive field. Barlow and Levick concluded from these and similar data that direction-of-motion response asymmetry was computed by numerous, small 'subunits', approximately 0.5° wide, replicated across the excitatory receptive field center. The mechanism for the motion asymmetry was postulated to be a delayed inhibitory effect propagated asymmetrically only in the null direction (see Figure 6).

Experiments by Wyatt and Daw (1975) and more recently, by Amthor and Grzywacz (1993) quantifed the properties of the inhibitory mechanism mediating directional selectivity in rabbit. The impulse response of the inhibition lasts between 0.5 and 1 second. In response to a long step, rather than a flash of light, the inhibition is sustained. Although the null-direction inhibition is strongest for short interslit distances (on the order of those postulated for Barlow and Levick's subunits), it is still quite effective for interslit distances larger than half the size of the excitatory receptive field center. In On–Off DS ganglion cells, null-direction inhibition is such that directional selectivity appears to be computed independently for light (On) and dark (Off) regions of moving objects or textures.

Amthor and Grzywacz (1993) also investigated whether the inhibitory interaction with excitation was linear, by determining the contrast dependency of the interaction. If the interaction between null-direction inhibition and excitation were linear, such as a subtraction (which would have to be followed later by some nonlinearity essential for the directional selectivity), increasing the contrast of the first (inhibiting) slit in a two-slit experiment, should cause a downward shift (subtraction) in the response versus contrast function of the second slit. On the other hand, nonlinear interactions could cause changes in the shape of the response versus contrast function. For example, a shunting-inhibition mechanism (Torre & Poggio, 1978) would tend to divide the response versus contrast function by a constant factor. The data showed that the contrast dependency of the null-direction inhibition is division-like, rather than subtraction-like.

Pharmacological studies in rabbit and turtle retina indicated that the inhibitory neurotransmitter gamma-aminobutyric acid (GABA) (via $GABA_A$ receptors) mediates null-direction inhibition in DS ganglion cells (Caldwell et al., 1978; Ariel & Adolph, 1985). In the presence of picrotoxin, a GABA antagonist, the null-direction responses of the DS ganglion cells increased to levels similar to the levels elicited by preferred-direction motion. Similar experiments using antagonists to glycine (such as strychnine) showed that glycine had little specific effect on directional selectivity in either the On–Off or On DS ganglion cells of the retina.

An important question about the mechanism of null-direction inhibition is its site of action: does the null-direction inhibition act on the ganglion cell itself or at some presynaptic site? A ganglion-cell mechanism might be evidenced by inhibitory conductances and perhaps hyperpolarizations during null-direction motions in intracellular recordings. Miller (1979) and Werblin (1970) reported such hyper-polarizations in On–Off DS ganglion cells of rabbit and mudpuppy respectively. However, other recordings in these and other species have not exhibited such hyperpolarizations. Amthor *et al.* (1989) observed small depolarizations during null-direction motion in On–Off DS ganglion cells of the rabbit. Similar depolarizations for null-direction motion have also been observed in experiments with turtle (Marchiafava, 1979) and frog (Watanabe & Murakami, 1984).

These studies indicate the possible existence of a direct inhibitory input onto the DS ganglion cell. However, they do not settle the question of whether the inhibitory input mediates directional selectivity or has a different role. In section 5.3 we will discuss recent patch-clamp data indicating that the latter is correct in turtle. Before that, we will describe detailed biophysical models of the inhibitory mechanism of retinal directional selectivity, which motivated the patch-clamp studies.

5.2 Shunting inhibition mechanism for directional selectivity

Torre and Poggio (1978) proposed a biophysical implementation of the nonlinear inhibition advanced by Barlow and Levick. Their rationale was based on retinal directional selectivity being elicited by motions spanning short distances almost anywhere inside the receptive field (see preceding section). Torre and Poggio suggested that this subunit-like behavior could be accounted for by each subunit corresponding to a branch of the DS ganglion cell's dendritic tree. To keep the computation constrained to each branch, they suggested that the inhibition mediating retinal directional selectivity works through a synapse that causes local changes of membrane conductance (shunting inhibition) and little hyperpolarization. To do so, the synapse would have its reversal potential near the cell's resting potential. To understand how such a synapse works, consider a patch of membrane receiving excitatory (g_e) and shunting inhibitory (g_i) synaptic conductances. Setting the resting and inhibitory reversal potentials to zero (without loss of generality), the voltage V obeys:

$$C\frac{dV(t)}{dt} + (g_e(t) + g_i(t) + g_{leak})V(t) = g_e(t)E_e + g_{leak}E_{leak} \qquad (6)$$

where C is membrane capacitance, g_{leak} is the membrane's leak conductance, and E_e and E_{leak} are reversal potentials of g_e and g_{leak} respectively. When g_i is large, that is, $g_i \gg g_e, g_{leak}$, then V falls rapidly towards the following equilibrium $(dV/dt \to 0)$ value:

$$V(t) \rightarrow \frac{g_e E_e + g_{leak} E_{leak}}{g_i} \tag{7}$$

which is nearly zero, because g_i is large. Therefore, this inhibition is division-like, rather than subtraction-like.

Torre and Poggio also pointed out that a shunting-inhibition mechanism for vertebrate retinal directional selectivity might be consistent with the quadratic nonlinearity observed in insects. Their argument was that for sufficiently low contrasts, one can neglect the higher order nonlinearities (equation 5), as in Taylor-series approximation.

5.3   Problems with previous models for retinal directional selectivity

The involvement of shunting inhibition in vertebrate retinal directional selectivity has been supported experimentally by the inhibitory mechanism mediating retinal directional selectivity in rabbit being division-like (see section 5.1). Moreover, as pointed out in section 5.1, there is evidence for a direct shunting inhibitory input onto the DS ganglion cell, as postulated by the Torre and Poggio model.

However, recent experiments did not support other aspects of the Torre and Poggio model. A quadratic approximation is not valid even for the smallest contrasts eliciting responses (Grzywacz et al., 1990). This was demonstrated by the failure of rabbit DS ganglion cells in the frequency-doubling and superposition-of-nonlinearities tests. (Previously, Grzywacz and Koch (1987) argued theoretically that the quadratic approximation for shunting inhibition would probably not be valid under physiological conditions.) Furthermore, a whole-cell patch-clamp experiment by Borg-Graham and Grzywacz (1992) casts doubt on whether the relevant inhibition acts on the DS ganglion cell of turtle. In whole-cell patch-clamp recordings, in which there was no ATP or Mg^{2+} in the electrode, direct inhibition of the ganglion cell was blocked. Under these conditions, ganglion cells maintained directionality despite lack of inhibition. Borg-Graham and Grzywacz concluded that the site of the nonlinear inhibitory mechanism is in an amacrine cell's dendrite, since its origin appears not to be in bipolar (Werblin, 1970; Marchiafava, 1979) or ganglion cells (Borg-Graham & Grzywacz, 1992), and one does not observe it in amacrine-cell somas (Werblin, 1970; Marchiafava, 1979). (The direct inhibition onto the DS ganglion cells in turtle probably mediates other functions than directional selectivity, such as gain control.)

Recent data also challenge the Barlow and Levick model. As pointed out in section 5.1, pharmacological evidence indicates that the inhibition mediating retinal directional selectivity is GABAergic via $GABA_A$ receptors. Smith et al. (1991, 1994) extended the analysis of the effect of picrotoxin on turtle retinal directional selectivity. They studied this effect as a function of contrast and speed of a continuously moving slit, and for apparent motions. Their data showed that picrotoxin does not eliminate directional selectivity in every single cell. In 28% of the DS ganglion cells, the preferred and null directions were actually reversed under picrotoxin. In 28% of the cells, directional selectivity was maintained despite saturating concentrations of picrotoxin. From these results, Smith et al. suggested that GABA does not mediate directional selectivity by being at the output of a synapse whose receptive field is asymmetric (as the Barlow and Levick model would suggest). Otherwise, GABA antagonists should eliminate directional selectivity and most certainly, should not reverse it.

Finally, the postulate of an exclusive inhibitory mechanism of retinal directional selectivity is not correct. Grzywacz and Amthor (1993) expanded earlier experiments of Barlow and Levick (1965) to find that if the spatio-temporal parameters of the stimulus are appropriate, then there is a strong preferred-direction facilitation. They also obtained evidence (based on apparent-motion protocols) that the facilitatory effect is not entirely due to non-specific mechanisms such as ganglion-cell spiking threshold. In addition, facilitation is not just dis-inhibition. When facilitatory and inhibitory effects were disentangled with appropriate control of stimulus parameters such as the use of steps, rather than flashes of light, Grzywacz and Amthor showed that the maximal strength of inhibition was only slightly larger (by a factor of about 1.30) than the maximal strength of facilitation across a large range of contrasts (10 to 70%). Maximal strength of inhibition (facilitation) is defined as the maximal reduction (increase) in the number of spikes illicited by the second frame of the apparent motion. Because facilitation and inhibition have comparable maxima under some conditions, one cannot neglect the contributions of facilitation to the properties of retinal directional selectivity.

5.4   A new model for retinal directional selectivity

Vaney (1990), and Borg-Graham and Grzywacz (1992) independently proposed new models for retinal directional selectivity that overcome the difficulties discussed in the previous section. Figure 7 illustrates some of the common points of the Vaney, and Borg-Graham and Grzywacz models. They are similar to the Reichardt model (Figure 3A) in that the spatial asymmetry is excitatory, that is, the inhibitory mechanism is spatially *symmetric*. However, different than that model, the nonlinearity is not quadratic, but rather inhibitory, as in the Barlow and Levick model, and of the shunting type, as in the Torre and Poggio model. Based on their patch-clamp results, Borg-Graham and Grzywacz suggested that the inhibition acts on an amacrine cell's dendrite. Based on anatomical findings, Vaney emphasized that this amacrine dendrite might belong to the cholinergic starburst amacrine cell. Because this cell exists in two subpopulations, one responding to light onset and another responding to light offset (Famiglietti, 1983, 1991), directional selectivity would be computed separately for light (On) and dark (Off) regions of moving objects or textures, this is consistent with available data (see section 5.1).

Computer simulations with this model show that it can generate retinal directional selectivity in the dendrites of an amacrine cell without leading to retinal directional selectivity in its soma (Borg-Graham & Grzywacz, 1992). In addition, simulations show that the model can account for reversals of preferred and null directions under picrotoxin as a result of saturation of the synapse between the amacrine dendrite and DS ganglion cell. Finally, the model can account for preferred-direction facilitation as, for example, synaptic facilitation. Such a facilitatory mechanism is consistent with an extension of facilitation outside the dendritic tree of the ganglion cell (Grzywacz & Amthor, 1993) without an extension of its excitatory receptive field (Amthor *et al.*, 1984; Amthor *et al.*, 1989; Yang & Masland, 1992).

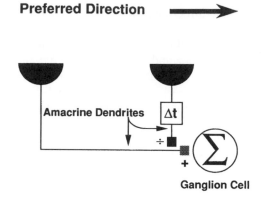

Figure 7   An amacrine-cell model for retinal directional selectivity. This model is similar to the Reichardt model in that the asymmetry is excitatory. But this model's nonlinearity is inhibitory, as in the Barlow and Levick model. Because it resulted from detailed retinal studies, one can identify in this model the underlying cellular processes. The ganglion cell sums ($\sum$) the inputs from several asymmetric excitatory amacrine dendrites, which in turn, are inhibited by shunting inhibitory (amacrine) processes.

6   CONCLUSIONS

We have argued that because of computational constraints, visual systems do not try to compute detailed and precise optic flows. Rather, they focus on the measurement of direction of local motion and estimate local speed roughly from the spatial and temporal properties of directionally selective neurons. To measure direction, directionally selective neurons must use a spatially asymmetric process and a nonlinearity. Although from a computational perspective, the latter should ideally be quadratic, neural mechanisms are not suitable for it for practical reasons. Inspection of retinal directional selectivity, an example of a system for which much anatomical, physiological, and pharmacological data are available, reveals that local direction measurement might be based on a division-like inhibitory process. In the retina, amacrine-cell dendrites might in some cases be the substrate of spatial asymmetry.

The general lesson is that to understand a visual computation, or more generally, a brain computation, it is at least useful, if not necessary, to use a multi-level approach combining the computational, behavioral, and neural perspectives.

ACKNOWLEDGMENTS

Support for this work came from a grant from the Office of Naval Research (N00014-91-J-1280), from grants from the National Eye Institute (EY08921) and Air Force Office of Sponsored Research (F49620-92-J0156), and an award from the Paul L. and Phyllis C. Wattis Foundation to N.M.G., from a grant from the National Eye Institute (EY05070) to F.R.A., from a Human Frontiers Science Program Organisation Long-term Fellowship to J.M.H., and from a core grant from the National Eye Institute to

Smith-Kettlewell (EY06883). We thank Dr Suzanne McKee for helpful comments on the manuscript.

REFERENCES

Adelson, E. H. & Bergen, J. R. (1985). Spatio-temporal energy models for the perception of motion. *J. Opt. Soc. Am. A*, **2**, 284–299.

Albrecht, D. G. & De Valois, R. L. (1981). Striate cortex responses to periodic patterns with and without the fundamental harmonics. *J. Physiol.*, **319**, 497–514.

Amthor, F. R. & Grzywacz, N. M. (1991). The nonlinearity of the inhibition underlying retinal directional selectivity. *Vis. Neurosci.*, **6**, 197–206.

Amthor, F. R. & Grzywacz, N. M. (1993). Inhibition in On-Off directionally selective ganglion cells in the rabbit retina. *J. Neurophysiol.*, **69**, 2174–2187.

Amthor, F. R., Oyster, C. W. & Takahashi, E. S. (1984). Morphology of ON–OFF direction selective ganglion cells in the rabbit retina. *Brain Res.*, **298**, 187–190.

Amthor, F. R., Takahashi, E. S. & Oyster, C. W. (1989). Morphologies of rabbit retinal ganglion cells with complex receptive fields. *J. Comp. Neurol.*, **280**, 97–121.

Andrews, B. W. & Pollen, D. A. (1979). Relationship between spatial frequency selectivity and receptive field profile of simple cells. *J. Physiol.*, **287**, 163–176.

Ariel, M. & Adolph, A. R. (1985). Neurotransmitter inputs of directionally sensitive turtle retinal ganglion cells. *J. Neurophysiol.*, **54**, 1123–1143.

Baker, C. L. & Braddick, O. J. (1982). Does segregation of differently moving areas depend on relative or absolute displacement? *Vision Res.*, **22**, 851–856.

Barlow, H. B. & Levick, W. R. (1965). The mechanism of directionally selective units in rabbit's retina. *J. Physiol.*, **178**, 477–504.

Barlow, H. B., Hill, R. M. & Levick, W. R. (1964). Retinal ganglion cells responding selectively to direction and speed of motion in the rabbit. *J. Physiol.*, **173**, 377–407.

Borg-Graham, L. J. & Grzywacz, N. M. (1992). A model of the direction selectivity circuit in retina: Transformations by neurons singly and in concert. In T. McKenna, J. Davis & S. F. Zornetzer (Eds.) *Single Neuron Computation*, pp. 347–375. Academic Press, Orlando, Florida, USA.

Bravo, M. J. & Watamaniuk, S. N. J. (1991). Speed segregation and transparency in random dot displays. *Invest. Ophthalmol. Vis. Sci.*, **33**, 1050.

Burden, R. L. & Faires, J. D. (1989). *Numerical Analysis*. PWS-Kent Publishing Company, Boston, MA.

Burr, D. C., Ross, J. & Morrone, M. C. (1986). Seeing objects in motion. *Proc. R. Soc. Lond. B*, **227**, 249–265.

Caldwell, J. H., Daw N. W. & Wyatt H. J. (1978). Effects of picrotoxin and strychnine on rabbit retinal ganglion cells: lateral interactions for cells with more complex receptive fields. *J. Physiol.*, **276**, 277–298.

Chubb, C. & Sperling, G. (1988). Drift-balanced random stimuli: a general basis for studying non-Fourier motion perception. *J. Opt. Soc. Am. A*, **5**, 1986–2006.

Daugman, J. G. (1985). Uncertainty relation for resolution in space, spatial frequency, and orientation optimized by two dimensional visual cortical filters. *J. Opt. Soc. Am. A*, **2**, 1160–1169.

Douglas, R. J. & Martin, K. A. C. (1992). Exploring cortical microcircuits: a combined anatomical, physiological, and computational approach. In T. McKenna, J. Davis & S. F. Zornetzer (Eds.) *Single Neuron Computation*, pp. 381–412. Academic Press, Orlando, Florida, USA.

Dow, B. M. (1974). Functional classes of cells and their laminar distribution in monkey visual cortex. *J. Neurophysiol.*, **37**, 927–946.

Emerson, R. C., Bergen, J. R. & Adelson, E. H. (1992). Directionally selective complex cells and the computation of motion energy in cat visual cortex. *Vision Res.*, **32**, 203–218.

Fahle, M. & Poggio, T. (1981). Visual hyperacuity: Spatio-temporal interpolation in human vision. *Proc. R. Soc. Lond. B*, **213**, 451–477.

Famiglietti E. V. (1983). On and Off pathways through amacrine cells in mammalian retina: the synaptic connections of 'starburst' amacrine cells. *Vision Res.*, **23**, 1265–1279.

Famiglietti, E. V. (1991). Synaptic organization of starburst amacrine cells in rabbit retina: Analysis of serial thin sections by electron microscopy and graphic reconstruction. *J. Comp. Neurol.*, **309**, 40–70.

Fennema, C. I. & Thompson, W. B. (1979). Velocity determination in scenes containing several moving objects. *Comput. Graph. Image Proc.*, **9**, 301–315.

Foley, J. M. (1980). Binocular distance perception. *Psychol. Rev.*, **87**, 411–434.

Gabor, D. (1946). Theory of communication. *J. Inst. Electr. Eng.*, **93**, 429–457.

Georgeson, M. A. & Shackleton, T. M. (1989). Monocular motion sensing, binocular motion perception. *Vision Res.*, **29**, 1511–1523.

Gibson, J. J. (1950). *The Perception of the Visual World*. Houghton-Mifflin, Boston, Mass.

Gottsdanker, R. M. (1956). The ability of human operators to detect acceleration of target motion. *Psychol. Bull.*, **53**, 477–487.

Grzywacz, N. M. & Amthor, F. R. (1993). Facilitation in On-Off directionally selective ganglion cells in the rabbit retina. *J. Neurophysiol.*, **69**, 2188–2199.

Grzywacz, N. M. & Koch, C. (1987). Functional properties of models for direction selectivity in the retina. *Synapse*, **1**, 417–434.

Grzywacz, N. M. & Poggio, T. (1990). Computation of motion by real neurons. In S. F. Zornetzer, J. L. Davis & C. Lau. (Eds.) *An Introduction to Neural and Electronic Networks*, pp. 379–403. Academic Press, San Diego, USA.

Grzywacz, N. M. & Yuille, A. L. (1990). A model for the estimate of local image velocity by cells in the visual cortex. *Proc. R. Soc. Lond. B*, **239**, 129–161.

Grzywacz, N. M. & Yuille, A. L. (1991). Theories for the visual perception of local velocity and coherent motion. In M. S. Landy & J. A. Movshon (Eds.) *Computational Models of Visual Processing*, pp. 231–252. MIT Press, Cambridge, MA.

Grzywacz, N. M., Amthor, F. R. & Mistler, L. A. (1990). Applicability of quadratic and threshold models to motion discrimination in the rabbit retina. *Biol. Cybernet.*, **64**, 41–49.

Harris, M. G. (1986). The perception of moving stimuli: a model of spatiotemporal coding in human vision. *Vision Res.*, **26**, 1281–1287.

Hausen, K. (1981). Monocular and binocular computation of motion in the Lobula Plate of the fly. *Verh. Dtsch. Zool. Ges.*, **74**, 49–70.

Heeger, D. J. (1987). A model for the extraction of image flow. *J. Opt. Soc. Am. A*, **4**, 1455–1471.

Heeger, D. J. (1993). Modeling simple cell direction selectivity with normalized, half-squared, linear operators. *J. Neurophysiol.*, **70**, 1885–1898.

Helmholtz, H. V. (1910/1962). *Physiological Optics.* (Vol. III), Dover, New York.

Hildreth, E. C. (1983). *The Measurement of Visual Motion.* MIT Press, Cambridge, MA.

Hildreth, E. C. (1992). Recovering Heading for Visually Guided Navigation. *Vision Res.*, **32**, 1177–1192.

Hochstein, S. & Shapley, R. M. (1976). Quantitative analysis of retinal ganglion cell classifications. *J. Physiol.*, **262**, 237–264.

Hoffman, D. D. & Bennett, B. M. (1985). The computation of structure from fixed-axis motion: nonrigid structures. *Biol. Cybernet.*, **51**, 293–300.

Hoffman, D. D. & Bennett, B. M. (1986). The computation of structure from fixed-axis motion: rigid structures. *Biol. Cybernet.*, **54**, 71–83.

Holub, R. A. & Morton-Gibson, M. (1981). Response of visual cortical neurons of the cat to moving sinusoidal gratings: Response-contrast functions and spatiotemporal integration. *J. Neurophysiol.*, **46**, 1244–1259.

Horn, B. K. P. & Schunck, B. G. (1981). Determining optical flow. *Artificial Intelligence*, **17**, 185–203.

Howard, I. P. (1961). An investigation of the satiation process in the reversible perspective of revolving skeletal shapes. *Q. J. Exp. Psychol.*, **9**, 19–33.

Hubel, D. H. & Wiesel, T. N. (1962). Receptive fields, binocular interaction and functional architecture in the cat's visual cortex. *J. Physiol.*, **1160**, 106–154.

Johnston, E. B. (1991). Systematic distortions of shape from stereopsis. *Vision Res.*, **31** 1351–1360.

Jones, J. P. & Palmer, L. A. (1987). An evaluation of the two-dimensional Gabor filter model of simple receptive fields in cat striate cortex. *J. Neurophysiol.*, **58**, 1233–1258.

Julesz, B. (1971). *Foundations of Cyclopean Vision*, University of Chicago Press, Chicago.

Koenderink, J. J. (1986). Optic flow. *Vision Res.*, **26**, 161–170.

Koenderink, J. J. & van Doorn, A. J. (1976). Local structure of movement parallax of the plane. *J. Opt. Soc. Am.*, **66**, 717–723.

Komatsu, H. & Wurtz, R. H. (1988). Relation and cortical areas MT and MST to pursuit eye movements. III. Interaction with full-field visual stimulation. *J. Neurophysiol.*, **60**, 621–644.

Lee, D. N. (1976). A theory of visual control of braking based on information about time-to-collision. *Perception*, **5**, 437–457.

Longuet-Higgins, H. C. (1984). The visual ambiguity of a moving plane. *Proc. R. Soc. Lond. B*, **223**, 165–175.

Longuet-Higgins, H. C. & Prazdny, K. (1980). The interpretation of moving retinal images. *Proc. R. Soc. Lond. B*, **208**, 385–397.

Maffei, L. & Fiorentini, A. (1977). Spatial frequency rows in the striate visual cortex. *Vision Res.*, **17**, 257–264.

Marchiafava, P. L. (1979). The responses of retinal ganglion cells to stationary and moving visual stimuli. *Vision Res.*, **19**, 1203–1211.

Marr, D. (1982). *Vision: A Computational Investigation into the Human Representation and Processing of Visual Information.* Freeman, San Francisco.

Marr, D. C. & Ullman, C. (1981). Directional selectivity and its use in early visual processing. *Proc. R. Soc. Lond. B*, **211**, 1881–1887.

Maturana, H. R., Lettvin, J. Y., McCulloch, W. S. & Pitts, W. H. (1960). Anatomy and physiology of vision in the frog (*Rana pipiens*). *J. Gen. Physiol.*, **43** (Suppl. 2), 129–171.

McKee, S. P. & Welch, L. (1985). Sequential recruitment in the discrimination of velocity. *J. Opt. Soc. Am. A*, **2**, 243–251.

McKee, S. P., Silverman, G. H. & Nakayama, K. (1986). Precise velocity discrimination despite random variations in temporal frequency and contrast. *Vision Res.*, **26**, 609–619.

Miles, F. A. & Wallman, J. (Eds.) (1993). *Visual Motion and Its Role in the Stabilization of Gaze*. Elsevier, Amsterdam.

Miller, R. F. (1979). The neuronal basis of ganglion-cell receptive-field organization and the physiology of amacrine cells. In F. O. Schmitt & F. G. Worden (Eds.) *The Neurosciences, Fourth Study Program*, pp. 227–245. MIT Press, Cambridge, Massachusetts.

Nakayama, K. (1985). Biological image motion processing: a review. *Vision Res.*, **25**, 625–660.

Nakayama, K. & Loomis, J. M. (1974). Optical velocity patterns, velocity-sensitive neurons, and space perception: a hypothesis. *Perception*, **3**, 63–80.

Nawrot, M. & Blake, R. (1991). The interplay between stereopsis and structure from motion. *Percept. Psychophys.*, **49**, 230–244.

Poggio, T. (1983). Visual Algorithms. In O. J. Braddick & A. C. Sleigh (Eds.) *Physical and Biological Processing of Images,* pp. 128–153, Springer-Verlag, Berlin.

Poggio, T. & Reichardt, W. E. (1973). Considerations on models of movement detection. *Kybernetik*, **13**, 223–227.

Poggio, T. & Reichardt, W. E. (1976). Visual control of orientation behaviour in the fly: Part II: Towards the underlying neural interactions. *Q. Rev. Biophys.*, **9**, 377–438.

Ratliff, F. (1965). *Mach Bands: Quantitative Studies on Neural Networks in the Retina*. Holden-Day, San Francisco.

Regan, D. (1986). Visual processing of four kinds of relative motion. *Vision Res.*, **26**, 127–145.

Regan, D. & Beverley, K. I. (1984). Figure-ground segregation by motion contrast and by luminance contrast. *J. Opt. Soc. Am. A,* **1**, 433–442.

Regan, D. M., Kaufman, L. & Lincoln, J. (1986). Motion in depth and visual acceleration. In K. R. Boff, L. Kaufman & J. P. Thomas (Eds.) *Handbook of Perception and Human Performance. Volume I. Sensory Processes and Perception*. Wiley, New York.

Reichardt, W. (1969) Movement perception in insects. In W. Reichardt (ed) *Processing of Optical Data by Organisms and Machines*, pp 465–493. Academic Press, London.

Reichardt, W. & Poggio, T. (1976). Visual control of orientation behaviour in the fly. Part I. A quantitative analysis. *Q. Rev. Biophys.*, **9**, 311–375.

Reichardt, W., Egelhaaf, M. & Schlögl, R. W. (1988). Movement detectors provide sufficient information for local computation of 2-D velocity field. *Die Naturwissenschaften*, **75**, 313–315.

Reid, R. C., Soodak, R. E. & Shapley, R. M. (1987). Linear mechanisms of directional selectivity in simple cells of cat striate cortex. *Proc. Natl Acad. Sci. USA,* **84**, 8740–8744.

Reid, R. C., Soodak, R. E. & Shapley, R. M. (1991). Directional selectivity and spatiotemporal structure of receptive fields of simple cells in cat striate cortex. *J. Neurophysiol.,* **66**, 505–529.

Rieger, J. H. & Lawton, D. T. (1985). Processing differential image motion. *J. Opt. Soc. Am. A,* **2**, 354–360.

Richards, W. (1985). Structure from stereo and motion. *J. Opt. Soc. Am. A,* **2**, 343–349.

Schmerler, J. (1976). The visual perception of accelerated motion. *Perception,* **5**, 167–185.

Sekuler, A. B. (1990). Motion segregation from speed differences: evidence for nonlinear processing. *Vision Res.,* **30**, 785–795.

Sekuler, A. B. (1992). Simple-pooling of undirectional motion predicts speed discrimination for looming stimuli. *Vision Res.,* **32**, 2277–2288.

Smith, R. D., Grzywacz, N. M. & Borg-Graham, L. (1991). Picrotoxin's effect on contrast dependence of turtle retinal directional selectivity. *Invest. Ophthalmol. Vis. Sci.,* **32**, 1263.

Smith, R. D., Grzywacz, N. M. & Borg-Graham, L. (1994). Complex effects of GABA$_A$ antagonists on turtle's retinal directional selectivity. Submitted for publication.

Sparks, D. L. & Mays, L. E. (1990). Signal transformation required for the generation of saccadic eye movements. *Ann. Rev. Neurosci.,* **13**, 309–336.

Tittle, J. S. & Braunstein, M. L. (1991). Shape perception from binocular disparity and structure-from-motion. In *Sensor Fusion III: 3-D Perception and Recognition, Proceedings of the SPIE,* **1383**, 225–234.

Tolhurst, D. J. & Movshon, J. A. (1975). Spatial and temporal contrast sensitivity of striate cortical neurons. *Nature,* **257**, 674–675.

Torre, V. & Poggio, T. (1978). A synaptic mechanism possibly underlying directional selectivity to motion. *Proc. R. Soc. Lond. B,* **202**, 409–416.

Ullman, S. (1979a). The interpretation of structure from motion. *Proc. R. Soc. Lond. B,* **203**, 405–426.

Ullman, S. (1979b). *Interpretation of Visual Motion.* MIT Press, Cambridge, Mass.

Ullman, S. (1984). Maximizing rigidly: the incremental recovery of 3-D structure from rigid and nonrigid motion. *Perception,* **13**, 255–274.

Vaina, L. M., Grzywacz, N. M. & LeMay, M. (1990). Structure from motion with impaired local-speed and global motion-field computations. *Neural Comp.,* **2**, 416–432.

Vaina, L. M., Grzywacz, N. M. & LeMay, M. (1992) Testing computational theories of motion discontinuities: a psychophysical study. In G. Sandini (ed) *Lecture Notes in Computer Science* (Vol 588), pp 222–226. Springer-Verlag, Berlin, Germany.

Vaney, D. I. (1990). The mosaic of amacrine cells in the mammalian retina. In N. Osborne & J. Chader (Eds.) *Progress in Retinal Research* (vol. 9), pp. 49–100. Pergamon Press, Oxford.

van Santen, J. P. H. & Sperling, G. (1985). Elaborated Reichardt detectors. *J. Opt. Soc. Am. A,* **2**, 300–320.

Verri, A., Girosi, F. & Torre, V. (1989). Mathematical properties of the two-dimensional motion field: From singular points to motion parameters. *J. Opt. Soc. Am. A,* **6**, 698–712.

Wallach, H. & O'Connell, D. N. (1953). The kinetic depth effect. *J. Exp. Psychol.*, **45**, 205–217.

Warren, W. H. & Hannon, D. J. (1988). Direction of self-motion is perceived from optical flow. *Nature*, **336**, 162–163.

Watamaniuk, S. N. J. (1992). Visible persistence is reduced by fixed-trajectory motion but not by random motion. *Perception*, **21**, 791–802.

Watamaniuk, S. N. J. & Duchon, A. (1992). The human visual system averages speed information. *Vision Res.*, **32**, 931–941.

Watanabe, S.-I. & Murakami, M. (1984). Synaptic mechanism of directional selectivity in ganglion cells of frog retina as revealed by intracellular recordings. *Jpn. J. Physiol.*, **34**, 497–511.

Watson, A. B. & Ahumada, A. J. (1985). Model of human visual motion sensing. *J. Opt. Soc. Am. A*, **2**, 322–342.

Werblin, F. S. (1970). Responses of retinal cells to moving spots: Intracellular recording in *Necturus maculosus*. *J. Neurophysiol.*, **33**, 342–350.

Werkhoven, P. & Koenderink, J. J. (1991). Visual processing of rotary motion. *Percept. Psychophys.*, **499**, 73–82.

Wheatstone, C. (1838). Contributions to the physiology of vision. Part the first. On some remarkable, and hitherto unobserved, phenomena of binocular vision. *Phil. Trans. R. Soc.*, **128**, 371–394.

Wyatt, H. J. & Daw, N. W. (1975). Directionally sensitive ganglion cells in the rabbit retina: Specificity for stimulus direction, size and speed. *J. Neurophysiol.*, **38**, 613–626.

Yang, G. & Masland, R. H. (1992). Direct visualization of the dendritic and receptive fields of directionally selective retinal ganglion cells. *Science*, **258**, 1949–1952.

Yuille, A. L. & Grzywacz, N. M. (1988). A computational theory for the perception of coherent visual motion. *Nature*, **333**, 71–74.

3

Motion Processing in the Primate Cerebral Cortex

R. J. Snowden

University of Wales College of Cardiff, UK

1   INTRODUCTION

The study of the visual cortex has led to the notion of specialization. Different parts of the brain seem to be involved in processing the visual image for different attributes (e.g. DeYoe & Van Essen, 1988). One of the central pillars of this view has been the identification of a 'motion area' – a small part of the brain whose function appears to be to analyse the temporal aspects of our world. Normally termed the middle temporal area (MT) due to its initial description in the New World monkey (Allman & Kaas, 1971), it is actually located in the superior temporal sulcus of Old World monkeys (and is often referred to as V5 – Zeki, 1974). Converging evidence strongly implicates its importance in motion perception. In recent years we have also come to appreciate that there may be other areas, particularly the medial superior temporal (MST) area, that may also have an important role in motion perception. The aim of this review will be to describe our current knowledge concerning these areas and to place them in the context of a 'motion pathway' – a linked set of visual areas through which the visual information concerning movement flows.

In attempting such a review the hardest thing is to know what not to review. I have therefore some disclaimers. The first is that I have confined the review to a few selected areas of the brain. The review revolves around area MT and considers the functional physiology of this area in relation to that of its major input area (V1), and in relation to its major output areas (medial superior temporal – MST and ventral intraparietal – VIP). This has meant leaving out many other areas that may, or indeed may not, play important roles in motion perception (for further discussion of some of these areas see Chapter 7 by Logothetis). Secondly, visual motion processing has been examined in many species. In this review I have tried to concentrate on evidence obtained from the Old World monkey. This has meant leaving out much work, particularly some interesting studies of the New World monkey. Much of what I shall claim has its counterpart in the New World monkey; however, due to its nocturnal lifestyle, there are sure to be some differences (e.g. the nocturnal monkeys have a smaller foveal/peripheral bias; Webb & Kaas, 1976). One major reason for concentrating upon the Old World monkey is its supposed similarity in visual function to that of humans (e.g. Siegel & Andersen, 1991). In a final section I review some of the studies that have suggested that the visual areas discussed here may have counterparts in the human brain. Thirdly, I have deliberately left out some important studies because of their more thorough treatment in other sections of this book (for example responses to plaid stimuli (see Chapter 8 by Wilson and Chapter 9 by Stoner and Albright), to stimuli defined by things other than luminance (see Chapter 7 by Logothetis), and pursuit eye movements (see Chapter 15 by Krauzlis)).

1.1   A special place for motion?

One of the techniques available to us to understand the importance of any particular part of the brain is to remove it and look for the changes in behaviour that accrue from this damage. Discrete lesions of area MT can be produced by the local injection of a neurotoxin. Newsome and Paré (1988) trained macaque monkeys to discriminate the direction of motion of a set of dots. To measure the animals' ability at this task a number of the moving dots could be given a random direction of motion, thus producing patterns in which the movement was only partially correlated from frame to frame (Figure 1A). In normal circumstances the animal is very good at this task and only requires as little as 3% correlation to discriminate one direction from its opposite (a figure similar to human performance on this task (Downing & Movshon, 1989)). If a small part of area MT was destroyed by the neurotoxin ibotenic acid, then the animal had great difficulties with the task (thresholds rose by 400–800%) if the stimulus was presented to the part of the visual field known to be served by the lesioned section of MT, but performed 'normally' if the stimulus was presented in another part of the visual field (Figure 1B). When other visual functions were tested (e.g. contrast sensitivity, colour perception) it was found that the lesion did not affect vision. Thus it is not that the animal cannot see the pattern, it seems to do so with all the acuity and sensitivity it always had, it simply cannot tell us in which direction the pattern moved. It could be described as having a 'motion scotoma' where area MT had been damaged (for further discussion of this and related lesion studies see Chapter 7 by Logothetis).

It seems that the integrity of MT is needed for motion perception. Further studies have demonstrated the sufficiency of its cells. Newsome et al. (1989) measured the required correlation for a cell to discriminate between two opposite directions of motion. They show that many MT cells can do this with at least the efficiency of the behaving animal given some simple assumptions. In addition the activity of these neurones was shown to be highly correlated with the response decision of the animal. Snowden et al. (1992) have shown that some MT neurones are capable of the same direction discrimination thresholds (approximately 1°) as human observers.

A final demonstration of MT's importance is that manipulation of the activity of

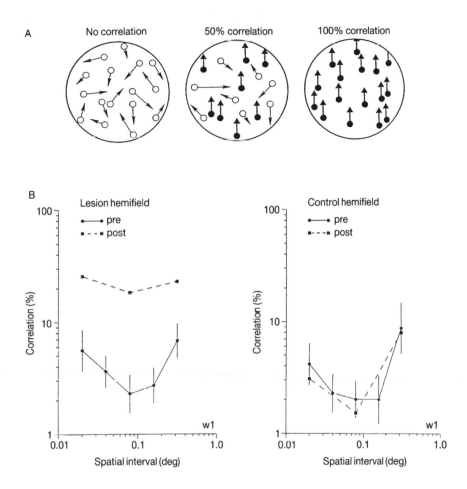

Figure 1   A. Representation of the stimuli used by Newsome and Paré (1988). The dark circles represent dots that are all moving in the same manner (upwards) whereas the open circles represent dots moving in a pseudorandom fashion. By varying the number of dots moving in the same manner (the correlation) a measure of performance can be obtained. B. Correlation thresholds for detecting the direction of motion at a number of displacement levels (speeds). The left panel depicts the performance of a macaque monkey both pre and post to a lesion of area MT, the right panel was for the hemifield which was not subject to a lesion. The animal's performance was strongly affected by the lesion in the experimental hemifield. Adapted from Newsome and Paré (1988).

small clusters of its cells can alter the animal's perceptual decision. Salzman *et al.* (1990) trained animals to signal their perceived direction of motion of a stimulus. They then stimulated a small cluster of cells within MT whose preferred direction of motion was one of the two possible answers the animal could give. They found that the animal was far more likely to signal that its perceived direction of motion was in the direction that the stimulated cells preferred. The implication of this is that the activity in

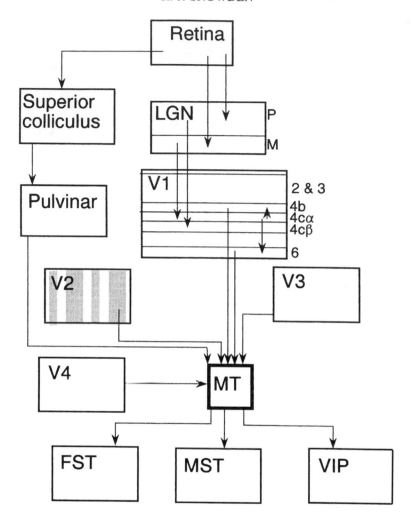

Figure 2 A representation of some of the known anatomical pathways through which information reaches and leaves area MT. See text for details. LGN: Lateral geniculate nucleus. P: Parvocellular division. M: Magnocellular division. MT: Middle temporal. FST: Fundus (or floor) of the superior temporal. MST: medial superior temporal. VIP: ventral intraparietal.

just a small cluster of MT cells is not only correlated with, but is causal to, the animal's behaviour and conceivably its perception of the direction of motion of a stimulus.

2　A PATHWAY FOR MOTION

Earlier I alluded to the notion of a motion pathway – a specialist route through which information about movement is analysed. Here I briefly outline some of its supposed structures and connections (see also Figure 2). The lateral geniculate nucleus (LGN) of the thalamus receives the output of retinal ganglion cells before itself projecting to the striate cortex. The lateral geniculate nucleus contains six layers of cells. Four of these have quite small cells (parvocellular) and receive information from small retinal ganglion cells. The other two layers have larger cells (magnocellular) and receive information from the larger retinal ganglion cells. The properties of the two (referred to from now as P and M) types of cell are somewhat different. P-cells have colour-opponent input in that their central zone is fed by one type of cone (e.g. long-wavelength) whereas their surround is fed by another type (e.g. middle-wavelength).[1] On the other hand the receptive fields of the M-pathway are larger and receive a mixed cone input. The question therefore of interest is: do both these groups of cells contribute equally to the eventual input into MT? Recent evidence suggests not. Maunsell *et al.* (1990) recorded from single cells and small clusters of cells in area MT whilst selectively blocking the response of either the M- or P-divisions of the lateral geniculate nucleus by chemical means. It was found that blocking the activity of M-cells caused profound and often complete loss of activity in MT, whilst blocking the activity of cells in the P-division rarely had an effect. It should, however, be noted that there were instances where blocking the P-division definitely did have an effect. Thus while it would be incorrect to state that area MT is driven solely by the M division, it does appear to be the major contributor.

The main projection of the lateral geniculate nucleus is into layer 4c of the striate cortex. Here our apartheid remains with the P-pathway connecting to layer $4c\beta$ and the M-pathway to layer $4c\alpha$ (Hubel & Wiesel, 1972). Layer $4c\alpha$ projects to layer 4b, which in turn is known to project directly to area MT (Shipp & Zeki, 1985a; Shipp & Zeki, 1989a). There is also a known connection between some cells in layer 6 of striate cortex and area MT (Fries *et al.*, 1985; Shipp & Zeki, 1989a). Layer 4b also projects to specific sites in area V2 (Shipp & Zeki, 1985), which in turn project to area MT (Shipp & Zeki, 1989b).

MT also receives a 'forward' input from area V3 (Ungerleider & Desimone, 1986). It should also be noted that MT feeds backwards to these same areas (Ungerleider & Desimone, 1986b; Shipp & Zeki, 1989a, 1989b). MT has 'intermediate'[2] connections with areas V4, V3A and the parieto-occipital area (PO). There is also an input to MT from the superior colliculus that comes through the pulvinar nucleus of the thalamus (Beneveto & Standage, 1983). Indeed, this latter pathway may be of great importance as lesioning the striate cortex by itself does not silence area MT (Rodman *et al.*, 1989). However, combined lesions of striate cortex and the superior colliculus do (Rodman *et al.*, 1990).

Area MT projects to several areas in the cortex. Most notable are the connections to the MST and the fundus superior temporal area (FST) (though see section 4.1) and the VIP. It is believed that these areas may also be heavily involved in motion perception (see section 4). There are also extensive feedback connections from these areas to area MT (Ungerleider & Desimone, 1986b). In turn areas MST and FST make numerous

---

[1] There is still some debate as to whether a single cone type feeds the surround or a mixture of cone types (see Reid & Shapley, 1992).
[2] Connections have been classified as 'forward', 'intermediate' and 'backward' according to which layers they originate from and terminate in; see, for instance, Maunsell and Newsome (1987).

connections in both the parietal and temporal cortex, and to the eye movement areas in the frontal cortex (Boussaoud et al., 1990).

This brief survey of the connections of area MT is, at best, incomplete in the sense that I have not tried to convey all the complexities of the connections that have already been charted, let alone the many connections and interrelations that have yet to be documented (for a more detailed review see Felleman and van Essen, 1991 and Chapter 7 of this book by Logothetis). Even given this brevity it is clear that MT plays a central role in such a motion pathway, that it has the potential to gather information about motion from many different sources, and its outpourings have consequences for many different visual functions.

3   THE MIDDLE TEMPORAL VISUAL AREA

3.1   Representation of the visual world

3.1.1   Topographic representation

In the macaque monkey area MT is located on the posterior bank and fundus of the superior temporal sulcus (Figure 3). It is normally characterized anatomically by an area of dense myelination in the lower layers. However, its precise definition has been the subject of some debate. Initial evidence accrued from studies that either looked at the degeneration effects of damage to various portions of V1 (Ungerleider & Mishkin, 1979), or using anterograde tracers injected into V1 (Rockland & Pandya, 1981). These studies reported (1) that there is a striate projection zone in the caudal part of the superior temporal sulcus, (2) that within this area there is an orderly topographic map of the visual field, (3) that only the central 20–30° of visual field are actually in the heavily myelinated zone, with the more peripheral representation extending medially from this zone. However, a later study (van Essen et al., 1981) suggested that the whole of the striate projection was within the heavily myelinated zone, and the earlier reports were due to accidental damage to area V2. Topographic maps have also been produced by locating the receptive field of individual cells within area MT (Gattass & Gross, 1981; Desimone & Ungerleider, 1986; Ungerleider & Desimone, 1986a; Maunsell & van Essen, 1987; Erickson et al., 1989). Some of the confusion seems to have occurred due to a rather strange irregularity in MT. Within the heavily myelinated zone there is indeed an orderly representation of both the foveal and mid-peripheral visual fields. As one crosses the medial border of the heavily myelinated zone there is also an abrupt discontinuity in the visual field representation. This second region has been termed MTp to distinguish it from the rest of MT, but is regarded as part of MT as both are needed to process the full output of both V1 and V2 (Ungerleider & Desimone, 1986a).

3.1.2   Receptive fields and magnification factor

It is a general rule of the visual system that as the number of synapses from the photoreceptors of the cells concerned increases, the receptive fields become larger. MT is no exception to this rule. The receptive fields of area MT cells are approx-

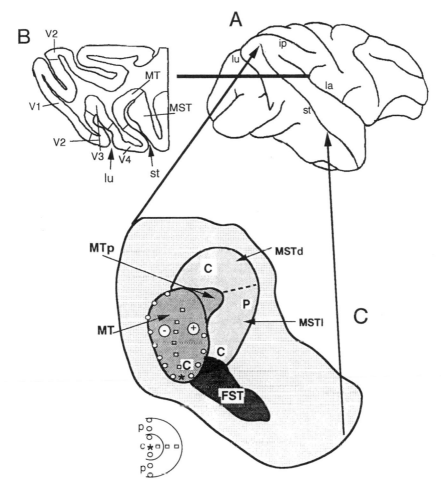

Figure 3   Position of motion areas in the macaque cortex. A. Location of sulci and gyri. lu is the lunate sulcus, st the superior temporal sulcus, ip the intraparietal sulcus and la the lateral sulcus. B. Horizontal section through the point marked by the bold line. The approximate position of some of the visual areas is indicated. C. A graphical representation of the unfolded superior temporal sulcus (between the points indicated by the arrows). C: Central visual field. P: Peripheral visual field.

imately 10 times the size of those in striate cortex at any given eccentricity (Gattass & Gross, 1981). Albright & Desimone (1987) measured the receptive field (RF) size of over 500 single cells in MT out to 25° eccentricity (Figure 4) and found an approximate relationship of:

$$\text{RF size} = 1.04° + 0.61 \text{ eccentricity}$$

with a scatter of RF size that was approximately one-third of the RF size. As mentioned above the field size of MT cells is approximately 10 times the equivalent of its V1 counterpart. However, if one only includes the direction selective cells of

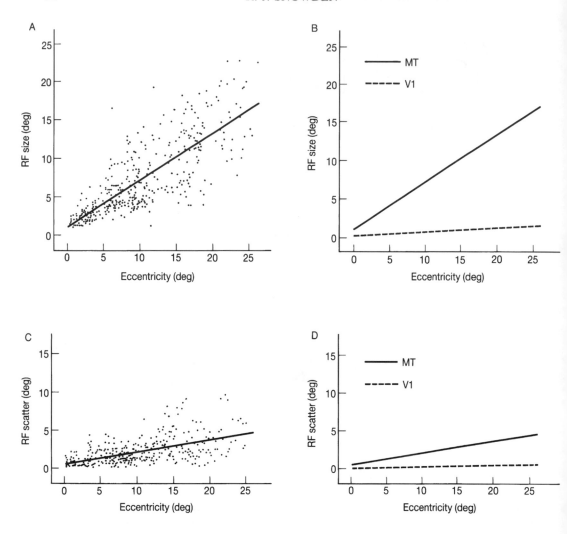

Figure 4   The size (A) and scatter (C) of receptive fields (RF) are plotted against eccentricity for area MT. Sections B and D compare the regression lines obtained for area MT with those from area V1 (Dow *et al.*, 1981). From Albright and Desimone (1987).

V1 then this figure reduces to around only three times (Mikami *et al.*, 1986). This is due to the directional selective cells of area V1 (which are mainly confined to upper layer 4, and layer 6 (Hawken *et al.*, 1988)) having larger receptive fields than the non-directional cells.

A second factor that should be considered is the magnification factor, the cortical distance corresponding to one degree of separation in the visual field. Estimates from the data of Gattass and Gross (1981) and Albright and Desimone (1987) suggest the relationship:

$$\text{Magnification factor} = 1.14 * \text{eccentricity}^{-0.76}$$

Thus there is about 1.14 mm/degree of cortex devoted to processing at an eccentricity of 1°. However, a more recent study places the figure for the central 1° at about 3.3 mm/degree (Erickson *et al.*, 1989). It is possible that the discrepancy could be due to the differing size of the animals. Maunsell and van Essen (1987) have shown a correlation between body weight and the size of MT when MT is defined on myelo-architecture. As it has been suggested that the heavily myelinated area in the superior temporal sulcus may actually be the foveal representation (see above), this suggests a more expanded foveal representation in larger animals, and could therefore account for the discrepancy noted above.

These estimates of cortical magnification can be compared to area V1. Though there is still some debate over the striate magnification factor, a comparable figure at 1° would be 6–8 mm/degree (Dow *et al.*, 1981), which means MT only devotes around one-fifth to one-third of the cortex that V1 does to this eccentricity. As one considers greater eccentricities this discrepancy reduces to around one-third at an eccentricity of 20° (Albright & Desimone, 1987). This conclusion appears to be at odds with the data of Erickson *et al.* (1989). They suggest that the ratio of the central fovea (the central 1°) to the rest of the area is around 0.10 for both MT and V1, suggesting the relative magnification of the foveal to peripheral visual field is preserved in the mapping of V1 on to MT. Again the discrepancy might be accounted for by the relative enlargement of the foveal representation in larger monkeys (see last paragraph). At this stage it would seem unwise to make any strong claims about differences in the rate of change of magnification factor with eccentricity in these two visual areas, and psychophysical studies that have made claims as to which visual area may govern performance on a particular task on the basis of how performance varies with eccentricity may be premature.

3.2 Directional selectivity

3.2.1 Occurrence

The most notable property of area MT is the very high proportion of directionally selective cells (Dubner & Zeki, 1971). Directional selectivity has been defined in many ways. Here I shall take the arbitrary value of a cell responding at five times the rate in its 'preferred' direction than it does in the direction 180° from the preferred direction (the antipreferred direction) when the stimulus is moving at the speed that produces the greatest response in the preferred direction. Various estimates have been made using a variety of different stimuli (Table 1). These studies estimate the proportion of directional selective cells to be around 84%. A comparison to area V1 is somewhat difficult as many studies of V1 attempt typically to search for direction-ally selective cells in order to make comparison with MT and hence produce a strong bias (e.g. Mikami *et al.*, 1986; Snowden *et al.*, 1992). Examining studies that appear to have far less bias (De Valois *et al.*, 1982; Hawken *et al.*, 1988) suggests a figure of 13–17%. Again, this comparison may be radically altered if we only consider those cells that project to area MT. The incidence of directional selectivity in these cells (i.e. those of layers 4b and 6) is far higher than those of V1 overall. Again estimating from the data of Hawken *et al.* on the incidence of directional selectivity in the upper layers

Table 3.1 Incidence of directional selectivity in area MT

Study	Stimulus	Measure	Mean DI[a]	% direction selective
Snowden et al. (1992)	Random dot pattern	1–N/P[b]	1.01	83
Rodman & Albright (1987)	Slits or single spots	1–N/P	1.01	78
Albright (1984)	slits	1–N/P	1.00	85
Albright (1984)	random dot patterns	1–N/P	1.05	69
Maunsell & van Essen (1983b)	slits	1–N/P	0.93	84
Mikami et al. (1986)	slits	1–N/P	0.98	80

[a] Directionality index. [b] N/P: response in null direction divided by response in preferred direction.

of layer 4 and from layer 6, the figure seems to be around 42%. While this still seems to be a long way from the figure of 84% for MT cells, findings that the projection from layer 4b of the V1 to MT is patchy (Shipp & Zeki, 1985a) suggest the possibility that there could be a sub-population of cells, which project to MT, that may have very similar directionally selective properties as MT cells themselves. This in turn may imply that area MT does not 'create' directional selectivity, but merely inherits it from its V1 afferents. This position is strengthened by a study (Movshon & Newsome, 1984) which stimulated area MT whilst recording from cells in area V1. Cells that were antidromically stimulated at short latency (suggesting a direct connection) were all (but one) directionally selective complex cells of layers 4b and 6. However, there are two other points that may speak against such a notion. Firstly, as mentioned in section 2 both Rodman et al. (1989) and Girard et al. (1992) have shown that the properties of MT cells change very little when the activity of V1 is eliminated – so either area MT creates its own properties from the input or the input coming from the superior colliculus might be sufficient (but this seems unlikely given the very different response properties of these two areas). Secondly, when considering apparent motion stimuli (sets of rapidly presented stationary frames) area MT is able to give a directionally selective response over a greater range of displacements (approximately three to four times) than is tolerated by V1 cells (Mikami et al., 1986). As the V1 cells are not giving a directional response at these large displacements this suggests that MT must be creating it for itself. Taken all together the most consistent explanation would be that MT may inherit many of its functional properties, but is also creating many of its own.

3.2.2 Organization

Early studies of MT showed no bias in the preferred direction over the population of cells (Maunsell & van Essen, 1983b), but did hint that cells that were near to each other physically were also near to each other in functional terms. This suspicion was consolidated by Albright *et al.* (1984) who recorded the preferred direction of motion of a succession of neurones on long electrode penetrations (Figure 5). Their finding was simple: the more vertical the electrode penetration was to the cortical surface the least change in preferred direction was observed from cell to cell. Such a result would occur if the cells were arranged in vertical columns where all cells within a column had the same preferred direction. This has been confirmed by labelling studies (Tootell & Born, 1990). Albright *et al.* also noticed that instead of a small change in preferred direction from one cell to the next there were occasionally shifts of around 180° and the frequency of these shifts was related to the angle of penetration. This suggests that an organization somewhat akin to that of Figure 6, where along one axis preferred direction of motion changes gradually, whereas in the orthogonal direction motions of opposing direction are encountered.

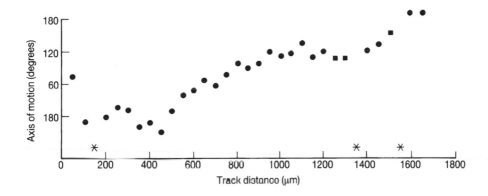

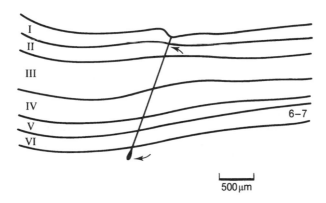

Figure 5 The preferred axis of motion of a number of cells is plotted against the distance of a penetration through area MT. Note that preferred axis of motion varies gradually from cell to cell. (Circles: preferred axis of motion for a unidirectional cell. Squares: preferred axis of motion for a bidirectional cell. Asterisks: A pandirectional cell.) From Albright *et al.* (1984).

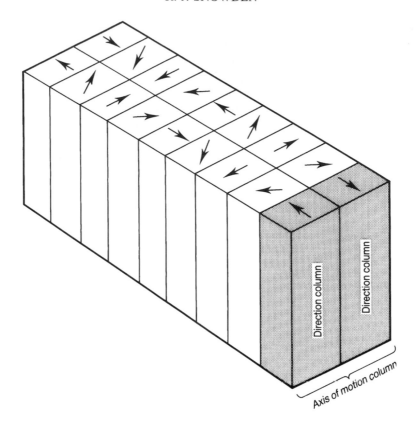

Figure 6 Representation of direction columns in area MT. Within any one column all the cells have the same preferred direction. As one moves in one direction the preferred direction of motion varies in small steps whereas in the orthogonal direction preferred direction of motion reverses. From Albright *et al.* (1984).

A second organization was noted by Albright (1989). He examined the preferred direction of motion of cells with respect to their position in the visual field. He found a bias in the population of cells such that there was an over-representation of motions away from the fovea. This only occurred for cells whose receptive fields were well away from the fovea; those near the fovea show no bias. Albright suggests that this over-representation may be due to the characteristic pattern of motion that occurs as an animal moves through space. If the head moves forward then the visual field expands away from the fovea (see Chapter 11 by M. Harris and Chapter 12 by Cumming).

3.3   Directional tuning

As well as considering a cell's response in the preferred and antipreferred direction we should also consider its response to any direction – its directional tuning. In considering directional tuning (as opposed to directional selectivity) one must be careful not to confuse tuning for the orientation of the stimulus with tuning for direction. If one were to move a bar perpendicular to its length then one could not be sure if changes in response were due to changes in the orientation or direction of motion of the stimulus. It would therefore seem that the best stimuli to use are ones that are isotropic in orientation, such as single spots or random dot patterns. The response of one neurone to such a stimulus is depicted in Figure 7. Albright (1984) has examined the tuning of MT cells for a variety of stimuli (slits, spots and random dot patterns). The tuning of a cell was found by measuring its response to many directions of motion and then fitting a Gaussian function through the points. The 'bandwidth' of the cell was then calculated by measuring the full width of the curve at half its maximum height. A very similar procedure was also used in considering the response of MT cells in an alert preparation to random dot patterns by Snowden *et al.* (1992) except they simply reported the standard deviation of the Gaussian function. For simplicity here I shall correct both studies to give half-width at half-height. Albright found fairly similar tuning for the three types of stimuli (means for spots = 52°; slits = 45°; random dots = 41°); however, the difference between the single spot and the other stimuli was significant. The variation (as expressed in the standard deviation of the above figures) was about 15°. The measurements by Snowden *et al.* find the tuning to be somewhat broader (mean = 52°, SD = 20.5, N = 30) than those of Albright.

3.4 Responses to the speed of stimuli

Moving stimuli not only have a direction of motion, but also have a speed of motion. This parameter too seems to be an important dimension to which MT neurones are sensitive. Rodman and Albright (1987) identified a number of types of response into which MT neurones may be grouped. Some cells (termed S1) have a tuning for speed in the preferred direction that is mirrored by a similar dependence in the antipreferred direction (Figure 8B). Others (S2 type) have a tuning in the antipreferred direction that is the mirror opposite in the preferred (i.e. a strong response in the preferred direction corresponds to a weak response in the antipreferred, Figure 8C), whilst a third type (termed NT) has tuning in the preferred direction but no tuning in the antipreferred (Figure 8A).

The study of Rodman and Albright used moving slits to probe sensitivity to speed. Unfortunately this type of stimulus confounds two possible reasons for the change in response. Any stimulus can be reconceptualized as composed of discrete components in the spatial and temporal frequency domains. Thus a moving slit has characteristic spatial and temporal frequencies. As the slit changes speed the temporal frequencies shift to higher frequencies; hence the change in response might be due to temporal frequency rather than speed. This is a particularly important point as (a) to success-fully track a stimulus we need a representation of speed, not temporal frequency, and (b) neurones of the striate cortex are tuned for temporal frequency not speed (Tolhurst & Movshon, 1975). Clearly one important elaboration that MT could perform is to move from a temporal frequency code to a speed code. To test this we need to show that the neurone gives its maximum response at the same speed of motion, regardless of changes in the spatial composition of the stimulus. Brief accounts of such an

A

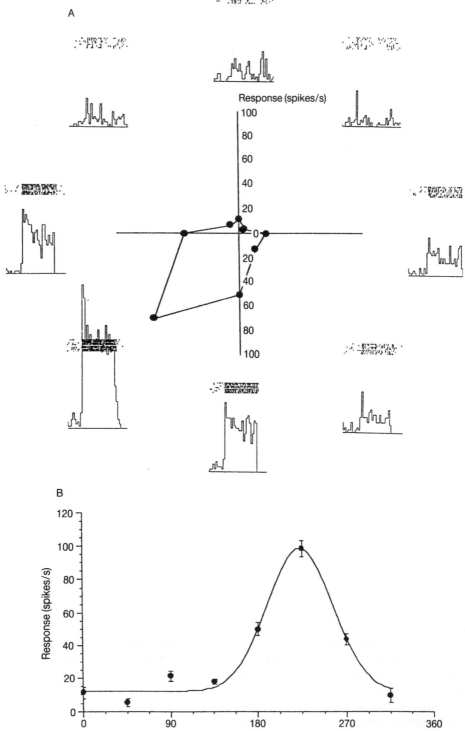

B

experiment have been reported (Newsome *et al.*, 1983; Movshon *et al.*, 1988). They found some evidence of a small number of neurones that did show a preferred speed rather than a preferred temporal frequency over a small range of spatial frequencies. While this evidence still needs substantial reproduction and elaboration it is suggestive that MT may well provide a 'pattern-independent' code for speed.

One theme of this review is to look to see what changes in response properties occur at each stage in our motion pathway. In the current context we know that many V1 cells are also tuned for speed (e.g. Orban *et al.*, 1986), so is there really any elaboration of response properties in area MT? One important difference might be in the range of speeds to which each area is sensitive. As mentioned in section 3.2.1, area MT can be directionally selective over a greater range of spatial displacements in an apparent motion stimulus (Mikami *et al.*, 1986). So if the temporal properties of the two areas were similar this should translate into area MT processing higher speeds than area V1. Mikami *et al.* (1986) also measured the greatest time delay between the successive frames of an apparent motion stimulus and found that the figure (mean approximately 100 ms) was similar in both areas. This 'greatest time delay' figure is, however, rather inappropriate when trying to consider the fastest speeds. As one increases the time delay for a given spatial displacement the stimulus becomes slower. The information one really needs is the minimum time over which the cells show their directional selectivity, but unfortunately I know of no data on this in any brain area. Mikami *et al.* do provide estimates of the maximal speed for smooth motion in both area V1 and MT. Perhaps not surprisingly, they find a positive correlation in both areas between the maximum spatial displacement and the max-imum speed. However, they also show a difference between V1 and MT above the already noted differences in maximal displacement. On average, if there are two cells, one from MT and one from V1, with matched maximal displacement tolerance then the V1 cell responds to speeds three to four times as great as the MT cell. Now reconsidering the earlier fact that the spatial displacement limit is three to four times larger in MT cells than V1 cells this suggests that the maximal velocity cut-off should be similar in MT and V1!

So far we have only considered the upper range of velocities; do the areas differ at slower speeds? From the speed tuning curves one can also produce an optimal velocity and this has been done for both V1 and MT cells (van Essen, 1985). This study does indeed show that MT cells have higher optimal speed than V1 cells. Another parameter of interest is a cell's sensitivity to motion, or its lower speed cut-off. Little evidence is available on this point but a recent study (Lagae *et al.*, 1993) suggests that both areas lose directional selectivity at about the same speed.

Many aspects of our vision vary with the eccentricity of viewing, and indeed many psychophysical studies of human motion perception show these variations (e.g. van de

---

Figure 7 A. Response histograms for an MT cell stimulated with several different directions of motion of a random dot pattern. The largest response occurs to movement down and to the left. B. The level of activity is plotted against direction of motion and the points have been fitted by a Gaussian function. Unpublished data of Snowden, Treue and Andersen; see Snowden *et al.* (1992) for details.

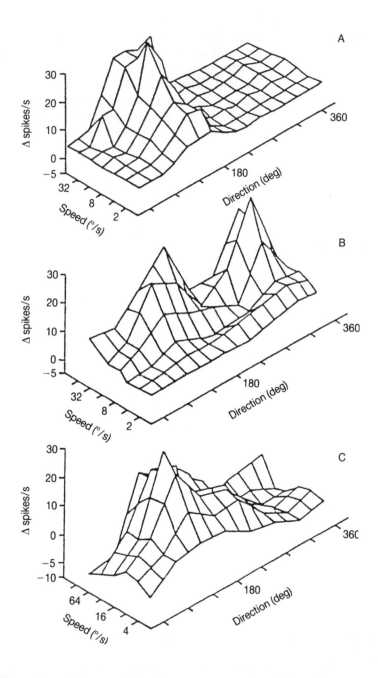

Figure 8   Three-dimensional plot showing the response of three MT cells to various directions of motion and speeds. A. This cell gives a strong response which is dependent on speed in the preferred direction but is not tuned for speed in the antipreferred direction (NT type). B. This cell gives responses in both preferred and antipreferred directions which have similar tuning for speed (S1 type). C. This cell is tuned for speed in the preferred direction but has reciprocal tuning in the antipreferred direction (S2 type). From Rodman and Albright (1987).

Grind *et al.*, 1983). Do MT cells show variation in their response to speed with eccentricity? Again it may depend upon what measure we choose to take. If we consider the optimal speed of movement for a cell there appears to be little (Maunsell & van Essen, 1983b) or no (Lagae *et al.*, 1993) variation in this parameter. On the other hand if we consider maximal speed then there appears to be a small increase with eccentricity (Lagae *et al.*, 1993) which is accompanied by a small increase in maximal spatial displacement of an apparent motion sequence (Mikami *et al.*, 1986). Once again, however, there appears to be a small increase in temporal dependency that may well nullify this change (see their Figure 7). The weak, or lack of, evidence for a change in the speed properties of MT is surprising given the psychophysical evidence for changes in motion processing with eccentricity – it may be time for a reappraisal of this evidence in human observers.

3.5 Invariance of response properties

So far I have considered directional selectivity, directional tuning and responses to speed for stimuli that were generally optimized for the cell concerned. Since real visual stimuli do not oblige us in this manner it is of interest to see how the cells cope with changes from optimal. The list of changes that one might consider is huge (e.g. pattern size, contrast, speed, position within receptive field, polarity, dot density, etc.) but I want to suggest here (with just a few examples) that the functional properties of MT cells remain invariant to other changes, the only difference being the overall level of activity in the cell (see also Chapter 7 by Logothetis). This mere scaling of response would allow any individual property (such as direction) to be easily extracted from a comparison of the activity over a number of cells.

Rodman and Albright (1987) considered the direction tuning of MT cells as a function of speed. Figure 9 illustrates the response of three cells to a variety of speeds. It is most notable that the peak of the curve (the cell's preferred direction) remains constant but the overall level of response varies. In their population of neurones they found no systematic changes in tuning bandwidth over a range of speeds from 1/16 of the optimal to 16/1 of the optimal. Lagae *et al.* (1993) have shown that the optimal speed, overall response and directional index are invariant to reversals in the polarity of a bar stimulus.

3.6 Interactions between directions of motion

So far we have only considered how cells respond to the motion of isolated elements. Any system that is to perform the many functions to which motion information may contribute (see Chapter 1) will have to do so when the visual field is cluttered with objects and the movement of those objects (caused by movement of the object or movement of the animal or its eyes).

A landmark study on this question was that of Allman *et al.* (1985a) in the New World monkey, which has been substantially reproduced in the Old World monkey (Tanaka *et al.*, 1986; Lagae *et al.*, 1989; Born & Tootell, 1992). In these studies it was noted that while most MT cells give a hearty response to small objects such as lines, or

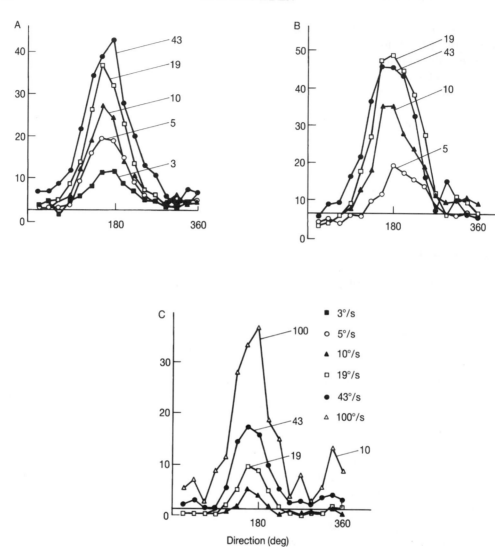

Figure 9 Directional tuning of three MT cells at various speeds. The changes in speed seem only to vary the response rate rather than the directional tuning. From Rodman and Albright (1987).

small patches of dots that are moving in the cell's preferred direction, many became silent (or much reduced in their response) if presented with a large array of moving dots. Similarly, if a bar is present moving across the cell's receptive field, then the response can be silenced by a background, which itself does not impinge on the receptive field of the neurone, moving in the same direction and speed. If the background motion is varied in some manner (such as changing its speed or direction) the effect is weakened, and can in some instances be reversed so as to enhance the response to the bar when the background is moving in a direction opposite to that of the bar. Two points are of great importance. Firstly, if the bar is not present then the background by itself fails to modulate the response of the cell concerned – it is beyond its 'classical receptive field' (Allman *et al.*, 1985b). Thus the neurone can be thought of as comparing its own stimulus with stimuli occurring in other parts of the visual field. Secondly, this effect seems distinct from other reported surround effects (e.g. McIlwain, 1964) that are non-specific in nature. The modulatory surrounds of MT are very specific in that they show maximal suppression when the background has the same *motion* properties as the centre. Indeed Tanaka *et al.* (1986) report one cell that gave a strong response to all directions of movement and was suppressed whenever the background was in the opposite direction to the centre direction regardless of the test direction (Allman *et al.*, 1985a also mention two cells with somewhat similar properties). It seems that the surround is a relativistic one serving to suppress or enhance the response to the centre (similar cells have been reported in both pigeon (Frost & Nakayama, 1983) and cat (Hammond & Smith, 1982)).

There is evidence that this surround suppression may well be a qualitative difference between MT and V1. Neurones of area V1 do not show suppression of this type (Orban *et al.*, 1987), though some V2 neurones do. In addition there appears to be some clustering of cells of this type with respect to the laminae of MT. Cells of this type are more prevalent outside of the layers receiving direct input from V1 (layers 4 and 6) (Tanaka *et al.*, 1986; Lagae *et al.*, 1989; Born & Tootell, 1992).

As a consequence of the interactions from beyond the classical receptive field a large field of moving dots should only drive a certain fraction of MT cells. Born and Tootell (1992) presented animals with such a stimulus after injecting the activity marker 2-deoxyglucose. On later processing for the labelled tissue they noticed that area MT was not uniformly stained; instead there appeared to be patches of light and darker stained tissue. This suggested that cells with antagonistic surrounds may be grouped together. Subsequent single cell recording confirmed that the areas of light staining did indeed contain cells with antagonistic surrounds whilst the darker stained patches contained cells with either no surround or a surround that increased the response of the cell as the stimuli became larger than the classical receptive field (Figure 10). They also demonstrate that electrode penetrations vertical to the cortical surface produce cells of a highly consistent type. Hence we should add this columnar organization to the columnar organization into direction of motion described in section 3.2.2.

What is the role of these types of cell? It is tempting to suggest that the cells that respond only to small objects could provide figure/ground segregation, whilst the large field cells could process information that relates to the animal's movements. Given the evidence presented below about the possible functions of area MST, and its possible divisions therein, it will be of interest to see if the dark and light patches within MT make differing connections.

Other common real-world visual stimuli are motion borders, and motion transparency (where two patterns move simultaneously in differing directions). Such scenes occur when an animal moves through a thicket or when the wind moves the leaves in a tree. The problem for the visual system is to assign the correct movement to the correct pattern. Snowden *et al.* (1991) investigated this by comparing the response of V1 and MT neurones to random dot patterns moving in the cell's preferred direction alone, or when it was embedded along with another motion. They noted that all MT

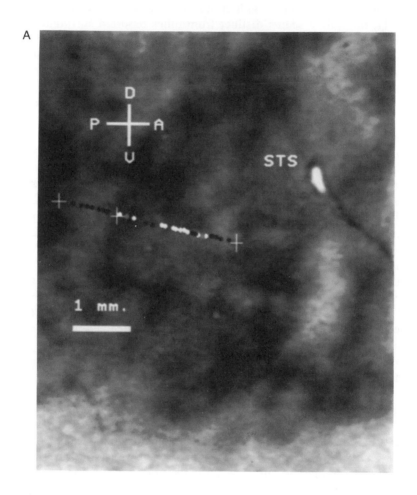

A

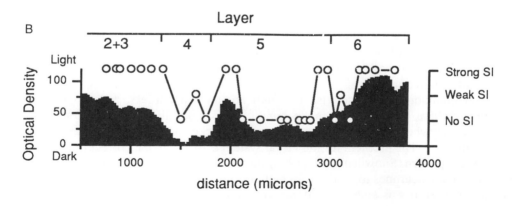

B

cells gave a reduced response when the preferred motion was embedded, whereas most V1 cells responded just as well. This result contrasts and complements those of Allman and colleagues in that Snowden's results show a reduced response to relativistic motion *within* the classical receptive field, and a weak response if a motion boundary falls within the receptive field (see also Britten & Newsome, 1990). Combining the two results we can see how a population of such cells would respond to a motion border. Cells whose receptive field falls upon the motion boundary would be suppressed due to differential motion within the classical receptive field, whereas those far from the boundary would be suppressed due to the non-classical receptive field suppression. Only those cells close to either side of the boundary could give strong signals. A second possibility is that the suppression of response caused by one direction upon another may be important in getting rid of noise signals that are inherent in any complex stimuli such as random dots (Snowden, 1989). The finding that damage to area MT degrades performance as soon as any noise is added to the stimulus (Newsome & Paré, 1988) (see also sections 1.1 and 5.1) would seem to support such a notion.

4 THE MEDIAL SUPERIOR TEMPORAL (MST) AREA

In the cortex anterior to MT directionally selective cells are also prevalent (Desimone & Ungerleider, 1986; Saito et al., 1986). Perhaps this is not surprising given that area MT projects heavily into this area. Indeed, it was the projection from MT that initially defined this area (Maunsell & van Essen, 1983a). The area does contain a topographic representation of the visual field but it is extremely crude. The area differs from MT in several respects: firstly the receptive fields are considerably larger than those at a comparable eccentricity in MT (Desimone & Ungerleider, 1986; Tanaka et al., 1986). Putting a figure on the difference is quite difficult. Studies that have quantitatively examined this issue have suffered several problems. Firstly, the definition of what is or is not MST is still under debate, and has changed several times. Therefore some studies may have included MTp as part of MST, and possible differences about the lateroventral borders of MST could have large consequences for these studies (see below). Secondly, as the receptive fields are quite large it is often hard to define their eccentricity. Most studies use the geometric mean of the receptive field to define eccentricity, but this may not be the 'functional centre'. Recent evidence has shown that the part of the field giving the strongest response is often some way from the geometric mean (Tanaka et al., 1993). Finally, there is the puzzling question of what

Figure 10 A. A section of area MT which has been stained for the activity marker 2-deoxyglucose. After injection of the marker the animal viewed a large field of moving random dots. The subsequent uptake of the marker is seen to be patchy in the sense that the tissue was not uniformly stained. B. The amount of suppression elicited from the surround of the classical receptive field (see text) is plotted for several cells on a penetration through the area of cortex displayed in A. The amount of staining (defined by the optical density) is also displayed. Note that the strongest suppression occurs when there is little staining. STS: superior temporal sulcus; D: dorsal; V: ventral; P: posterior; A: anterior. From Born and Tootell (1992).

to use to map the receptive field. The two 'classic' methods (first, finding the limits of where a cell responds to a small object; second, increasing the size of the object until no more summation is observed) may be quite inappropriate. For instance, some cells actually decrease their response as the object is made bigger, while others will only respond to large objects!

4.1 One, two or more areas?

Defining area MST has been difficult due to the lack of myeloarchitectural boundaries or other anatomical markers. We have had to rely on functional physiology to make distinctions.

The most dorsal and medial portion most adjacent to MT have receptive fields near the fovea, as one moves more ventral the receptive fields are more peripheral (see Figure 3C). Just anterior to foveal MT the receptive fields are once again more foveal. This particular zone has been thought to be part of FST (Desimone & Ungerleider, 1986; Boussaoud *et al.*, 1990) due to the lower incidence of directionally selective cells and possible differences in its projections. Others have found its incidence of directional selectivity to be similar to the rest of MST (Komatsu & Wurtz, 1988) and have therefore included it as part of this area. The dorsal portion (MSTd) appears to have response properties distinct from the lateral portion (MSTl) and the disputed area has response properties that are in accord with MSTl. For the purpose of this review I shall regard the disputed zone as belonging to MSTl, but acknowledge the current uncertainty.

4.2 MSTd

One of the most striking findings is that some cells respond to relative motion within their receptive fields. Some cells are selective for a pattern rotating in a particular direction, some respond to an expanding pattern and others to a contracting pattern (Saito *et al.*, 1986; Tanaka & Saito, 1989; Duffy & Wurtz, 1991a, 1991b). Some cells also respond to simple translations of the pattern. These movements are those occurring during an animal's movement (see Chapter 11 by M. Harris), and this leads to the hypothesis that these cells are important for optic flow. As the whole visual field undergoes this motion when an animal moves we should expect these cells to prefer a large stimulus to a small one. Several studies have shown that these cells are generally far more responsive to large fields of moving dots than to small fields or isolated elements (Komatsu & Wurtz, 1988; Tanaka & Saito, 1989; Duffy & Wurtz, 1991a).

Optic flow patterns have often been described according to a mathematical breakdown into an expansion/contraction component, a rotational component and a shear component (e.g. Koenderink, 1986; see also Chapter 11 by M. Harris). The finding that there are cells responsive to two of these components is suggestive that this might be what MSTd is attempting. However, this type of analysis into 'basis functions' is rarely observed in the brain (for an instance where this may happen see Simpson, 1984). For instance, MT could encode direction of translation by having just two groups of cells, say one for vertical axis and one for horizontal axis, but it has instead neurones tuning for many (all?) directions. Could MST have neurones tuned to many (all) possible optic flow patterns? If we consider an expansion/contraction pattern (Figure 11) we can see that each dot changes its radial component but not its angular one, whereas a rotation does not change its radial component but does change its angular component. Between the axes there are other stimuli that spiral clockwise or

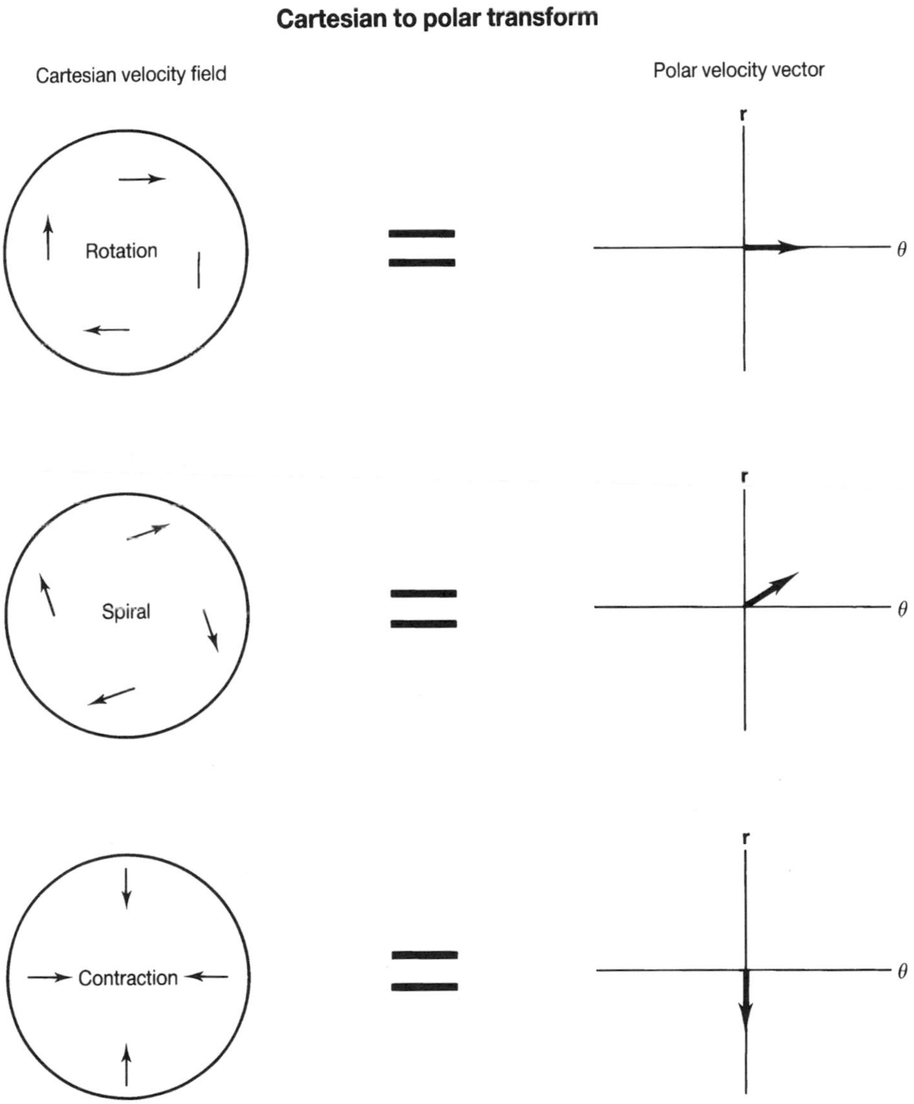

Cartesian to polar transform

Figure 11   The relationship between Cartesian and polar coordinate systems. In the upper plot a clockwise rotation is redescribed as a change in angle (θ) but no change in radius (r). A contraction (bottom) has a negative change in r, but no change in θ. A spiral has a change both in r and in θ.

anticlockwise, in or out. Are there neurones that prefer a spiral stimulus? Graziano *et al.* (1994) presented such stimuli and found neurones that were tuned for spiral stimuli (Figure 12A), indeed their incidence was similar to cells tuned to the major axes (Figure 12B). From this they conclude that MSTd is not breaking the optic flow into

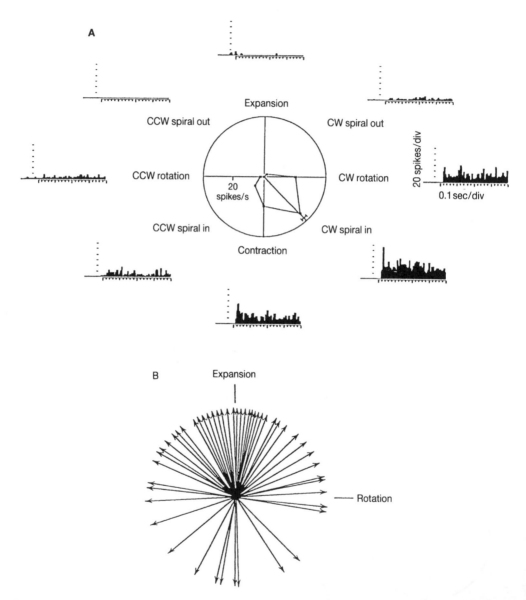

Figure 12   A. The response of a MSTd cell to various stimuli. The centre plots the response in a polar coordinate system (see Figure 11). Note the cells give the strongest response to a stimulus that spirals clockwise and contracts. B. The preferred stimulus for a population of MSTd cells is plotted in polar coordinates. Note that there are cells tuned for many different directions of motion. CW: clockwise; CCW: counter clockwise. From Graziano *et al.* (1994).

basic functions but operates in polar space just like MT operates in Cartesian space. This finding might explain the 'double-component' (those that respond to, say, both a contraction and a rotation) cells of Duffy and Wurtz (1991b). These cells may actually be tuned to a spiral and thus give a non-optimal response to the contraction and the rotation.

Given that MSTd receives its major input from MT, how does its preference for complex motions arise from simple translational motion inputs? The most simple model would be to arrange MT afferents in an orderly manner within the neurone's receptive field. To produce an 'expansion' cell we should feed an MT neurone with an upward preference to the top of the receptive field, bottom to bottom, left to the left and right to the right, etc. (Figure 13). A crucial test of this model would be to use a fairly small stimulus and measure the cell's response in different parts of the receptive field. Such a test is shown in Figure 14A (from Graziano *et al.*, 1994). The cell was presented with either a clockwise or anticlockwise rotation in various parts of its receptive field. The cell gave exactly the same preference for clockwise motion everywhere the test patterns were positioned (Figure 14B). A similar pattern of results is also shown for an expansion cell (Figure 14C). Graziano *et al.* found that all the cells they tested gave this position invariant response strongly refuting the simple model of Figure 13 (see also Duffy & Wurtz, 1991b). At the moment it is unclear just how MST cells produce their selectivity, though there are some interesting ideas emerging (Tanaka *et al.*, 1989; Sereno *et al.*, 1991).

Figure 13   A possible model of how selectivity to expanding stimuli may be produced. Each subsection is sensitive to linear motion, and the preferred linear motion varies systematically across the receptive field. On this model there is also a change in the preferred speed with cells near the edge of the receptive field preferring greater speeds.

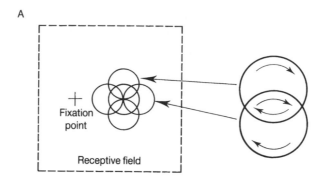

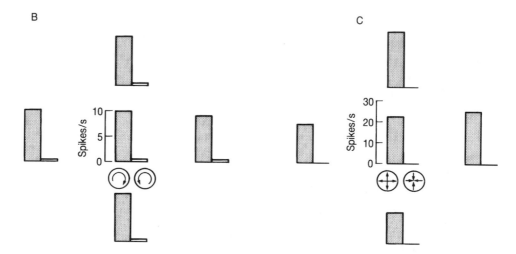

Figure 14   A. The properties of MSTd cells were tested at five different positions within the receptive field of each cell. The five positions were arranged in an overlapping clover-leaf pattern. This arrangement means on some trials movement is in one direction in a particular part of the receptive field, whereas on other trials it would be in the opposite direction. B. The response of a cell which preferred clockwise rather than anticlockwise rotation. The cell maintains its preference at all five positions tested. C. A similar test for a cell which preferred expanding rather than contracting patterns. From Graziano *et al.* (1994).

4.3   MSTl

Cells in MSTl appear to have very different properties to those of MSTd. These cells do not seem to be sensitive to the optic flow type of pattern. Interestingly, Tanaka *et al.* (1993) found that some cells respond very well to small patterns in motion, but then decrease their response as the stimulus increases in size (Figure 15, cell 1)!

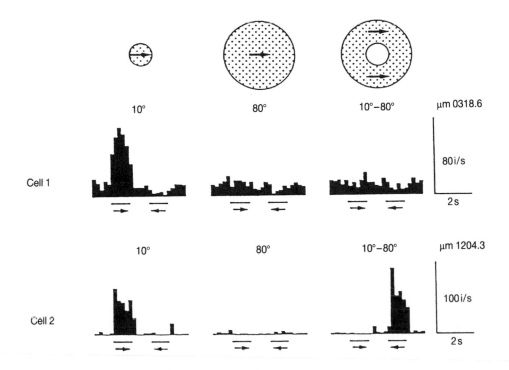

Figure 15   The responses of two cells from area MST1 to various stimuli. The first stimulus was a moving random dot pattern of diameter 10°, the second one of 80°, and the third the pattern of 80° but with the central 10° occluded. Cell 1 responds well to the small pattern but not to the large patterns. Cell 2 responds well to the small pattern but not to the large pattern. This cell also gives a strong response to the large pattern when the centre is occluded. However, it does so when the pattern is moving in the opposite direction to its preferred direction for the small pattern (i: impulses). From Tanaka *et al.* (1993).

Note this is the opposite to the type of response found in MSTd. If MSTd is processing large optic flow patterns, then it is tempting to think that MSTl is doing the complementary job of analysing the movements of objects relative to their surroundings (figure/ground segregation). A remarkable finding supports this interpretation. Some cells that are directionally selective for small fields but lose this directionally selective response for large fields, can be made directionally selective by placing a small stationary occluding object in front of the large pattern. However, the preferred direction for the pattern is now the opposite of its preferred direction for a small stimulus (Figure 15, cell 2). The result could be thought of as signalling a 'relativistic' motion. A small object moving to the right on a stationary background has, by definition, a rightward component relative to its background. A stationary object with a background moving to the left also has a rightward component relative to its background.

5   A HUMAN MOTION PATHWAY?

5.1   The effects of brain damage

If the human brain possesses structures and functions similar to those described for the macaque monkey we should occasionally find that they have malfunctioned due to damage or disease. Such damage is rarely selective and interpretation of results obtained from psychophysical studies on such patients should be treated with caution.

Zihl *et al.* (1983) reported on a woman who appeared to have lost her ability to sense motion following a bilateral cerebral vascular lesion (see also Chapter 7 by Logothetis). The patient appeared to have normal vision in other respects (e.g. colour vision). Such a deficit has been labelled cerebral akinetopsia (Zeki, 1991). Quantitative testing of this patient has revealed that her deficit strongly resembles that of the MT lesioned monkeys. Baker *et al.* (1991) used similar techniques to those of Newsome and Paré (1988) (see section 1.1). They found that the patient could determine the direction of motion under a limited set of conditions but this residual performance was eliminated by small amounts of incoherence in the stimulus and even if some of the dots were kept stationary whilst the rest moved. In section 3.6 I suggested one possible function of area MT was to eliminate noise and segregate motions. If indeed a human homologue of MT has been damaged then we should expect to see the patient having difficulties when the motion information is degraded or when more than one motion exists – the pattern of results reported (see also Vaina, 1989). Of course the damage accrued by this patient and others of a similar nature (Vaina *et al.*, 1990) is unlikely to just affect a human MT; there must be a substantial probability that human MST and other adjacent structures are also damaged. So we are unsure just what is causing the problems and this may explain some of the many discrepancies reported from patient to patient. An illustration of this point comes from some recent studies that have shown abnormal motion perception in subjects with cerebellar lesions (Ivry & Diener, 1991; Nawrot & Rizzo, 1992).

5.2   Visualization of visual function

The functional fractionation of the human brain can also be revealed by sophisticated imaging techniques whilst the subject views a specially prepared stimulus that exclusively activates (hopefully) the mechanism(s) one wishes to examine. Zeki *et al.* (1991) used positron emission tomography (PET) to measure regional blood flow while subjects viewed either a colourful Mondrian pattern or a pattern of moving dots. These stimuli activated several areas of the brain in common, which were taken to be coincident with previous estimates of human V1 and V2. The Mondrian (colour) pattern also activated a region of the lingual and fusiform gyri that was not affected by the presentation of the moving dots. Likewise the moving dot pattern activated a region near the parietal/occipital junction that was not activated by the Mondrian pattern. The implication is clear. Within the human cortex there is a fractionation of the visual image. Whether the area revealed by the motion stimulus is a 'human MT' is still a matter of some speculation. If human cortex is similar to that of the macaques such a stimulus should only drive a sub-population of MT cells, but should also drive most MSTd cells and VIP cells. Another PET study using the technique of presenting the subjects with the same stimulus from trial to trial but getting them to make judgements on different attributes of that stimulus (e.g. its colour or movement) seems to give a somewhat different pattern of activity in the human brain (Corbetta *et al.*, 1990). Imaging techniques such as PET and others that are still in their infancy offer a great chance for us to begin to explore the human cortex in much greater detail.

REFERENCES

Albright, T. D. (1984). Direction and orientation selectivity of neurons in visual area MT of the macaque. *J. Neurophysiol.*, **52**, 1106–1130.

Albright, T. D. (1989). Centrifugal directional bias in the middle temporal visual area (MT) of the macaque. *Vis. Neurosci.*, **2**, 177–188.

Albright, T. D. & Desimone, R. (1987). Local precision of visuotopic in the middle temporal area (MT) of the macaque. *Exp. Brain Res.*, **65**, 582–592.

Albright, T. D., Desimone, R. & Gross, C. G. (1984). Columnar organization of directionally selective cells in visual area MT of the macaque. *J. Neurophysiol.*, **51**, 16–31.

Allman, J. M. & Kaas, J. (1971). A representation of the visual field in the caudal third of the temporal gyrus of the owl monkey (*Aotus trivirgatus*). *Brain Res.*, **31**, 84–105.

Allman, J., Miezin, F. & McGuiness, E. (1985a). Direction- and velocity-specific responses from beyond the classical receptive field in the middle temporal visual area (MT). *Perception*, **14**, 105–126.

Allman, J., Miezin, F. & McGuiness, E. (1985b). Stimulus specific responses from beyond the classical receptive field. *Ann. Rev. Neurosci.*, **8**, 407–430.

Baker, C. L. J., Hess, R. F. & Zihl, J. (1991). Residual motion perception in a 'motion-blind' patient: assessed with limited lifetime random dot stimuli. *J. Neurosci.*, **11**, 454–461.

Beneveto, L. A. & Standage, G. P. (1983). The organization of projections of the retinorecipient and nonretinorecipient nuclei of the pretectal complex and layers of the superior colliculus to the lateral pulvinar and medial pulvinar in the macaque monkey. *J. Comp. Neurol.*, **271**, 307–336.

Born, R. T. & Tootell, R. B. H. (1992). Segregation of global and local motion processing in primate middle temporal area. *Nature*, **357**, 497–499.

Boussaoud, D., Ungerleider, L. G. & Desimone, R. (1990). Pathways for motion analysis: cortical connections of the medial superior temporal and fundus of the superior temporal visual areas of the macaque. *J. Comp. Neurol.*, **296**, 462–495.

Britten, K. H. & Newsome, W. T. (1990). Responses of MT neurons to discontinuous motion. *Invest. Ophthalmol. Vis. Sci. Suppl.*, **31**, 238.

Corbetta, M., Miezin, F. M., Dobmeyer, S., Shulman, G. L. & Petersen, S. E. (1990). Attention modulation of neural processing of shape, color, and velocity in humans. *Science*, **248**, 1556–1559.

De Valois, R. L., Yund, E. W. & Hepler, H. (1982). The orientation and direction selectivity of cells in macaque visual cortex. *Vision Res.*, **22**, 531–544.

Desimone, R. & Ungerleider, L. G. (1986). Multiple visual areas in the caudal superior temporal sulcus of the macaque. *J. Comp. Neurol.*, **248**, 164–189.

DeYoe, E. A. & Van Essen, D. C. (1988). Concurrent processing streams in monkey visual cortex. *Trends Neurosci.*, **11**, 219–226.

Dow, B. M., Snyder, A. Z., Vautin, R. G. & Bauer, R. (1981). Magnification factor and receptive field size in foveal striate cortex of the monkey. *Exp. Brain Res.*, **44**, 213–228.

Downing, C. & Movshon, J. A. (1989). Spatial and temporal summation in the detection of motion in stochastic random dot displays. *Invest. Ophthalmol. Vis. Sci. Suppl.*, **30**, 72.

Dubner, R. & Zeki, S. M. (1971). Response properties and receptive fields of cells in an anatomically defined region of the superior temporal sulcus in the monkey. *Brain Res.*, **35**, 528–532.

Duffy, C. J. & Wurtz, R. H. (1991a). Sensitivity of MST neurons to optic flow stimuli. I. A continuum of response selectivity to large-field stimuli. *J. Neurophysiol.*, **65**, 1329–1345.

Duffy, C. J. & Wurtz, R. H. (1991b). Sensitivity of MST neurons to optic flow stimuli. II. Mechanisms of response selectivity revealed by small-field stimuli. *J. Neurophysiol.*, **65**, 1346–1359.

Erickson, R. G., Dow, B. M. & Snyder, A. Z. (1989). Representation of the fovea in the superior temporal sulcus of the macaque monkey. *Exp. Brain Res.*, **78**, 90–112.

Felleman, D. J. & van Essen, D. C. (1991). Distributed hierarchical processing in the primate cerebral cortex. *Cerebral Cortex*, **1**, 1–47.

Fries, W., Keizer, K. & Kuypers, H. G. J. M. (1985). Large layer VI cells in macaque striate cortex (Meynert cells) project to both superior colliculus and prestriate visual area V5. *Exp. Brain Res.*, **58**, 613–616.

Frost, B. J. & Nakayama, K. (1983). Single visual neurons code opposing motion independent of direction. *Science*, **220**, 744–745.

Gattass, R. & Gross, C. G. (1981). Visual topography of striate projection zone (MT) in posterior superior temporal sulcus of the Macaque. *J. Neurophysiol.*, **46**, 621–638.

Girard, P., Salin, P. A. & Bullier, J. (1992). Response selectivity of neurons in area MT of the macaque monkey during reversible inactivation of area V1. *J. Neurophysiol.*, **67**, 1437–1446.

Graziano, M. S. A., Andersen, R. A. & Snowden, R. J. (1994). Tuning of MST neurons to spiral stimuli. *J. Neurosci.*, **14**, 54–67.

Hammond, P. & Smith, A. T. (1982). On the sensitivity of complex cells in feline striate cortex to relative motion. *Exp. Brain Res.*, **47**, 457–460.

Hawken, M. J., Parker, A. J. & Lund, J. S. (1988). Laminar organization and contrast sensitivity of direction-selective cells in the striate cortex of the Old World monkey. *J. Neurosci.*, **8**, 3541–3548.

Hubel, D. H. & Wiesel, T. N. (1972). Laminar and columnar distribution of geniculocortical fibres in macaque monkey. *J. Comp. Neurol.*, **146**, 421–450.

Ivry, R. B. & Diener, H. C. (1991). Impaired velocity perception in patients with lesions of the cerebellum. *J. Cogn. Neurosci.*, **3**, 355–366.

Koenderink, J. J. (1986). Optic flow. *Vision Res.*, **26**, 161–180.

Komatsu, H. & Wurtz, R. H. (1988). Relation of cortical areas MT and MST to pursuit eye movements. I. Localization and visual properties of neurons. *J. Neurophysiol.*, **60**, 580–603.

Lagae, L., Gulyas, B., Raiguel, S. & Orban, G. A. (1989). Laminar analysis of motion information processing in macaque V5. *Brain Res.*, **496**, 361–367.

Lagae, L., Raiguel, S. & Orban, G. A. (1993). Speed and direction selectivity of macaque middle temporal neurons. *J. Neurophysiol.*, **69**, 19–39.

Maunsell, J. H. R. & Newsome, W. T. (1987). Visual processing in monkey extra-striate cortex. *Ann. Rev. Neurosci.*, **10**, 363–401.

Maunsell, J. H. R. & van Essen, D. (1983a). The connections of the middle temporal area (MT) and their relationship to a cortical hierarchy in the macaque monkey. *J. Neurosci.*, **3**, 2563–2586.

Maunsell, J. H. R. & van Essen, D. C. (1983b). Functional properties of neurons in middle temporal visual area of the macaque monkey. I. Selectivity for stimulus direction, speed, and orientation. *J. Neurophysiol.*, **49**, 1127–1147.

Maunsell, J. H. R. & van Essen, D. C. (1987). Topographic organization of the middle temporal visual area in the macaque monkey: Representational biases and the relationship to callosal connections and myeloarchitectonic boundaries. *J. Comp. Neurol.*, **266**, 535–555.

Maunsell, J. H. R., Nealey, T. A. & DePriest, D. D. (1990). Magnocellular and parvocellular contributions to responses in the middle temporal visual area (MT) of the macaque monkey. *J. Neurosci.*, **10**, 3323–3334.

McIlwain, J. T. (1964). Receptive fields of optic tract axons and lateral geniculate cells: Peripheral extend and barbiturate sensitivity. *J. Neurophysiol.*, **27**, 1154–1173.

Mikami, A., Newsome, W. T. & Wurtz, R. H. (1986). Motion selectivity in macaque visual cortex. II. Spatiotemporal range of directional interactions in MT and V1. *J. Neurophysiol.*, **55**, 1328–1339.

Movshon, J. A. & Newsome, W. T. (1984). Functional characteristics of striate cortical neurons projecting to MT in the macaque. *Soc. Neurosci. Abstracts*, **10**, 933.

Movshon, J. A., Newsome, W. T., Gizzi, M. S. & Levitt, J. B. (1988). Spatio-temporal tuning and speed sensitivity in macaque cortical neurons. *Invest. Ophthalmol. Vis. Sci. Suppl.*, **29**, 327.

Nawrot, M. & Rizzo, M. (1992). Abnormal motion perception with human cerebellar lesions. *Invest. Ophthalmol. Vis. Sci. Suppl.*, **33**, 1130.

Newsome, W. T. & Parén, E. B. (1988). A selective impairment of motion perception following lesions of the middle temporal visual area (MT). *J. Neurosci.*, **8**, 2201–2211.

Newsome, W. T., Gizzi, M. S. & Movshon, J. A. (1983). Spatial and temporal properties of neurons in macaque MT. *Invest. Ophthalmol. Vis. Sci. Suppl.*, **24**, 106.

Newsome, W. T., Britten, K. H. & Movshon, J. A. (1989). Neuronal correlates of a perceptual decision. *Nature*, **341**, 52–54.

Orban, G. A., Kennedy, H. & Bullier, J. (1986). Velocity sensitivity and direction selectivity of neurons in area V1 and V2 of the monkey: Influence of eccentricity. *J. Neurophysiol.*, **56**, 462–480.

Orban, G. A., Gulyas, B. & Spileers, W. (1987). A moving noise background modulates responses to moving bars of monkey V2 cells but not of monkey V1 cells. *Invest. Ophthalmol. Vis. Sci. Supp.*, **28**, 197.

Reid, R. C. & Shapley, R. M. (1992). Spatial structure of cone inputs to receptive fields in primate lateral geniculate nucleus. *Nature*, **356**, 716–718.

Rockland, K. S. & Pandya, D. N. (1981). Cortical connections of the occipital lobe in the rhesus monkey: interconnections between areas 17, 18, 19 and superior temporal sulcus. *Brain Res.*, **212**, 249–270.

Rodman, H. R. & Albright, T. D. (1987). Coding of visual stimulus velocity in area MT of the macaque. *Vision Res.*, **27**, 2035–2048.

Rodman, H. R., Gross, C. G. & Albright, T. D. (1989). Afferent basis of visual responses in area MT of the macaque. I. Effects of striate cortex removal. *J. Neurosci.*, **9**, 2033–2050.

Rodman, H. R., Gross, C. G. & Albright, T. D. (1990). Afferent basis of visual response properties in area MT of the macaque: II. Effects of superior colliculus removal. *J. Neurosci.*, **10**, 1154–1164.

Saito, H.-A., Yukie, M., Tanaka, K., Hikosaka, K., Fukada, Y. & Iwai, E. (1986). Integration of direction signals of image motion in the superior temporal sulcus of the macaque monkey. *J. Neurosci.*, **6**, 145–157.

Salzman, C. D., Britten, K. H. & Newsome, W. T. (1990). Cortical microstimulation influences perceptual judgements of motion direction. *Nature*, **346**, 174–177.

Sereno, M. E., Zhang, K. & Sereno, M. I. (1991). How position-independent detection of sense of rotation or dilation is learned by a Hebb rule. *Soc. Neurosci. Abstracts*, **17**, 441.

Shipp, S. & Zeki, S. (1985a). Segregated output to area V5 from layer 4B of macaque monkey striate cortex. *J. Physiol. (Lond.)*, **369**, 32P.

Shipp, S. & Zeki, S. (1985b). Segregation of pathways leading from area V2 to areas V4 and V5 of macaque monkey visual cortex. *Nature*, **315**, 322–325.

Shipp, S. & Zeki, S. (1989a). The organization of connections between areas V5 and V1 in macaque monkey visual cortex. *Eur. J. Neurosci.*, **1**, 309–332.

Shipp, S. & Zeki, S. (1989b). The organization of connections between areas V5 and V2 in macaque monkey visual cortex. *Eur. J. Neurosci.*, **1**, 333–354.

Siegel, R. M. & Andersen, R. A. (1991). The perception of structure from visual motion in monkey and man. *J. Cogn. Neurosci.*, **2**, 306–319.

Simpson, J. I. (1984). The accessory optic system. *Ann. Rev. Neurosci.*, **7**, 13–41.

Snowden, R. J. (1989). Motions in orthogonal directions are mutually suppressive. *J. Opt. Soc. Am. A*, **6**, 1096–1101.

Snowden, R. J., Treue, S., Erickson, R. E. & Andersen, R. A. (1991). The response of area MT and V1 neurons to transparent motion. *J. Neurosci.*, **11**, 2768–2785.

Snowden, R. J., Treue, S. & Andersen, R. A. (1992). The response of neurons in area V1 and MT of the alert rhesus monkey to moving random dot patterns. *Exp. Brain Res.*, **88**, 389–400.

Tanaka, K. & Saito, H. (1989). Analysis of motion of the visual field by direction, expansion/contraction, and rotation cells clustered in the dorsal part of the medial superior temporal area of the macaque monkey. *J. Neurophysiol.*, **62**, 626–641.

Tanaka, K., Hikosaka, K., Saito, H.-A., Yukie, M., Fukada, Y. & Iwai, E. (1986). Analysis of local and wide-field movements in the superior temporal visual areas of the macaque monkey. *J. Neurosci.*, **6**, 134–144.

Tanaka, K., Fukada, Y. & Saito, H. (1989). Underlying mechanisms of the response specificity of the expansion/contraction and rotation cells in the dorsal part of the medial superior temporal area of the macaque monkey. *J. Neurophysiol.*, **62**, 642–656.

Tanaka, K., Sugita, Y., Moriya, M. & Saito, H.-A. (1993). Analysis of object motion in the ventral part of the medial superior temporal area of the macaque visual cortex. *J. Neurophysiol.*, **69**, 128–142.

Tolhurst, D. J. & Movshon, J. A. (1975). Spatial and temporal contrast sensitivity of striate cortical neurones. *Nature*, **257**, 674–675.

Tootell, R. B. H. & Born, R. T. (1990). Patches and directional columns in primate area MT. *Invest. Ophthalmol. Vis. Sci. Suppl.*, **31**, 238.

Ungerleider, L. G. & Desimone, R. (1986a). Cortical connections of the visual area MT in the macaque. *J. Comp. Neurol.*, **248**, 190–222.

Ungerleider, L. G. & Desimone, R. (1986b). Projections to the superior temporal sulcus from the central and peripheral field representations of V1 and V2. *J. Comp. Neurol.*, **248**, 147–163.

Ungerleider, L. G. & Mishkin, M. (1979). The striate projection zone in the superior temporal sulcus of *Macaca mulatta*: Location and topographic organization. *J. Comp. Neurol.*, **188**, 347–366.

Vaina, L. M. (1989). Selective impairment of visual motion interpretation following lesions of the right occipito-parietal area in humans. *Biol. Cybernetics*, **61**, 347–359.

Vaina, L. M., Lemay, M., Bienfang, D. C., Choi, A. Y. & Nakayama, K. (1990). Intact 'biological motion' and 'structure from motion' perception in a patient with impaired motion mechanisms: A case study. *Vis. Neurosci.*, **5**, 353–369.

van de Grind, W. A., van Doorn, A. J. & Koenderink, J. J. (1983). Detection of coherent movement in peripherally viewed random-dot patterns. *J. Opt. Soc. Am. A*, **73**, 1674–1683.

van Essen, D. C. (1985). Functional organization of primate visual cortex. In A. Peters & E. G. Jones (Eds.) *Cerebral Cortex*. Plenum, New York.

van Essen, D. C., Maunsell, J. H. R. & Bixby, J. L. (1981). The middle temporal visual area in the macaque: myeloarchitecture, connections, functional properties and topographic organization. *J. Comp. Neurol.*, **199**, 293–326.

Webb, S. V. & Kaas, J. H. (1976). The sizes and distribution of ganglion cells in the retina of the owl monkey, *Aotus trivirgatus. Vision Res.*, **16**, 1247–1254.

Zeki, S. M. (1974). Functional organization of a visual area in the posterior bank of the superior temporal sulcus of the rhesus monkey. *J. Physiol. (Lond.)*, **236**, 549–573.

Zeki, S. (1991). Cerebral akinetopsia. *Brain*, **114**, 811–824.

Zeki, S., Watson, J. D. G., Lueck, C. J., Friston, K. J., Kennard, C. & Frackowiak, R. S. J. (1991). A direct demonstration of functional specialization in the human visual cortex. *J. Neurosci.*, **11**, 641–649.

Zihl, J., Cramon, D. V. & Mai, N. (1983). Selective disturbance of movement vision after bilateral brain damage. *Brain*, **106**, 313–340.

4

The Psychophysics of Motion Perception

Suzanne P. McKee and Scott N. J. Watamaniuk

Smith-Kettlewell Eye Research Institute, San Francisco, USA

1 INTRODUCTION

In physical terms, motion is defined as a change in space over time. The human visual system has a highly precise code for both spatial and temporal dimensions. Since motion information is implicitly present in the spatial and temporal codes, there had been a question traditionally about the need to invoke special mechanisms[1] to encode motion. At this date, the physiological evidence for special motion mechanisms in primates is so overwhelming that this question seems a quaint remnant of scientific history. Unfortunately, physiological evidence does not solve the psychophysicist's problem. In psychophysical experiments, there is no guarantee that human 'motion' judgments are limited or determined by the motion system just because the stimulus is moving. Any study of motion tests the ingenuity of the psychophysicist to prove that the judgment is indeed determined by a motion-specific mechanism, rather than by spatial or temporal mechanisms. Over the years, a number of techniques have been developed to deal with this problem. These techniques include motion thresholds measured at contrast threshold, thresholds measured after adaptation to moving stimuli, thresholds measured with stimuli that obscure consistent position and/or time information, and thresholds measured concurrently for motion, time and space.

---

[1] Mechanism is the term used by psychophysicists to refer to the physiological entity responsible for detecting a target, or discriminating a difference between targets. It commonly refers to a single neural unit (Graham, 1989), or perhaps to a small pool of neural units with nearly identical tuning for particular stimulus dimensions, including retinal position.

VISUAL DETECTION OF MOTION
ISBN 0–12–651660–X

2 TECHNIQUES FOR ISOLATING MOTION MECHANISMS

2.1 Motion at contrast threshold

In many experimental conditions, the direction of motion can be coarsely identified at contrast threshold. For example, if the percentage correct for detecting a drifting vertical grating is compared to the percentage correct for identifying the drift direction (left or right) at the same contrast, detection and identification are equally good, at least at low spatial frequencies (Levinson & Sekuler, 1975; Watson *et al.*, 1980; Derrington & Henning, 1993). At contrast threshold, a target can be detected only by a mechanism which is optimally stimulated by the characteristics of the target; all other mechanisms will have a smaller signal. So, if observers can identify the direction of motion as easily as they detect the stimulus, the detecting mechanism must convey information about direction – it must be specifically 'labeled' for direction of motion (Watson & Robson, 1981). Unlike their results with low spatial frequency (SF) gratings, Watson *et al.* (1980) found that high spatial frequency gratings, drifting at low temporal frequencies (TF), were not labeled for direction at contrast threshold (Figure 1). This result implies that contrast detection for targets moving at slow velocities (low TF/high SF) is mediated by directionally *non-selective* mechanisms.

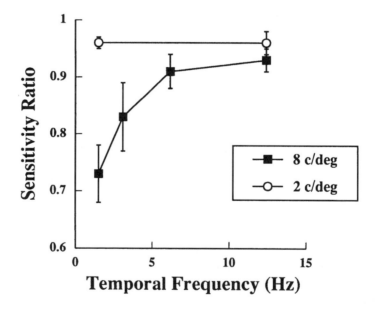

Figure 1 Ratio of threshold for detection of a moving grating and for identifying its direction of motion, as a function of temporal frequency. The data, taken from Watson *et al.* (1980) were averaged over three subjects; the standard errors were estimated from the variance among the three subjects. The sensitivity ratio shows that mechanisms detecting faster speeds (high TF/low SF) also encode direction at threshold, while mechanisms detecting slower speeds (low TF/high SF) are not direction selective.

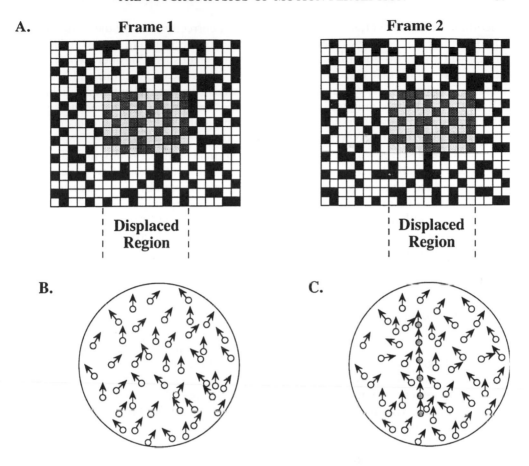

Figure 2 A. Diagram of dense random dot kinematogram. Central region of dots in frame 1 is displaced in frame 2, with new dots filling in. A displaced region is shaded differently to identify it; there is no difference in shading in experimental kinematograms. B. Sparse random dot kinematogram in which each dot can move in a different direction from frame to frame. If the range of directions is small, the display will appear to flow in the mean direction, e.g. up for this diagram. C. Single 'signal dot' is moved in a consistent direction for many frames (six in diagram), while background 'noise' dots move at random in directions chosen from a range of 160 degrees.

It does not mean that there are no direction-selective motion mechanisms at slow velocities, only that they are less sensitive than non-selective mechanisms.

One drawback to using gratings is that judgments about direction are confined to the two opposite directions perpendicular to the bars of the grating. To determine the number of independent directions available at contrast threshold, Ball *et al.* (1983) used the detection–identification paradigm for a moving random dot stimulus. The stimulus consisted of a sparse pattern of bright dots (8 dots/deg^2), with each dot moving in a direction chosen from a 30 degree range centered on a mean direction (Figure 2B). Observers judged which of two mean directions were presented as a

function of stimulus contrast. The observer cannot correctly identify the same proportion of stimuli that he or she detects, unless the mean directions are spaced sufficiently
far apart that they do not stimulate overlapping mechanisms. Consider two direction
mechanisms – one that responds maximally to the vertical direction, and a second
tuned to 30 degrees left of vertical. On most trials, the vertically-tuned mechanism
will detect a stimulus moving in a vertical direction and the other mechanism will
have insufficient signal to respond. But, if the tuning of the two mechanisms overlaps
significantly, the one tuned 30 degrees off vertical will occasionally detect the vertical
target and the observer will then report that the vertical stimulus is moving left. Using
this paradigm, Ball *et al.*, found that observers could identify direction as readily as
they detected the stimulus only when mean directions differed by 120 degrees or more.
Other studies have produced similar estimates of the breadth of directional tuning
(Marshak & Sekuler, 1979; Mather, 1980). These results show only that human motion
mechanisms are very broadly-tuned for direction. Different procedures, which will
be described below, are required to estimate the *number* of different directional
mechanisms in the human visual system.

2.2 Motion adaptation

After staring for a few seconds at a stimulus moving in a given direction, a
stationary target viewed subsequently may appear to move in the opposite direction
for a short time. This phenomenon, first described in the 19th century, is called the
motion aftereffect (Wohlgemuth, 1911). Arguing from physiological evidence for
directionally-selective mechanisms, Sekuler and Ganz (1963) proposed that motion
mechanisms tuned to the direction of the moving stimulus had been selectively
fatigued or adapted by prolonged exposure to motion in one direction. A stationary
target stimulates all motion mechanisms equally, but the selective adaptation had
significantly reduced the sensitivity of one group creating a local imbalance. The
residual response of the unadapted mechanisms to the stationary target produced the
appearance of illusory motion in the direction opposite to the adapted direction.
Sekuler and Ganz reasoned, that if directionally-selective mechanisms existed in the
human visual system, the contrast threshold for targets moving in the adapted direction
might be higher than the threshold for the opposite direction. They found selective
elevation for grating targets moving in the same direction as the adapting stimulus,
when compared to thresholds for gratings moving in the opposite direction. Since the
spatial and temporal characteristics of the test gratings for the two directions were
identical, directionally-selective adaptation, like directional labelling at contrast
threshold, was compelling evidence that the moving target was detected by a
motion-specific mechanism.

   The strength of the motion aftereffect can be enhanced by using a moving or
flickering display as the test stimulus, rather than the traditional stationary pattern
(Hiris & Blake, 1992; Nishida & Sato, 1993). Hiris and Blake (1992) reported that a
sparse display of dots moving at random in all directions was biased by selective
directional adaptation. After adaptation, the random direction display appeared to flow
opposite the direction of the adapting stimulus. Increasing the proportion of dots
moving in one direction will also cause the randomly-moving dots to appear to flow

in that direction (Williams & Sekuler, 1984; Newsome & Pare, 1988). Hiris and Blake showed that the bias induced by selective adaptation was indistinguishable from the physical bias created by changing the proportion of dots in the display moving in a single direction – another indication that the motion aftereffect can be used to select a population of human motion units tuned for a particular direction.

Regan and Beverley (1978a, 1978b, 1979) and Regan (1986) showed that complex motions, such as motion-in-depth, looming (changing size), rotation and shearing, can also be selectively adapted (see also Chapter 12 by Cumming). For example, thresholds for detecting an oscillation in overall size were elevated by prolonged exposure to size oscillations, but thresholds for lateral oscillations of the same figure were little affected by size oscillations (Regan & Beverley, 1978b) The results from each of these adaptation studies may argue for a specific class of neural mechanisms tuned to each pattern of motion (Beverley & Regan, 1973), or they may instead reflect a type of contingent learning in which regular association of particular stimulus dimensions affects subsequent sensitivity and perception (Wyatt, 1974). Certainly the time course for the initial adaptation, and for its extinction, are considerably longer than the simple motion aftereffects described above (Regan & Beverley, 1978a).

2.3   Obscuring position information

Motion measurements made after selective adaptation, or at contrast threshold, are useful in exploring human motion processing, but psychophysicists also want to study direction and speed processing under conditions in which the targets are highly visible. To obscure information about position in these conditions, many studies have used random dot targets. If a region of a dense random dot pattern is displaced laterally from frame-to-frame, it will appear to move (Anstis, 1970; Julesz, 1971; Lappin & Bell, 1972; Braddick, 1974). This type of display, known as a random dot kinematogram, is diagrammed in Figure 2A. Subjects can be asked to judge the shape or location of the displaced region to verify that they can detect its presence. It is often said that a random dot kinematogram isolates the motion system because there is no position information present in the display. Strictly speaking, this claim is not correct. Although the camouflaged region cannot be detected in any one frame, there is an abundance of position information in the display. The problem is that the observer has no way of identifying which subset of position cues constitute the 'signal', and so must rely on some distinction between 'signal' and 'noise' to segment the display. Almost any distinction will do (luminance, color, size, flicker, texture, orientation). The local change in contrast or luminance produced by displacing the region could be detected by a mechanism that is not selective for motion. Additional evidence is required to demonstrate that this judgment is specifically mediated by the motion system, and fortunately such evidence is available. For one thing, subjects can identify the direction of the motion of the camouflaged region in addition to specifying its shape and location (Baker & Braddick, 1982). Moreover, this judgment is strongly dependent on the time between frames and the size of the displacement – constraints that are likely properties of motion-specific mechanisms (Braddick, 1974).

Sparse random dot displays, such as the one diagrammed in Figure 2B, are particularly useful for studying direction discrimination (Ball *et al.*, 1983;

Watamaniuk *et al.*, 1989), and global flow (Williams & Sekuler, 1984; Williams *et al.*, 1991). Suppose that a single dot, moving vertically, were used as the stimulus for measuring direction discrimination. On each trial, the direction of the dot could be presented either moving slightly right or slightly left of vertical, and the experimenter could use the observer's judgments of left or right to determine the minimum detectable difference in direction. Is this a motion judgment or an implicit judgment of orientation based on the path traced out by a moving dot? Recently, Westheimer and Wehrhahn (1993) demonstrated that direction judgments based on the motion of a single dot are actually mediated by a mechanism sensitive to static orientation, rather than direction of motion. In the random dot display, each dot can move on a slightly different path since the direction of motion can be chosen randomly from a narrow range on every frame; Figure 2B shows a single frame of vectors chosen from a 90 degree range of directions. Consistent information about orientation or direction cannot be obtained from the motion of any individual dot. To determine the mean direction of the display, the motion system must average the motion vectors. Watamaniuk *et al.* (1989) found that direction discrimination for displays composed of a narrow range of directions (about 30 degrees) was identical to direction discrimination for displays in which all dots moved in the same direction. The dots in these displays 'jump' in apparent or stroboscopic motion, rather than in continuous motion, so if all the dots from all frames were presented simultaneously, rather than sequentially, observers would see nothing except a field of dots with no defined orientation. It seems likely that direction discrimination for these displays is based on motion, rather than orientation.

In most experimental studies of direction or speed with highly visible targets, the basis of the judgment is much less certain. The best experimental strategy is to make comparable measurements of position, orientation or timing thresholds using similar stimuli. If the speed or direction thresholds are better than the thresholds for the confounding variables of orientation, distance or duration, the measured thresholds are undoubtedly limited by motion-sensitive mechanisms. An alternative strategy is to show that the variables that affect motion thresholds have different effects on static judgments of position, thereby strengthening the case for motion specificity. For example, adapting to motion in one direction would probably not affect orientation judgments, but may degrade direction discrimination for the adapted direction.

3 SAMPLED MOTION

Many of the techniques described above use sampled motion, rather than 'real' continuous motion, to study human motion processing. In fact, most modern electronic visual displays (movies, television, computer graphics) present a sequence of static pictures in rapid succession. Nevertheless, the motion in these displays appears continuous. The human visual system has limited spatial and temporal resolution, so sampled motion that exceeds these resolution limits will necessarily be indiscriminable from continuous motion. Human spatial resolution for static sinusoidal gratings is about 60 cycles per degree (cpd); higher spatial frequency gratings are blurred together into a single homogeneous field by the optics of the eye. The temporal resolution limit is between 50 and 100 Hz; a light flickering at a faster rate is indistinguishable from a continuously-presented light.[2] The question is whether these stringent sampling rates (> 60 cpd and > 100 Hz) are necessary to produce an adequate stimulus for the human motion system. What rate of sampled motion is indistinguishable from continuous motion?

3.1   Equating continuous and sampled motion

Watson *et al.* (1983) used a forced choice procedure to measure the *temporal* sampling frequency that produced motion which was indistinguishable from continuous motion. They found that the sampling frequency depended on the spatial characteristics of the stimulus and the speed of the motion. A low frequency sinusoidal grating (1 cpd) moving slowly (8 deg/s) could be sampled at a relatively low rate of 45 Hz, but a thin bright line moving at the same speed had to be sampled at a much faster rate (over 100 Hz). Even higher sampling rates were required for a line moving at faster velocities. It is easy to see why speed affects the perception of continuity at a fixed frame rate – the faster the target moves, the larger the spatial step between frames, so the spaces between successive positions become visible. Using sinusoidal grating targets drifting at 12 Hz, Burr *et al.* (1986) measured the *spatial* sampling rate at which continuous and sampled motion were indistinguishable. The tolerable 'jump' size varied with spatial frequency from a minimum of about 30 arc s (120 samples/deg) at high spatial frequencies (30 cpd) to about 6 degrees at low spatial frequencies (0.07 cpd).

Most electronic systems cannot present complex pictures at temporal rates of 100 Hz or spatial rates of 120 samples/degree. Thus, it might seem surprising that human observers are largely unaware of the discrepancy between continuous and sampled motion in visual displays. Part of the explanation may be that features in these displays seldom move at high velocities, and that the electronic media (cameras, phosphors, etc.) blur the images somewhat, so that discontinuities are harder to discern. There could be another reason why sampled motion is so readily tolerated – the spatial resolution of the motion system may be considerably below the resolution of the visual system as a whole (Morgan, 1992). Static spatial mechanisms could detect that the sampled motion was not continuous, but motion mechanisms would have an adequate signal. However, even if sampled motion provides an adequate signal for the motion system, why would observers ignore the information supplied by the higher-resolution static mechanisms? Perhaps, spatial discontinuities are masked or obscured by the rich textures of natural images, or perhaps, sampled motion images are cognitively acceptable facsimiles of real motion much as electronically-reproduced color and depth are acceptable – not quite real, but close enough.

3.2   Sampled motion in studies of human motion processing

Practical considerations aside, sampled motion, often called apparent motion, has been a valuable tool in psychophysical studies. The question is not whether the observer can

---

[2] The critical flicker fusion rate depends on the eccentricity of the target (Tyler, 1981, 1985; McKee & Taylor, 1984).

discriminate sampled from continuous motion by any means, but rather whether a particular sampling rate is an adequate stimulus for some types of human motion judgments, and what that reveals about the subsystem responsible for the particular judgment. Scientific interest in apparent motion began well over a century ago when Exner (1875) showed that two electric sparks flashed in rapid succession in different places produced a sensation of motion. For the next half-century, most of the research on this phenomenon examined how much time and space could separate briefly-presented targets without destroying the illusion of motion, and how these variables interacted with other visual dimensions, such as target luminance. The results were codified into a set of 'laws', known as Korte's laws (Graham, 1965). Generally, observers reported a sensation of motion when targets were separated by as much as 10 degrees and by 200–300 ms.

With the appearance of the first quantitative models for physiological motion detectors (Hassenstein & Reichardt, 1956; Reichardt, 1959), studies of sampled motion took on a special significance. The properties of these models are described elsewhere in this volume (see Chapter 2 by Grzywacz *et al.* and Chapter 5 by Mather), but a 'generic' motion detector is drawn in Figure 3. It consists of a pair of spatially-separated receivers connected so that the signal from one receiver is delayed with respect to the other. If a stimulus moves continuously with the appropriate speed and direction, the signals from both receivers will arrive simultaneously at some common junction producing an enhanced direction-specific response. For sampled motion, the space and/or time between the samples could be too large to produce a direction-specific response in any size of motion detector. Thus, sampled motion can be used to

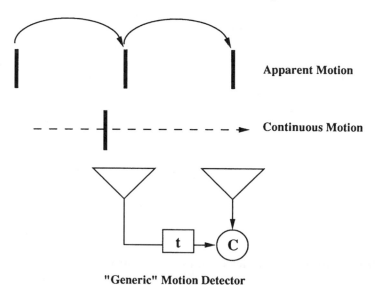

Apparent Motion

Continuous Motion

"Generic" Motion Detector

Figure 3 'Generic' motion detector consists of two receivers connected by a delay line (t in diagram) so that the signal from the first receiver arrives at common junction simultaneously with the signal from the second receiver, if speed of moving target is appropriate. A target hopping in apparent motion will produce the same stimulation as continuous motion if hop size and time between hops are appropriate for the motion detector.

determine the largest spatial and temporal interval that will stimulate these postulated detectors.

This type of reasoning has produced a spate of papers using sampled motion to determine the spatial and temporal requirements for human motion processing. As noted above, the older apparent motion studies, based on two briefly-presented targets, suggested that the spatial and temporal separation between targets could be very large indeed. A number of more recent studies have cast doubt on this conclusion. For a sequence of many targets, Kolers (1972) found that the spacing between target presentations needed to produce the best sensation of motion was much smaller (14 arc min) than for the pair of targets used in the classical studies. Similarly, Sperling (1976) showed that the time between presentations for best apparent motion was much shorter for a sequence than for a pair. Even more damaging were the results from studies using random dot kinematograms such as the one shown in Figure 2A. Note that the camouflaged region is displaced laterally in one 'jump' from frame 1 to frame 2 – it moves in apparent motion. Braddick (1974) measured the spatial displacement that produced a correct identification of the camouflaged shape. He found that the maximum displacement, now widely known as 'd_{max}', was about 15 arc minutes in the fovea, independent of the size of the elements in the target. Larger displacements produced an incoherent motion percept which did not permit observers to identify the shape. Subsequent work demonstrated that there were strong constraints on the time between frames as well. Observers could not correctly identify the direction of motion if the time between frames was longer than about 80 ms (Morgan & Ward, 1980; Baker & Braddick, 1985a).

How do we reconcile these results with the classical studies? Braddick suggested that there were two motion processes mediating apparent motion. The first, which he called the 'short range' process, depended on the responses of low-level motion detectors of the type shown in Figure 3, and was responsible for the segregation of the camouflaged form in dense random dot kinematograms. The second 'long-range' process, which perhaps depended on high-level cognitive processes, produced the sensation of motion associated with the single pair of targets used in the classical studies. In addition to the large discrepancy between their tolerable temporal and spatial intervals, these two processes differed in other ways (Braddick, 1974; Anstis, 1978, 1980). Long-range apparent motion (> 1 deg between samples) did not produce motion aftereffects, and short-range motion did not operate dichoptically, i.e. the camouflaged region of the random dot stimulus could not be detected when the two frames were presented to alternate eyes.

d_{max} was initially interpreted as representing the span separating the two receivers in the low-level human motion detector. However, it was soon apparent that d_{max} was not a fixed value. For one thing, it increased with retinal eccentricity (Baker & Braddick, 1982, 1985b; Koenderink et al., 1985; Van de Grind et al., 1986). Since the spatial dimensions of the receptive fields of visual neurons increase with eccentricity, the change in the size of d_{max} seemed to reflect a general property of the underlying neural substrate – motion detectors were scaled along with other visual functions such as acuity (Van de Grind et al., 1993). More puzzling was the effect of blurring or spatial filtering. Chang and Julesz (1983) showed that d_{max} increased when the random dot kinematograms were filtered so that there were no high spatial frequencies in the image. This result was confirmed in a number of studies that used narrow-band

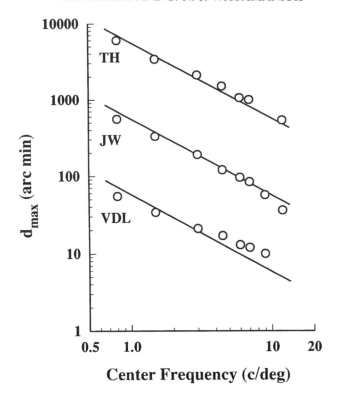

Figure 4 d_{max} plotted as a function of center frequency of narrow-band filtered random dot kinematogram (filtered version of Figure 2A) for three subjects; upper two data sets displaced vertically by 1 log unit each for clarity. Maximum displacement scales with spatial frequency. Data from Bischof and DiLollo (1990).

filtering of the random dot kinematograms. As shown in Figure 4, d_{max} varies systematically with the spatial frequency content of the stimulus – the lower the center frequency, the larger the displacement limit (Chang & Julesz, 1983; Cleary & Braddick, 1990a; Bischof & DiLollo, 1990). Why would removing the high frequencies from the target increase the span of the motion detectors?

It has been frequently suggested that d_{max} is set by the characteristics of the random dot stimulus, rather than by the spatial properties of human motion detectors (Lappin & Bell, 1976; Cavanagh & Mather, 1989; Cavanagh, 1991; Morgan, 1992; Morgan & Fahle, 1992). To understand why the limit might be set by the stimulus, consider the 'correspondence problem' diagrammed in Figure 5A and 5B. It shows schematically a one-dimensional strip of dots from a random dot kinematogram; the central eight dots in the first frame are displaced to the right in the second frame. If the displacement is small (< 15 arc min) and the time brief (< 80 ms), observers report the coherent motion percept diagrammed in Figure 5A, rather than the incoherent motion diagrammed in Figure 5B. How does the brain decide which dots to match in the two frames, since there are many potential matches for every dot?

One possibility is that the brain performs an operation, akin to cross-correlation, that minimizes the number of mismatches between frames (Lappin & Bell, 1976). In simplistic terms, this amounts to matching each dot in the first frame to the nearest

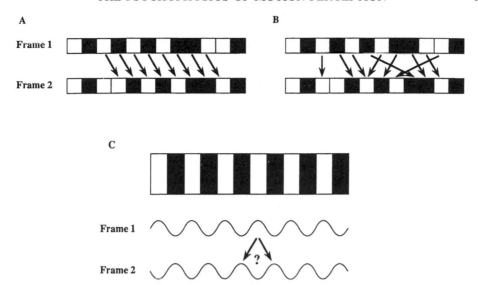

Figure 5 A and B. Diagram of correspondence problem in random dot kinematogram. Human observers report seeing consistent motion diagrammed in A rather than chaotic random motion diagrammed in B. C. Diagram of correspondence problem for sinusoidally varying grating targets, showing that a shift of 180 degrees of phase is ambiguous.

similar dot in the second frame. If the random dot display is blurred (low-pass filtered), the dots tend to form clumps – effectively larger elements – so that the nearest match is more distant, and the tolerable displacement larger. However, if the brain were using a cross-correlation strategy, then d_{max} for an unblurred target would be limited by element size and density, but d_{max} is nearly independent of element size (elements < 10 arc min), and is not affected by density (Braddick, 1974; Baker & Braddick, 1982). To explain these findings, Morgan (1992) and Morgan and Fahle (1992) proposed that a coarse spatial filter precedes all motion detection in the human visual pathways; this coarse filter would blur the dots neurally into clumps, so that element size would not affect d_{max} until element size exceeded a large value. Their idea is difficult to reconcile with Anderson and Burr's psychophysical evidence (1987) for very small motion-detecting mechanisms (2 arc min), but, when contrast effects are incorporated to explain density independence, it does provide a good account of the d_{max} results.

A different explanation for d_{max} comes from considering the responses of motion units to the random dot kinematogram. Most contemporary motion models assume that motion units are built out the same scaled spatial filters[3] that are thought responsible

---

[3] Spatial filter is an engineering term used to describe the visual sensitivity of a localized group of similar neural units to the spatial frequency spectrum. When plotted on a spatial axis, the sensitivity profile of a spatial filter resembles the sensitivity profile of a simple cortical unit, e.g. the 'Mexican Hat' function. Current spatial models assume that every region of the visual field is served by a family of different sized filters, tuned to different spatial frequency ranges. The exact composition of the family varies with retinal eccentricity. In the 'generic' motion unit, the receivers shown in Figure 3 are replaced by two overlapping bandpass filters, optimally tuned to the same spatial frequency range.

for other aspects of human spatial vision (see Chapter 2 by Grzywacz *et al.* and Chapter 5 by Mather). These filtered units respond to a limited spatial frequency range and are generally most sensitive to a sinusoidal grating of a characteristic frequency. The 'correspondence problem' in apparent motion is particularly simple for a sinusoidal grating. If a grating is displaced by 180 degrees of phase (see luminance profiles in Figure 5C), the direction of motion is completely ambiguous. If it is displaced by more than 180 degrees, it will appear to move in the opposite direction (aliasing) as it should, e.g. + 270 degrees is equivalent to − 90 degrees. Therefore, d_{max} for a grating must be less than 180 degrees of phase.

A random dot display is a broadband stimulus, containing many spatial frequencies, so many different-sized filters will respond to the moving region. To a first approximation, each motion unit should respond correctly to any displacement of the random dot stimulus that does not exceed one half-cycle of its characteristic frequency. A given displacement may be below the limit for some units, and beyond it for other units which will then signal motion in the opposite direction. Given these competing signals from different units, which units determine the measured psychophysical limit? d_{max} is between 15 and 20 arc minutes, corresponding to half-cycles of between 2 and 1.5 cpd. It is thus unlikely that the limit is set by the motion units with the smallest receptive fields (high spatial frequency sensitivity), unless there are no motion units tuned to spatial frequencies beyond 2 cpd. Perhaps units tuned to 1.5 cpd produce the strongest motion signals and so determine the maximum displacement. But if so, why does d_{max} increase when the random dots are filtered or blurred to remove all frequencies above 1.8 cpd? From their studies on the effect of selective filtering, Cleary and Braddick (1990b) concluded that the signals from units tuned to higher frequencies selectively masked or inhibited lower frequency units.

There was another curious outcome from the Cleary–Braddick studies – the measured d_{max} for their bandpass-filtered stimuli was considerably greater than a half-cycle of the center frequency of their filters (Cleary & Braddick, 1990a). Faced with this experimental conundrum, Bischof and DiLollo (1991) proposed an imaginative synthesis to explain the d_{max} limit. They suggested that the limit was set by the average horizontal spatial frequency of all image components, each weighted by the contrast sensitivity function of the human observer. When the horizontal components of diagonal features were included in their calculations, d_{max} never exceeded the half-cycle limit. Their proposal implies that the brain takes a linear sum of all the contrast energy in one direction. No inhibition from high spatial frequencies is required to explain the difference between blurred and unfiltered images, because the average spatial frequency of a blurred image is lower than the average for an unblurred image.

Which explanation for d_{max} is right? Is the human motion system very low pass (Morgan & Fahle, 1992)? Do high frequency signals inhibit low frequency signals (Cleary & Braddick, 1990b)? Or is the limit set by the linear sum of signals in one direction (Bischof & DiLollo, 1991)? More studies are needed to determine which is the best explanation since all are adequate to explain much of the data available on d_{max}.

One thing is clear. The old conceptual dichotomy between short-range and long-range apparent motion is threadbare. Instead of a simple two-mechanism system, there appears to be a continuum of motion detectors, responsive to a wide range of different spans (Turano & Pantle, 1985; Burr *et al.*, 1986; Bischof & DiLollo, 1990; Cavanagh,

1991). Turano and Pantle (1985) found that the maximum displacement that produced a motion aftereffect depended on the spatial frequency content of the stimulus. Carney and Shadlen (1993) have recently shown that direction discrimination is possible for a dichoptically-presented random dot display, although it is difficult or impossible to identify the shape of the camouflaged form in these dichoptic displays; their measurements indicate that this dichoptically-defined motion is indeed mediated by a 'short-range' process.

To summarize, most of the characteristics that distinguished long and short range motion have disappeared under close experimental scrutiny. Nevertheless, there are apparent motion phenomena that are not explained by current filter models of motion processing (Chubb & Sperling, 1988; Cavanagh *et al.*, 1989; Victor & Conte, 1990; Pantle, 1992; Boulton & Baker, 1993a,b). Some researchers have suggested a new dichotomy based on first- and second-order motion detectors (see Chapter 6 by Smith), but there may still be some room for the old cognitive 'long-range' system. It remains to be seen whether current psychophysical and physiological studies can explain all these phenomena by simple mechanisms of early visual processing.

4   BASIC PSYCHOPHYSICAL DATA ON EARLY MOTION PROCESSING

Although the d_{max} studies focused on the capacity of the brain to use motion to segregate figure and ground, the fundamental information supplied by the early motion system is image velocity – the local speed and direction of moving features. In the next sections, we will review studies of the best human performance in judging direction and speed, as well as which variables affect these judgments.

4.1   The minimum motion threshold

The minimum motion threshold refers to the smallest displacement (distance traversed) that can be reliably detected by a human observer. Generally, observers are required to judge whether the target moved left or right during each test trial. Provided that the moving test target is presented near a stationary reference target, the smallest detectable displacement is less than 10 arc seconds in the fovea (Tyler & Torres, 1972; Westheimer, 1978, 1979; Legge & Campbell, 1981). This referenced or relative motion threshold is a hyperacuity,[4] comparable in magnitude to the static positional hyperacuities, e.g. vernier acuity. Two factors indicate that the relative motion threshold is indeed mediated by a motion mechanism. First, when the separation

---

[4] Hyperacuity refers to the ability to detect a change in spatial location with a precision that is much lower than comparable measurements of resolution (ordinary visual acuity) at the tested retinal location. For example, in the fovea, the best hyperacuity thresholds are 2–6 arc seconds, while resolution acuity is roughly 30 arc seconds.

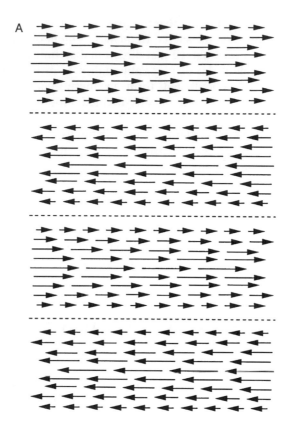

Figure 6A   Diagram of motion vectors showing continuous displacement of adjacent regions of a random dot kinematogram. The length of each vector shows the speed of the movement. Vertically, the vectors trace out a sinusoidal oscillation of the random dot stimulus.

between the test and reference targets is greater than about 10 arc minutes, the threshold for an abrupt displacement is better than the increment threshold for judging the distance separating a static pair of targets (Westheimer, 1979; Legge & Campbell, 1981). Second, the motion threshold depends on the speed of the change (Tyler & Torres, 1972). Nakayama and Tyler (1981) used an oscillating random dot target to study minimum motion thresholds in isolation from position cues. The oscillations were sinusoidal in form, producing a shearing motion between adjacent regions of the random dots (Figure 6A). Thresholds were measured as the minimum detectable amplitude of the sinusoidal motion (the excursion between peak and trough). Like a static vernier target, this shearing pattern is self-referencing because adjacent regions move in opposite directions. However, the minimum motion threshold for this random dot pattern occurs at peak-to-trough separations ranging from 1 to 3 degrees, whereas the threshold for detecting a deviation from straightness measured with a sinusoidally-varying static line (periodic vernier acuity) reaches a minimum at a peak-to-trough

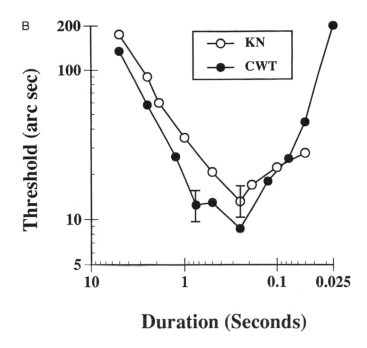

Duration (Seconds)

Figure 6B The minimum detectable displacement measured as a function of the duration of the displacement in one direction before reversal to opposite direction, i.e. half oscillation. The detectable displacement depends on the speed of the movement, as well as the spatial extent. Data from Nakayama and Tyler (1981).

separation of 15 arc minutes. The dependence of threshold displacement on the temporal frequency of the oscillations, and hence the speed of the change, is shown in Figure 6B.

Relative motion thresholds are much higher at peripheral retinal loci than in the fovea. The eccentricity function for relative motion is identical to the eccentricity function for other hyperacuities, when measured with line targets scaled in length for eccentricity (McKee et al., 1990). Like other hyperacuity thresholds, relative motion thresholds rise more rapidly with increasing retinal eccentricity than grating resolution acuity (McKee & Nakayama, 1984; Levi et al., 1985).

Minimum motion thresholds measured without an adjacent stationary reference target are called absolute motion thresholds. In the fovea, the absolute motion threshold is much higher than the relative motion threshold, but in the periphery, the absolute and relative motion thresholds are very nearly the same (Tyler & Torres, 1972; Levi et al., 1984). Because of this difference in the effect of a reference, absolute motion thresholds show a very shallow rise with eccentricity (Johnson & Leibowitz, 1974; Levi et al., 1984). Figure 7 shows thresholds for relative and absolute motion as a function of eccentricity for brief (150 ms) targets presented in the parafovea (McKee et al., 1990). One explanation for this reference effect is that the brain cannot monitor involuntary eye movements with hyperacuity precision, so, in

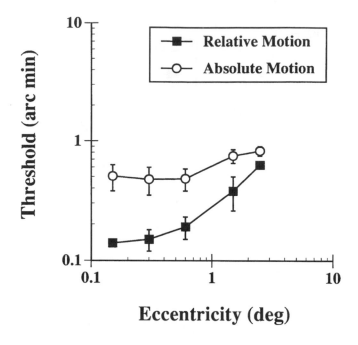

Figure 7　Minimum detectable displacement as a function of target line eccentricity. Relative motion judgments made with stationary reference line present in field of view. Absolute motion judgments made without reference line. Absolute and relative thresholds are similar at eccentric loci. Data from McKee *et al.* (1990).

the absence of a stationary reference, the brain is unable to distinguish between small motions of the target and motions of the eye. Random oculomotor fluctuations constitute an additive source of noise, so their effect depends on the inherent sensitivity of the tested retinal locus. In the fovea, this noise swamps the highly precise motion signal, but in the periphery, oculomotor noise is comparable or smaller than the smallest motion signal that the neural substrate can detect. Apparently, the human motion system uses a differential signal to distinguish object motion from the noise generated by the oculomotor system. See Chapter 14 by L. Harris for a discussion of this issue.

The effect of contrast on the minimum motion threshold depends on the type of target used to measure the threshold. Contrast has a relatively small effect on thresholds measured with grating targets. Nakayama and Silverman (1985) found that the displacement threshold for 2 cpd grating reached its asymptotic value at a contrast of 2–3%. However, displacement thresholds for edge targets improve over a much larger contrast range. Lee *et al.* (1993) found that parafoveal thresholds for edge targets reached asymptote at 20–40% contrast, while Mather (1987) found that foveal displacement thresholds improved as the square root of contrast over the whole range from 1 to 80% contrast. Interestingly, contrast increment thresholds, measured as a Weber fraction ($\Delta C/C$), also improve with the square root of contrast (Legge & Foley, 1980). If contrast increment and motion displacement thresholds are limited by the same source of noise, e.g. noise in the ganglion cells of the magnocellular pathways as

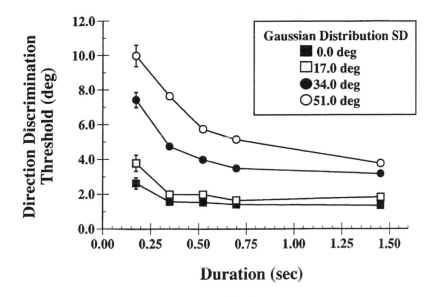

Figure 8 Direction discrimination threshold for mean direction of sparse random dot kinematograms such as one shown in Figure 2B. Individual dots move in directions chosen at random from a range specified by the standard deviation of Gaussian distribution of directions. For narrow ranges of directions (SD = 17 degrees), the threshold is the same as when dots all move in same direction. Direction discrimination improves with increasing duration.

Lee *et al.* have proposed, then these two thresholds should show a similar dependence on contrast.[5] Motion thresholds based on edge displacement appear to confirm this prediction.

4.2 Direction discrimination

Ball and Sekuler (1979) and Watamaniuk *et al.* (1989) used high contrast random dot patterns, similar to the one diagrammed in Figure 2B, to measure direction discrimination. As shown in Figure 8, angular shifts of 1–2 degrees can be detected if the distribution of directions in the display is narrow (SD = 20 degrees or less), and target duration is greater than about 300 ms (Watamaniuk *et al.*, 1989). These results indicate that the motion system integrates direction signals within broadly-tuned mechanisms for a substantial period of time. Studies from the Sekuler laboratory indicate that these directionally-selective mechanisms have a bandwidth of roughly 60 degrees at half-height, and that 12 such mechanisms evenly distributed around the clock can account for human direction discrimination (Watamaniuk *et*

---

[5] For a more complete explanation of the relationship between contrast and hyperacuity thresholds see Klein and Levi (1985), Bowne (1990) or McKee (1991).

al., 1989; Watamaniuk & Sekuler, 1992; Williams *et al.*, 1991). DeBruyn and Orban (1988) found comparable precision in direction discrimination using a random dot pattern of the type shown in Figure 2A. In their display, all components of the display moved in the same direction. Control experiments showed that subjects could not use orientation or position information to judge direction. DeBruyn and Orban found the optimum direction discrimination at speeds ranging from 4 to 64 deg/s.

Although the directional sensitivity of the mechanisms responding to the random dot 'flow' fields shown in Figure 2B is quite good, their spatial resolution is very coarse. We found that blurring the flow field with ground glass (cut-off spatial frequency < 5 cpd), so that individual dots in the display were not resolved, had no effect on the precision of direction discrimination. Fine scale mechanisms, tuned to higher spatial frequencies, might be more sensitive to changes in direction than the coarse mechanisms that respond to these flow fields. Directional-selectivity is related indirectly to orientational tuning because the optimum tuning for direction is perpendicular to the optimum tuning for orientation. Orientation discrimination for static line targets, presumably mediated by fine scale mechanisms, is about 0.5 degree (Westheimer *et al.*, 1976), which argues that direction discrimination could be somewhat more precise at the finest scales.

Recently, Watamaniuk (1992) used a different approach to obscure static positional information. The test target was a single dot moving (or 'hopping' in apparent motion) on a straight trajectory in the midst of a background of dots in random apparent motion; a diagram of this stimulus is shown Figure 2C. Subjects were asked whether the trajectory moved to the left or right of vertical from trial to trial. The single test dot would be invisible if the individual dots were not resolved so this judgment probably depends on a moderately-high spatial frequency. To demonstrate that trajectory direction discrimination depended on motion sensitivity, rather than on orientation, Watamaniuk changed the motion of the dots in the noise background so that range of directions was confined to 160 degrees. With this change, the background noise appeared to move in the mean direction of the range (Williams & Sekuler, 1984). Direction discrimination thresholds were elevated when the background noise moved in the same direction as the trajectory, but were not affected by background motion in the opposite direction. Since static judgments of orientation would be insensitive to the motion of the background, this result is strong evidence that the trajectory direction is detected by a motion mechanism. Subjects were able to judge the direction of the trajectory with a precision of 2 degrees, comparable to the best performance with the flow fields. Apparently, directional-sensitivity at fine scales is not much better than at coarser scales.

Under certain circumstances, the direction of motion may be misperceived. Anstis (1970) showed that if the contrast of the moving region in a random dot kinematogram was reversed in the second frame, the region appeared to move in the opposite direction. This effect, known as 'reverse phi', is predicted by contemporary motion models, but there are some misperceptions which are not so readily explained. Derrington and Henning (1987) found that a moving sinusoidal grating superimposed on a low frequency stationary grating appeared to move in the opposite direction for brief presentations (< 100 ms). Cormack *et al.* (1992) reported that a moving feature presented against a textured background appeared to move in a direction oblique to the orientation of the background contours, when viewed peripherally.

4.3 Speed discrimination

In a typical speed discrimination experiment, the observer is shown a standard speed and a test speed, and is required to judge whether the test speed is faster or slower than the standard – a procedure known as the Method of Constant Stimuli. The percentage of stimuli judged 'faster' is plotted as a function of the test speed, and a cumulative normal curve is fitted to the data. The increment threshold for speed (ΔV) is estimated from the slope of the fitted curve, while the perceived speed (or bias) is determined from the stimulus corresponding to the 50% point on the curve. If the objective is to learn how precisely the observer can judge speed, i.e. what is the smallest increment threshold, the experimenter usually provides feedback about the correctness of the observer's responses. In studies of perceived speed, feedback is not given because the experimenter wants to know what standard speed matches a test speed.

The Weber fraction for speed ($\Delta V/V$) is the smallest proportional difference between two speeds that the human observer can judge reliably. Although the best Weber fraction in the literature is 0.015 (Pantle, 1978),[6] most studies report values between 0.04 and 0.08 (Bourdon, 1902; Brown, 1961; McKee, 1981; Nakayama, 1981; Pasternak, 1987; Turano & Pantle, 1989; Watamaniuk & Duchon, 1992). Orban and colleagues reported that the Weber fraction was roughly constant at 0.06–0.08 for speeds ranging from 4 to 64 deg/s; they found higher values for both slower and faster speeds (Orban et al., 1984; DeBruyn & Orban, 1988). This level of precision is largely independent of the stimulus pattern used for the measurements; the same Weber fractions have been reported for moving lines, dense random dot kinematograms (Figure 2A), sparse flow fields (Figure 2B), and drifting sinusoidal gratings.

Despite the evident agreement among studies, psychophysical measurements of speed discrimination are a methodological nightmare because so many dimensions covary with target speed. Consider the moving line pictured in Figure 9. If target duration is kept fixed, the observer can judge speed on the basis of the distance traversed by the target. If the distance traversed is fixed, the observer can make the judgment on the basis of duration. Mandriota et al. (1962) compared speed discrimination for exactly these two conditions (fixed duration and fixed distance) to a condition in which the distance traversed (and hence the duration) was randomized from trial to trial. The Weber fractions were about 0.05 for the fixed duration condition, 0.10 for the fixed distance condition, and over 0.15 for the randomized distance condition. Their results raised serious questions about whether human observers possessed an independent, precise signal for speed.

In judging speed, how much do observers rely on the spatial or temporal information that necessarily covaries with speed? Under the best of circumstances,

---

[6] Pantle's stimulus duration was exceptionally long (2 seconds), but it is not obvious why a long duration should improve performance.

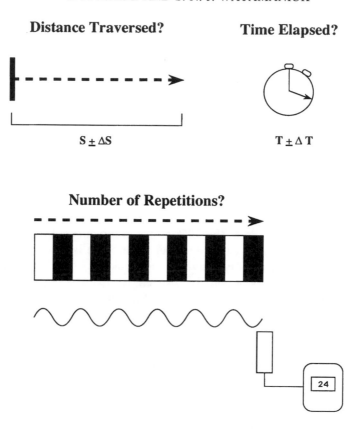

Figure 9 Diagram of confounding factors in speed discrimination (see text). S refers to distance, T refers to time.

speed discrimination is not nearly as precise as judgments of distance, for which the Weber fractions are typically in the 0.02–0.03 range (Burbeck, 1987; Klein & Levi, 1987). However, judgments of duration or temporal asynchrony are generally poorer than speed judgments, yielding Weber fractions ranging from 0.06 to 0.20 (McKee & Taylor, 1984; Orban *et al.*, 1984). McKee (1981) directly compared duration judgments to speed judgments for targets moving a *fixed* distance. She demonstrated that the speed judgments were more precise than the duration judgments – strong evidence that target speed was not inferred from the spatial and temporal extents of the target. Lappin *et al.* (1975) reached the same conclusion from similar measurements. McKee also found that when target distance was randomized, observers were initially very confused and made many errors in judging speed, but that with practice and feedback, they learned to ignore the random variations in distance, and to judge speed with the same precision as when the distance was fixed. Recently, Katz *et al.* (1990) found that briefly-presented targets appear to move faster than targets moving at the same speed presented for longer durations. The effect of duration on the perceived velocities may account for the initial confusion of McKee's observers, who subsequently learned to compensate for this effect. Given these considerations, the best strategy for measuring speed discrimination with aperiodic targets is to use a fixed distance for the traverse, and a relatively long duration (~ 500 ms). If there is any doubt about the basis of the judgment, concurrent measurements of duration discrimination should be made under the same stimulus conditions.

Repetitive targets, such as drifting sinusoidal gratings, add their own headaches to studies of speed discrimination. Changes in speed introduce subtle changes in grating contrast, particularly at high drift rates.[7] More significantly, observers can, in principle, judge speed by the number of bars moving past a fixed location (see diagram at the bottom of Figure 9), or more generally, by the temporal frequency (Smith & Sherlock, 1957). Does the observer use speed or temporal frequency to judge grating speed? If the spatial frequency of a grating is changed, the temporal frequency of a grating must also be changed to maintain the same speed (speed = temporal frequency/ spatial frequency). By manipulating both the spatial and temporal frequency of a grating, quite large changes in temporal frequency can be introduced without altering target speed. Can an observer still respond precisely to small changes in speed, if large random variations in temporal frequency are introduced by large random changes in spatial frequency?

McKee et al. (1986) measured speed discrimination using gratings of either five or nine different spatial frequencies ranging from 0.5 to 1.5 cpd. The different spatial frequencies were presented at random from trial to trial with appropriate changes in temporal frequency to produce the same mean speed of 5 deg/s ($\pm \Delta V$). The observer's task was to judge, without feedback, whether the test target was moving faster or slower than the mean. If an observer could only discriminate speed on the basis of temporal frequency, then his or her threshold would have been substantially degraded by these random variations. In fact, the Weber fractions, estimated from the data pooled from all spatial frequencies, were only slightly elevated (to ~ 0.07). Smith and Edgar (1991) replicated this study for a wider range of speeds, and noted that the loss of precision produced by spatial frequency randomization depends to some degree on the particular speed, although randomization never increased thresholds by more than a factor of two, even in the worst case. McKee et al. (1986) also obtained separate psychometric functions for each of the spatial frequencies in their randomization study, and used the 50% point on each function to determine the perceived speed associated with each temporal/spatial frequency pair. As shown in Figure 10, the perceived speed was fairly close to the mean speed of 5 deg/s for all the frequencies. Nevertheless, there is a systematic deviation apparent in the data; low temporal frequencies look somewhat slower than high temporal frequencies. This type of bias has been noted in other studies (Diener et al., 1976). However, Smith and Edgar (1990) showed that these perceptual biases depend jointly on both the spatial and temporal frequency of the target, and cannot be explained on the basis of temporal frequency alone.

The results from the McKee et al. study do not provide conclusive evidence that observers judge grating speed on the basis of speed rather than temporal frequency.

---

[7] The contrast or luminance of aperiodic targets may also covary with target speed. Whether these changes in contrast and luminance are sufficient to be useful as a cue to speed is unknown.

	ΔV/V V = 10 deg/sec	ΔTF/TF Drifting TF = 10 Hz	ΔTF/TF Counterphase TF=10 Hz
KN	0.08 ± 0.005	0.08 ± 0.005	0.17 ± 0.01
SM	0.07 ± 0.004	0.11 ± 0.008	0.14 ± 0.01

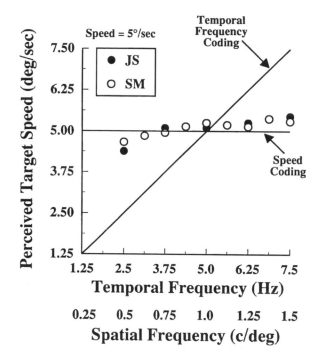

Figure 10 Upper box shows the Weber fractions for speed discrimination (ΔV/V) for drifting gratings, temporal frequency discrimination (ΔTF/TF) for drifting gratings, and temporal frequency discrimination (ΔTF/TF) for counterphase gratings, all measured with random variations in target spatial frequency. Weber fraction for speed is somewhat better than Weber fractions for temporal frequency.

Lower graph shows perceived speed, estimated from the 50% point on psychometric functions, for different spatial-temporal frequency pairs that each together specified a speed of 5 deg/s. Different TF/SF pairs were randomly chosen from trial to trial and the observer's task was to judge whether the test speed was faster or slower than the mean speed. No feedback was provided. Observers were responding on the basis of target speed, not target temporal frequency. Data taken from McKee *et al.* (1986).

For example, the observers could have measured the temporal frequency, and then indirectly estimated target speed using the cue provided by the target spatial frequency – an 'unconscious calculation' of speed from TF/SF. More compelling evidence for the primacy of speed comes from direct measurements on the precision of flicker judgments. Pasternak (1987) showed that the Weber fractions for counterphase flicker were systematically higher, by about a factor of two, than the Weber fractions for judging the drift rate of a sinusoidal grating of the same contrast and spatial frequency. Mandler (1984) also found that the Weber fractions for uniform flicker were about 0.10 for temporal frequencies ranging from 1 to 10 Hz, again roughly twice the speed of the Weber fraction. These results show that an observer cannot use a temporal frequency 'count', either directly or indirectly, as the basis of a speed judgment, because the speed judgments are more precise.

Smith and Edgar (1991) asked observers to judge the temporal frequency of drifting gratings when the speed was randomly varied from trial to trial – the inverse of the McKee *et al.* experiment. With practice, their observers learned to discriminate temporal frequency as well as speed in some conditions, but, on average, their temporal frequency judgments with random variations in speed were less precise than speed judgments made with random variations in temporal frequency. The box at the top of Figure 10 shows the Weber fractions for speed discrimination, and for temporal frequency discrimination for both drifting and counterphase gratings, all measured when the spatial frequency of the target was randomized from trial to trial (McKee *et al.*, 1986). Although temporal frequency and speed signals probably depend on some common site in the early visual processing, subsequent motion processing seems to provide the human observer with slightly better access to the speed signal.

Thresholds for speed discrimination are independent of target contrast above a contrast of 5% (Turano & Pantle, 1989; McKee *et al.*, 1986). Since many spatial and some motion thresholds improve with increasing contrast, speed discrimination must be limited by an internal source of noise occurring at a neural level beyond the earliest stage of visual processing (Bowne, 1990; Watamaniuk & Duchon, 1992). Contrast does affect perceived speed somewhat, particularly for targets of different contrast presented side-by-side (Thompson, 1982). The higher contrast target looks faster than the lower contrast target (Stone & Thompson, 1992).

Speed discrimination for slow speeds (< 10 deg/s) is only precise in the central visual field, where the Weber fraction for speed reaches a minimum at speeds of 3–10 deg/s and then rises again at speeds exceeding about 70 deg/s. As the target is moved to more eccentric positions, the speed discrimination function reaches a minimum at increasingly faster speeds. As shown in Figure 11, the minimum is reached at 10–15 deg/s at 20 degrees eccentricity, and at 20–30 deg/s at 40 degrees eccentricity. This increase follows the increase in the minimum angle of resolution also found with increasing eccentricity (McKee & Nakayama, 1984; Van de Grind *et al.*, 1983, 1986; Koenderink *et al.*, 1985). Thus, the minimum speed producing precise discrimination at any retinal locus is probably determined by the smallest local receptive field, which in turn limits the size of the span in the 'generic motion detector' (see Figure 3). Retinal eccentricity also influences perceived speed. Targets appear to move more slowly in the periphery (Tynan & Sekuler, 1982; Johnston & Wright, 1986).

Human observers are exceptionally *insensitive* to target acceleration (Gottsdanker, 1956; Brown, 1961; Schmerler, 1976). The Weber fraction for judging instantaneous changes in speed is roughly five times the Weber fraction for judging the difference between speeds presented in isolation (Snowden & Braddick, 1991). Apparently, the

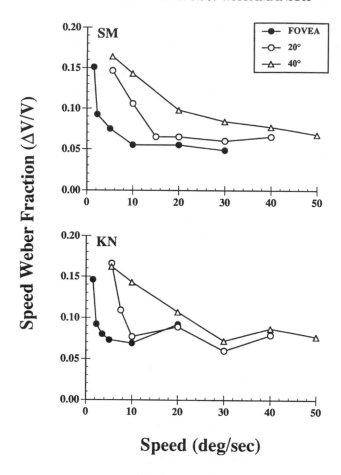

Figure 11 Weber fractions for speed discrimination as a function of mean speed, measured at three different eccentricities (fovea, 20 degrees in lower visual field, and 40 degrees in lower visual field). The curves reach a minimum at different mean speeds depending on the retinal locus. Observers can only respond precisely to slow speeds presented in the central visual field. Data taken from McKee and Nakayama (1984).

motion system must integrate for a substantial period of time (100–200 msec) to obtain a precise speed signal, an operation which obscures acceleration, as well as other higher derivatives of velocity (Nakayama, 1985; Werkhoven *et al.*, 1992). In short, the motion system averages the local speed signals to reach a Weber fraction of roughly 0.05 (Watamaniuk & Duchon, 1992).

5   CONCLUDING REMARKS

Many computational models of motion processing have appeared in the last ten years. Psychophysical measurements are the only way to test which of these models is appropriate for human motion processing, although neurophysiological studies in primates can supply important information about the 'physiological hardware' responsible for the motion signals. Whatever the model, a psychophysicist must have something to measure, and in studies of motion, that comes down to measurements of direction and speed. In this chapter, we have summarized the basic data on these fundamental measurements, as well as provided some guidance on how these measurements are best made. Local estimates of direction and speed are the raw material used by the brain for subsequent elaboration into the global flow fields that guide our actions and define our surroundings. Thus, these simple psychophysical measurements are good indicators of the quality (signal/noise ratio) of the signals that are available for use in higher order motion processing.

ACKNOWLEDGEMENT

Work on this chapter was supported by AFOSR grant F49620-92-J0156.

REFERENCES

Anderson, S. J. & Burr, D. C. (1987). Receptive field size of human motion detection units. *Vision Res.*, **27**, 621–635.

Anstis, S. M. (1970). Phi movement as a subtraction process. *Vision Res.*, **10**, 1411–1430.

Anstis, S. M. (1978). Apparent movement. In R. Held, H. W. Leibowitz & H. L. Teuber (Eds.) *Handbook of Sensory Physiology*, pp. 655–673. Springer-Verlag, New York.

Anstis, S. M. (1980). The perception of apparent movement. *Phil. Trans. Roy. Soc. London B*, **290**, 153–168.

Baker, C. L. & Braddick, O. J. (1982). The basis of area and dot number effects in random dot motion perception. *Vision Res.*, **22**, 1253–1259.

Baker, C. L. & Braddick, O. J. (1985a). Temporal properties of the short-range process in apparent motion. *Perception*, **14**, 181–192.

Baker, C. L. & Braddick, O. J. (1985b). Eccentricity-dependent scaling of the limits for short-range apparent motion perception. *Vision Res.*, **25**, 803–812.

Ball, K. & Sekuler, R. (1979). Masking of motion by broadband and filtered directional noise. *Percept. Psychophys.*, **26**, 206–214.

Ball, K., Sekuler, R. & Machamer, J. (1983). Detection and identification of moving targets. *Vision Res.*, **23**, 229–238.

Beverley, K.I. & Regan, D. (1973). Evidence for the existence of neural mechanisms selectively sensitive to the direction of movement in space. *J. Physiol.*, **235**, 17–29.

Bischof, W. F. & DiLollo, V. (1990). Perception of directional sampled motion in relation to displacement and spatial frequency: Evidence for a unitary motion system. *Vision Res.*, **30**, 1341–1362.

Bischof, W. F. & DiLollo, V. (1991). On the half-cycle displacement limit of sampled directional motion. *Vision Res.*, **31**, 649–660.

Boulton, J. & Baker, C. L. (1993a). Different parameters control motion perception above and below a critical density. *Vision Res.*, **33**, 1803–1811.

Boulton, J. & Baker, C. L. (1993b). Dependence on stimulus onset asynchrony in apparent motion: Evidence for two mechanisms. *Vision Res.*, **33**, 2013–2019.

Bourdon, B. (1902). *La Perception Visuelle de l'Espace*. Reinwald, Paris.

Bowne, S. F. (1990). Contrast discrimination cannot explain spatial frequency, orientation or temporal frequency discrimination. *Vision Res.*, **30**, 449–462.

Braddick, O. (1974). A short-range process in apparent motion. *Vision Res.*, **14**, 519–527.

Brown, R. H. (1961). Visual sensitivity to differences in velocity. *Psychol. Bull.*, **58**, 89–103.

Burbeck, C. A. (1987). Position and spatial frequency in large-scale localization judgments. *Vision Res.*, **27**, 417–427.

Burr, D., Ross, J. & Morrone, C. (1986). Smooth and sampled motion. *Vision Res.*, **26**, 643–652.

Carney, T. & Shadlen, M. N. (1993). Dichoptic activation of the early motion system. *Vision Res.*, **33**, 1977–1995.

Cavanagh, P. (1991). Short-range vs. long-range motion: Not a valid distinction. *Spatial Vision*, **5**, 303–309.

Cavanagh, P. & Mather, G. (1989). Motion: The long and the short of it. *Spatial Vision*, **4**, 103–129.

Cavanagh, P., Arguin, M. & von Grünau, M. (1989). Interattribute apparent motion. *Vision Res.*, **29**, 1197–1204.

Chang, J. J. & Julesz, B. (1983). Displacement limits for spatial frequency filtered random-dot cinematograms in apparent motion. *Vision Res.*, **23**, 1379–1385.

Chubb, C. & Sperling, G. (1988). Drift-balanced random stimuli: A general basis for studying non-Fourier motion perception. *J. Opt. Soc. Am. A*, **5**, 1986–2006.

Cleary, R. & Braddick, O. J. (1990a). Direction discrimination in narrow-band filtered kinematograms. *Vision Res.*, **30**, 303–316.

Cleary, R. & Braddick, O. J. (1990b). Masking of low frequency information in short-range apparent motion. *Vision Res.*, **30**, 317–327.

Cormack, R., Blake, R. & Hiris, E. (1992). Misdirected visual motion in the peripheral visual field. *Vision Res.*, **32**, 73–80.

DeBruyn, B. & Orban, G. A. (1988). Human velocity and direction discrimination measured with random dot patterns. *Vision Res.*, **28**, 1323–1335.

Derrington, A. M. & Henning, G. B. (1987). Errors in direction-of-motion discrimination with complex stimuli. *Vision Res.*, **27**, 61–75.

Derrington, A. M. & Henning, G. B. (1993). Detecting and discriminating the direction of motion of luminance and colour gratings. *Vision Res.*, **33**, 799–811.

Diener, H. C., Wist, E. R., Dichgans, J. & Brandt, Th. (1976). The spatial frequency effect on perceived velocity. *Vision Res.*, **16**, 169–176.

Exner, S. (1875). Experimentelle Untersuchungen der einfachsten psychischen Processe. *Pflügers Arch. Physiol.*, **62**, 423–432.

Gottsdanker, R. M. (1956). The ability of human operators to detect acceleration of target motion. *Psychol. Bulletin*, **53**, 477–487.

Graham, C. H. (1965). *Vision and Visual Perception*, Chapter 20, pp. 575–588. John Wiley, New York.

Graham, N. V. S. (1989). *Visual Pattern Analyzers*, p. 26. Oxford University Press, New York.

Hassenstein, B. & Reichardt, W. (1956). Functional structure of a mechanism of perception of optical movement. *Proceedings I of International Congress of Cybernetics, Namur*, pp. 797–801.

Hiris, E. & Blake, R. (1992). A new perspective on the visual motion aftereffect. *Proc. Nat. Acad. Sci., USA*, **89**, 9025–9028.

Johnson, C. A. & Leibowitz, H. W. (1974). Practice, refractive error, and feedback as factors influencing peripheral motion thresholds. *Percept. Psychophys.*, **15**, 276–280.

Johnston, A. & Wright, M. J. (1986). Matching velocity in central and peripheral vision. *Vision Res.*, **26**, 1099–1109.

Julesz, B. (1971). *Foundations of Cyclopean Perception*. University of Chicago Press, Chicago.

Katz, E., Gizzi, M. S., Cohen, B. & Malach, R. (1990). The perceived speed of an object is affected by the distance it travels. *Perception*, **19**, 387.

Klein, S. A. & Levi, D. M. (1985). Hyperacuity thresholds of 1 sec: Theoretical predictions and empirical validation. *J. Opt. Soc. Am. A*, **2**, 1170–1190.

Klein, S. A. & Levi, D. M. (1987). Position sense of the peripheral retina. *J. Opt. Soc. Am. A*, **4**, 1543–1553.

Koenderink, J. J., van Doorn, A. J. & van de Grind, W. A. (1985). Spatial and temporal parameters of motion detection in the peripheral visual field. *J. Opt. Soc. Am. A*, **2**, 252–259.

Kolers, P. A. (1972). *Aspects of Motion Perception*, p. 37. Pergamon Press, Oxford.

Lappin, J. S. & Bell, H. H. (1972). Perceptual differentiation of sequential visual patterns. *Percept. Psychophys.*, **12**, 129–134.

Lappin, J. S. & Bell, H. H. (1976). The detection of coherence in moving random-dot patterns. *Vision Res.*, **16**, 161–168.

Lappin, J. S., Bell, H. H., Harm, O. J. & Kottas, B. (1975). On the relation between time and space in the visual discrimination of velocity. *J. Exp. Psychol.: Human Perception and Performance*, **1**, 383–394.

Lee, B. B., Wehrhahn, C., Westheimer, G. & Kremers, J. (1993). Macaque ganglion cell responses to stimuli that elicit hyperacuity in man: Detection of small displacements. *J. Neurosci.*, **13**, 1001–1009.

Legge, G. E. & Campbell, F. W. (1981). Displacement detection in human vision. *Vision Res.*, **21**, 205–213.

Legge, G. E. & Foley, J. M. (1980). Contrast masking in human vision. *J. Opt. Soc. Am.*, **70**, 1458–1471.

Levi, D. M., Klein, S. A. & Aitsebaomo, P. (1984). Detection and discrimination of the direction of motion in central and peripheral vision of normal and amblyopic observers. *Vision Res.*, **24**, 789–800.

Levi, D. M., Klein, S. A. & Aitsebaomo, P. (1985). Vernier acuity, crowding and cortical magnification. *Vision Res.*, **25**, 963–978.

Levinson, E. & Sekuler, R. (1975). The independence of channels in human vision selective for direction of movement. *J. Physiol.*, **250**, 347–366.

McKee, S. P. (1981). A local mechanism for differential velocity detection. *Vision Res.*, **21**, 491–500.

McKee, S. P. (1991). The physical constraints on visual hyperacuity. In J. J.

Kulikowski, V. Walsh & I. J. Murray (Eds.) *Vision and Visual Dysfunction*, **5**, *The Limits of Vision*. Macmillan Press, London.

McKee, S. P. & Nakayama, K. (1984). The detection of motion in the peripheral visual field. *Vision Res.*, **24**, 25–32.

McKee, S. P. and Taylor, D. G. (1984). Discrimination of time: comparison of foveal and peripheral sensitivity. *J. Op. Soc. Am. A*, **1**, 620–627.

McKee, S. P., Silverman, G. H. & Nakayama, K. (1986). Precise velocity discrimination despite random variations in temporal frequency and contrast. *Vision Res.*, **26**, 609–619.

McKee, S. P., Welch, L., Taylor, D. G. & Bowne, S. F. (1990). Finding the common bond: Stereoacuity and the other hyperacuities. *Vision Res.*, **30**, 879–891.

Mandler, M. B. (1984). Temporal frequency discrimination above threshold. *Vision Res.*, **24**, 1873–1880.

Mandriota, F. J., Mintz, D. E. & Notterman, J. M. (1962). Visual velocity discrimination: Effects of spatial and temporal cues. *Science*, **138**, 437–438.

Marshak, W. M. & Sekuler, R. (1979). Mutual repulsion between moving visual targers. *Science*, **205**, 379–382.

Mather, G. (1980). The movement aftereffect and a distribution–shift model for coding the direction of visual movement. *Perception*, **9**, 379–382.

Mather, G. (1987). The dependence of edge displacement thresholds on edge blur, contrast and displacement distance. *Vision Res.*, **27**, 1631–1637.

Morgan, M. J. (1992). Spatial filtering precedes motion detection. *Nature, Lond.*, **335**, 344–346.

Morgan, M. J. & Fahle, M. (1992). Effects of pattern element density upon displacement limits for motion detection in random binary luminance patterns. *Proc. Roy. Soc., Lond. B*, **248**, 189–198.

Morgan, M. J. & Ward, R. (1980). Conditions for motion flow in dynamic visual noise. *Vision Res.*, **20**, 431–435.

Nakayama, K. (1981). Differential motion hyperacuity under conditions of common image motion. *Vision Res.*, **21**, 1475–1482.

Nakayama, K. (1985). Higher order derivatives of the optical velocity vector field: Limitations imposed by biological hardware. In D. Ingle, M. Jeannerod & D. Lee (Eds.) *Brain Mechanisms and Spatial Vision*. Martinus Nijhoff, Holland.

Nakayama, K. & Silverman, G. H. (1985). Detection and discrimination of sinusoidal grating displacements *J. Opt. Soc. Am. A*, **2**, 267–274.

Nakayama, K. & Tyler, C. W. (1981). Psychophysical isolation of movement sensitivity by removal of familiar position cues. *Vision Res.*, **21**, 427–433.

Newsome, W. T. & Pare, E. B. (1988). A selective impairment of motion perception following lesions of the middle temporal area (MT). *J. Neurosci.*, **8**, 2201–2211.

Nishida, S. & Sato, T. (1993). Two kinds of motion aftereffect reveal different types of motion processing. *Invest. Ophthalmol. Vis. Sci. Suppl.*, **34**, 1363.

Orban, G. A., de Wolf, J. & Maes, H. (1984). Factors influencing velocity coding in the human visual system. *Vision Res.*, **24**, 33–39.

Pantle, A. (1978). Temporal frequency response characteristics of motion channels measured with three different psychophysical techniques. *Percept. Psychophys.*, **24**, 285–294.

Pantle, A. (1992). Immobility of some second-order stimuli in human peripheral vision. *J. Opt. Soc. Am. A*, **9**, 863–867.

Pasternak, T. (1987). Discrimination of differences in speed and flicker rate depends on directionally-selective mechanisms. *Vision Res.*, **27**, 1881–1890.

Regan, D. (1986). Visual processing of four kinds of relative motion. *Vision Res.*, **26**, 127–145.

Regan, D. & Beverley, K. I. (1978a). Illusory motion in depth: After effect of adaptation to changing size. *Vision Res.*, **18**, 209–212.

Regan, D. & Beverley, K. I. (1978b). Looming detectors in the human visual pathway. *Vision Res.*, **18**, 415–422.

Regan, D. & Beverley, K. I. (1979). Binocular and monocular stimuli for motion in depth: changing disparity and changing-size feed the same motion-in-depth stage. *Vision Res.*, **19**, 1331–1342.

Reichardt, W. (1959). Autocorrelation and the central nervous system. In W. A. Rosenblith (Ed.) *Sensory Communication*, pp. 303–318. MIT Press, Cambridge.

Schmerler, J. (1976). The visual perception of accelerated motion. *Perception*, **32**, 931–941.

Sekuler, R. W. & Ganz, L. (1963). Aftereffect of seen motion with a stabilized retinal image. *Science*, **139**, 419–420.

Smith, A. T. & Edgar, G. K. (1990). The influence of spatial frequency on perceived temporal frequency and perceived speed. *Vision Res.*, **30**, 1467–1474.

Smith, A. T. & Edgar, G. K. (1991). The separability of temporal frequency and velocity. *Vision Res.*, **31**, 321–326.

Smith, O. W. & Sherlock, L. (1957). A new explanation of the velocity-transposition phenomenon. *Am. J. Psychol.*, **70**, 102–105.

Snowden, R. J. & Braddick, O. J. (1991). The temporal integration and resolution of velocity signals. *Vision Res.*, **31**, 907–914.

Sperling, G. (1976). Movement perception in computer-driven visual displays. *Behav. Res. Meth. Instr.*, **8**, 144–151.

Stone, L. S. & Thompson, P. (1992). Human speed perception is contrast dependent. *Vision Res.*, **32**, 1535–1549.

Thompson, P. (1982). Perceived rate of movement depends on contrast. *Vision Res.*, **22**, 377–380.

Turano, K. & Pantle, A. (1985). Discontinuity limits for the generation of visual motion aftereffects with sine- and square-wave gratings. *J. Opt. Soc. Am. A*, **2**, 260–266.

Turano, K. & Pantle, A. (1989). On the mechanism that encodes the movement of contrast variations: velocity discrimination. *Vision Res.*, **29**, 207–221.

Tyler, C. W. (1981). Specific deficits of flicker sensitivity in glaucoma and ocular hypertension. *Invest. Ophthalmol. Vis. Sci.*, **20**, 204–212.

Tyler, C. W. (1985). Analysis of visual modulation sensitivity. II. Peripheral retina and the role of photoreceptor dimensions. *J. Opt. Soc. Am. A*, **2**, 393–398.

Tyler, C. W. & Torres, J. (1972). Frequency response characteristics for sinusoidal movement in the fovea and periphery. *Percept. Psychophys.*, **12**, 232–236.

Tynan, P. D. & Sekuler, R. (1982). Motion processing in peripheral vision: Reaction time and perceived velocity. *Vision Res.*, **22**, 61–68.

Van de Grind, W. A., Koenderink, J. J. & Van Doorn, A. J. (1983). Detection of coherent movement in peripherally viewed random-dot patterns. *J. Opt. Soc. Am.*, **73**, 1674–1683.

Van de Grind, W. A., Koenderink, J. J. & Van Doorn, A. J. (1986). The distribution of human motion detector properties in the monocular visual field. *Vision Res.*, **26**, 797–810.

Van de Grind, W. A., Koenderink, J. J., Van Doorn, A. J., Milders, M. V. & Voerman, H. (1993). Inhomogeneity and anisotropies for motion detection in the monocular visual field of human observers. *Vision Res.*, **33**, 1089–1107.

Victor, J. D. & Conte, M. M. (1990). Motion mechanisms have only limited access to form information. *Vision Res.*, **30**, 289–301.

Watamaniuk, S. N. J. (1992). Simultaneous direction information from global flow and a local trajectory component. *Invest. Ophthalmol. Vis. Sci. Suppl.* **33**, 1050.

Watamaniuk, S. N. J. & Duchon, A. (1992). The human visual system averages speed information. *Vision Res.*, **32**, 931–941.

Watamaniuk, S. N. J. & Sekuler, R. (1992). Temporal and spatial integration in dynamic random-dot stimuli. *Vision Res.*, **32**, 2341–2347.

Watamaniuk, S. N. J., Sekuler, R. & Williams, D. W. (1989). Direction perception in complex dynamic displays: the integration of direction information. *Vision Res.*, **29**, 47–59.

Watson, A. B. & Robson, J. G. (1981). Discrimination at threshold: Labelled detectors in human vision. *Vision Res.*, **21**, 1115–1122.

Watson, A. B., Thompson, P. G., Murphy, B. J. & Nachmias, J. (1980). Summation and discrimination of gratings moving in opposite directions. *Vision Res.*, **20**, 341–347.

Watson, A. B., Ahumada, A. & Farrell, J. E. (1983). The window of visibility: A psychophysical theory of fidelity in time-sampled visual motion display. *NASA Technical Paper 2211*.

Werkhoven, P., Snippe, H. P. & Toet, A. (1992). Visual processing of optic acceleration. *Vision Res.*, **32**, 2313–2329.

Westheimer, G. (1978). Spatial phase sensitivity for sinusoidal grating targets. *Vision Res.*, **18**, 1073–1074.

Westheimer, G. (1979). The spatial sense of the eye. *Invest. Ophthalmol. Vis. Sci.*, **18**, 893–912.

Westheimer, G. & Wehrhahn, C. (1993). Discrimination of direction of motion in human vision. *J. Neurophysiol.*, **71**, 33–37.

Westheimer, G., Shimamura, K. & McKee, S. P. (1976). Interference with line-orientation sensitivity. *J. Opt. Soc. Am.*, **66**, 332–338.

Williams, D. W. & Sekuler, R. (1984). Coherent global motion percepts from stochastic local motions. *Vision Res.*, **24**, 55–62.

Williams, D. W., Tweten, S. & Sekuler, R. (1991). Using metamers to explore motion perception. *Vision Res.*, **31**, 275–286.

Wohlgemuth, A. (1911). On the aftereffect of seen movement. *Br. J. Psychol., Monographs Supplement*, **1**.

Wyatt, H. J. (1974). Singly and doubly contingent after-effects involving color, orientation and spatial frequency. *Vision Res.*, **14**, 1185–1194.

Part 3
Inputs to Local Motion Detectors

5

Motion Detector Models: Psychophysical Evidence

George Mather

University of Sussex, UK

1 INTRODUCTION

A variety of image attributes can support motion perception. Psychophysical evidence indicates that separate sub-systems analyse the motion of first-order attributes (defined by variations in local intensity) and second-order attributes (defined by variations in textural attributes) (Chubb & Sperling, 1988; Cavanagh & Mather, 1989; Mather & West, 1993). Processing of second-order motion is discussed by Smith in Chapter 6. This chapter will review psychophysical research concerning the visual processes which detect first-order motion. The next section briefly outlines the spatiotemporal frequency content of motion displays; the third section summarizes a class of motion detector model which encodes motion on the basis of Fourier energy; and the final section will assess psychophysical evidence claiming to shed light on the specific model implemented in human vision.

2 INFORMATION AVAILABLE FOR MOTION DETECTION

Moving images can be represented as three-dimensional intensity distributions, where two axes are space (x and y) and the third axis is time (t). We will consider stimuli in which motion and spatial variation is restricted to the x axis, partly on the grounds of simplicity, and partly because the local processes discussed in this chapter can only signal the component of velocity orthogonal to local contour (additional processing is required to recover the full velocity vector. See Chapter 8 by Wilson). Such motion can be represented by two-dimensional (2D) plots in which the horizontal axis corresponds to space (x) and the vertical axis corresponds to time (t). Figure 1 (left) illustrates several so-called xt plots of motion stimuli (Fahle & Poggio, 1981; Adelson

VISUAL DETECTION OF MOTION
ISBN 0–12–651660–X

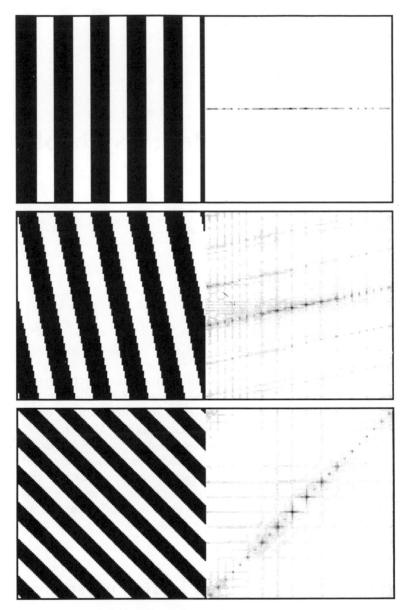

Figure 1   Information available in motion displays. Left: motion displays represented as *xt* plots. The horizontal axis represents *x* position, and the vertical axis represents time (increasing downward). Right: spatiotemporal frequency spectra of the *xt* plots on the left. The horizontal axis represents spatial frequency and the vertical axis represents temporal frequency, both increasing from zero at the centre of the plot. The top row depicts a stationary square wave (vertical contours) and its spatiotemporal Fourier spectrum (energy is spread along the horizontal, or zero temporal frequency axis). The middle and bottom rows depict a rightward drifting square wave and its spectrum, at two velocities. Frequency components lie along a tilted line in frequency space, with angle of tilt specifying velocity (ratio of spatial frequency to temporal frequency).

& Bergen, 1985): the top plot depicts a stationary square wave grating, and therefore contains vertical contours (no change in x position over time); rightwards drift creates contours in the plot which are tilted anticlockwise from vertical (middle and bottom), with velocity reflected in the slope of the contours. Leftwards motion would create contours tilted clockwise from vertical.

Following Watson *et al.*'s (1986) elegant analysis, it is illuminating to consider these motion stimuli in the frequency domain. Figure 1 (right) presents spatiotemporal Fourier spectra of the xt plots in Figure 1 (left). The vertical axis of the spectrum represents temporal frequency, and the horizontal axis represents spatial frequency, with the origin at the centre of the plot. Thus, stationary contours in the xt plot yield spatial frequency components which are spread out along the horizontal (zero temporal frequency) axis, as in the top plot. If the contours begin to move at a specific velocity, each spatial frequency component will take on a particular temporal frequency and, since temporal frequency is given by the product of velocity and spatial frequency, each spatial frequency component will slide vertically up or down the temporal frequency axis so that it falls on a tilted line in the spatiotemporal frequency spectrum (at any one velocity, higher spatial frequencies have higher temporal frequencies). The slope of this line specifies velocity, as is obvious by inspection of Figure 1 (a more rigorous analysis is available in Watson *et al.*, 1986). Thus directional motion confines energy to a line falling across two antisymmetric quadrants of the spectrum: leftward motion produces energy in the top-left and bottom-right quadrants of the spectrum, and rightward motion produces energy in the top-right and bottom-left quadrants.

3   MODELS OF MOTION DETECTION

Motion detectors can be constructed which detect the spatiotemporal signature of movement by sampling small antisymmetric regions of the frequency spectrum. Equivalently, they can be viewed as detecting oriented structure in xt space. There are a variety of ways to construct them, and a number of specific models have been proposed. The same basic principle of operation underlies all models, but they differ in terms of detailed structure. A clear understanding of the similarities and differences between models is essential if we are to draw meaningful conclusions from the psychophysical data.

3.1   Similarities between models

3.1.1   Local spatiotemporal filtering

Early visual processing begins with local spatiotemporal filtering, by means of neural receptive fields. For example, sustained or X-type retinal ganglion cells have a band-pass spatial response, and a low-pass temporal response (Enroth-Cugell & Robson, 1966). Such receptive fields are symmetrical and non-oriented about the x and t axes of an xt plot, and would obviously not offer a directionally selective response to motion. The general strategy in all spatiotemporal filtering models of motion detection is to combine the outputs of such local component filters to construct a direction-selective receptive field, which is elongated along a particular axis in space–time. The angle given by the axis of elongation specifies the receptive field's preferred velocity.

3.1.2 Paired inputs in approximate quadrature

How can one construct a receptive field which is oriented in space–time, using non-oriented filters as components? Minimally, such a receptive field can be constructed from a pair of input receptive fields having spatiotemporal weighting functions which are offset relative to each other, so that in combination they collect energy along a specific axis in space–time (see also Grzywacz *et al.* in Chapter 2, section 4.1). A common thread running through all detector models is the use of quadrature relationships between component receptive fields to introduce a spatiotemporal offset.

Consider first how quadrature relationships can be used to introduce a spatial offset in receptive field structure between two component fields. The Gabor function offers a convenient and simple way to characterize spatial receptive field structure, since it consists of a sinewave multiplied by a Gaussian 'window'. Cortical receptive fields can be accurately modelled in this way (Marcelja, 1980; Pollen & Ronner, 1981; though see Hawken & Parker, 1987). It also offers an optimal compromise between spatial precision and frequency resolution (see Grzywacz *et al.*, Chapter 2). An offset in the spatial sensitivity profiles of two such receptive fields, centred on coincident retinal locations, can be implemented by phase-shifting the sinewave component of the Gabor function in one field relative to the function in the other field. A phase shift of one-quarter cycle, or 90°, will convert an even-symmetric receptive field (cosine phase) into an odd-symmetric one (sine phase). Figure 2 (top) illustrates two Gabor functions exhibiting this spatial 'quadrature' relationship. A phase shift of 90° prevents spatial aliasing, because it is scaled to the spatial frequency response of the receptive field. In general, any pair of receptive fields which exhibit even and odd symmetry about the same spatial axis can be said to approximate a quadrature pair.

Imagine a horizontally drifting sinewave, sampled by two receptive fields differing in spatial phase or position by a quarter of a spatial cycle along the x axis (a quadrature pair). Image intensity at each receptive field position will fluctuate over time sinusoidally as the sinewave passes over and, because of the spatial offset, the temporal modulation at the two positions will be offset by a quarter of a *temporal* cycle. If the sinewave drifts rightward, then the temporal modulation at the lefthand sampling position will 'lead' the modulation at the righthand position (e.g. a dark peak will reach the lefthand position a quarter-cycle before it reaches the righthand position). If the sinewave drifts leftward, then modulation at the righthand sampling position will lead. A direction-selective response can be constructed by introducing a relative temporal offset (delay) between responses of the two receptive fields. The net response of each receptive field is given by the product of its spatial response and its temporal response. If the temporal responses also form a quadrature pair, so that the righthand field's response leads the lefthand field's by a quarter temporal cycle (i.e. lefthand response is relatively delayed), then the two responses will be temporally in-phase for rightward motion, but 180° out-of-phase for leftward motion. If the temporal response of the lefthand field leads the right, then the two responses will be in phase for leftward motion. As in the case of spatial quadrature, the precise form of the

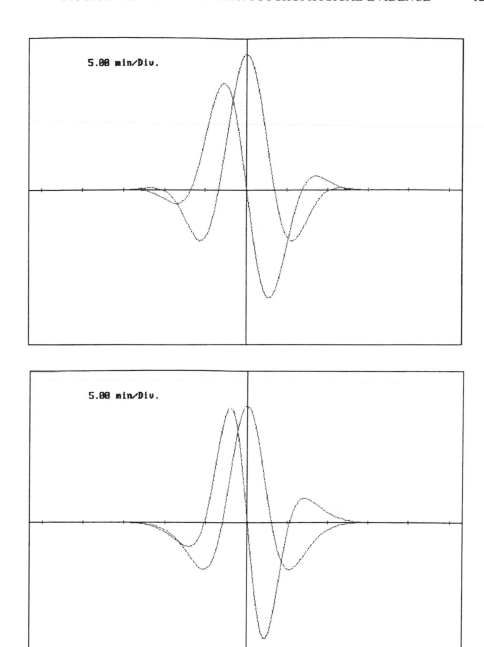

Figure 2   Quadrature pairs of spatial filters. Top: a quadrature pair of Gabor filters. One filter has even symmetry (central peak), and the other has odd symmetry (peak on the left), created by shifting the sinewave component of the filter response through 90 degrees of phase. Bottom: a difference of Gaussian's spatial filter (central peak), and its spatial derivative (peak on the left). Note the similarity with the Gabor pair.

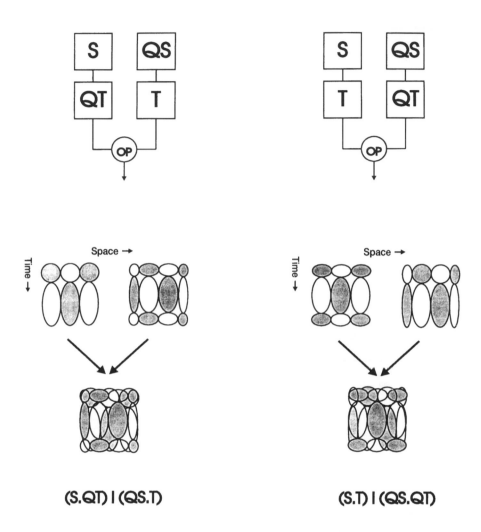

Figure 3 Basic motion detector. Top: a basic motion detector consists of two paths. Each path contains a spatial filter and a temporal filter, and filters in the two paths form quadrature pairs, S and QS, and T and QT. Responses in the two paths are combined (OP) to create a direction selective output. Preferred direction depends on the particular form of the spatial and temporal filters.

Middle and bottom: examples of spatiotemporal receptive fields forming a basic detector, taken from Watson and Ahumada's linear motion sensor. Component spatiotemporal filters are shown in the middle row. Reading from the left, the first filter (S.QT) has an even-symmetrical spatial response and odd-symmetrical temporal response; the second (QS.T) is odd over space and even over time; the third (S.T) is even over both space and time; and the fourth (QS.QT) is odd over both space and time. When superimposed, as shown in the bottom row, the pairs of component filters form receptive fields containing regions which are aligned for leftward motion (shown on the left) or for rightward motion (shown on the right). In a faithful version of the linear motion sensor, the superimposition would be implemented as addition.

temporal responses is not crucial, as long as they exhibit (or approximate) temporal quadrature.

Thus direction-specific responses can be constructed by combining the outputs of paired component receptive fields which exhibit spatial and temporal quadrature. The basic detector, illustrated in Figure 3 (top), consists of two paths, each containing a spatial filter and a temporal filter. The pairs of filters in the two paths are in spatial and temporal quadrature (S vs. QS; and T vs. QT). The outputs of the two paths are combined (OP) to generate a direction-specific response. A number of specific models for motion detection have been proposed, which all implement this basic scheme. Models differ in terms of implementation details. For example, some models propose linear combination of responses in the two paths, whereas others propose non-linear combination. The next section briefly reviews the idiosyncracies of different models.

3.2   Differences between models

3.2.1   Linear motion sensor

This detector is perhaps the purest form of quadrature detector. It consists of two paths, which Watson and Ahumada (1985) call 'main' and 'quadrature': they each contain a spatial filter (S and QS, respectively, in spatial quadrature), and a temporal filter (T and QT, respectively, in temporal quadrature), as depicted in Figure 3 (top). The spatiotemporal response of each path is given by the product of its spatial response and its temporal response. The outputs of the two paths are summed to create a linear spatiotemporally oriented receptive field. Watson and Ahumada used a low-pass temporal filter and bandpass (Gabor) spatial filter, but argue that the precise form of the filters is not important as long as they form quadrature pairs in the two paths. The middle and bottom rows of Figure 3 depict the two detectors in the top row in terms of the component spatiotemporal filters used by Watson and Ahumada. Consider first the lefthand detector and pair of component filters. The leftmost filter in the middle row has an even spatial response and an odd temporal response, corresponding to the (S.QT) path of the detector. The second filter in the middle row has an odd spatial response and even temporal response, corresponding to the (QS.T) path. In the bottom row of the figure the two components have been overlaid to show how the positive and negative areas of response combine to create a spatio-temporal receptive field which is tuned to leftward motion (clockwise tilt in xt space). In a similar way, the righthand detector in Figure 3 is tuned to rightward motion (anticlockwise tilt) because of the particular pattern of excitatory and inhibitory regions in its component filters.

Since the detector is linear, its output oscillates in time at the same temporal frequency as the image components within its spatial frequency pass band. Watson and Ahumada exploit this oscillation to estimate image velocity.

3.2.2   Energy detector

Adelson and Bergen's (1985) 'energy' detectors are also constructed using quadrature pairs of spatial and temporal filters. They implement quadrature detectors using four

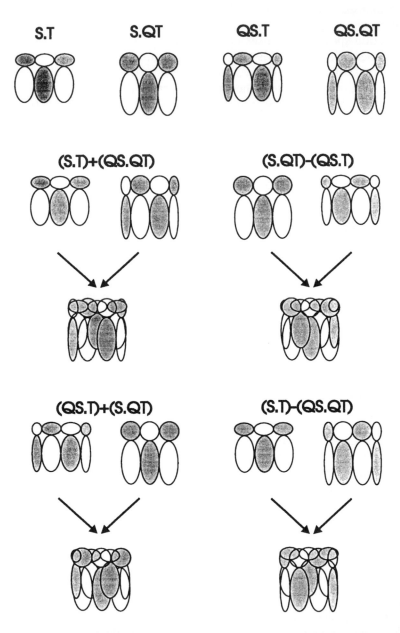

Figure 4 Adelson and Bergen's energy detector. The top row depicts component spatiotem-poral receptive fields similar to those in Figure 3, but based on the weighting functions suggested by Adelson and Bergen. The remaining rows show how these components can be combined in pairs to create four direction-selective receptive fields, two tuned to rightward motion (anticlockwise tilt, row 3), and two tuned to leftward motion (clockwise tilt, row 5). Rightward 'energy' is computed by squaring the outputs of the two rightward detectors, and then summing their responses. Leftward energy involves identical operations on the outputs of the leftward detectors.

psychophysically plausible component receptive field profiles, a quadrature pair of spatial filters and a quadrature pair of temporal filters. These filters are shown in Figure 4, though Adelson and Bergen argue that no significance should be attached to these particular filter functions. They note that, given such pairs of filters, it is possible to construct four different component spatiotemporal fields consisting of pairwise products of spatial and temporal response: (S.T), (QS.T), (S.QT), (QS.QT). Figure 4 (top row) depicts the four non-oriented spatiotemporal fields, which vary in terms of the particular pattern of excitatory and inhibitory regions they contain (note the slight difference compared to the filters in Figure 3). It is possible to create two direction-specific detectors oriented for rightward motion, and two oriented for leftward motion, by taking appropriate sums and differences between components. The two rightward detectors are ((S.T) + (QS.QT)) and ((S.QT) − (QS.T)) as illustrated in Figure 4 (rows 2 and 3). The two leftward detectors are: ((QS.T) + (S.QT)) and ((S.T) − (QS.QT)) as illustrated in Figure 4 (rows 4 and 5). Rows 3 and 5 in Figure 4 depict the four detectors (note that subtraction is equivalent to addition if one of the two spatio-temporal components involved is reversed in 'polarity', or phase-shifted through 180° so that positive areas become negative and vice versa).

The four individual detectors have basically the same form as Watson and Ahumada's, as Adelson and Bergen acknowledged (compare Figures 3 and 4). Each (linear) detector's response depends on the contrast polarity of the moving stimulus; the sign of its response 'will depend on the sign of the stimulus contrast, so that a black bar and a white bar moving in the same way will give inverted responses' (Adelson & Bergen, 1985, p. 290). (Of course, a rightward detector will not respond to a leftward moving bar, and vice versa, whatever its polarity. Further, in a physiologically plausible system responses cannot invert completely, so individual cells will exhibit preferences for particular stimulus polarities.)

Unlike Watson and Ahumada, Adelson and Bergen view the polarity sensitivity of detector responses as undesirable, and remove its effects by first squaring each detector's output and then summing the outputs of each pair of detectors tuned to the same direction. The resulting summed response forms their energy detector, and is taken as a measure of local motion energy in the range of frequencies to which the detector is tuned. The linear receptive field concept, as depicted in Figure 4, is only applicable up to the squaring operation. Adelson and Bergen allow for the possibility of motion opponency by including an optional final stage in which the arithmetic difference is taken between rightward and leftward energy responses. However they do not view opponency as an essential feature of their model.

3.2.3 Elaborated Reichardt Detector

The output of van Santen and Sperling's (1984, 1985) Elaborated Reichardt Detector (ERD) is given by the difference between the time-averaged responses of two 'subunits' tuned to opposing directions. Each subunit in the ERD is basically a quadrature detector, since it consists of two paths containing quadrature pairs of spatial and temporal filters. The outputs of the two quadrature paths are multiplied to yield a directional response from the subunit. (van Santen and Sperling (1985) offer addition of quadrature path outputs followed by squaring, as an alternative to multiplication, to emphasize similarities with the previous two detectors.) van Santen and

Sperling argue that the precise form of the spatial and temporal filters is not crucial, and give examples which use quadrature pairs of spatial Gabor filters and temporal low-pass filters. The most distinctive features of the ERD are the temporal averaging of subunit outputs, and the subtraction of oppositely-tuned subunit outputs. The basic motion detector in Figure 3 can be extended to form an ERD by first adding a 'time averaging' box to the output of each detector, and then bringing together the outputs of the two detectors in a box which computes the difference between their outputs. However, van Santen and Sperling play down these two features. They argue that some form of temporal averaging is needed in some situations 'to obtain simple mathematical results', but 'there are many applications in which even this weaker form of averaging is not needed' (1985, p. 302). Averaging is included in their model, they argue, primarily to provide a single real number to predict response strength in experiments, but they concede that it is not possible to determine how much integration occurs in the detector and how much at higher levels. As for opponency, van Santen and Sperling state that 'one can imagine a scheme in which subunits tuned to rightward and to leftward motion have no direct ''physical'' connection with each other and occur in pairs only in the formal sense that for every leftward-tuned detector we can find elsewhere in the system a rightward-tuned detector that is its complete mirror image' (1984, p. 470). They demonstrate that in a wide range of conditions (e.g. well-behaved quadrature subunits) this independent scheme is equivalent to the opponent subtraction scheme. The underlying close relation between the ERD and the previous two detectors is clear. van Santen and Sperling show how the output of the Watson–Ahumada detector is equivalent to the ERD's if 'linking' assumptions are added to the former which involve squaring of detector output followed by temporal integration and subtraction. Similarly, they argue that the variant of Adelson and Bergen's detector which includes opponency is fully equivalent to the ERD: the outputs of the two detectors differ only by a multiplicative constant. Emerson *et al.* (1992) point out that although the output of the ERD is equivalent to the final output of the opponent energy detector, the earlier non-opponent stages of the energy detector have no equivalent in the ERD.

van Santen and Sperling (1985) prefer to view the linear motion sensor and energy detector as 'special cases' of the ERD, but it is equally justifiable to view the ERD as equivalent in status to the other detectors, since all are variants of a general quadrature detector. The ERD includes unique processing stages involving temporal averaging and opponent subtraction.

3.2.4   Gradient detector

The gradient detector was originally put forward as a model for human vision by Marr and Ullman (1981), and versions have been described in Harris (1986) and in Mather (1987, 1990). It exploits the fact that a moving intensity edge generates spatial and temporal changes in image intensity which jointly specify direction and speed. Marr and Ullman proposed that the detector contains two channels, S and T, which differ in their spatial and temporal responses. The S channel computes spatial gradient and the T channel computes temporal gradient. Motion information can be derived from a comparison of spatial and temporal gradients. The signs of the gradients are sufficient to specify direction, and the magnitudes specify velocity. If we assume that both S and

T channels contain spatiotemporal filters, then the filter functions in the two channels must have the following relationships: the spatial response in the S channel is the spatial derivative of the spatial response in the T channel, and the temporal response in the T channel is the temporal derivative of the temporal response in the S channel. Marr and Ullman proposed that the nonlinear combination (e.g. product) of S and T responses specifies direction, but other rules are also possible (e.g. addition of S and T, followed by squaring, to emphasize similarities with other models).

The gradient detector exploits quadrature relationships by virtue of the fact that the filters in the two channels are in (at least approximate) spatial and temporal quadrature. Consider first the spatial responses of the S and T channels. If the T channel has an even-symmetrical spatial response (e.g. difference-of-Gaussians (DoG) profile), its spatial derivative in the S channel will have an odd-symmetrical spatial response, so the two responses will be in spatial quadrature. Figure 2 (bottom) shows a difference-of-Gaussians profile and its spatial derivative. In the case of the Gabor functions frequently used in other models, a phase shift of 90° from cosine phase to sine phase is equivalent to spatial differentiation, since the derivative of a sine function is a cosine and vice versa. Note that the quadrature pair of Gabor filters in Figure 2 (top) looks very similar to the DoG and its spatial derivative in Figure 2 (bottom). Similarly, if one channel has an odd-symmetrical *temporal* response, its temporal derivative in the other channel will be even-symmetrical, and therefore its approximate quadrature partner. Thus the basic motion detector in Figure 3 (top) can be described as a gradient detector if we consider the QS component filter to be the spatial derivative of the S filter, and the QT component filter to be the temporal derivative of the T filter. Marr and Ullman speculated that X-type ganglion cells could mediate the S channel, and Y-type ganglion cells could mediate the T channel.

3.3 Summary

A remarkable feature of these motion detector models is that, despite superficial differences, the underlying principles of operation are so similar. Several key features are common to all, including:

● the use of paired input channels;
● the presence of quadrature relationships between filter functions in the two channels;
● the concept of a spatiotemporal receptive field (until the intrusion of non-linearities);
● the presence of non-linearities (except for the linear sensor).

Given such overlap between models, it is unlikely that psychophysical data will conclusively reject an entire model, so it may be more realistic to ask: 'What model offers the best heuristic description of the system?' If a particular detail of a model is not supported by data (e.g. opponent-processing in ERDs), then the heuristic value of the model is diminished, even if other details of the model survive unscathed, and an alternative model is to be preferred.

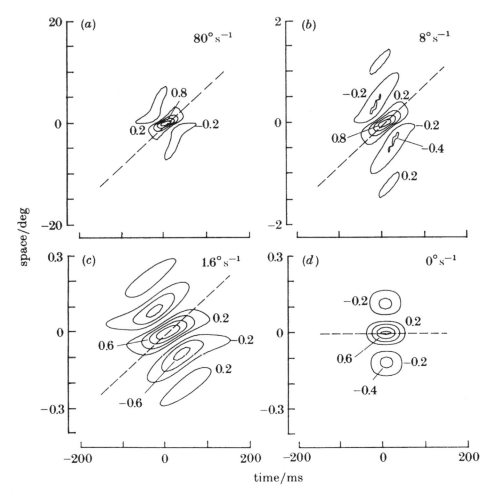

Figure 5   Human motion detecting receptive fields estimated from masking data by Burr *et al.* (1986). Note that, in these graphs, space is plotted vertically and time horizontally. Graphs (a), (b) and (c) were obtained using test gratings drifting at 8 Hz; graph (d) was obtained using a temporal frequency of 0.3 Hz (almost stationary). All receptive fields contains sub-regions of alternating polarity, but only the three fields estimated at high temporal frequency are oriented in space–time.

4   PSYCHOPHYSICAL EVIDENCE

A very large number of experimental studies have examined the response properties of low-level motion detectors. Many of these studies are described in other chapters (e.g. McKee and Watamaniuk, Chapter 4). This review will concentrate on studies which address the following questions:

- Do motion detectors have spatiotemporally limited responses, consistent with the receptive field concept?
- Is there evidence that receptive fields are tuned to quadrature?
- Is there evidence for non-linearities in detector responses?
- Are oppositely-tuned detectors yoked in opponent-process fashion?

4.1 Spatiotemporal tuning

Clear psychophysical evidence for spatiotemporally oriented receptive fields comes from Burr *et al.* (1985). They estimated receptive field structure by measuring contrast sensitivity to movement in the presence of masking gratings at various spatial and temporal frequencies. Figure 5 shows their estimate of receptive field shape in *xt* space. There is also convincing physiological evidence that the receptive fields of cortical direction-specific cells are tuned to spatiotemporal orientation (Emerson *et al.*, 1987; McLean & Palmer, 1989; Emerson *et al.*, 1992).

Spatiotemporal tuning has been studied extensively using random dot kinematograms (RDKs), which are assumed to offer no motion cues based on segmented features and shapes. A simple RDK contains two frames of random dots, presented in temporal sequence. At least some of the dots in the second frame are identical to those in the first, except for the addition of a coherent spatial offset in one direction or its opposite. The offset creates an impression of motion when the kinematogram is animated, which is usually quantified in terms of the percentage of correct responses in a direction discrimination task. Early data indicated that there were strict spatial and temporal limits on the subjects' ability to perform the discrimination (e.g. Braddick, 1973, 1974; Lappin & Bell, 1976). Performance typically fell to 80% correct or less for spatial displacements above about 0.25° (Dmax), and for temporal inter-frame intervals above about 80 ms (Tmax). These limits have been interpreted in terms of the structural properties of Reichardt detectors. The spatial limit can be related to the spatial offset between input sub-units, and the temporal limit can be related to the detector's temporal delay filter (see, for example, Baker & Braddick, 1985; Sekuler *et al.*, 1990). However, recent evidence indicates that the spatial and temporal limits of discrimination in RDKs can be attributed to physical properties of the stimulus, and tell us little about the structural properties of motion detectors. The evidence centres on the consequences of two simple manipulations. First, when the random dots are spatially blurred, or low-pass filtered, the displacement limit Dmax increases, roughly in proportion with the filter space constant (beyond some threshold level of blur; as can be seen in Figure 6 (left), which is replotted from Morgan and Mather, 1994) (see also Chang & Julesz, 1983; Cleary & Braddick, 1990; Morgan, 1992). Second, when the onset and offset of each frame in the RDK is temporally smoothed (so that the dots appear and disappear gradually instead of suddenly), the temporal limit Tmax also increases in proportion with the filter time constant beyond a certain time constant (see Figure 6 (right), replotted from Mather and Tunley, 1993).

Two alternative interpretations have been proposed for these effects, a 'structural' account and an 'informational' account. The structural explanation assumes that motion detectors can be divided into groups tuned to different ranges of spatial frequency and temporal frequency. The optimal spatial displacement and temporal

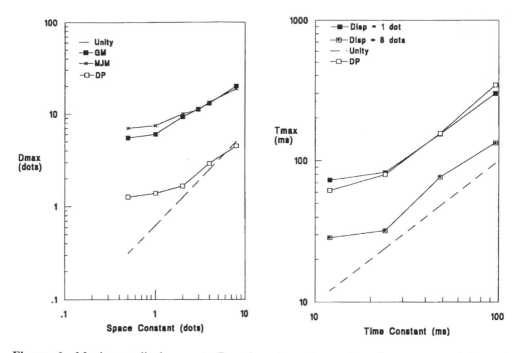

Figure 6 Maximum displacement (Dmax) and maximum inter-frame temporal interval (Tmax) supporting direction discrimination in two-frame random dot patterns. The left graph shows Dmax (in multiples of dot width) as a function of the space constant of a spatial Gaussian (i.e. low-pass) filter applied to each frame of the pattern. Data for two subjects are shown, replotted from Morgan and Mather (1994). Actual dot size was 4.5 min arc, and frame duration was 72 ms. The curve labelled DP plots Dmax computed from spatiotemporal Fourier transforms of two-frame random element arrays; vertical location of this curve is arbitrary (see text for details). The right graph shows Tmax (in ms) as a function of the time constant of a Gaussian temporal smoothing filter applied to the onset and offset of each frame in the pattern. Data for two frame-to-frame displacements are shown, and represent the mean performance of six observers, replotted from Mather and Tunley (1993). Dot size was 2 min arc, and frame duration was 384 ms (chosen to allow each pattern to reach maximum contrast even for filter values giving the most gradual frame onset and offset; the same pattern of results was obtained for shorter frame durations). Curve DP plots computed Tmax, calculated as in the other graph.

interval for each detector depends on its preferred spatial frequency and temporal frequency. For example, a detector tuned to high spatial frequencies prefers short displacements, and a detector tuned to high temporal frequencies prefers brief inter-frame intervals. In spatially broadband patterns, displacements beyond the limit of high-frequency channels generate incoherent responses in these detectors which interfere with the coherent responses in lower frequency channels. Consequently, Cleary and Braddick (1990) argue that 'performance is limited by the output of the highest frequency channel activated'. Low-pass spatial filtering removes high-frequency motion responses, and therefore permits the greater Dmax values supported by low-frequency channels. A similar explanation can be devised for the effects of temporal smoothing, which removes incoherent responses generated in high temporal frequency detectors at long inter-frame intervals.

According to the informational explanation, Dmax and Tmax are set by the physical properties of the display. Thus, Bischof and Di Lollo (1990), Morgan (1992), and Eagle and Rogers (1991) argue that Dmax is determined by the mean separation between adjacent elements in the 'neural image' of the pattern. Once the between-frame spatial displacement exceeds half the mean separation between elements, then the discrimination task becomes ambiguous. According to this view, low-pass spatial filtering extends Dmax because it increases the mean separation between pattern elements (though only when the externally applied filter exceeds the space constant of the visual system's internal filter). In other words, Dmax is a sampling effect. The consequences of sampling for Dmax and Tmax become clearer if we consider the pattern in the spatiotemporal frequency domain. Discrete temporal sampling in RDKs (inherent in successively viewed static frames) introduces spurious energy into the spatiotemporal frequency spectrum of the pattern, in the form of repeating replicates of the pattern's energy. The replicas are spaced at intervals on the temporal frequency axis given by the inter-frame interval, and at intervals on the spatial frequency axis given by the between-frame displacement (Figure 7). Since random dot patterns are broadband, the sampling replicates cover a wide range of frequencies, and tend to overlap with the signal energy to generate alias signals at inappropriate velocities (e.g. in the opposite direction to the pattern's actual displacement). The overlap becomes more severe as the sampling interval over space or time increases, since the replicates move closer to the signal in Fourier space, eventually corrupting the signal so much that discrimination performance falls below the criterion set for Dmax or Tmax. Low-pass spatial filtering or temporal smoothing extends Dmax or Tmax because it ameliorates the corrupting effect of sampling artefacts at high spatiotemporal frequencies. Figure 6 also shows plots of Dmax and Tmax computed from 'directional power' (DP), a simple physical measure of the Fourier energy available for direction discrimination. DP is computed as the ratio of summed energy in the leftward quadrants of the transform to summed energy in the rightward quadrants, and has been used in several studies to estimate the Fourier-based signal in motion displays (e.g. Dosher *et al.*, 1989; Nishida & Sato, 1992; Boulton & Baker, 1993). Dmax and Tmax values were computed from DP as follows. First DP was computed for two-frame arrays of random black–white elements, as a function of between-frame displacement and inter-frame interval (during the interval between frames all elements became grey, at mean luminance). In agreement with psychophysical data, DP declined as displacement and inter-frame interval increased (due to the intrusion of sampling replicates). Under the assumption that discrimination performance depends on DP, an arbitrary 'threshold' level of DP was selected, and the displacement (or inter-frame interval) which just reached this threshold value was calculated. These computed Dmax and Tmax values are shown in Figure 6; they increase as the low-pass space or time constant rises, as expected from the sampling account and in agreement with the psychophysical data.

There are points of contact between the structural and informational explanations, in that both attribute Dmax and Tmax to visual responses generated by high spatio-temporal frequencies. The structural explanation posits the existence of detectors tuned to narrow bands of spatial and temporal frequency, whereas the informational explanation does not require assumptions about detector properties but restricts itself

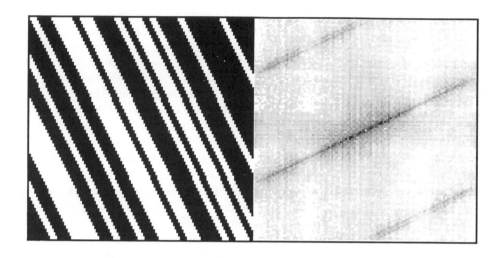

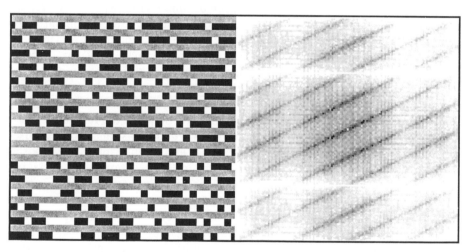

Figure 7 Sampling effects in spatially random patterns. Upper plots: an *xt* plot of a rightward drifting array of random black–white elements, and its spatiotemporal Fourier transform. Contours in the *xt* plot are tilted anticlockwise, and energy in the transform is spread along a tilted line specifying drift velocity. A motion detector collecting energy in the rightward quadrants of the transform (i.e. top right and bottom left) would give a strong response (cf. Figure 1). Lower plots: a sampled version of the upper *xt* plot, and its spatiotemporal Fourier transform. The array of elements is shown occupying a series of static positions separated by blank grey inter-frame intervals (the conventional way of presenting random dot kinematograms). Discrete sampling introduces replicates of the original signal in the Fourier transform. Energy spills over into the leftward quadrants (i.e. top left and bottom right), offering alias signals for leftward tuned detectors.

to a consideration of the Fourier energy available in the stimulus. To this extent the informational account is more parsimonious. Morgan and Mather (1994) presented further data on spatial blur effects which they claim cannot be accommodated by the structural explanation. They measured Dmax for two-frame RDKs in which only one frame was spatially low-pass filtered while the other remained broadband. According to the structural explanation, Dmax should match the value obtained when both frames are *low-pass*, since blurring just one frame should be sufficient to remove high-frequency motion responses. The informational explanation, on the other hand, predicts that Dmax will match the value obtained when both frames are *broadband*, because filtering only one frame does not completely remove the high spatial frequency components in the pattern which are responsible for the ambiguities at large displacements. Morgan and Mather found that Dmax matched the value obtained when both frames are broadband, supporting the informational explanation.

The debate about the explanation for variations in Dmax and Tmax warns against drawing inferences about spatiotemporal tuning properties on the basis of such data. An alternative response measure has been advocated by Koenderink and his colleagues (e.g. van Doorn & Koenderink, 1982a, 1982b). They manipulate the proportion of dots in a RDK which are correlated between frames, to find the proportion ('signal-to-noise ratio') required for observers to reach a criterion level of performance in direction discrimination. Some of the data obtained using this technique will be described in detail below.

In summary, despite misgivings about data from RDKs, there is strong psychophysical and physiological evidence that early motion detectors are tuned to spatiotemporal orientation.

4.2   Quadrature tuning

4.2.1   Sampled sinewaves

A number of studies have used discontinuously moving sinewaves, searching for evidence of quadrature tuning in motion detectors. Nakayama and Silverman (1985) measured contrast sensitivity for phase-shifting sinewave gratings. Subjects were given two brief views of a sinewave, the second view shifted by a fixed phase angle relative to the first, and required to report the direction of shift. Sensitivity was maximal for phase shifts of 90°, consistent with detection by quadrature-based detectors. Baker *et al.* (1989) measured motion after-effects (MAEs) generated by adaptation to phase-shifting sinewaves and found that, across a range of frequencies, the most effective adapting stimuli involved shifts of 0.2 cycles at most, somewhat less than the optimal phase shift found by Nakayama and Silverman.

Watson (1990) reviewed these two studies, and demonstrated that there are critical differences between the two-frame displays of Nakayama and Silverman (1985) and the multi-frame displays of Baker *et al.* (1989). He argued that the latter displays were contaminated by sampling artefacts introduced by a low television (TV) refresh rate (see the previous section), and concluded that the linear motion sensor would not predict any specific optimum phase shift. Clean predictions can be made only for two-frame displays, such as Nakayama and Silverman's. Watson considered two alternatives:

1   An energy model, in which direction is given by the difference between the squared magnitudes of left and right detector outputs ('energies'), as in Adelson and Bergen's detector.
2   A magnitude model, in which direction is given by the difference between the magnitudes of left and right detector outputs at the temporal frequency showing the greatest magnitude, as in Watson and Ahumada's detector.

Both schemes predict an optimum at 0.25 cycle displacements, but the magnitude model fitted the phase-angle function obtained by Nakayama and Silverman more closely than the energy model, at least by visual inspection. However, more data would be needed to establish this difference as statistically reliable.

A number of more recent studies measuring direction discrimination performance for two-frame sine-wave displays have also found optimal performance at displacements of a quarter-cycle (Cleary, 1990; Derrington & Cox, 1992; Morgan & Cleary, 1992b), though Boulton and Hess (1990) obtained optimum performance somewhat below a quarter-cycle.

4.2.2 Four-stroke apparent motion

When a static shape is replaced by an exact copy of itself, shifted slightly in position, an impression of movement is evoked in the direction of the spatial shift. Anstis and Rogers (1975) reported that when the contrast polarity of the display is reversed, so that objects brighter than the background become darker (and vice versa), the direction of apparent motion also reverses. Sato (1989) compared conventional RDKs with RDKs in which contrast is reversed between frames, and found that Dmax and other measures were similar for the two types of display, except for the reversal in apparent direction. Anstis and Rogers (1986) elaborated on the original reversed apparent motion effect to create a so-called four-stroke stimulus which appears to move unidirectionally, yet contains only contours which oscillate in position and synchronously reverse contrast polarity. Figure 8 (left) illustrates a simple variant as an *xt* display. In frames 1 and 2 (as identified on the left), a bright bar against a grey background shifts right; in frames 2 and 3 the bar shifts back, but reverses in polarity to become dark; in frames 3 and 4 the dark bar shifts right; in frames 4 and 5 the bar shifts back but reverses in polarity (frame 5 is identical to frame 1, so the stimulus repetitively cycles through four frames). Each transition generates an impression of rightward motion. A number of other displays which generate surprising impressions of motion basically constitute variants of this effect (e.g. Gregory & Heard, 1982; Anstis & Mather, 1985; Freeman *et al.*, 1991). The display in Figure 8 (right) is a variant containing continuous, rather than discrete variation in each bar's intensity as a function of time.

Notice that the temporal luminance modulation applied to the righthand bar in the displays of Figure 8 lags behind the modulation applied to the lefthand bar by one-quarter of a temporal cycle, to create short segments of oriented structure in *xt* space. The effectiveness of these stimuli is therefore at least consistent with detection by spatiotemporally oriented motion receptive fields tuned to quadrature. Anstis and

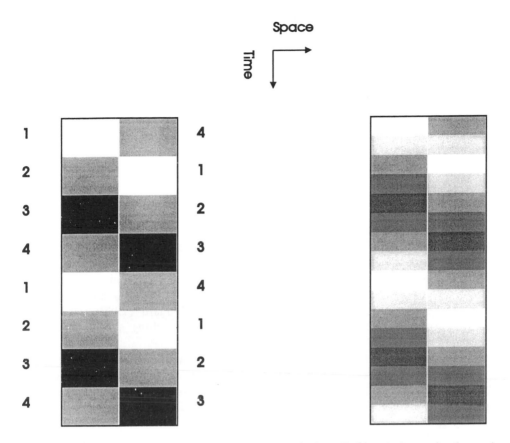

Figure 8 Examples of four-stroke apparent motion displays. Lefthand plot: a simple version in which a bright bar shifts right (frame 1 – frame 2); then shifts back and reverses contrast (frame 2 – frame 3); then shifts right again (frame 3 – frame 4); and finally returns to its original position and polarity (frame 1). Righthand plot: a more sophisticated version in which bar intensity varies more continuously over time. In both displays, modulation at the righthand bar position lags behind modulation at the lefthand position by one-quarter of a temporal cycle.

Rogers (1975, 1986) prefer to explain the effects by 'spatial smoothing'. They examined the spatial location of contours in their displays, after filtering by physiologically plausible receptive fields, and found contours which shift in the direction reported (see also Mastebroek & Zaagman, 1988). The quadrature and smoothing accounts are closely related. Anstis and Rogers' explanation basically amounts to the observation that after filtering by spatiotemporally non-oriented receptive fields the xt plot contains oriented structure corresponding to the perceived direction. They leave open the question of how the oriented structure is detected. The explanation based on quadrature takes one further step in arguing that this structure is extracted by detectors with elongated receptive fields, constructed from the symmetrical receptive fields as described earlier. Both Marr and Ullman (1981) and Adelson and Bergen (1985) showed that their detector models can account for reversed apparent motion. Anstis and Rogers (1986) point out that almost any motion model which successfully detects real motion should sense motion in their stimuli (see also Mather, 1989).

So, four-stroke stimuli are consistent with motion detection by spatiotemporally oriented receptive fields tuned to quadrature, but they cannot offer conclusive proof. More generally, there is now sufficient agreement in the available data for us to conclude that motion detectors rely on (at least approximate) spatiotemporal quadrature.

4.3   Non-linearities

4.3.1   Response compression

Many studies of human motion perception have found severe non-linearities in contrast response. For example, Keck *at al.* (1976) found that motion after-effect velocity did not vary with adapting contrasts above 0.03. Johnston and Wright (1985) reported that the lower threshold for motion detection was constant above a contrast of 0.05. Nakayama and Silverman (1985) reported saturation in a direction discrimination task above contrasts of 0.02. Derrington and Goddard (1989) reported that discrimination performance actually deteriorates after reaching a peak at contrasts of about 0.05.

A marked non-linearity in contrast response is the inevitable consequence of nonlinear interactions between the two paths in the motion detector, as mentioned by Grzywacz *et al.* in Chapter 2: a small increase in stimulus contrast will result in a large increase in detector response, which may therefore reach saturation very rapidly at relatively low contrasts. There is physiological evidence that response compression arises at the level of cortical motion detecting cells, and cannot be attributed to the response properties of component filters (e.g. magnocellular lateral geniculate nucleus (LGN) cells). Sclar *et al.* (1990) compared the contrast response of LGN cells with the responses of cortical V1 and middle temporal (MT) cells. Cortical response functions were twice as steep as LGN response functions.

Other evidence for non-linear interactions can be found in van Santen and Sperling (1984). Their display contained a row of abutting vertical bars. Each bar's intensity was modulated sinusoidally, and modulations in adjacent bars were 90° out of temporal phase to offer a signal for motion detectors. In one experiment, van Santen and Sperling set the modulation depth of adjacent bars to different values, and found that discrimination performance depended on the product of the two modulation depths. Combinations of modulation depth which yielded the same product produced similar discrimination performance. Morgan and Cleary (1992a) reported similar effects, and related them to Reichardt detector properties.

4.3.2   Polarity sensitivity

A number of experiments have reported effects which indicate that motion detectors are selectively sensitive to the spatial polarity and temporal polarity of intensity change.

4.3.2.1 *Adaptation effects*

Anstis (1967) was the first to report that after adaptation to a uniform field which repetitively grows gradually brighter, a steady uniform field shows a negative after-effect, appearing to grow dimmer (the so-called ramp after-effect). He also reported that spatial luminance gradients appear to move after adaptation to temporal luminance change. This effect was followed up by Anstis in 1990, when he reported MAEs generated by adaptation to luminance change. The apparent direction of the MAE was determined jointly by the temporal polarity of the ramp after-effect, and the spatial polarity of the stationary edge onto which the ramp after-effect was projected, so that the same ramp after-effect could generate MAEs moving in opposite directions when projected onto edges with opposite spatial polarities. Similar adaptation effects were also reported by Moulden and Begg (1986). Anstis concluded that 'neural representations of luminance change must provide an input into the neural pathways that signal motion'. The same conclusion was reached by Mather *et al.* (1991), who used a very different technique to investigate polarity specificity. Subjects adapted to a counterphase grating containing two oppositely-moving sawtooth components, selected so that both generated the same sign of luminance change over time (e.g. repetitive gradual brightening). In a test counterphase containing one component with the same temporal sign as the adapting stimulus and another with the opposite sign, only the direction given by the unadapted sign was visible. Mather *et al.* also found spatial polarity specific adaptation to motion, using an analogous procedure.

4.3.2.2 *Flashed edges*

When a briefly presented static edge is replaced by a uniform field, an impression of motion is evoked, the direction of which depends on the polarity of the edge and on the mean luminance of the uniform field (Mather, 1984; Moulden & Begg, 1986). For example if a vertical edge, dark on the left and bright on the right, is replaced by a bright uniform field, naive observers report a fleeting impression of leftward motion just as the edge disappears. If the uniform field is dark, rightward motion is seen. If an edge of opposite spatial polarity is used, reported directions reverse. The effects can be explained in terms of polarity-specific motion detectors: the conjunction of spatial polarity given by the edge, and temporal polarity given by the transition to the uniform field specifies direction.

These polarity selective motion effects indicate that the inputs to motion detectors are half-wave rectified, with positive and negative portions of spatial and temporal response arriving at different detectors. It is interesting to note that direction-specific simple cortical cells tend to show polarity sensitivity, but complex cells do not (e.g. Goodwin & Henry, 1975; Emerson *et al.*, 1987).

4.4 Opponent processing

van Doorn and Koenderink (1982a, 1982b) used RDKs that were divided into spatial slices or temporal slices. Spatial slicing involved division of the pattern into thin spatial strips, with dots in immediately adjacent strips moving in opposite directions. Temporal slicing involved patterns which repetitively reversed direction at fixed temporal intervals (Figure 9). van Doorn and Koenderink (1982a, 1982b) found that

G. MATHER

Space →

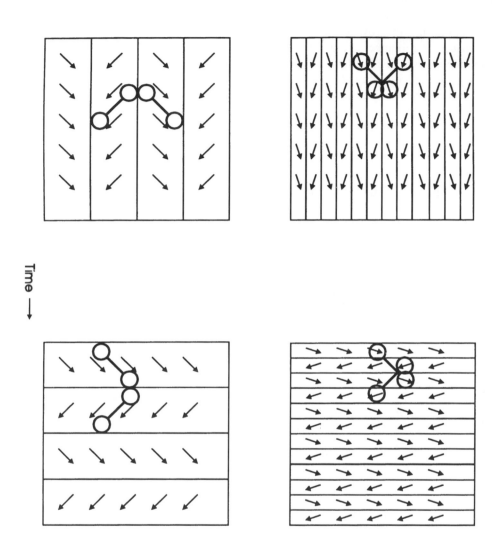

Time →

Figure 9 Space–time plots of sliced random dot kinematograms used by van Doorn and colleagues. Top plots: the pattern is divided into spatial slices (columns) containing dots moving in opposite directions (arrows), with either thick slices (left) or thin slices (right). Bottom plots: the pattern is divided into temporal slices (rows), so that the dot pattern repeatedly reverses in direction over time (arrows), with either thick slices or thin slices. Dipoles represent spatiotemporally elongated receptive fields of motion detectors tuned to rightward and leftward movement. Note that detectors with non-overlapping receptive fields can respond in the case of thick slices, but in the case of thin slices, receptive fields tuned to opposite directions overlap. A system based on opponent connections between overlapping detectors tuned to opposite directions would produce no net directional response to thin slices, yet observers perceive transparent motion.

at very narrow slice widths (both spatial and temporal) subjects reported perceptual transparency, in which the two directions seemed to be overlaid. At wide slice widths the alternating spatial or temporal structure of the pattern became apparent. At intermediate widths motion perception was severely impaired. According to van Doorn and Koenderink, their results rule out gradient-based detectors but can be explained by Reichardt detectors. They reasoned that Reichardt detectors are 'bi-local' (i.e. have two input receptive fields) whereas gradient detectors are 'local' (i.e. have one input receptive field), and 'the mere existence of the transparency phenomenon seems hard to explain on the basis of local rather than bilocal models' (Koenderink et al., 1985). The argument runs as follows: perceptual alternation occurs when the duration (or width) of each slice is greater than the detector's temporal delay constant (or sub-unit separation), so that detectors tuned to opposite directions respond alternately to each slice (Figure 9, left). Transparency takes over when the duration (or spatial width) of each slice is so small relative to the temporal delay constant (or sub-unit separation) that detectors responding to opposite directions are active simulta-neously at the same retinal location (Figure 9, right). They reasoned that a 'local' detector could never mediate transparency because at narrow stripe widths the opposing directions would both excite the same detector, and so cancel each other out. This may be true, but as we have seen in section 2, both Reichardt *and* gradient detectors can be constructed from paired input paths, exhibiting spatiotemporal quadrature, so both can be considered bilocal. Indeed transparency actually rules out the Reichardt detector: at narrow stripe widths the simultaneous signals in detectors tuned to opposite directions would cancel out at the opponent stage. van Doorn and colleagues were the first to reach this conclusion, but preferred to describe their favoured model as 'a unidirectional version of the Reichardt-detector' (van de Grind et al., 1992), i.e. without the opponent stage. However, it seems inappropriate to apply the term 'Reichardt detector' to a system which does not incorporate opponent-processing, since this is perhaps the most characteristic feature of the Reichardt model.

The earliest reported empirical 'demonstration' of opponent-processing in motion perception is the motion after-effect (apparent motion seen in a stationary pattern, opposite to the direction of a moving adapting stimulus (Sutherland, 1961)). The illusion indicates that perceived direction depends on the relative response of detec-tors tuned to different directions, and does not imply any special status for oppositely-tuned detectors (Mather, 1980). In fact other psychophysical data are inconsistent with opponent-processing, or at least do not require it. For example at threshold, detectors tuned to opposite directions are independent (e.g. Watson et al., 1980; Stromeyer et al., 1984). In addition, careful analyses of the responses of cortical direction-specific cells yield no evidence of opponent processing (Emerson et al., 1987, 1992).

4.5   Summary and conclusions

The psychophysical data can be summarized as follows:

1   There is strong evidence that motion detecting receptive fields are spatio-temporally elongated, and are tuned to approximate spatiotemporal quadrature.

2 Detector responses exhibit major non-linearities, namely severe response compression, and polarity sensitivity.
3 There is no evidence to support opponent-process connections between detectors.

What conclusions can we draw about the relative merits of alternative models? Non-linearities rule out a faithful implementation of the linear motion sensor. There is also psychophysical and physiological evidence against the opponency which is crucial to ERDs. Two models are left, the energy detector and the gradient detector. At present, there is no evidence to conclusively rule out a particular characteristic feature of either model, so neither can be rejected. Gradient detectors appear to map on to the family of polarity-sensitive detectors in the middle stage of the energy model (preceding the square-and-sum operation that removes polarity sensitivity). As already discussed, there is psychophysical evidence for polarity sensitivity in motion detectors, and it is tempting to identify these detectors with polarity-sensitive simple cortical cells. Further, Emerson et al. (1987, 1992) have identified the later polarity-independent stage of the energy model with complex cortical cells. Thus, the energy and gradient models may offer different descriptions of parts of the same mechanism. The former emphasizes spatiotemporal energy, and may be useful for understanding motion phenomena in the frequency domain (e.g. sampling effects; the fluted-square-wave illusion. See examples in Adelson and Bergen, 1985). The latter emphasizes spatio-temporal intensity gradients, and is useful for understanding polarity-specific effects (e.g. Moulden & Begg, 1986; Mather, 1987). The choice between the two models presently rests on heuristic value.

ACKNOWLEDGEMENTS

This work was supported by the Medical Research Council and the Science and Engineering Research Council.

REFERENCES

Adelson, E. H. & Bergen, J. R. (1985). Spatiotemporal energy models for the perception of motion. J. Opt. Soc. Am., A2, 284–299.
Anstis, S. M. (1967). Visual adaptation to gradual change of luminance. Science, 155, 710–712.
Anstis, S. M. (1990). Motion aftereffects from a motionless stimulus. Perception, 19, 301–306.
Anstis, S. M. & Mather, G. (1985). Effects of luminance and contrast on direction of ambiguous apparent motion. Perception, 14, 167–179.
Anstis, S. M. & Rogers, B. J. (1975). Illusory reversal of visual depth and movement during changes of contrast. Vision Res., 15, 957–961.
Anstis, S. M. & Rogers, B. J. (1986). Illusory continuous motion from oscillating positive-negative patterns: implications for motion perception. Perception, 15, 627–640.

Baker, C. L. & Braddick, O. J. (1985). Temporal properties of the short-range process in apparent motion. *Perception*, **14**, 181–192.

Baker, C. L., Baydala, A. & Zeitouni, N. (1989). Optimal displacement in apparent motion. *Vision Res.*, **29**, 849–859.

Bischof, W. F. & Di Lollo, V. (1990). Perception of directional sampled motion in relation to displacement and spatial frequency: evidence for a unitary motion system. *Vision Res.*, **30**, 1341–1362.

Boulton, J. C. & Baker, C. L. (1993). Dependence on stimulus onset asynchrony in apparent motion: evidence for two mechanisms. *Vision Res.*, **33**, 2013–2019.

Boulton, J. C. & Hess, R. F. (1990). The optimal displacement for the detection of motion. *Vision Res.*, **30**, 1101–1106.

Braddick, O. J. (1973). The masking of apparent motion in random-dot patterns. *Vision Res.*, **13**, 355–369.

Braddick, O. J. (1974). A short-range process in apparent motion. *Vision Res.*, **14**, 519–527.

Burr, D. C., Ross, J. & Morrone, M. C. (1986). Seeing objects in motion. *Proc. Roy. Soc. Lond.*, **B227**, 249–265.

Cavanagh, P. & Mather, G. (1989). Motion: the long and short of it. *Spatial Vision*, **4**, 103–129.

Chang, J. J. & Julesz, B. (1983). Displacement limits for spatial frequency filtered random-dot cinematograms in apparent motion. *Vision Res.*, **23**, 1379–1385.

Chubb, C. & Sperling, G. (1988). Drift-balanced random stimuli: a general basis for studying non-Fourier motion perception. *J. Opt. Soc. Am.*, **A5**, 1986–2007.

Cleary, R. (1990). Contrast dependence of apparent motion. *Vision Res.*, **30**, 463–478.

Cleary, R. & Braddick, O. J. (1990). Masking of low frequency information in short-range apparent motion. *Vision Res.*, **30**, 317–327.

Derrington, A. M. & Cox, M. J. (1992). Errorless performance in a two-frame apparent motion task using high contrast stimuli, a failure to replicate Cleary (1990). *Vision Res.*, **32**, 2191–2193.

Derrington, A. M. & Goddard, P. A. (1989). Failure of motion discrimination at high contrasts: evidence for saturation. *Vision Res.*, **29**, 1767–1776.

Dosher, B. A., Landy, M. S. & Sperling, G. (1989). Kinetic depth effect from optic flow – I. 3D shape Fourier motion. *Vision Res.*, **32**, 1789–1813.

Eagle, R. A. & Rogers, B. J. (1991). Maximum displacement (Dmax) as a function of density, patch size, and spatial filtering in random-dot kinematograms. *Invest. Ophthalmol. Vis. Sci. (Suppl.)*, **32**, 893.

Emerson, R. C., Citron, M. C., Vaughn, W. J. & Klein, S. A. (1987). Nonlinear directionally selective subunits in complex cells of cat striate cortex. *J. Neurophysiol.*, **58**, 33–65.

Emerson, R. C., Bergen, J. R. & Adelson, E. H. (1992). Directionally selective complex cells and the computation of motion energy in cat visual cortex. *Vision Res.*, **32**, 203–218.

Enroth-Cugell, C. & Robson, J. (1966). The contrast sensitivity of retinal ganglion cells in the cat. *J. Physiol.*, **187**, 512–552.

Fahle, M. & Poggio, T. (1981). Visual hyperacuity: spatiotemporal interpolation in human vision. *Proc. R. Soc. Lond.*, **B 213**, 415–477.

Freeman, W. T., Adelson, E. H. & Heeger, D. J. (1991). Motion without movement. *Computer Graphics,* **25**, No. 4, 27–30.

Goodwin, A. W. & Henry, G. H. (1975). Direction selectivity of complex cells in a comparison with simple cells. *Neurophysiology,* **38**, 1524–1540.

Gregory, R. L. & Heard, P. (1982). Visual dissociations of movement, position, and stereo depth: some phenomenal phenomena. *Q. J. Exp. Psychol.,* **35**, 217–237.

Harris, M. G. (1986). The perception of moving stimuli: a model of spatiotemporal coding in human vision. *Vision Res.,* **26**, 1281–1287.

Hawken, M. J. & Parker, A. J. (1987). Spatial properties of neurons in the monkey striate cortex. *Proc. Roy. Soc. Lond.,* **B231**, 251–288.

Johnston, A. & Wright, M. J. (1985). Lower thresholds of motion for gratings as a function of eccentricity and contrast. *Vision Res.,* **25**, 179–185.

Keck, M. J., Palella, T. D. & Pantle, A. (1976). Motion aftereffect as a function of the contrast of sinusoidal gratings. *Vision Res.,* **16**, 187–191.

Lappin, J. S. & Bell, H. H. (1976). The detection of coherence in moving random dot patterns. *Vision Res.,* **16**, 161–168.

McLean, J. & Palmer, L. A. (1989). Contribution of linear spatiotemporal receptive field structure to velocity selectivity of simple cells in area 17 of cat. *Vision Res.,* **29**, 675–679.

Marcelja, S. (1980). Mathematical description of the responses of simple cortical cells. *J. Opt. Soc. Am.,* **70**, 1297–1300.

Marr, D. & Ullman, S. (1981). Directional selectivity and its use in early visual processing. *Proc. R. Soc. Lond.,* **B 211**, 151–180.

Mastebroek, H. A. K. & Zaagman, W. H. (1988). Apparent movements induced by luminance modulations: a model study. *Perception,* **17**, 667–680.

Mather, G. (1980). The movement after-effect and a distribution shift model of direction coding. *Perception,* **9**, 379–392.

Mather, G. (1984). Luminance change generates apparent movement: implications for models of directional specificity in the human visual system. *Vision Res.,* **24**, 1399–1405.

Mather, G. (1987). The dependence of edge displacement thresholds on edge blur, contrast, and displacement distance. *Vision Res.,* **27**, 1631–1637.

Mather, G. (1990). Computational modelling of motion detectors: responses to two-frame displays. *Spatial Vision,* **5**, 1–14.

Mather, G. & Tunley, H. (1993). Temporal filtering enhances motion detection. *Perception,* **22**, S31.

Mather, G. & West, S. (1993). Evidence for second-order motion detectors. *Vision Res.,* **33**, 1109–1112.

Mather, G., Moulden, B. & O'Halloran, A. (1991). Polarity specific adaptation to motion in the human visual system. *Vision Res.,* **31**, 1013–1019.

Morgan, M. J. (1992). Spatial filtering precedes motion detection. *Nature,* **355**, 344–346.

Morgan, M. J. & Cleary, R. (1992a). Effects of contrast substitutions upon motion detection in spatially random patterns. *Vision Res.,* **32**, 639–643.

Morgan, M. J. & Cleary, R. (1992b). Ambiguous motion in a two-frame sequence. *Vision Res.,* **32**, 2195–2198.

Morgan, M. J. & Mather, G. (1994). Motion discrimination in two-frame sequences with differing spatial frequency content. *Vision Res.*, **34**, 197–208.

Moulden, B. & Begg, H. (1986). Some tests of the Marr–Ullman model of movement detection. *Perception*, **15**, 139–155.

Nakayama, K. & Silverman, G. H. (1985). Detection and discrimination of sinusoidal grating displacements. *J. Opt. Soc. Am.*, **A2**, 267–274.

Nishida, S. & Sato, T. (1992). Positive motion after-effect induced by bandpass-filtered random-dot kinetograms. *Vision Res.*, **32**, 1635–1646.

Pollen, D. A. & Ronner, S. F. (1981). Phase relationship between adjacent simple cells in the visual cortex. *Science*, **212**, 1409–1411.

Sato, T. (1989). Reversed apparent motion with random patterns. *Vision Res.*, **12**, 1749–1758.

Sclar, G., Maunsell, J. H. R. & Lennie, P. (1990). Coding of image contrast in central visual pathways of the macaque monkey. *Vision Res.*, **30**, 1–10.

Sekuler, R., Anstis, S. M., Braddick, O. J., Brandt, T., Movshon, J. A. & Orban, G. (1990). The perception of motion. In L. Spillman & J. S. Werner (Eds.) *Visual Perception: The Neurophysiological Foundations*. Academic Press, San Diego.

Stromeyer, C. F., Kronauer, R. E., Madsen, J. C. & Klein, S. A. (1984). Opponent-movement mechanisms in human vision. *J. Opt. Soc. Am.*, **A1**, 876–884.

Sutherland, N. S. (1961). Figural after effects and apparent size. *Quart. J. Exp. Psychol.*, **13**, 222–228.

van Doorn, A. J. & Koenderink, J. J. (1982a). Temporal properties of the visual detectability of moving spatial white noise. *Exp. Brain Res.*, **45**, 179–188.

van Doorn, A. J. & Koenderink, J. J. (1982b). Spatial properties of the visual detectability of moving spatial white noise. *Exp. Brain Res.*, **45**, 189–195.

van de Grind, W. A., Koenderink, J. J. & Doorn, A. J. van (1992). Viewing distance invariance of movement detection. *Exp. Brain Res.*, **91**, 135–150.

van Santen, J. P. & Sperling, H. G. (1984). Temporal covariance model of human motion perception. *J. Opt. Soc. Am.*, **A1**, 451–473.

van Santen, J. P. & Sperling, H. G. (1985). Elaborated Reichardt detectors. *J. Opt. Soc. Am.*, **A2**, 300–320.

Watson, A. B. (1990). Optimal displacement in apparent motion and quadrature models of motion sensing. *Vision Res.*, **30**, 1389–1393.

Watson, A. B. & Ahumada, A. J. (1985). Model of human visual-motion sensing. *J. Opt. Soc. Am.*, **A2**, 322–342.

Watson, A. B., Thompson, P. G., Murphy, B. J. & Nachmias, J. (1980). Summation and discrimination of gratings moving in opposite directions. *Vision Res.*, **20**, 341–347.

Watson, A. B., Ahumada, A. J. & Farrell, J. E. (1986). Window of visibility: a psychophysical theory of fidelity in time-sampled visual motion displays. *J. Opt. Soc. Am.*, **A3**, 300–307.

6

The Detection of Second-order Motion

Andrew T. Smith

Royal Holloway, University of London, UK

1 INTRODUCTION

Typically, objects are either lighter or darker than their backgrounds. It therefore follows that when an object moves relative to its background, the image cast on an observer's retina contains moving luminance discontinuities: at the edge of the object is a luminance step and this step changes its position over time. The way in which such motion is detected is the subject of the preceding chapters. The discussion contained therein is in most cases based on the assumption, whether explicit or implicit, that the motion to be detected is defined in such terms. However, it is also possible for image motion to be defined by the change in position of objects or sub-regions which differ from their backgrounds in other ways, such as colour, texture, contrast or flicker rate.

1.1 Definitions

Consider a point in an image. There are only two dimensions in which such a point source can vary: intensity and wavelength. A description of the spatiotemporal variations in intensity and wavelength contained in an image specifies the first-order characteristics of the image. All other properties of the image are derived from these two primary properties. For example, one familiar derived property is contrast. Contrast gives a measure of the variability of luminance across space and has no meaning in relation to a point source; a contrast value is derived from some set of luminance values. If local contrast values are defined for all regions of an image and the image is then described in terms of spatiotemporal variations in local contrast, that description specifies one of many possible second-order characteristics of the image. This class of second-order characteristic will become more familiar later in the chapter. Another derived property is orientation. Orientation specifies the spatial configuration of a set of luminance values. The way in which local orientation

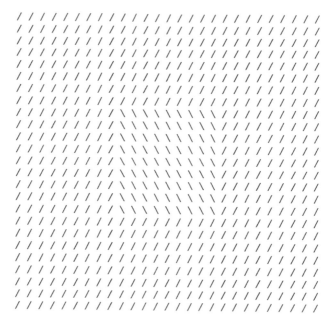

Figure 1   A second-order square defined by orientation.

changes across space (or time) is a second-order characteristic. Figure 1 shows an image in which orientation varies across space and this variation gives rise to second-order spatial structure (a square). Because the elements are equally dense in all areas, the mean (locally-averaged) luminance is the same for all parts of the image and so there is no first-order square in the image.

   Thus, a second-order characteristic of an image is one which describes spatial and/ or temporal variations in the value of some parameter which is derived from either luminance or wavelength (usually the former). In principle it is also possible to define third- and subsequent-order characteristics of an image. Since the distinction between second-order and higher-order characteristics may be of little importance for the purposes of this chapter, the term 'second-order' will be used loosely to mean 'second- and higher-order' or 'not first-order'. When spatial structure arises from second-order characteristics the structure is referred to as second-order structure (or form), when motion arises from second-order characteristics the motion is referred to as second-order motion. The term 'second-order motion' was coined by Cavanagh and Mather (1989).

1.2   Some examples of second-order motion

The most comprehensive published descriptions of second-order motion patterns are those of Chubb and Sperling (1988, 1989). These authors have described a variety of such patterns, some of which will briefly be described here for illustrative purposes. A distinction may be drawn between images which contain both first-order and second-order motion cues and those which contain only second-order cues. Images of the latter category are perhaps the most useful, from an experimental point of view, because they may be expected to isolate detection mechanisms that are sensitive to second-order motion, if such exist. Chubb and Sperling introduced the term 'drift-balanced' to describe an animated image sequence in which there is no *net* directional motion energy (moving luminance-domain Fourier energy; see Chapter 2 by Grzywacz *et al.*), i.e. such first-order motion energy that exists is equal in opposite directions. Thus a second-order motion sequence may contain much moving first-order motion energy but any motion perceived is purely second-order provided the image sequence is drift-balanced. The significance of drift-balance is that motion, although often visible to an observer, is invisible to a linear, energy-based motion-detection system, which requires net directional motion energy in order to respond. Second-order motion is often also called 'non-Fourier motion' for this reason (see, for example, Chapter 8 by Wilson).

One type of drift-balanced stimulus involves contrast-modulated noise. If an image consisting simply of two-dimensional static noise is multiplied by a one-dimensional (horizontal) sinewave, the result is a noise pattern whose contrast is modulated sinusoidally in the horizontal dimension (see Figure 2c). It has the appearance of a vertical grating (Figure 3, top left) but the grating is defined by variations in contrast, not luminance. If the horizontal position of the multiplying sinewave is now incremented on each frame of an animation sequence, while the noise itself remains stationary, motion of the 'grating' is clearly perceived. The way in which the image changes over time is illustrated as a space–time plot in Figure 3 (top right). Chubb and Sperling (1988) have shown formally that this image is drift-balanced. The pattern is also 'micro-balanced', i.e. it remains drift-balanced after arbitrary spatial and temporal filtering (such as that imposed by early processing in the visual system).

Another example of second-order motion involves temporal frequency-modulated noise. Take a pattern of two-dimensional noise like that in Figure 3 (lower left). Imagine that each noise pixel reverses its polarity (changes from black to white or vice versa) at some temporal frequency which is different for different pixels. If the temporal frequency of reversal varies sinusoidally in the horizontal dimension, the image again has the appearance of a vertical grating, but the grating is defined by spatial variation in flicker rate, not contrast. If the horizontal position of the sinewave used to modulate flicker rate is now incremented on each frame of an animation sequence while, again, the noise itself remains stationary, motion of the grating is perceived. This image too has been shown by Chubb and Sperling (1988) to be drift-balanced. One frame of the sequence is illustrated in Figure 3 (lower left): it appears simply as noise. Its change in appearance over time is shown in Figure 3 (lower right; shown as a squarewave modulation for clarity).

A third example of second-order motion, also based on modulated noise and containing elements of both the above examples, is as follows. In the two examples above, either contrast or flicker rate varies across space to give spatial structure which can then be made to move. If, instead, all pixels flicker at the same rate, but the depth of flicker (extent of change in luminance) is modulated sinusoidally in one dimension, again a grating is seen. If the modulating waveform moves while the noise pixels remain stationary, second-order motion results. Again, the image is drift-balanced.

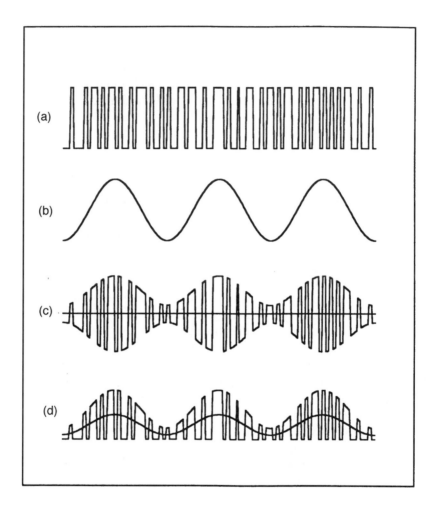

Figure 2 Luminance profiles showing the method of construction of contrast-modulated noise and the effect of rectification upon it. Each trace represents a horizontal slice through an image and shows luminance as a function of spatial position. (a) Visual noise formed by assigning each pixel to be either light or dark at random. (b) A sinusoid. (c) Contrast-modulated noise formed by taking the product of waveforms (a) and (b) and re-scaling. Also shown (straight line) is the mean luminance of the image locally averaged over a patch of pixels: contrast waxes and wanes symmetrically about a constant mean luminance value. (d) The result of half-wave rectifying (and re-scaling) the image represented in (c), together with the mean luminance profile of the transformed image. The mean luminance now varies in the horizontal dimension. The waveform describing this variation has the same form (though not amplitude) as that in (b), i.e. the representation of the image now contains Fourier energy at the spatial frequency of the modulating sinusoid.

Figure 3 Sample frames of two second-order motion sequences (left) and the space–time plots describing the associated motion (right). *Top left*: the contrast of a sample of two-dimensional noise is modulated sinusoidally in the horizontal dimension. *Top right*: a space–time plot in which the horizontal axis represents the horizontal dimension in space and the vertical axis represents time. Thus each row of pixels represents a horizontal slice through an image like that shown top left at one moment in time, and consists of noise whose contrast is modulated sinusoidally. Succeeding rows represent the same slice at different times. As time progresses (top to bottom on the plot) the contrast modulation drifts leftwards (reflecting leftward motion) while the noise sample itself is static. *Bottom left*: one frame of a second-order motion sequence in which the frequency of polarity-reversal of noise pixels is modulated by a drifting squarewave. The image appears simply as binary 2D noise. *Bottom right*: a space–time plot describing variations in the pattern shown bottom left over time. The flicker rate of any pixel can be seen by following it down its column (representing time). This flicker rate varies in the horizontal dimension: for simplicity the variation is shown as a squarewave alternation between 0 Hz (no reversal) and polarity reversal on every image update. The waveform drifts leftwards over time, giving leftward second-order motion.

1.3   Theoretical significance of second-order motion

Second-order motion of many different types is readily perceived (e.g. Julesz, 1971; Ramachandran *et al*., 1973; Petersik *et al*., 1978; Anstis, 1980). Since the incidence of pure second-order motion (i.e. unaccompanied by first-order motion) in natural images is probably rather low, it is pertinent to ask why the visual system is sensitive to it. Two possible answers spring to mind. Firstly, second-order motion might be of greater biological significance than is apparent on casual reflection and consequently a mechanism might have evolved specifically to detect it. Secondly, it might be that our motion detection systems are designed to detect first-order motion but, by the nature of their design, they detect second-order motion incidentally. If the former is true then sensitivity to second-order motion becomes an important topic in its own right. If the latter, then the study of sensitivity to second-order motion is likely to shed light on standard motion mechanisms and perhaps provide critical tests of computational models of motion detection. Either way, therefore, it is a topic that merits attention.

The question of how many motion detection mechanisms we have and to what they respond has been debated in a different context for some time. A distinction has been drawn (see Chapter 4 by McKee and Watamaniuk and Chapter 5 by Mather) between motion detection based on tracking the positions of features and that based on a low-level mechanism operating directly on image intensity. This distinction was first made explicit by Braddick (1974, 1980) who proposed the existence of two motion-detection systems, one based on each strategy, which he referred to as 'long-range' and 'short-range' mechanisms. This proposal may now be seen as a key development in the subject and met with widespread, though not universal, support (see Cavanagh and Mather, 1989; Petersik, 1991 for commentaries). Detailed computational models of both feature-based and intensity-based motion-detection have been developed. The most explicit feature-tracking model is that of Ullman (1979), who developed a 'minimal mapping theory' which provides algorithms for computing the likely correspondences between low-level 'tokens' such as edges and corners that are detected at different times. Several models in which motion detection is based directly on image intensity without reference to features have been published (see Chapter 2 by Grzywacz *et al*.). Perhaps the most influential is that of Adelson and Bergen (1985) who developed the concept of 'motion energy' models, in which motion is detected by spatiotemporal filters that are oriented in x–t (space–time) space, without the need for identifying features and the temporal correspondences between them. Gradient models (Marr & Ullman, 1981), in which spatial and temporal intensity gradients at a single location are computed and used to detect motion, provide another variant on the intensity-based motion computation principle.

Since second-order motion stimuli can be regarded as containing features that move, a simple and parsimonious explanation of the visibility of second-order motion patterns is that they are detected by a correspondence-based system. Cavanagh (1991, 1992) has recently developed a modification of the long-range/short-range distinction. He distinguishes between active and passive motion-detection processes. In this scheme the distinction between feature-based and energy-based detection is preserved; the passive process describes low-level, intensity-based operations while the active process operates by attentive tracking of features or objects. The high level, active process can operate on features however they are defined, so that the distinction between first- and second-order is of little importance for this mechanism. All that matters is that the image contains visible features. This view provides the basis for a version of the possibility outlined earlier that second-order motion is detected incidentally and is of no great significance in its own right.

However, Chubb and Sperling (1988, 1989) have shown theoretically that in many cases a simple non-linear transformation (such as rectification) of the luminance profile of the image, in combination with spatial and temporal filtering, is sufficient to convert second-order motion to first-order motion, which may then be analysed by means of standard energy-based motion detection, obviating the need to encode global features explicitly. An intuitive feel for how rectification achieves this conversion may be gained from Figure 2d. Rectification is physiologically plausible: the responses of 'on-centre' and 'off-centre' pathways approximate half-wave rectification and their sum could in principle be used to provide full-wave rectification. If this is how second-order motion is detected, then one of three (mutually exclusive) conclusions must be drawn. Firstly, the 'long-range' system might not be a high-level correspondence-based process at all, but might reflect simple (though gross) non-linear transformations of the image followed by energy-based motion detection similar to that used for short-range motion. This is the view that was taken by Chubb and Sperling (1988). Secondly, the short-range or low-level system might not be as Adelson and Bergen conceptualized it. Instead of simply detecting motion energy in a linear representation of the image it might precede energy detection with some non-linear operation which allows it to detect both first-and second-order motion. Thirdly, both linear energy and feature-based mechanisms might exist as originally conceived but in addition a third mechanism, involving non-linear transformation followed by energy detection, might exist. Each of these three conclusions has important implications and so it is important to know which is correct. A view on this question will be expressed later in the chapter (section 2.6).

1.4   Models of second-order motion detection

Most of the prevailing models of motion detection are models of first-order motion and simply do not address the issue of second-order motion. The work of Chubb and Sperling (1988) provides an outline of a model which deals with both types of motion, but no more. Their notion of two parallel energy detection systems, one involving a luminance non-linearity, has recently been developed further by Wilson et al. (1992). This model is more comprehensive and has considerable predictive power concerning the visibility of both first- and second-order motion (see Chapter 8 by Wilson). In this model, two parallel, low-level, intensity-based motion-detection pathways exist. In one, the detection of motion energy is preceded only by bandpass spatial frequency filtering. In the other, the image is also filtered and the output of the filter is then squared before being filtered again at a different scale (a lower spatial frequency). The effect of this is that low spatial frequency modulations in a high-frequency carrier (such as those in the image in Figure 3 (top left)) are passed by the second channel while luminance modulations at the same spatial frequency are not. A key feature of the model is that the outputs of the two motion-detection systems are then pooled to give a single motion signal. An outline of the model is shown in Chapter 8, Figure 7. This model is currently the most comprehensive model of motion detection which accounts for both first- and second-order motion.

Zanker (1993) has pointed out that the Reichardt correlation principle of motion detection (Reichardt, 1961) may be applied to signals other than raw luminance (as in the original Reichardt detector) or filtered luminance signals (as in the motion-energy model) and has developed a model of the detection of one specific type of second-order motion. His model involves two layers of Reichardt detectors. The first is conventional and gives a first-order motion signal. In the second, the outputs of two spatially separated first-layer detectors are compared, with a delay. The result is that detectors in the second layer are sensitive to discontinuities in motion, and to the motion of these discontinuities. Thus, sensitivity to moving boundaries defined by differences in image motion of the elements on the two sides of the boundary (a second-order characteristic) is achieved using two serial stages. This is not a general model of second-order motion. It could, in principle, be extended to other types of second-order motion, but this would involve a proliferation of classes of Reichardt detector, one for each class of second-order input.

An entirely different approach to the need to model the detection of both first- and second-order motion has been taken by Johnston et al. (1992). They have argued that it is unnecessary, from a theoretical point of view, to invoke two separate mechanisms for the two types of motion and they have developed a single model which deals effectively with both. The model is based on the gradient principle (Marr & Ullman, 1981; see Chapter 5 by Mather) in which at a given point in space the temporal luminance gradient is divided by the spatial luminance gradient to give an estimate of local motion speed and direction. The essence of the model of Johnston et al. is that the sum of a series of temporal derivatives is divided by the sum of a series of spatial derivatives (giving sensitivity to a set of spatial scales) to give the local motion estimate. Johnston et al. have shown that the model gives appropriate responses to a variety of moving images, both first- and second-order. Grzywacz (1992) has also developed a single model of motion detection which deals with both types of motion. These models have the benefit of parsimony, but have not yet been tested extensively against empirical data.

In summary, only a few models of the detection of second-order motion have been published and there is no consensus as to whether it is appropriate to model the detection of first- and second-order motion separately or together. This is an empirical question and much of the remainder of the chapter will be devoted to empirical studies bearing on that question.

2   EMPIRICAL STUDIES OF SENSITIVITY TO SECOND-ORDER MOTION

In this section, the psychophysical and physiological literature concerning the sensitivity of the visual system to second-order motion will be reviewed.

2.1   Detection sensitivity

Several investigators have measured the sensitivity of the human visual system to second-order motion stimuli. Smith *et al.* (1994) measured contrast sensitivity for simple detection, orientation detection and direction detection. The image used was the contrast-modulated two-dimensional (2D) noise pattern illustrated in Figures 2c and 3 (top left). The objective was to see whether direction of motion can be detected at detection threshold, as it can for first-order gratings under most conditions, though not at very low temporal or high spatial frequencies (Watson *et al.*, 1980; Green, 1983). The results (Figure 4) showed that thresholds for identifying the orientation and direction of the grating are identical under most conditions: if the spatial structure of the second-order pattern is visible then its direction of motion is detectable. The similarity in the detection of first- and second-order motion patterns could be interpreted as reflecting detection by a common mechanism, but it might also reflect independent detection by mechanisms using a common detection principle. Other authors have also noted similarities in the detection of first- and second-order motion patterns. Nishida (1993) has recently demonstrated a striking example. He developed a second-order checkerboard pattern in which the checks were either grey or made up of

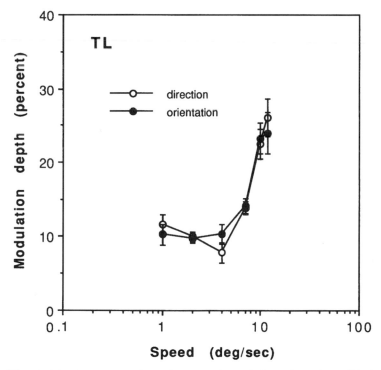

Figure 4   Thresholds, for one subject, for detection of the orientation (filled circles) and direction (hollow circles) of a sinusoidally (1 c/deg) contrast-modulated two-dimensional static noise field, as a function of the drift speed of the modulating waveform. Reproduced from Smith *et al.* (1994).

(A)

RDK (Random Dot Kinematogram)

(B)

RWK (Random Window Kinematogram)

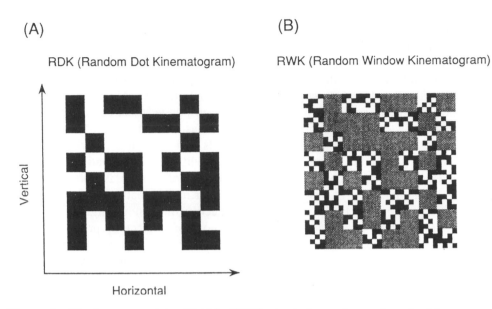

Vertical

Horizontal

Figure 5 The images used by Nishida (1993). A. A first-order random dot kinematogram comprising black and white checks. B. A second-order random dot pattern in which the 'dots' consist of either patches of small black and white checks or uniform grey checks of the same mean luminance. Reproduced with permission from Nishida (1993).

a number of smaller, black and white checks to give the same mean luminance as the grey checks (Figure 5). Such patterns were then used to produce random dot kinematograms (RDKs). When the check type was reversed between updates, motion was seen in the reverse direction, just as for a luminance RDK in which luminance polarity is reversed (Anstis, 1970). In the evoked potential realm, Victor and Conte (1992b) have reported that first- and second-order motion elicit indistinguishable visual evoked potentials.

A number of relevant experiments have been conducted using either contrast-modulated (amplitude-modulated or AM) gratings or beats. An AM grating is similar in some ways to contrast-modulated noise (section 1.2). An example of an AM grating is that introduced by Henning *et al.* (1975), which comprises the sum of three sine gratings of the same orientation (say vertical) but different spatial frequencies (say 9, 10 and 11 cycles per degree (c/deg)). This has the appearance of, and is mathematically equivalent to, a 10 c/deg grating whose contrast is modulated sinusoidally in the horizontal dimension with a periodicity of 1 c/deg, the 'difference frequency'. Indeed, the image can be generated by multiplying a 10 c/deg carrier by a 1 c/deg envelope as well as by addition of three components; the two methods are equivalent. To produce drift of the modulation but not the carrier, the envelope is drifted (using the multiplication method) while the carrier remains static, just as for contrast-modulated noise. Using the addition method, the 9 c/deg and 11 c/deg components are made to drift in opposite directions with equal temporal frequencies while the 10 c/deg component remains static. Again the two methods are equivalent. Thus, the image is approximately drift balanced because one component moves in each direction. Strictly speaking it is not drift balanced because analysis at 9 c/deg will reveal first-order motion energy with a clear direction, as will analysis at 11 c/deg (though in the opposite direction). But since the bandwidth of spatial filtering in the visual system is quite broad, there will be heavy overlap in the mechanisms detecting the two components and so within one filter pass-band (centred, say, at 10 c/deg) the image can be regarded as drift balanced. A beat pattern is the same as an AM grating but without the centre frequency and has a similar appearance. Moving beats are also approximately, though not strictly, drift balanced.

AM gratings and beats have been used mainly because of the ease with which they can be generated and have few obvious advantages over contrast modulation of spatially broadband carriers. Nonetheless, they have generated some interesting results. Badcock and Derrington (1985) noted that the detection of motion of a 30 c/deg grating is made much easier by the addition of a static 28 c/deg grating (which results in a moving beat) suggesting that a mechanism other than first-order motion energy detection is used. Derrington and Badcock (1985) reported that temporal modulation does not improve detectability of a 1 c/deg beat, as it does a 1 c/deg grating, suggesting detection by mechanisms with different temporal frequency sensitivities. They also measured the highest and lowest drift speeds at which direction could be identified correctly and found the range to be considerably more limited for beats than for gratings of the same spatial frequency, particularly at low contrasts. This result is consistent with the recent suggestion of the same authors (Derrington et al., 1993) that beat motion is detected by a mechanism which has poorer temporal acuity than that which detects first-order motion. The evidence on which this claim is based is that the direction of motion of a beat cannot be detected at durations below 200 ms (Figure 6). At very brief durations most of the energy is at high temporal frequencies, which would be invisible to a mechanism sensitive only to low frequencies. Cropper (1994) has recently shown that speed discrimination is affected differently by contrast for gratings and beats. Reducing contrast (to 0.5 log units above threshold) results in a marked impairment in performance for beats, not seen with gratings.

In summary, comparisons of the detectability of first- and second-order motion yield some close similarities (suggesting the use of common processing principles) but also some clear differences (suggesting the involvement of different detection mechanisms).

2.2 Peripheral sensitivity to second-order motion

Several researchers have noted anecdotally that second-order motion is, at least in some cases, invisible in the periphery, raising the possibility that it may be a feature of central vision only. But rather few studies have considered this question in detail. Solomon and Sperling (1991) have reported that sensitivity to second-order motion (in terms of the highest detectable spatial frequency) declines with eccentricity rather more rapidly than for first-order motion, although they find that motion is, in fact, visible in the periphery (12 deg) for low spatial frequencies. Arimura and Nishida (1992) reported that direction discrimination performance for 'random-window

A. T. SMITH

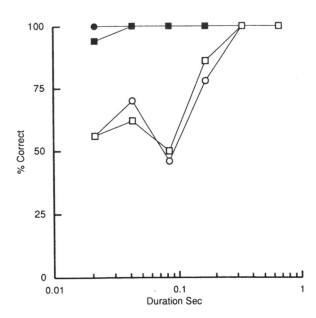

Figure 6 Performance of two subjects (circles and squares) in a direction identification task for moving gratings (solid symbols) and beats (hollow symbols) as a function of duration. Direction can be detected at all durations for gratings but only above about 120 ms for beats. Reproduced with permission from Derrington *et al.* (1993).

kinematograms' (like those used by Nishida (1993); see section 2.1) declines faster with eccentricity than does performance with first-order random dot kinematograms, and suggest that this may reflect greater attentional demands in the case of second-order motion.

Pantle (1992) reported that a contrast-modulated grating in which the modulation envelope drifts while the carrier remains static appears stationary when presented peripherally. In his experiments the spatial structure of the contrast modulation was detectable but its motion was not. The same result was found for flicker-defined second-order gratings and for several other types of second-order motion stimulus. Pantle suggested that second-order motion detection may be mediated by a mechanism that serves central vision only. However, his experiments were all conducted at a single eccentricity (8 deg) and using a single, low drift rate (1 deg/s). In a more comprehensive study using contrast-modulated noise, Smith *et al.* (1994) measured direction-sensitivity for each of a range of eccentricities, using a drift speed of 4 deg/s. They found that second-order motion is detectable at least up to 20 deg eccentricity and that the rate at which direction sensitivity declines with eccentricity is extremely similar to that found with first-order gratings (Figure 7a). They reconciled their results with those of Pantle as follows. At very low speeds, direction detection performance diverges rapidly from orientation performance (as is the case for first-order gratings). Sensitivity to second-order motion is relatively poor at the best of times (100% modulation depth is typically only about 10 times detection threshold) and so, as eccentricity is increased, direction thresholds soon reach 100% modulation depth and the direction cannot be identified, even though the orientation may still be visible. Thus Pantle's result reflects the fact that his stimulus was below direction-detection threshold because of its low speed and his conclusion does not apply to higher speeds.

In summary, second-order motion is less visible in the periphery than first-order motion, but this appears to be because sensitivity is lower in the first place and so declines to threshold at a more modest eccentricity, not because sensitivity declines at a faster rate with eccentricity. This situation is illustrated graphically in Figure 7b.

2.3 Adaptation to second-order motion

Several studies have shown that adaptation to second-order motion does not give rise to a motion aftereffect. Adaptation to stereo-defined motion (Anstis, 1980), to a horizontal travelling wave comprising the ends of a set of vertical lines (Anstis, 1980) and to 'long-range' motion (Anstis & Moulden, 1970) have all been reported to produce little or no aftereffect. Derrington and Badcock (1985) adapted subjects to drifting beats and found no effect on the temporal frequency thresholds for detecting motion in the same and opposite directions. Mather (1991) devised a second-order stimulus consisting of a bar that steps continuously in one direction but whose polarity (black or white) is determined randomly on each trial. This means that the polarity changes on 50% of updates, giving reversed first-order motion, but not in the remaining updates, giving forward motion. Thus, when the stimulus is integrated over time, motion energy is equal in both directions and the motion perceived is second-order. Adapting to this stimulus produced some motion aftereffect, but much less than adapting to first-order motion (a drifting bar of fixed polarity). The authors of all these studies interpreted their results as evidence for the existence of separate mechanisms for the detection of first-order and second-order motion, the second-order motion mechanism being less susceptible to adaptation.

However, studies of adaptation to second-order motion using paradigms other than the conventional motion aftereffect have yielded a rather different result. Turano (1991) used the technique of threshold elevation. She adapted subjects to AM gratings in which the modulation drifted and the carrier remained static and found substantial direction-specific threshold elevation. Moreover, direction-specific threshold elevation for the detection of sine gratings was obtained following adaptation to AM gratings, and vice versa. This transfer between image types led Turano to postulate that they are detected by a common mechanism.

von Grunau (1986) succeeded in obtaining a motion aftereffect following adaptation to a 'long-range' stimulus (a grating patch which moved in discrete 4 deg steps with a 50 ms blank interval between frames) by using a counterphasing test grating in place of the conventional static test pattern. McCarthy (1993) used a similar test procedure to study motion aftereffects with smoothly-drifting second-order adaptation patterns. He adapted subjects to a drifting contrast-modulation of a grating and tested for a motion aftereffect using a grating whose contrast was modulated with a counterphase waveform (the sum of two oppositely-drifting sinusoids). He found that the aftereffect measured in this way was as strong for adaptation to drifting contrast modulations as for drifting luminance modulations (sine gratings). Ledgeway (1994) independently obtained the same result using a 2D noise carrier and, in addition, demonstrated that

a

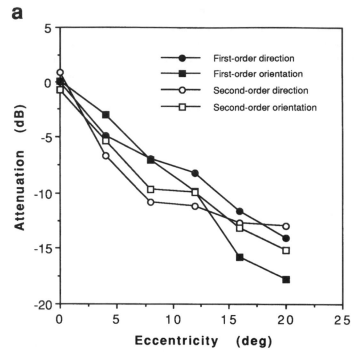

b

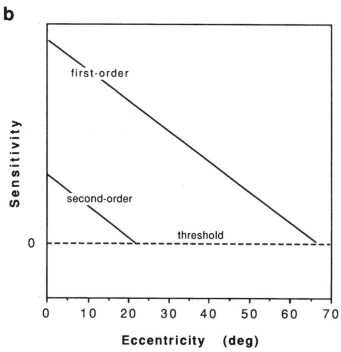

cross-over adaptation effects occur (adaptation to second-order motion gives an aftereffect in a first-order test pattern and vice versa). He used stimuli that were equated for visibility across the two types by presenting them at the same multiple of detection threshold. The magnitude of the effect, measured by varying the relative amplitudes of the two sinusoids so as to null it, was as large in cross-over conditions as in those where the same stimulus type was used for adapting and testing. Ledgeway and Smith (1994b) have also reported motion aftereffects with second-order adaptation, but using the more conventional approach of measuring aftereffect duration. Again, strong aftereffects were observed using a dynamic test pattern (a 2D noise field contrast-modulated by a counterphase grating), even though little or no aftereffect was visible in a static test pattern following identical adaptation. Again, cross-over effects were found that were as large as those resulting from adapting and testing with the same stimulus type (Figure 8).

The important difference between static and dynamic test patterns may be that static patterns contain positional cues which may, at some level, signal 'no motion'. Aftereffects will then be visible only if sufficiently strong to override these cues. Dynamic patterns do not contain such cues and so provide a more sensitive test pattern which will yield visible aftereffects for lower adaptation strengths. It must be remembered that sensitivity to contrast-modulation is low compared with sensitivity to luminance modulation. In fact Ledgeway and Smith (unpublished observations) found that adaptation to conventional sine gratings, equated for visibility (approx. 3% contrast) with their second-order patterns, also elicit aftereffects in dynamic but not static test patterns. Thus, when effective contrast or visibility is equated, results for motion aftereffects (like those for threshold elevation) are very similar for first-order and second-order adaptation. However, this does not necessarily imply a common detection mechanism. Indeed, Nishida and Sato (1993) have presented evidence that it is possible simultaneously to produce aftereffects in opposite directions by adapting to a pattern which has first-order components in one direction and second-order components in the other. When a static test pattern was used, an aftereffect was seen in the direction opposite to the first-order adaptation direction. But the aftereffect seen in a dynamic test pattern was often in the direction opposite to the second-order adaptation direction, suggesting adaptation in a separate motion mechanism. The cross-over of aftereffects between the two image types found by Turano (1991), Ledgeway (1994) and Ledgeway and Smith (1994b) could reflect a pooling of motion signals following separate detection, rather than common detection.

Figure 7 (a) Threshold sensitivity for the detection of the orientation (squares) and direction (circles) of a drifting sine grating (filled symbols) and a sinusoidally contrast-modulated two-dimensional static noise carrier (hollow symbols), as a function of viewing eccentricity, for one observer. Sensitivity at each eccentricity is shown relative to foveal sensitivity. Redrawn from the data of Smith *et al.* (1994). (b) Diagram illustrating the effects of eccentricity on the visibility of first- and second-order motion stimuli. For first-order stimuli, sensitivity declines with eccentricity, reaching zero at some large eccentricity. For second-order motion, visibility declines with eccentricity at the same rate as for first-order motion, but nonetheless reaches zero at a much more modest eccentricity because of lower foveal sensitivity.

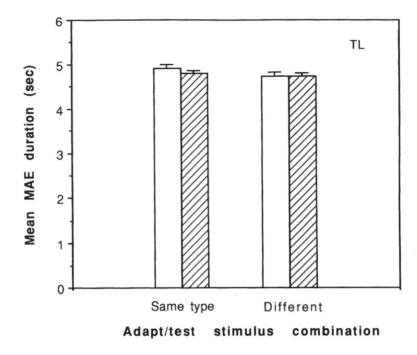

Figure 8 Motion aftereffect (MAE) duration following adaptation to either first-order (white bars) or second-order (hatched bars) motion tested with directionally-ambiguous first-order or second-order test patterns. The test patterns were either of the same type (both first-order or both second-order) or of different types. All patterns were presented at the same multiple of their independent detection thresholds. Redrawn from Ledgeway and Smith (1994b).

2.4 Second-order motion and colour

Another image characteristic in terms of which motion may be defined is colour. If the colour of an otherwise homogeneous image is varied sinusoidally in one dimension, without modulating luminance, a grating is produced (an equiluminant, or chrominance, grating). If this pattern moves, its motion is visible, although the percept may be impoverished in various ways compared with perception of luminance motion. As stated in section 1.1, colour is a first-order attribute. However, there are numerous suggestions in the literature that motion is supported by chrominance less effectively than by luminance and so first-order luminance and first-order chrominance cannot be regarded as equivalent stimuli for a common first-order mechanism. Thus, either colour makes a limited contribution to the standard, first-order luminance motion system or else chrominance motion is detected by a separate, less sensitive, system.

Cavanagh and Anstis (1991) found that it is possible to null chrominance-defined motion by adding luminance motion in the opposite direction. Superficially, this is suggestive of a common luminance–chrominance motion-detection mechanism, but it could also reflect summation of motion signals detected separately at a lower level. Chromatic motion sensitivity does have some aspects in common with first-order

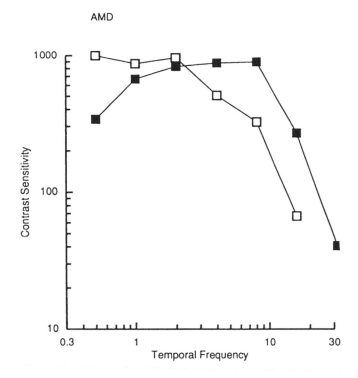

Figure 9 Thresholds for detecting the direction of motion of drifting luminance gratings (filled squares) and colour gratings (hollow squares), as a function of drift temporal frequency. Spatial frequency is 1 cycle/deg. Reproduced with permission from Derrington and Henning (1993).

(luminance) motion sensitivity. For example, direction can be identified at contrasts close to detection threshold over much of the spatiotemporal range, although estimates of just how close vary (Lindsey & Teller, 1990; Cavanagh and Anstis, 1991; Mullen & Boulton, 1992; Derrington & Henning, 1993), and chromatic motion patterns are capable of generating motion aftereffects (Mullen & Baker, 1985). However, as discussed in sections 2.1 and 2.3, these results have both recently been shown to apply also to second-order luminance motion patterns.

In other respects, chrominance and luminance motion differ perceptually. The perceived speed of a chromatic grating is reduced compared with that of a luminance grating (Cavanagh *et al.*, 1984). Perceived speed depends on contrast in both cases (Stone & Thompson, 1992; Mullen & Boulton, 1992) but the difference persists even when appropriate controls for the effect of contrast are used (Henning and Derrington, 1994). Mullen and Boulton (1992) have reported that, except at very high contrasts, colour-defined motion appears jerky while luminance motion appears smooth. Derrington and Henning (1993) report that the temporal contrast sensitivity profile is different for colour and luminance gratings (Figure 9).

Recent studies suggest that, in some respects, perception of chromatic gratings resembles that of second-order luminance stimuli more than first-order luminance.

Firstly, the temporal contrast sensitivity function, well-known to fall-off at low temporal frequencies for luminance gratings (e.g. Kulikowski & Tolhurst, 1973), does not show this behaviour for either beats (Derrington & Badcock, 1985) or chrominance gratings (Derrington & Henning, 1993; Figure 9). Secondly, Cropper (1994) has shown that, at low contrast, speed discrimination performance for drifting chrominance gratings is worse than for luminance gratings but similar to performance with beats. Similarly, again at low contrast levels, the lower threshold of motion (lowest speed at which direction can be correctly identified) for chromatic gratings has been shown to be much worse than that for luminance gratings but comparable to that for beats (Cropper & Derrington, 1994). The same authors found that at low contrast the direction of motion of a chrominance grating is not detectable at any drift speed for short (less than 120 ms) stimulus durations, as is also the case for beats but not for luminance gratings. This suggests that the mechanism detecting chrominance motion has the same limited temporal resolution that appears to affect the second-order motion system. These studies raise the possibility that colour-defined motion might be detected by a 'second-order' motion mechanism, even though strictly speaking it is first-order. If so, our models of second-order motion will require adjustment to incorporate this fact. In both the above studies, the behaviour of chrominance gratings resembled first-order luminance more closely at high contrast levels, leading the authors to suggest a colour input to each of two motion systems.

2.5   Involvement in kinetic depth

When an observer views, on a 2D screen, an image comprising a set of moving dots whose motion paths are projections of rigid 3D motion, a striking impression of 3D motion is seen. A simple example is a circular patch of dots which all move in the same direction but whose speeds vary as if they were painted on a rotating sphere. Such an image is readily perceived as a sphere; indeed it can be extremely difficult to perceive it as a set of dots moving in a plane. This phenomenon is usually called the kinetic depth effect (Wallach & O'Connell, 1953).

Landy et al. (1991) have addressed the question of whether second-order motion can support the kinetic depth effect. They used several types of second-order motion pattern. One was a pattern of dots, each of which consisted of black and white checks, presented on a grey background of the same mean luminance (contrast modulation). Another was a pattern of dynamic noise dots on a static noise back-ground (flicker rate modulation). Performance in a task in which subjects had to identify the shape seen and specify its direction of motion was always poor and was often close to zero. Prazdny (1986) found rather better performance using simulated 3D objects made from sparsely-sampled 'wires' defined by static-on-dynamic noise. Thus, kinetic depth can be supported to a limited extent by second-order motion, but first-order motion provides a very much more effective basis. Whether the difference is qualitative, reflecting two separate motion mechanisms only one of which drives kinetic depth effectively, or simply reflects the sensitivity differences discussed in connection with adaptation studies in section 2.3, is unknown.

2.6   Features or motion energy?

As stated in section 1.3, second-order motion might, in principle, be detected by tracking features, presumably at a high level. Alternatively, detection could be based on a non-linear transformation of the image, in combination with spatial and temporal filtering, followed by standard motion energy analysis (Chubb & Sperling, 1988). Several of the studies reviewed in sections 2.1 to 2.3 are relevant to the question of which theory is correct, in that they show either similarities or differences between first-order and second-order motion, suggesting that the energy principle believed to be used for first-order motion is or is not, respectively, used for second-order motion. It will be evident that this indirect approach to the question has not resolved the issue. In this section, empirical studies bearing more directly on this question are considered.

Support for the use of motion energy in second-order motion detection comes from the study by Nishida (1993) described previously (section 2.1). He devised a stimulus (see Figure 5) consisting of a regular array of checks, each of which either was grey or else itself consisted of a matrix of smaller black and white checks, giving the same mean luminance as the large grey checks. He used these patterns to generate second-order random-dot kinematograms: on successive frames the entire pattern was shifted rigidly to give a perception of (second-order) motion. He found that if he reversed the check type (grey checks become patterned and vice versa) between updates, motion was seen in the opposite direction to that in which the pattern was moved. This phenomenon, termed reversed phi motion, is well known in the first-order domain (Anstis, 1970; Sato, 1989) and is predicted by motion energy models. Feature tracking (which in any case would not be favoured by the dense array of checks used by Nishida) would either match features despite the contrast change (giving veridical direction detection) or else fail to make matches (giving no motion) and so does not predict reversed direction perception. This logic applies equally for first- and second-order stimuli. Nishida's study therefore provides strong evidence for the use of motion energy in the detection of second-order motion.

Werkhoven et al. (1993) have also presented evidence that second-order motion is detected using the energy principle. They employed a rotating ring (Figure 10) made of patches of grating, designed to give two competing motion paths: homogeneous (between patches of the same spatial frequency) and heterogeneous (between patches of different spatial frequencies). The spatial phases of the gratings in the patches were randomized, so that motion could not be detected on the basis of luminance energy. Clearly, feature matching predicts motion between patches of the same spatial frequency. However, it was found that the heterogeneous path could often dominate. They suggested that this dominance reflects the activity of a single 'texture grabber' which rectifies the image and then analyses motion energy, irrespective of the spatial scale of the pattern or texture from which it arises. The relative strength of the two competing motion paths depended on contrast in a way predicted by such a system and the authors found no need to posit even a minor role for feature matching.

Smith (1994) has also obtained direct evidence for the use of the motion energy strategy in second-order motion detection, though not to the exclusion of feature matching. He conducted two experiments suggesting that both strategies may be used. In the first, an image was devised in which the two strategies predict perception of opposite directions of motion. The image was 2D noise whose contrast was

Figure 10 A ring comprising alternating patches of high and low spatial frequency. Reproduced with permission from Werkhoven *et al.* (1993).

modulated in the horizontal dimension by a $3f + 4f$ waveform (the sum of two sinusoids whose frequencies are in the ratio $3:4$). In this image the features, like the modulating waveform, repeat with a period of f (1 c/deg). The modulating waveform was then drifted horizontally in discrete steps of $\frac{1}{4}$ of a period of f (0.25 deg) while the noise remained static, to give second-order motion. If detection is based on feature-matching, motion should be perceived in the correct direction. However, if the image is rectified, energy is introduced at $3f$ and at $4f$. The $4f$ component is effectively static, so energy detection will be based on the $3f$ component, which will alias, giving reversed motion. A similar logic was applied to first-order stimuli by Georgeson and Harris (1990). Smith (1993) found (Figure 11) that motion was unambiguously seen in the reversed direction, suggesting strongly that detection employs the energy strategy. However, if a 60 ms interstimulus interval (isi) (containing unmodulated noise) was introduced between frames, motion was perceived in the veridical direction. This suggests that the energy strategy is normally used for detection of second-order motion but the feature-matching strategy is used in circumstances unfavourable for energy detection. In a second experiment, thresholds were measured for detecting the direction of drift of a sinusoidal contrast modulation. A mask was then introduced which was designed to disrupt the features (bars) of the drifting second-order grating without greatly affecting the motion energy in the (rectified) image. The mask consisted of a 1D random contrast modulation of the 2D noise, which produced vertical second-order bars of random width. These were stationary and of fixed contrast. Direction identification thresholds were measured with and without the mask. Successive frames of the display were presented either without an isi or separated by an isi of 60 ms during which the 2D noise carrier and the mask (if used) were present but the sinusoidal modulation was not. With the isi, performance

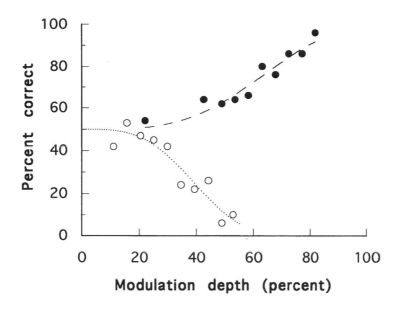

Figure 11 Performance of one subject on a direction judgement task as a function of the amplitude of a $3f + 4f$ contrast modulation which moves continuously in one direction in steps of one-quarter of a period of f (noise remains static). At low contrast, performance is at chance. At higher contrasts, with no interstimulus interval (isi) between updates (hollow circles), direction is consistently seen in the incorrect direction (0% correct); with a 60 ms isi (filled circles) direction is perceived in the correct direction. Redrawn from the data of Smith (1993).

was severely disrupted by the presence of masking features, suggesting the use of a feature-based mechanism. Without the isi, performance was much less affected, suggesting the use of motion energy.

Victor and Conte (1990) also suggest the involvement of both energy-based and feature-based mechanisms, with a strong dominance of energy. They measured the ability of observers to detect various types of second-order motion pattern. They found that moving texture boundaries defined by element size and/or flicker are easily detectable. Boundaries defined by higher-order characteristics of the texture structure, which contain features but which will not activate a mechanism of the type proposed by Chubb and Sperling, are much less visible, but are not invisible. They interpret their results as evidence for both a sensitive energy-based mechanism and a relatively insensitive feature-based mechanism.

Smith and Ledgeway (1994) dissociated energy-based and feature-based second-order motion detection in an adaptation experiment. In experiments discussed earlier (section 2.3), they showed that adaptation to second-order motion can elicit a motion aftereffect provided a dynamic test pattern is used. An interesting additional feature of their work is that when a 60 ms isi was used between updates of the adaptation pattern, the motion aftereffect was abolished. It seems, then, that adaptation occurs in the energy-based second-order system but the original assertion (Anstis, 1970) that feature-based motion does not cause aftereffects is upheld.

2.7   The role of early non-linearities

Having asserted (section 2.6) that second-order motion is detected using the motion
energy principle, at least in most circumstances, we must consider the nature of the
process which gives rise to motion energy in the representation of an image which
contains no net directional energy. In fact, any simple brightness non-linearity
(compressive, expansive, rectifying, etc.) applied at any stage of processing prior to
motion detection, will introduce net motion energy in the most commonly-used type of
second-order motion: drifting contrast-modulations. For example, a brightness nonlinearity occurring in the retina will introduce 'distortion products' in the Fourier
spectrum of the image which will thereafter appear as first-order components and will
be indistinguishable from those that exist in the image. Several investigators have
claimed the existence of such early distortion products (see, for example, MacLeod *et
al.*, 1992). Perhaps, therefore, the visibility of motion defined by contrast modulations
has a trivial explanation: rather than using a specialized second-order motion system,
such motion might be detected because, in the neural representation of the image, it
appears as first-order motion. If this were all there is to second-order motion detection,
the whole topic would be no more than a red herring. A related concern is the role of
distortion products arising in the display. Any brightness non-linearity in the display
will have the same effect as one in the brain and may lead to moving first-order
distortion products being mistaken for second-order motion. Even when the greatest
care is taken over linearization of the display, it can be difficult to be certain that
linearity has been achieved.

The detection of some types of second-order stimuli, particularly those defined by
spatial variations in temporal parameters, cannot be explained in such terms because a
brightness non-linearity is not sufficient to create net motion energy (Chubb &
Sperling, 1989). However, some of the studies reviewed in sections 2.1 to 2.6 are
vulnerable, when considered in isolation, to this artefactual explanation. In general all
studies which show similarities in the detection of luminance- and contrast-defined
motion (e.g. Turano, 1991; McCarthy, 1993; Smith *et al.*, 1994) raise this suspicion in
the mind. However, the early distortion product hypothesis predicts that contrast-
modulations, at least, will behave like luminance modulations in *all* respects, because
that is what they are on this view. In fact, as well as similarities there are some clear
differences in the processing of the two types of stimulus (see section 2.1). One of the
most direct and compelling demonstrations that the detection of contrast modulations
cannot be attributed to early non-linearities involves beats. Derrington and Badcock
(1986) have shown that sensitivity to beats is best when both components of the beat
have the same amplitude; when the amplitude of one component of such a beat is
increased, sensitivity decreases. This is inconsistent with detection of beats using early
distortion products: the amplitude of any distortion product is expected to be proportional to the product of the component amplitudes and so should increase when this
manipulation is applied, raising sensitivity. The same authors have also shown that it
is not possible to null the motion of a drifting beat by adding a luminance modulation
calculated to null any distortion product (Badcock & Derrington, 1989).

Thus, distortion products arising from an early non-linearity are not sufficient to
explain second-order motion detection. A distinction must be drawn, however,
between early non-linearities, which are usually regarded as imperfections in a

quasi-linear system, and non-linear transformations that are applied later, perhaps as a deliberate ploy. The scheme proposed by Chubb and Sperling (1988) is in the latter category, but nonetheless the motion signals that arise from it may properly be regarded as distortion products.

2.8   Separate or common energy mechanisms?

The studies reviewed in section 2.6, together with the considerations discussed in section 2.7, suggest that the low-level strategy outlined by Chubb and Sperling (1988) is, indeed, implemented in the visual system, but that we can also fall back on a high-level process of tracking features. Recall that the Chubb and Sperling strategy is to create first-order motion energy where none exists in the image, and then detect it using standard first-order motion analysis. Given strong evidence for a mechanism that detects such first-order motion energy in (transformed) second-order stimuli, taken together with the close qualitative similarities in the sensitivity of the visual system to the two types of pattern (see sections 2.1 to 2.3) in many respects, it is natural to wonder whether it is necessary to invoke two energy mechanisms, one for each type of image, or whether both types of motion might be detected by a common mechanism. This question was raised earlier (section 1.3). Johnston et al. (1992) have shown that a single mechanism is computationally feasible. In this section the question of whether it is suggested by empirical studies is addressed.

A few studies, some of them described in section 2.1, have demonstrated differences in the way in which sensitivity to first-order and second-order patterns varies as the spatial and temporal parameters of the image are changed, suggesting separate detection. However, other studies have shown similarities, consistent with processing by a common mechanism. One way to reconcile these findings is to propose the existence of two mechanisms with many properties in common (reflecting the fact that both use the same detection principle) but with one or two critical differences, in particular temporal acuity (Derrington & Henning, 1993). But such studies of sensitivity address the question only indirectly and should not, perhaps, be regarded as definitive evidence for separate mechanisms.

One approach that has been taken to the question of whether first- and second-order motion signals are represented in the same pathway has been to ask whether it is possible for a drifting first-order grating of one orientation to be integrated with a drifting second-order grating of another orientation (Figure 12) to yield a coherently moving plaid. If not, then separate motion mechanisms are implicated. Stoner and Albright (1992) conducted an experiment to answer this question and found that coherence is indeed obtained, provided that the contrasts of the two patterns are similar. However, this finding does not point unambiguously to a common motion detection mechanism for two reasons. Firstly, Victor and Conte (1992a) have failed to find evidence for coherence of mixed plaids. They attribute the difference in results between the two studies to differences in the instructions given to, and criteria used by, the two groups of subjects. Secondly, coherence of mixed plaids might in any case reflect separate detection of the two components followed by a subsequent pooling of motion signals from both pathways, as in the model of Wilson et al. (1992).

Indeed, two other lines of evidence provide compelling and direct evidence for

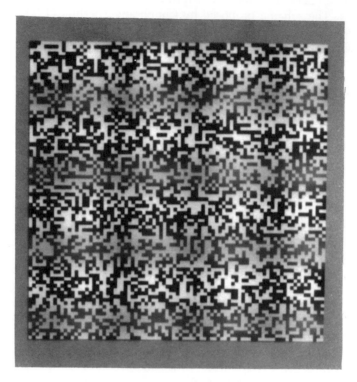

Figure 12 A plaid (the sum of two sine gratings of different orientations) in which one component is first-order (a conventional luminance-modulated grating, vertically-oriented) and the other is second-order (a horizontal contrast-modulated grating).

separate detection of first- and second-order motion. Firstly, two studies (Mather & West, 1993; Ledgeway & Smith, 1994a) have been presented that are based on the logic that if a single mechanism is responsible for the detection of both types of motion then that mechanism should be capable of integrating frames of mixed type presented in a single animation sequence. Mather and West (1993) presented two-frame random dot kinematograms in which the dots were defined either by first- or second-order characteristics. Direction identification performance was measured as a function of the size of the displacement of the dots between frames. When both frames were of the same type, direction could be detected accurately over a considerable range of displacements, but when they were of different types performance was at chance levels even for very small displacements (Figure 13a), suggesting that the two frames could not be integrated. Ledgeway and Smith (1994a) required observers to judge the direction of a multi-frame motion sequence in which the display alternated between a second-order sine grating and a first-order sine grating. All frames contained the same static, 2D noise sample. In frame n the noise was multiplied by a 1 c/deg, vertical sine grating to produce a second-order contrast-modulated image. In frame $(n + 1)$ the sine grating was displaced some fraction of its spatial period and was added to, rather than multiplied by, the noise to give a first-order image. On frame

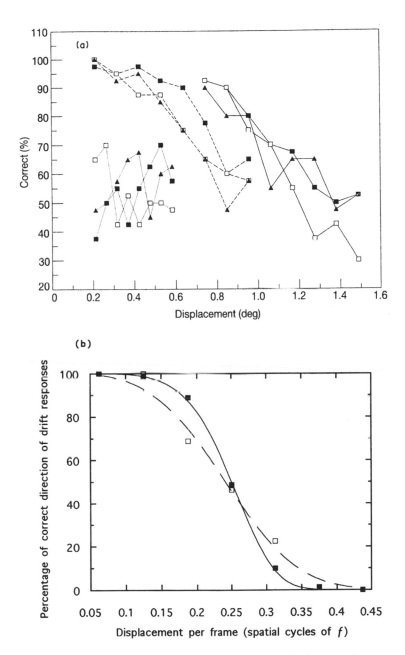

Figure 13 (a) Performance on a direction-identification task for two-frame random dot motion sequences of three types: both frames first-order dots (solid lines), both second-order (dashed lines) and one frame of each type (dotted lines). The different symbols show data for three different subjects. Reproduced with permission from Mather and West (1993). (b) Performance on a direction-identification task for a grating which alternates between first- and second-order on successive frames of the animation sequence, as a function of the size of the displacement between frames. Redrawn from the data of Ledgeway and Smith (1994a).

(n + 2) the grating was displaced again and was multiplied by the noise, and so on. They found that for displacements of 0.25 cycles per frame (0.5 cycles between images of the same type) no consistent direction was perceived (Figure 13b). This suggests the involvement of two separate mechanisms, each seeing only frames of one type (a 0.5 cycle step is ambiguous). For larger displacements, motion in the incorrect direction was consistently reported, again suggesting that images of like type only were integrated (step sizes between 0.5 and 1.0 cycles give aliasing).

Secondly, Harris and Smith (1992) have made a very simple observation that is hard to reconcile with a common detection mechanism. They found that second-order motion (contrast-modulated noise) does not elicit optokinetic nystagmus (OKN). OKN is the name given to the reflexive eye movements that are normally made in response to movement of the entire image with the object of stabilizing the image on the retina during movements of the eye or head. In primates, visual input to the OKN system is known to be dominated by cortical motion-detection mechanisms (see Chapter 14 by L. Harris and Chapter 15 by Krauzlis for a full discussion of eye movements elicited by motion and the visual input that drives them). It is hard to conceive of a single motion-detection mechanism which does not distinguish between first- and second-order inputs but which nonetheless drives OKN only when its input is first-order. Separate detection mechanisms are clearly implicated.

2.9 Physiological studies

An important source of evidence concerning the means of detection of second-order motion is the literature on the responses of single neurones to such motion (see also Chapter 7 by Logothetis). The first such study was by Albrecht and DeValois (1981). They recorded the responses of cells in the primary visual cortex of both cats and monkeys to drifting AM gratings. They found that such stimuli do not excite cells tuned to the modulation frequency. In contrast, Derrington (1987) recorded the responses of feline lateral geniculate nucleus (LGN) cells to drifting AM gratings and found that the response had a component at the modulation frequency. Derrington suggested that this component of the response reflects the presence of distortion products which arise post-receptorally in or before the LGN. He reconciled his results with those of Albrecht and DeValois by speculating that these distortion products might subsequently be removed by subtraction of the responses of on-centre and off-centre cells. Clearly, this does not provide a basis for the detection of second-order motion. However, Zhou and Baker (1993) have recently recorded responses of neurones in areas 17 and 18 of cat visual cortex to static high spatial frequency gratings with a drifting contrast modulation. They obtained responses to the contrast modulation in about 40% of neurones. Significantly, spatial frequency tuning was not the same for contrast modulations as for luminance modulations (Figure 14), ruling out any explanation in terms of simple distortion products. They interpret their findings as support for two parallel motion pathways that are subsequently integrated, as suggested by Wilson *et al.* (1992).

Physiological responses to second-order motion stimuli have also been studied in neurones in primate area MT (an area specialized for motion detection; see Chapter 3 by Snowden). Albright (1992) found that 87% of cells in MT responded to a second-

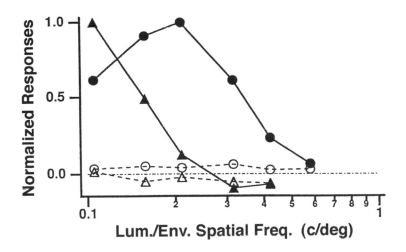

Figure 14 Responses of a neurone in cat visual cortex to drifting gratings (triangles) and drifting contrast modulations (circles) as a function of the spatial frequency of the stimulus. Solid symbols show responses to motion in the preferred direction, hollow circles responses to motion in the null direction. Reproduced with permission from Zhou and Baker (1993).

order motion pattern in which a bar defined by dynamic noise drifted across a static noise background. These neurones also responded to luminance-defined bars and the mean direction tuning was the same for both types of bar. A recent study (O'Keefe *et al.*, 1993) suggests that neurones in MT often respond to contrast-modulated noise but only if their receptive fields are near the fovea (see also section 2.2). Olavarria *et al.*, (1992) also obtained strong responses of MT neurones to texture-defined bars, although their stimuli were not drift-balanced. These results are consistent both with a common detection mechanism and with parallel mechanisms that converge in MT.

Thus, the available physiological data, when results for cats (Zhou & Baker, 1993) and primates are taken together, are suggestive of parallel motion pathways which (in primates) converge in area MT. But the data are sparse, and do not rule out a common detection mechanism. At present, psychophysical data are more telling in this respect.

2.10 Utility of second-order motion

The studies reviewed in the preceding sections have used images in which motion is defined only by second-order characteristics. In natural images, second-order motion is frequently present but is almost invariably accompanied by first-order motion. We should perhaps, therefore, think of the detection of second-order motion as a supplement, rather than an alternative, to the use of first-order motion. What is the evidence for such supplementary use? Derrington *et al.* (1992) have provided an example of the use of second-order motion in situations where first-order motion cues are also present. They constructed a conventional plaid comprising two sine gratings of the same spatial frequency but different orientations. Both components were then drifted in discrete jumps of 3/8 cycle of their period. This causes the periodic contrast variations of the plaid to alias even though the components are not under-sampled. Observers judged the overall direction of the plaid. In some circumstances, observers reported motion in the direction opposite that which is expected on the basis of the components, suggesting that the second-order characteristics were used in determining direction. This outcome would result either from squaring the image and detecting motion energy or from treating the contrast variations as features and tracking them, but either way the experiment shows that second-order motion can be used in the presence of first-order motion cues, and can even dominate where the latter are weak. Similarly, Derrington and Badcock (1992) found that direction discrimination performance for plaids is slightly better (in terms of the lowest speed at which direction can be identified) than for its components, suggesting that second-order information feeds into the computation of motion alongside first-order characteristics. Gorea and Lorenceau (1991) have also argued in favour of an influence of second-order characteristics, alongside first-order, in determining plaid perception. The idea of a supplementary, refining role for second-order motion will be raised again in Chapter 8 by Wilson.

3 CONCLUSIONS

In this chapter, second-order motion has been defined, the current computational models of its detection have been discussed and the empirical evidence concerning the sensitivity of the visual system to it has been reviewed. It is concluded that second-order motion is normally detected using the method outlined by Chubb and Sperling (1988), or something rather similar. In this process the image is filtered, rectified and then subjected to standard motion energy analysis. The bulk, at least, of the available evidence suggests that this is a separate process from the motion energy analysis that is applied to first-order images. That is, motion energy analysis is applied to the image twice, in parallel, each application revealing motion defined in different ways. Currently, the most comprehensive, plausible model of this parallel system is that of Wilson et al. (1992), which is discussed in more detail in Chapter 8. In addition to this dual motion energy system, second-order motion can also be detected by tracking the positions of second-order spatial structures, or features, over time. At present there is little reason to invoke a separate second-order feature-tracking mechanism; it seems more appropriate to think of a single mechanism, perhaps post-attentive (Cavanagh, 1992), that can track features however they are defined.

Several issues remain unresolved. Chief among these, perhaps, is the degree of commonality in the detection of second-order motion of different types. Only contrast-defined motion has been studied extensively and those studies that have used other types of second-order motion have paid little attention to this question. There is also a need for more extensive neurophysiological studies of the sensitivity of single neurones in the visual pathway, particularly in primates, to second-order motion.

ACKNOWLEDGEMENTS

The author is supported by grants from the Science and Engineering Research Council and the Medical Research Council. He is grateful to Dr David Badcock for helpful comments on an earlier draft of this chapter.

REFERENCES

Adelson, E. H. & Bergen, J. R. (1985). Spatiotemporal energy models for the perception of motion. *J. Opt. Soc. Am. A*, **2**, 284–299.

Albrecht, D. G. & DeValois, R. L. (1981). Striate cortex responses to periodic patterns with and without the fundamental harmonics. *J. Physiol. (Lond.)*, **319**, 497–514.

Albright, T. D. (1992). Form-cue invariant motion processing in primate visual cortex. *Science*, **255**, 1141–1143.

Anstis, S. M. (1970). Phi movement as a subtraction process. *Vision Res.*, **10**, 1411–1430.

Anstis, S. M. (1980). The perception of apparent movement. *Phil. Trans. R. Soc. Lond. B*, **290**, 153–168.

Anstis, S. M. & Moulden, B. P. (1970). Aftereffect of seen movement: evidence for peripheral and central components. *Q. J. Exp. Psychol.*, **22**, 222–229.

Arimura, K. & Nishida, S. (1992). Simultaneous detection of second-order motions in two regions. *Invest. Ophthalmol. Vis. Sci.*, **33**, 1142.

Badcock, D. R. & Derrington, A. M. (1985). Detecting the displacement of periodic patterns. *Vision Res.*, **25**, 1253–1258.

Badcock, D. R. & Derrington, A. M. (1989). Detecting the displacement of spatial beats: no role for distortion products. *Vision Res.*, **29**, 731–739.

Braddick, O. (1974). A short-range process in apparent motion. *Vision Res.*, **14**, 519–527.

Braddick, O. J. (1980). Low-level and high-level processes in apparent motion. *Phil. Trans. R. Soc. Lond. B*, **290**, 137–151.

Cavanagh, P. (1991). Short-range vs long-range motion: not a valid distinction. *Spatial Vision*, **5**, 303–309.

Cavanagh, P. (1992). Attention-based motion perception. *Science*, **257**, 1563–1565.

Cavanagh, P. & Anstis, S. (1991). The contribution of color to motion in normal and color-deficient observers. *Vision Res.*, **31**, 2109–2148.

Cavanagh, P. & Mather, G. (1989). Motion: The long and short of it. *Spatial Vision*, **4**, 103–129.

Cavanagh, P., Tyler, C. W. & Favreau, O. E. (1984). Perceived velocity of moving chromatic patterns. *J. Opt. Soc. Am. A*, **1**, 893–899.

Chubb, C. & Sperling, G. (1988). Drift-balanced random stimuli: a general basis for studying non-Fourier motion perception. *J. Opt. Soc. Am. A*, **5**, 1986–2006.

Chubb, C. & Sperling, G. (1989). *Second-order Motion Perception: Space/time Separable Mechanisms*, pp. 126–138. IEEE Computer Society Press, Washington DC.

Cropper, S. J. (1994). Velocity discrimination in chromatic gratings and beats. *Vision Res.*, **34**, 41–48.

Cropper, S. J. & Derrington, A. M. (1994). Motion of chromatic stimuli: first-order or second-order? *Vision Res.*, **34**, 49–58.

Derrington, A. M. (1987). Distortion products in geniculate X-cells: a physiological basis for masking by spatially modulated gratings? *Vision Res.*, **27**, 1377–1386.

Derrington, A. M. & Badcock, D. R. (1985). Separate detectors for simple and complex patterns? *Vision Res.*, **25**, 1869–1878.

Derrington, A. M. & Badcock, D. R. (1986). Detection of spatial beats: non-linearity or contrast increment detection? *Vision Res.*, **26**, 343–348.

Derrington, A. M. & Badcock, D. R. (1992). Two-stage analysis of the motion of 2-dimensional patterns, what is the first stage? *Vision Res.*, **32**, 691–698.

Derrington, A. M. & Henning, G. B. (1993). Detecting and discriminating the direction of motion of luminance and colour gratings. *Vision Res.*, **33**, 799–812.

Derrington, A. M., Badcock, D. R. & Holroyd, S. A. (1992). Analysis of the motion of 2-dimensional patterns: evidence for a second-order process. *Vision Res.*, **32**, 699–707.

Derrington, A. M., Badcock, D. R. & Henning, G. B. (1993). Discriminating the direction of second-order motion at short stimulus durations. *Vision Res.*, **33**, 1785–1794.

Georgeson, M. A. & Harris, M. G. (1990). The temporal range of motion sensing and motion perception. *Vision Res.*, **30**, 615–619.

Gorea, A. & Lorenceau, J. (1991). Directional performances with moving plaids: component-related and plaid-related processing modes coexist. *Spatial Vision*, **5**, 231–251.

Green, M. (1983). Contrast detection and direction discrimination of drifting gratings. *Vision Res.*, **23**, 281–289.

Grzywacz, N. M. (1992). One-path model for contrast-independent perception of Fourier and non-Fourier motions. *Invest. Ophthalmol. Vis. Sci.*, **33**, 954.

Harris, L. R. & Smith, A. T. (1992). Motion defined exclusively by second-order characteristics does not evoke optokinetic nystagmus. *Vis. Neurosci.*, **9**, 565–570.

Harris, M. G. (1986). The perception of moving stimuli: A model of spatiotemporal coding in human vision. *Vision Res.*, **26**, 1281–1287.

Henning, G. B. & Derrington, A. M. (1994) Speed, spatial-frequency and temporal-frequency comparisons in luminance and colour gratings. *Vision Res.*, **34**, 2093–2101.

Henning, G. B., Hertz, B. G. & Broadbent, D. E. (1975). Some experiments bearing on the hypothesis that the visual system analyses spatial patterns in independent bands of spatial frequency. *Vision Res.*, **15**, 887–897.

Johnston, A., McOwen, P. W. & Buxton, H. (1992). A computational model of the analysis of some first-order and second-order motion patterns by simple and complex cells. *Proc. R. Soc. Lond. B*, **250**, 297–306.

Julesz, B. (1971). *Foundations of Cyclopean Perception*. University of Chicago Press, Chicago.

Kulikowski, J. J. & Tolhurst, D. J. (1973). Psychophysical evidence for sustained and transient detectors in human vision. *J. Physiol. (Lond.)*, **232**, 149–162.

Landy, M. S., Dosher, B. A., Sperling, G. & Perkins, M. E. (1991). The kinetic depth effect and optic flow – II. First- and second-order motion. *Vision Res.*, **31**, 859–876.

Ledgeway, T. (1994). Adaptation to second-order motion results in a motion aftereffect for directionally-ambiguous test stimuli. *Vision Res.* (in press).

Ledgeway, T & Smith, A. T. (1994a). Evidence for separate motion-detecting mechanisms for first- and second-order motion in human vision. *Vision Res.* (in press).

Ledgeway, T & Smith, A. T. (1994b). The duration of the motion aftereffect following adaptation to first- and second-order motion. *Perception* (in press).

Lindsey, D. T. & Teller, D. Y. (1990). Motion at isoluminance: discrimination/ detection ratios for moving isoluminant gratings. *Vision Res.*, **30**, 1751–1761.

McCarthy, J. E. (1993). Directional adaptation effects with contrast modulated stimuli. *Vision Res.*, **33**, 2653–2662.

MacLeod, D. I. A., Williams, D. R. & Makous, W. (1992). A visual nonlinearity fed by single cones. *Vision Res.*, **32**, 347–363.

Marr, D. & Ullman, S. (1981). Directional selectivity and its use in early visual processing. *Proc. R. Soc. Lond. B*, **211**, 151–180.

Mather, G. (1991). First-order and second-order visual processes in the perception of motion and tilt. *Vision Res.*, **31**, 161–167.

Mather, G. & West, S. (1993). Evidence for second-order motion detectors. *Vision Res.*, **33**, 1109–1112.

Mullen, K. T. & Baker, C. L. (1985). A motion aftereffect from an isoluminant stimulus. *Vision Res.*, **25**, 685–688.

Mullen, K. T. & Boulton, J. C. (1992). Absence of smooth motion perception in color vision. *Vision Res.*, **32**, 483–488.

Nishida, S. (1993). Spatiotemporal properties of motion perception for random-check contrast modulations. *Vision Res.*, **33**, 633–646.

Nishida, S. & Sato, T. (1993). Two kinds of motion aftereffect reveal different types of motion processing. *Invest. Ophthalmol. Vis. Sci.*, **34**, 1363.

Olavarria, J. F., DeYoe, E. A., Knierim, J. J., Fox, J. M. & VanEssen, D. C. (1992). Neural responses to visual texture patterns in middle temporal area of the macaque monkey. *J. Neurophysiol.*, **68**, 164–181.

O'Keefe, L. P., Carandini, M., Beusmans, J. M. H. & Movshon, J. A. (1993). MT neuronal responses to 1st- and 2nd-order motion. *Soc. Neurosci. Abst.*, **19**, 1283.

Pantle, A. (1992). Immobility of some second-order stimuli in human peripheral vision. *J. Opt. Soc. Am. A*, **9**, 863–867.

Petersik, J. T. (1991). Comments on Cavanagh and Mather (1989): Coming up short (and long). *Spatial Vision*, **5**, 291–301.

Petersik, J. T., Hicks, K. I. & Pantle, A. J. (1978). Apparent movement of successively generated subjective figures. *Perception*, **7**, 371–383.

Prazdny, K. (1986). Three-dimensional structure from long-range apparent motion. *Perception*, **15**, 618–625.

Ramachandran, V. S., Rao, V. M. & Vidyasagar, T. R. (1973). Apparent movement with subjective contours. *Vision Res.*, **13**, 1399–1401.

Reichardt, W. (1961). Autocorrelation, a principle for the evaluation of sensory information by the central nervous system. In W. A. Rosenblith (Ed.) *Sensory communication*. Wiley, New York.

Sato, T. (1989). Reversed apparent motion with random dot patterns. *Vision Res.*, **29**, 1749–1758.

Smith, A. T. (1994). Correspondence-based and energy-based detection of second-order motion in human vision. *J. Opt. Soc. Am. A*, **11**, 1940–1948.

Smith, A. T. & Ledgeway, T. (1994). Adaptation to energy-based but not feature-based second-order motion (in preparation).

Smith, A. T., Hess, R. F. & Baker, C. L. J. (1994). Direction identification thresholds for second-order motion in central and peripheral vision. *J. Opt. Soc. Am. A*, **11**, 506–514.

Solomon, J. A. & Sperling, G. (1991). Can we see 2nd-order motion and texture in the periphery? *Invest. Ophthalmol. Vis. Sci.*, **32**, 714.

Stone, L. S. & Thompson, P. (1992). Human speed perception is contrast dependent. *Vision Res.*, **32**, 1535–1549.

Stoner, G. R. & Albright, T. D. (1992). Motion coherency rules are form-cue invariant. *Vision Res.*, **32**, 465–475.

Turano, K. (1991). Evidence for a common motion mechanism of luminance-modulated and contrast-modulated patterns: selective adaptation. *Perception*, **20**, 455–466.

Ullman, S. (1979). *The Interpretation of Visual Motion*. MIT Press, Cambridge.

Victor, J. D. & Conte, M. M. (1990). Motion mechanisms have only limited access to form information. *Vision Res.*, **30**, 289–301.

Victor, J. D. & Conte, M. M. (1992a). Coherence and transparency of moving plaids composed of Fourier and non-Fourier gratings. *Percept. Psychophys.*, **52**, 403–414.

Victor, J. D. & Conte, M. M. (1992b). Evoked potential and psychophysical analysis of Fourier and non-Fourier motion mechanisms. *Vis. Neurosci.*, **9**, 105–123.

von Grunau, M. W. (1986). A motion aftereffect for long-range stroboscopic apparent motion. *Percept. Psychophys.*, **40**, 31–38.

Wallach, H. & O'Connell, D. N. (1953). The kinetic depth effect. *J. Exp. Psychol.*, **45**, 205–217.

Watson, A. B., Thompson, P. G., Murphy, B. J. & Nachmias, J. (1980). Summation and discrimination of gratings moving in opposite directions. *Vision Res.*, **20**, 341–347.

Werkhoven, P., Sperling, G. & Chubb, C. (1993). The dimensionality of texture-defined motion: a single channel theory. *Vision Res.*, **33**, 463–486.

Wilson, H. R., Ferrera, V. P. & Yo, C. (1992). A psychophysically motivated model for two-dimensional motion perception. *Vis. Neurosci.*, **9**, 79–97.

Zanker, J. M. (1993). Theta motion: a paradoxical stimulus to explore higher order motion extraction. *Vision Res.*, **33**, 553–569.

Zhou, Y.-X. & Baker, Jr, C. L. B. (1993). A processing stream in mammalian visual cortex neurons for non-Fourier responses. *Science*, **261**, 98–100.

7
Physiological Studies of Motion Inputs

Nikos K. Logothetis
Baylor College of Medicine, Houston, USA

1 INTRODUCTION

We conceive motion to be the change of position of an object over time. In the same way, one could think of visual motion perception as a memory-based evaluation of successive loci of the retinal stimulus, thereby implying that visual motion is derived from the combination of other simpler visual dimensions. Yet, experimental work shows that biological systems possess a neural machinery, which is selectively tuned to basic properties of motion, such as the direction or the speed of a moving object.

The existence of a specialized neural motion system is hardly surprising given the multiplicity and the importance of the different perceptual roles that visual motion can play. Besides detecting moving objects, which is certainly of great biological import- ance, motion is essential for image segmentation, for the reconstruction of the third dimension, and for the perception of self-motion. In addition, motion information is used by the oculomotor system to foveate a moving object, and also to stabilize the image of the world on the retina during autokinesis. Only a few degrees of retinal slip can cause a deleterious effect on our visual acuity equivalent to nearly two to three degrees of myopia, showing the importance of gaze-holding eye movements for pattern vision.

The different roles of motion probably require that a variety of computations be carried out by different neural circuits. The input to all kinds of processing, however, is likely to be the mechanism that senses the direction and the speed of a moving stimulus.

This chapter selectively deals with this neural mechanism in the primate visual system. It does not provide a comprehensive review of all the work on the physio- logical mechanisms underlying the perception of motion, which is very extensive indeed, but (1) briefly describes the anatomy of the motion system, and the physio- logical properties of the cells in some of the structures that are pertinent to motion analysis, (2) examines the responses of the motion-processing cells to stimuli defined by cues other than luminance, and (3) discusses the behavioral deficits following selective lesions in some stations of the motion system. For a comprehensive review of biological motion processing the reader is referred to Nakayama (1985). For earlier reviews of lesions in structures related to motion analysis see Newsome and Wurtz (1988), Schiller and Logothetis (1990), Andersen and Siegel (1990), Merigan (1991), and Merigan and Maunsell (1993).

2   A SPECIALIZED NEURAL MECHANISM FOR THE ANALYSIS OF VISUAL MOTION

One of the most convincing demonstrations of the existence of a separate motion processing machinery in the visual system is the well-known motion aftereffect. After viewing one direction of motion for several minutes, people experience illusory motion in the opposite direction. This phenomenon, studied in detail by Wohlgemuth at the beginning of the century (Wohlgemuth, 1911), is also known as the waterfall illusion since it is often experienced after looking at a waterfall and then fixating on any stationary part of the surrounding landscape (Addams, 1964). In other words, it involves apparent motion without any position change, and it has been therefore taken as strong evidence for the existence of two separable visual mechanisms, one for spatial and one for motion processing (Gregory, 1966).

A related phenomenon is the well-studied direction-specific adaptation. Exposure for several minutes to an upward-moving grafting of high contrast produces a twofold increase in the detection threshold for upward motion, while the ability to see downward motion remains the same (Sekuler & Ganz, 1963). Since both the upward and the downward motion of the grating involve stimulation of the same spatial locations, direction-specific adaptation is also considered strong evidence for the existence of a specialized mechanism in the visual system that detects image motion.

Motion blindness is the most striking demonstration of behavioral isolation of such a mechanism in man. Patients with damage in the posterior cortex have normal visual capacities, such as acuity, contrast detection, color vision, saccadic eye movements, and flicker perception, while they seem completely unable to perceive continuous motion (see below). Instead, their perception is fragmented into a sequence of snapshots in which objects seem to change position abruptly.

How then is such a specialized processing instantiated in the visual system? In many species single unit recordings have revealed a great number of neurons in various stages of the visual system that respond selectively to the direction or speed of moving objects. The range of stimuli eliciting such responses can vary from bars or dots of light to complex patterns defined either by luminance or by other visual cues. In the primate, multiple visual areas seem to form an extensively interconnected network, whose components probably underlie different aspects of motion processing. The following discussion relates to the physiological properties of cells in some stages of motion processing.

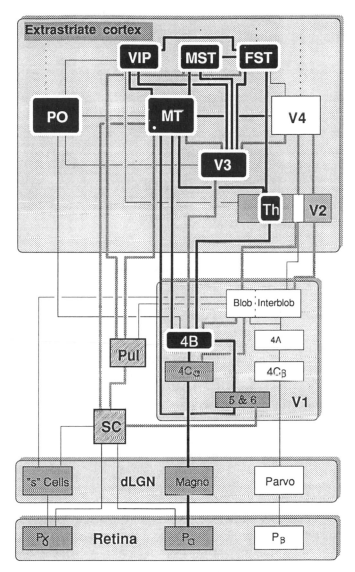

Figure 1 Visual areas related to motion processing. Black rectangles correspond to areas with high proportions of direction-selective neurons. White rectangles are areas mostly associated with form and color vision. Black, thick lines show the commonly accepted motion pathways, while gray thick lines show some possible alternative routes. Dashed thin lines show projections to other areas that are not considered here. SC, superior colliculus; Pul, pulvinar; Th, cytochrome oxidase staining thick stripes of area V2; MT, middle temporal visual; PO, parieto-occipital area; MST, middle superior temporal; FST, floor of superior temporal; VIP, ventral intraparietal; dLGN, dorsal lateral geniculate nucleus.

3 ORGANIZATION OF THE PRIMATE MOTION SYSTEM

Figure 1 shows schematically the components of the primate visual system wherein cells show sensitivity to image motion. The diagram does not include those structures that are involved in the control of eye movements. Vision is undoubtedly a sensory-motor event, and eye movements are not only crucial for image stabilization and orienting behavior, but also play an essential role in the perceptual judgment of motion, depth, size, and self-motion (Douglas *et al.*, 1993). Nonetheless, the issue of visuomotor integration extends beyond the scope of this chapter, and therefore I will concentrate on what is customarily called the sensory part of the visual motion system. The loci of some extrastriate areas involved in visual motion processing are illustrated in Figure 2.

3.1 Retinogeniculostriate system

The retinogeniculostriate system has been covered in Chapter 3 by Snowden. This section will therefore briefly outline the major points of interest.

3.1.1 Retina

The ganglion cell layer of the retina contains several spatially intermixed cell types that differ in their morphologies and their central projections (Boycott & Waessle, 1974; Leventhal *et al.*, 1981). In the primate the major classes of ganglion cells are the larger-sized P_α or A cells (10%), the smaller-sized P_β or B cells (80%), and a third heterogeneous group of small neurons (10%), reported as 'rarely encountered cells' (Schiller & Malpeli, 1977), W-like, or P_γ neurons (Perry & Cowey, 1984; Perry *et al.*, 1984).

The two major classes have distinctly different physiological properties (Gouras, 1969; De Monasterio & Gouras, 1975; Schiller & Malpeli, 1977; Kaplan & Shapley, 1986; Purpura *et al.*, 1988). The P_α neurons have high conduction velocities, transient responses, no color selectivity, and high contrast sensitivity, while the P_β cells have lower conduction velocities, sustained responses, low contrast sensitivity, and color opponency. Small differences are also found in the size of the receptive field centers of the P_α and P_β cells (Blakemore & Vital-Durand, 1986; Crook *et al.*, 1988; Derrington & Lennie, 1984), although measuring responses to gratings of high spatial frequency (Crook *et al.*, 1988) revealed similar spatial resolution for both cell classes at any eccentricities. Primate ganglion cells lack any overt directional selectivity. While most cells respond well to drifting gratings, they respond equally well for all directions.

Recently, Kolb and colleagues (1992) applied Golgi techniques to postmortem specimens of human retina to demonstrate the existence of ganglion cells that send dendritic processes into both sublaminae (a and b) of the inner plexiform layer, where the cells may receive direct bipolar input from both type a (Off) and type b (On) bipolar cells. These bistratified cells resemble the complex ganglion cells of the rabbit retina, which give both (On) and (Off) responses (On–Off cells), and which are

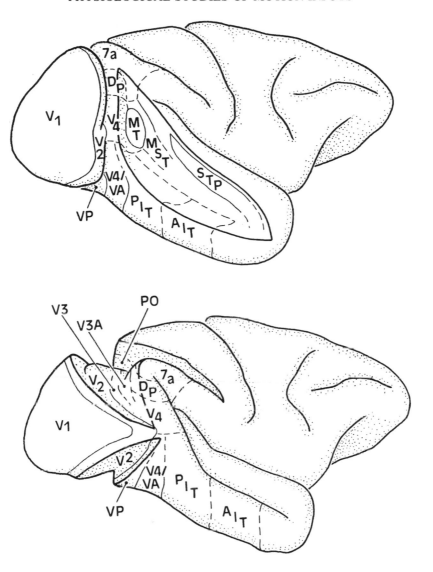

Figure 2 The locations of some visual areas related to motion processing. The superior temporal sulcus has been opened in the upper half and the lunate, parieto-occipital, and inferior occipital sulci in the lower half of the figure, to show the areas burried within the sulci. AIT, anterior inferior temporal cortex; PIT, posterior inferior temporal cortex; VA, ventral anterior; MT, middle temporal; MST, medial superior temporal; STP, superior temporal polysensory; VP, ventral posterior; PO, parieto-occipital. Reproduced with permission from Maunsell and Newsome (1987).

directionally selective (Barlow & Levick, 1965; Wyatt & Daw, 1975; Amthor *et al.*, 1984, 1989). Nevertheless, this type of cell is not commonly reported in the nonhuman primate. Thus whether such bistratified cells are directionally selective, and what central targets they project is currently unknown.

3.1.2 Dorsal lateral geniculate nucleus

More than 90% of the ganglion cells project to the dorsal lateral geniculate nucleus (dLGN), which in turn projects to striate cortex (V1). The dLGN of the Old World monkey is a laminated thalamic structure with six clearly discernible cellular layers, numbered from 1 to 6, beginning with the ventromedial hilar region. As described in Chapter 3, the LGN contains two major subdivisions the magnocellular (laminae 1 and 2) and the parvocellular (laminae 3 to 6) layers, with each subdivision receiving input from one class of ganglion cells (Figure 1); the magnocellular layers from the P_α cells and the parvocellular layers from P_β cells (Leventhal *et al.*, 1981; Perry *et al.*, 1984).

Scattered within the interlaminar zones, and within the 'S' layers of dLGN, just ventral to the magnocellular layers, is a third group of small (*koniocellular*) neurons (Campos-Ortega & Hayhow, 1970; Giolli & Tigges, 1970; Kaas *et al.*, 1978), which receive both retinal and collicular inputs (Campos-Ortega & Hayhow, 1970; Giolli & Tigges, 1970; Kaas *et al.*, 1978). Its retinal input originates from the P_γ ganglion cells (Itoh *et al.*, 1982) that are known to project to the superior colliculus (SC) as well (Schiller & Malpeli, 1977).

The response properties of the parvocellular and magnocellular dLGN neurons can be predicted from their retinal predecessors (Kaplan & Shapley, 1982; Derrington & Lennie, 1984; Derrington *et al.*, 1984). The parvocellular cells have sustained responses, color opponency, low contrast sensitivity, high spatial resolution, low temporal resolution, and low conduction velocity. The magnocellular units, on the other hand, have transient responses, no overt color opponency, high contrast sensitivity (Shapley *et al.*, 1981; Sclar *et al.*, 1990), large receptive fields, high temporal resolution (Derrington & Lennie, 1984), and high conduction velocity (Kaplan & Shapley, 1982; Schiller & Malpeli, 1978). Interestingly, despite the differences in conduction speeds, the transmission times between retina and cortex are very similar for both cell types (Lennie *et al.*, 1990). Both parvo- and magnocellular neurons respond strongly to moving gratings, but only a small fraction of either type has a weak directional preference (Lee *et al.*, 1979), which in the case of parvocellular neurons depends on the wavelength of the stimulus (Lee *et al.*, 1979).

3.1.3 Striate cortex

The spatial segregation observed in the dLGN is largely maintained in striate cortex with the magnocellular neurons projecting to layer $4C_\alpha$ and the parvocellular neurons to layer $4C_\beta$ (Hubel & Wiesel, 1972). Tangential sections of macaque V1, stained for cytochrome oxidase activity (Wong-Riley, 1979), reveal a pattern of regularly spaced blobs, best seen in layers 2 and 3 but also visible in layers 5 and 6. Neurons in layer $4C_\beta$ send their axons to the supragranular layers 2 and 3 in either the blob or the interblob regions (Lund, 1973; Lund & Boothe, 1975), while neurons in layer $4C_\alpha$ project to layer 4B.

A third, albeit less prominent, pathway relays through the koniocellular layers of dLGN directly to the supragranular layers of V1 (Casagrande & Debruyn, 1982; Diamond *et al.*, 1985). In the macaque this projection terminates deep in layers 2 and 3 within the cytochrome blobs (Livingstone & Hubel, 1982).

Hubel and Wiesel (1968) were the first to report direction selectivity in some complex cells of the anesthetized monkey. Subsequent work in the behaving monkey showed the existence of rapidly adapting units, with pronounced directional selectivity and broad speed tuning (Wurtz, 1969). About one-third of the striate cells are directionally selective (De Valois *et al.*, 1982; Schiller *et al.*, 1976; Foster *et al.*, 1985; Burkhalter & Van Essen, 1986; De Yoe & Van Essen, 1985; Felleman & Van Essen, 1987; Desimone & Schein, 1987; Ferrera *et al.*, 1993; Albright *et al.*, 1984; Albright, 1984; Maunsell & Van Essen, 1983c; Lagae *et al.*, 1993; Desimone & Ungerleider, 1986; Colby *et al.*, 1993; Bender, 1982; Goldberg & Wurtz, 1972; Snowden *et al.*, 1992). However, in striking contrast to the distribution of directional cells in cat striate cortex, those in the monkey are distributed nonuniformly and are most commonly found in the upper portions of layer 4, i.e. $4C_\alpha$, 4B, and 4A, and in layer 6 (Dow, 1974; Hawken *et al.*, 1988). Moreover, directional units have a patchy horizontal distribution within 4B. For further discussion of the properties of these cells the reader is referred to Chapter 3 by Snowden.

3.2   Tectopulvinar system

3.2.1   *Superior colliculus*

In the macaque, about 10% of the retinal ganglion cells project to the superior colliculus (Perry & Cowey, 1984), with terminals in the upper half of the superficial gray layer (Hendrickson *et al.*, 1970; Tigges & Tigges, 1970; Hubel *et al.*, 1975). Studies using antidromic activation of ganglion cells (Schiller & Malpeli, 1977) have shown that the population of ganglion cells projecting to the tectum consists mainly of P_γ cells, and a small proportion of P_α cells. Collicular cells that receive retinal input have short latencies and their responses persist after reversible inactivation of the magnocellular layers of LGN (Schiller *et al.*, 1979), or ablation or cooling of the striate cortex (Schiller *et al.*, 1974; Marrocco, 1978).

Cortical projections to the superficial layers of superior colliculus originate from many visual areas, including areas V1 (Wilson & Toyne, 1970; Ogren & Hendrickson, 1976; Graham, 1982), V2, V3, V4 and middle temporal (MT) (Fries, 1984). The projection from striate cortex arises from layer V (Lund *et al.*, 1975; Fries, 1984) and from the solitary cells of Meynert in layer VI (Fries & Distel, 1983; Fries, 1984); the extrastriate corticotectal projections originate only in layer V. Both V1 and V2 cells send fibers that terminate most densely in the superficial layers (Graham, 1982). Area MT sends a moderate projection to the ipsilateral superior colliculus (Maunsell & Van Essen, 1983b; Ungerleider *et al.*, 1984), which extends through the lower one-half of the superficial gray and the optic layer.

Cells in the superficial layers respond vigorously to visual stimuli; they have spatially restricted receptive fields consisting of a central activating region surrounded by a suppressive zone (Lane *et al.*, 1971; Schiller & Koerner, 1971) and the majority receive binocular inputs. While directional selectivity in superior colliculus cells appears to be very prominent in many lower species, it is limited in the primate (Schiller & Koerner, 1971; Cynader & Berman, 1972; Goldberg & Wurtz, 1972; Marrocco & Li, 1977). Approximately 10% of the units in the superficial layers respond preferentially to one direction of movement with no response or inhibition to motion in the opposite direction (Goldberg & Wurtz, 1972). Cells respond to a wide range of stimulus speeds and some neurons do not react to static stimuli at all. The optimal stimulus is usually an edge oriented perpendicularly to the direction of movement, but most neurons respond also to moving small square stimuli.

Recently, cells in the superficial layers of the colliculus were reported to respond strongly to relative motion (Davidson & Bender, 1991). In this study, the stimulus was a small spot, the target, within the receptive field of a collicular neuron, surrounded by a large random dot background pattern covering the area outside the receptive field. In agreement with previous studies, most cells in the superficial gray layer were found to be nonselective for the direction or the speed of the target, when plotted with stationary background. However, more than 90% of the neurons responded select-ively when stimulus and background differed in direction of motion by an angle of at least 60 degrees. Their response was independent of the absolute direction of either the stimulus or the background. This sensitivity, however, was considerably reduced by corticotectal tract lesion, suggesting a cortical contribution to the relative motion selectivity in colliculus.

3.2.2   Pulvinar

The pulvinar is a large nuclear mass forming the most caudal portion of the thalamus. It is customarily subdivided into several nuclei based on cytoarchitectonic and physiological criteria. The monkey pulvinar is divided into at least three large nuclei: the medial, the lateral, and the inferior pulvinar. Although all three nuclei receive some input from the superior colliculus (SC), the inferior pulvinar is the primary target of the colliculo-pulvinar projection, which originates in the superficial layers of SC (Harting et al., 1973; Benevento & Fallon, 1975a; Partlow et al., 1977). Other subcortical afferents of interest are those originating in the retina (Campos-Ortega et al., 1970; Itaya & Van Hoesen, 1983), as well as those arising from the parvocellular layers of the dLGN (Trojanowski & Jacobson, 1975b).

The cortico-pulvinar projection originates in layer 5 of striate and layers 5 and 6 of extrastriate cortex. The frontal and temporal cortices project to the medial pulvinar, while the occipital and parietal areas project to inferior and lateral pulvinar (Ogren & Hendrickson, 1976; Trojanowski & Jacobson, 1977; Ogren & Hendrickson, 1979; Ungerleider et al., 1983). Area MT is reciprocally connected to the inferior and lateral pulvinar nuclei (Standage & Benevento, 1983; Maunsell & Van Essen, 1983b). MT neurons send axons only to that part of the lateral pulvinar which, along with the inferior pulvinar, makes up a single, complete visual field representation. The inferior pulvinar projects to the striate cortex, while the lateral pulvinar (Lund et al., 1975; Ogren & Hendrickson, 1976) projects mainly to the extrastriate cortices. The medial pulvinar projects to the temporal and frontal cortices (Trojanowski & Jacobson, 1975a; Benevento & Fallon, 1975b; Benevento & Rezak, 1976), and to area 7a of the parietal lobe (Ungerleider et al., 1984; Asanuma et al., 1985; Colby et al., 1983).

The anterior portion of pulvinar contains multiple retinotopically organized areas. In particular, mapping studies have shown the existence of a complete, first-order representation of the contralateral field in a region that includes all of the cytoarchi-tectonically defined inferior pulvinar, with the lower field represented dorsomedially and the upper field ventrolaterally (Bender, 1981; Allman et al., 1972). The lateral pulvinar contains both retinotopically (Bender, 1981) and nonretinotopically (Bender, 1981; Petersen *et al.*, 1985) organized regions. The retinotopically organized regions receive strong projections from superior colliculus and cortex and project back to the cortex. Most cells in the pulvinar nuclei have well-defined receptive fields, varying from 1 to 5 degrees, are binocular, and respond to either static or moving stimuli (Benevento & Miller, 1981; Bender, 1982). Similar to collicular cells, some neurons react only to moving stimuli with no response to static stimulation. Neurons have been also reported that have color-opponent response characteristics (Felsten *et al.*, 1983).

The majority of the inferior pulvinar cells are tuned for orientation or direction (Bender, 1982). The response magnitude of the orientation-selective units depends on the axis of movement and the sharpness of tuning often depends on the stimulus length. For many pulvinar neurons, responses evoked by motion in the preferred direction can be more than two times that in the null direction. Most cells respond well to a relatively large range of speeds, with about 50% of them tuned to speeds greater than 32 degrees per second.

In the lateral portion of the caudal pole of the pulvinar, which has been shown to lack retinotopic organization, directional cells were found that respond to stimulus movement towards or away from the animal. Such neurons have large, unflanked receptive fields, and about one-half of them respond well to both monocular and binocular stimulation (Benevento & Miller, 1981).

Lesions of the superior colliculus have almost no effect on the visual response of the orientation- or direction-selective pulvinar cells. However, they do affect cells having non-oriented receptive fields (Bender, 1983). Striate cortex lesions, on the other hand, have a profound effect on the inferior pulvinar cell-responses, showing that inferior pulvinar cells are critically dependent on descending input from visual cortex (Bender, 1983).

3.3 Extrastriate areas

3.3.1 Areas V2, V3 and V4

Cells in the thick stripes of V2 tend to be selective for disparity, orientation or direction. Direction-selective cells comprise about 8 to 16% of V2 neurons (Zeki, 1978). The majority of V2 cells seem to respond best to slow movement (Orban *et al.*, 1986). Recordings in the alert behaving monkey showed that some V2 neurons can be activated significantly more by stimulus movement than by motion of the retinal image of a stationary stimulus resulting from self-induced eye movements (Galletti *et al.*, 1988).

Physiological studies of V3 reported a high percentage (about 40%) of cells selective for orientation or direction (Felleman & Van Essen, 1987). Some cells show a striking, complex property of direction selectivity for multiple directions. Most cells (more than 90%) are tuned to the speed of the moving stimulus. Both V3 and the thick stripes of V2 project to areas MT, medial superior temporal (MST), floor of the superior temporal sulcus (FST) and ventral intraparietal (VIP), areas with a high incidence of directionally selective cells.

Area V4 was initially considered to be specialized almost exclusively for the analysis of color (Zeki, 1973, 1977). More recently, however, it has been shown that neurons in this area respond to a variety of stimulus attributes including color, orientation, and direction (Desimone & Schein, 1987; Desimone *et al.*, 1985; Schein *et al.*, 1982). About 13% of V4 cells respond at least three times as well, and 24% at least two times as well to stimulus movement in the preferred direction, compared with a stimulus moving in the null direction, which is not markedly different from the proportion of cells with an equivalent directionality in V1 (Desimone & Schein, 1987; Ferrera *et al.*, 1993).

Experiments, in which the magnocellular or the parvocellular layers of dLGN were reversibly inactivated by injecting lidocaine or gamma-aminobutyric acid (GABA), have shown that direction selectivity in V4 is not significantly altered by the blockade of either pathway (Figure 3). This is in striking contrast to the strong effects of magnocellular layer inactivation on the responses of MT neurons (Maunsell *et al.*, 1990), and implies that the limited direction selectivity observed in area V4 does not necessarily require the M pathway. Area V4 is interconnected with MT, but responses in MT were consistently reduced, if not entirely eliminated, when the magnocellular subdivision of dLGN was inactivated.

3.3.2 The middle temporal visual area

Area MT or V5 was first described by Dubner and Zeki (1971) as an area outside the striate cortex of the macaque monkey, which is specialized for visual motion. MT contains a complete representation of the contralateral visual field (Zeki, 1974; Gattass & Gross, 1981; Van Essen *et al.*, 1981). Receptive field size increases linearly with eccentricity, and it is about ten times larger than in striate cortex (Albright & Desimone, 1987), suggesting a considerable convergence of inputs. The area is organized in a columnar fashion with blocks of cortical tissue containing neurons with similar directional selectivity (Albright, 1983; Albright *et al.*, 1984).

The connections of the retinogeniculate system to area MT have been described in Chapter 3. The projection from V1 and V2 originates in specific laminae or tangential subdivisions. V1 input is received from layers 4B and 6, which are the sites of directionally selective cells in this area (Lund *et al.*, 1975; Maunsell & Van Essen, 1983b; Ungerleider & Desimone, 1986; Shipp & Zeki, 1989), whereas V2 input originates from the cytochrome oxidase thick stripes (De Yoe & Van Essen, 1985; Shipp & Zeki, 1985), that in turn receive a direct projection from layer 4B (Hubel & Livingstone, 1987). Hence, whether directly or indirectly through V2, the input to MT seems to be dominated by the M pathway.

Strong physiological evidence for an almost exclusive M contribution in MT was recently provided by experiments, in which the responses of MT units were studied while selectively blocking the activity in the magnocellular or parvocellular layers of dLGN (Maunsell *et al.*, 1990). Whereas parvocellular blockade had minor effects on the activity of MT neurons, inactivation of magnocellular neurons produced complete or nearly complete block of MT responses (Figure 3), suggesting a predominance of the M pathway in the geniculostriate input to this area.

MT projects to various subcortical structures such as the ventral lateral geniculate

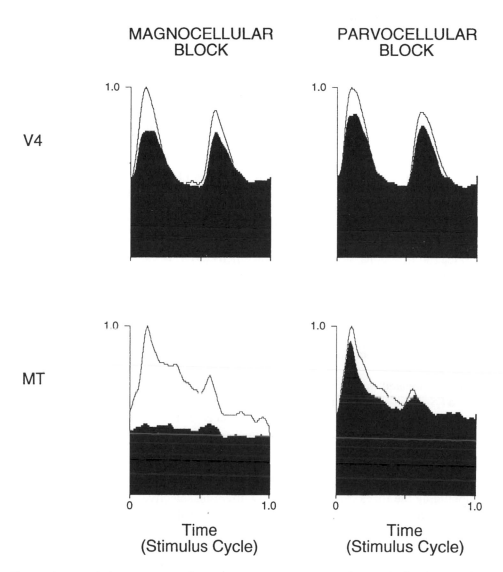

Figure 3   Population responses from all V4 and MT sites before and after inactivation of either parvocellular or magnocellular layers of dLGN. White histograms are responses before and black after the injection. Histograms from individual sites are normalized to the same peak pre-injection response and to a time of 1.0 stimulus cycle, and subsequently averaged together. The resulting mean histogram was smoothed with a Gaussian kernel. Adapted from Maunsell *et al.*, 1990; Ferrera *et al.*, 1993.

nucleus, the reticular nucleus of the thalamus, the claustrum, the putamen, the caudate nucleus, and the pontine nuclei. As described above, MT is also reciprocally connected to the inferior and lateral pulvinar nuclei (Standage & Benevento, 1983; Maunsell & Van Essen, 1983). Extensive reciprocal connectivity has been shown between MT and many cortical areas, which include the primary visual cortex, V2, V3, V3A, V4, V4t, the MST, the lateral intraparietal (LIP), VIP, the FST, and the frontal eye fields (FEF) (Zeki, 1969; Cragg, 1969; Maunsell & Van Essen, 1983b; Montero, 1980; Ungerleider & Mishkin, 1979; Van Essen *et al.*, 1981; Ungerleider & Desimone, 1986; Felleman & Van Essen, 1991).

A striking characteristic of MT is its high proportion of neurons that are selective to stimulus motion (Dubner & Zeki, 1971; Zeki, 1974; Maunsell & Van Essen, 1983a, 1983c; Albright, 1983). More than 90% of the cells of this area are directionally selective, and most of them show selectivity to the stimulus orientation when tested with stationary, flashed bars (Maunsell & Van Essen, 1983c; Albright, 1984). The response of the neurons to their preferred direction can be up to 10 times that to their nonpreferred or null direction (Maunsell & Van Essen, 1983c). Interestingly, however, the tuning of MT cells is not sharper than that of striate neurons (Albright, 1984). Area MT does not seem to be concerned with the fine tuning of its afferents from other directionally selective areas, the cells of which show direction selectivity. Instead, most MT as well as other superior temporal sulcus (STS) cells seem to elaborate their receptive field structure, responding to more complex stimuli that require integration of local motion vectors or integration of the output of directional spatial frequency channels. A detailed description of MT responses to complex stimuli is given by Snowden in Chapter 3, and Stoner and Albright in Chapter 9.

3.3.3   Area MST, FST and VIP

Area MST is located medial to MT occupying most of the bottom half of the upper bank of STS and a small proportion of the adjacent floor, while FST lies anterior to MT in the fundus of the sulcus. Both areas receive a direct projection from MT (Desimone & Ungerleider, 1986), and project to posterior parietal cortex. They are also connected to the pulvinar, mainly to its medial subdivision, to the striatum, and to the claustrum (Boussaoud *et al.*, 1992).

MST has a crude topographical organization, with cells that have large receptive fields including the fovea. Most are strongly direction selective, and they respond best to large moving stimuli, such as a pattern of random dots (Van Essen *et al.*, 1981; Desimone & Ungerleider, 1986; Komatsu & Wurtz, 1988). In addition, over 90% of MST neurons were found to be sensitive to the disparity of the visual stimulus (Roy *et al.*, 1992). Some neurons respond strongly to rotating or isotropically expanding or contracting patterns of motion, as well as to frontal parallel planar motion (Tanaka & Saito, 1989; Saito *et al.*, 1986; Duffy & Wurtz, 1991a, 1991b), and their response magnitude is invariant with respect to the size, shape or contrast of the pattern. Such neurons might contribute to the analysis of the large-field optic flow stimulation generated as an observer moves through the visual environment.

Area MST plays an important role in both the initiation and the maintenance of smooth pursuit. In addition, the fact that a subset of visually responsive, directional cells discharge during smooth pursuit of a small target in an otherwise dark room, even if the pursuit target is turned off transiently, suggests that MST also receives signals from extraretinal sources (Newsome *et al.*, 1988; Sakata *et al.*, 1983). Lesions in this area produce deficits in both pursuit eye movements and in optokinetic nystagmus (Dursteler & Wurtz, 1988). Monkeys with MST lesions were unable to match eye speed to target speed or to correctly estimate the amplitude of the saccade needed to acquire the target to compensate for target motion. Moreover, the lesioned animals were unable to pursue smoothly after the target was foveated.

Area FST has no retinotopical organization. Receptive field size is large, usually including the fovea. About 32% of the cells in FST show directional selectivity, mainly responding to complex object motion.

Area VIP, which probably coincides with a projection zone described by Maunsell and Van Essen (1983), contains neurons with response properties very similar to those reported for area MT. In fact, about 80% of VIP neurons respond at least twice (mean = 9.5 times) as well to stimulus movement in the preferred direction as compared to stimulus movement in the null direction (Colby *et al.*, 1993). Many neurons in area VIP are tuned to the speed of the stimulus, with a speed range from 10 to 320 deg/s.

4 RESPONSES TO NONLUMINANCE CUES

4.1 Responses to second-order stimuli

Moving stimuli fall into two broad classes on the basis of the spatial characteristics of the defining cues (Cavanagh & Mather, 1989). First-order stimuli are those that can be solely defined by their luminance or color difference from the background, while second-order stimuli are defined by the frequency with which specific combinations of intensity or color values occur for pairs of spatial locations (Cavanagh & Mather, 1989). Typical stimuli of the first class are patterns defined by luminance contrast. Second-order stimuli, on the other hand, are patterns in which the figures can be only perceived through changes in texture, motion, or disparity.

Although traditional motion models are unable to account for the detection of a second-order stimulus (Chubb & Sperling, 1988), both stimulus types can be equally detected by the motion system, and they are likely to share the same basic phenomena (Cavanagh & Mather, 1989).

In a recent study Albright (1992) examined the responses of MT neurons to one type of second-order stimulus, termed drift-balanced or non-Fourier motion stimulus (Chubb & Sperling, 1988). The stimulus has the appearance of twinkling dots that drift across a background of identical but static texture. MT neurons exhibited similar directional tuning for either the luminance or the second-order stimuli. Thus, neurons in this area can respond to patterns defined by other than luminance cues. Similar experiments have been performed with stimuli defined by static texture (Van Essen *et al.*, 1989, 1991) and isoluminant colors.

4.2 Chromatic input to motion system

4.2.1 Motion processing at isoluminance

The contribution of color information to motion perception has been a subject of recent debate, mainly because of the apparent separation of the pathways believed to underlie color and motion processing.

As discussed above, the P and M pathways remain largely segregated from the retina through the striate cortex. In cortex, the P pathway contributes mainly in the extrastriate visual areas specialized to form and color processing (Merigan & Maunsell, 1993), while the M pathway sends a direct projection to the area MT. Based on the physiological properties of cells at various levels of these processing streams and on the contributions that each stream makes in different extrastriate areas, it has been suggested that each pathway underlies the processing of different visual attributes. In particular, the P pathway is thought to play a central role in color vision and in the perception of high spatial frequency patterns, whereas the M pathway, referred to by some investigators as the luminance channel (Livingstone & Hubel, 1988; Livingstone, 1988; Shapley, 1990), is thought to play an important role in the perception of low contrast stimuli, depth and visual motion.

The hypothesis that motion and color are analyzed by separate neural systems the former of which is insensitive to color, and the latter to motion velocity, appeared to explain early psychophysical results, which showed that apparent motion normally visible in color random dot kinematograms disappeared when the dots had the same luminance with their background (Gregory, 1977; Ramachandran & Gregory, 1978). Moreover, when drifting sinusoidal color gratings were presented on a color display, a dramatic decrease of the perceived velocity occurred at and around isoluminance; for some spatial and temporal frequency combinations the drifting gratings could be even perceived as 'stopped' (Cavanagh et al., 1984).

Recent single unit recordings in the magnocellular layers of dLGN, however, show that M neurons do react to isoluminant color transitions with frequency-doubled, unsigned responses (Schiller & Colby, 1983; Logothetis et al., 1990; Lee et al., 1989). Although this type of response could be useful for the detection of temporal changes, it is unclear whether it can provide the motion system with unambiguous direction information.

4.2.2   Responses of MT cells to chromatic stimuli

The response of MT cells to isoluminant chromatic stimuli has been studied in different laboratories (Lee et al., 1989; Logothetis et al., 1990; Saito et al., 1989; Gegenfurtner et al., 1994; Charles et al., 1993). We have recorded from MT cells stimulated with chromatic gratings modulated along different directions in a cone contrast space (Cole et al., 1990; Eskew et al., 1990; Charles et al., 1993), based on the Smith and Pokorny cone fundamentals (Smith & Pokorny, 1975). The axes of this space determine changes in individual cone excitation normalized to the cone excitation caused by the background. The vertical axis depicts M cone contrast ($\Delta M/M$), and the horizontal axis depicts L cone contrast ($\Delta L/L$). Modulation approximately along the $45°$ axis represents luminance modulation of the stimulus, whereas along the $-45°$ axis represents isoluminant chromatic modulation (Figure 4). A chromatic grating consists of the superposition of two component gratings, generated by modulating sinusoidally the red and green phosphors of the monitor. The chromaticity of such a grating is determined by the amplitude and phase difference of the component gratings. For example, a yellow luminance grating is generated by an inphase modulation, while a red–green isoluminant grating by an counterphase

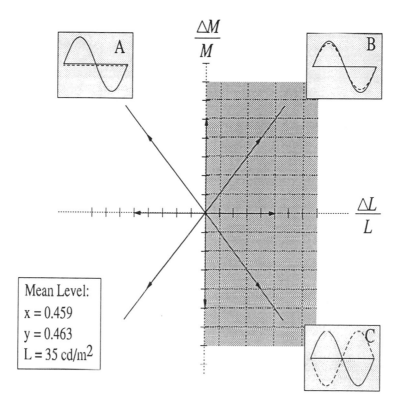

Figure 4 Cone contrast space (see text).

modulation of the red and green phosphors. Isoluminance was determined in psycho-physical experiments using the minimum motion technique (Anstis & Cavanagh, 1983).

Cells in MT were found to respond to red–green isoluminant stimuli. Figure 5 shows the responses of one MT cell to drifting color and luminance gratings. The gratings for each condition had identical velocities, directions and spatial frequencies. The top row represents the cell activity to sinusoidal gratings in which L and M cones are modulated 180° out of phase with respect to each other (chromatic modulation). The lower row shows the response to sinusoidal gratings in which the L and M cones are modulated in phase (luminance modulation). Each column of the histograms shows a different level of cone modulation. The response of most MT cells to isoluminant gratings was found to increase as the cone modulation increased. For the cell in Figure 5, the largest response was to an isoluminant red–green grating with a 16% L and M out of phase cone modulation. The directional tuning of most MT cells remained the same for different luminance contrasts of the chromatic gratings (Figure 6).

The results of these experiments demonstrate that MT is capable of processing information based on differences in wavelength as well as luminance. This is in agreement with recent psychophysical experiments that show a clear contribution of color to the motion system (Cavanagh & Anstis, 1991).

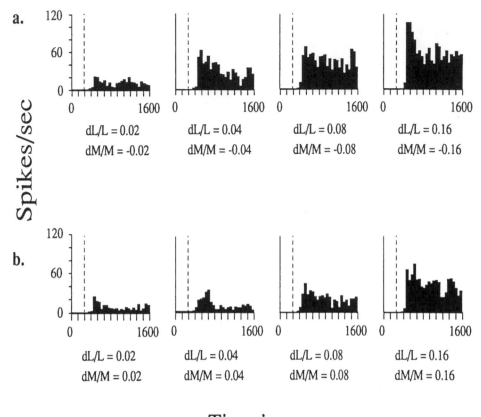

Figure 5 Response of an MT cell to luminance and chrominance contrast. (a) Cell responses to an antiphase L and M cone modulation. (b) Cell responses to an inphase L and M modulation. Cone contrasts are equal for each column and increase from left to right (2, 4, 8, and 16%). Bin width is 5 ms and there are ten trials for each condition.

What mechanisms mediate this response to chromatic borders? Electrophysiology has shown that the isoluminance point at which dLGN neurons give a minimum response varies from cell to cell (Logothetis *et al.*, 1990). This implies that no matter what the ratio of luminances between two colors, some cells will always respond. A motion response to pure chromatic stimuli could therefore be attributed to variation of the isoluminance point among different neurons. Nevertheless, psychophysical evaluation of this scatter in isoluminance points has shown that it accounts for only a small portion of the overall contribution of color to motion (Cavanagh & Anstis, 1991).

Could the unsigned response of the magnocellular neurons, discussed above, provide an input to the motion system? An obvious test for such an hypothesis is a color grating that jumps by 90°. While this would be a powerful motion stimulus for a system selective for the polarity of contrast (either luminance or color contrast), it

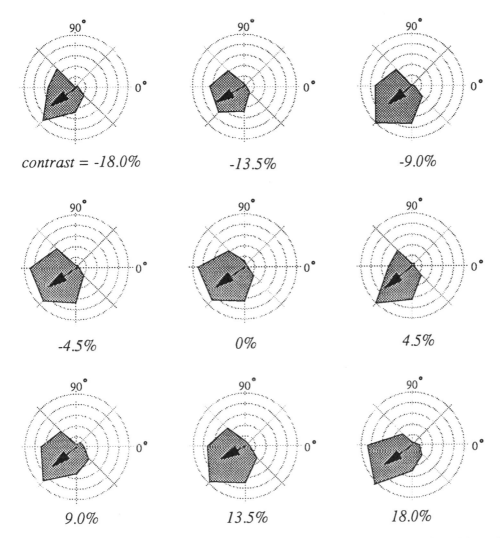

Figure 6 Direction tuning of an MT cell tested for nine luminance contrast values. The red modulation was constant and the green modulation varied from — 18.0% to + 18.0%.

would be an ambiguous motion stimulus (a jump could be seen 90° to the left or to the right) to any mechanism that only saw the color transitions and was indifferent to the direction of color change (Logothetis, 1991). Human observers report visible motion for 90 deg jumps (Cavanagh & Anstis, 1991), suggesting an opponent-color input to the motion system.

Dobkins and Albright (personal communication) recorded the activity of MT neurons using a similar stimulus. In their experiments monkeys viewed an 'apparent motion' stimulus in which the chromatic gratings underwent contrast-reversal each time they were displaced in a given direction. For small displacements MT cells responded to the motion of the chromatic borders disregarding the chromatic sign information. For large displacements, however, in which the chromatic borders provide ambiguous information, the direction selectivity of the neurons was determined by the motion of the signed chromatic borders.

5 SHAPE FROM MOTION

5.1 Sensitivity to motion-defined borders

The visual system can exploit a variety of different sources of information for the perception of shape. An object's boundary, for example, may be defined by a change in luminance contrast, chromatic contrast, or texture contrast. However, even when objects match their background in all the above attributes, figure–ground segregation may be achieved when the object or the observer is in motion. The boundaries of the object are, in this case, defined by motion contrast.

The capacity of the visual system to process shape defined by motion contrast can be studied using random dot patterns, whereby dots within a rectangular area of the display move in one direction, while dots in the complementary display area move in a different direction.

Psychophysical experiments using such displays showed that the sensitivity of human observers for orientation discrimination, for shape discrimination, and for vernier acuity is very similar for a dotted bar defined entirely by motion-sensitive elements and for a dotted bar whose boundary was determined by luminance contrast (Regan, 1986, 1989; Regan & Hamstra, 1991). Obviously, the visual system can efficiently use motion information for extracting the shape of an object. Given that form and movement processing are at least partly segregated, how is it that motion information is made directly available to those visual areas specialized in form analysis?

5.2 Responses of V4 cells to motion contrast

Recently cells have been reported in V4 of the behaving monkey that respond to the orientation of a grating defined through relative motion (Charles *et al.*, 1993). Figure 7 shows the responses of such a neuron to a luminance-contrast (A), and a motion-contrast (B) grating. The cell gave responses of almost equal magnitude for both types of gratings for all tested orientations (Figure 7C).

Figure 8 shows the distribution of the orientation selectivity of 89 units collected from one of the tested monkeys. The selectivity index (SI) was defined as the mean vector for the four tested angles weighted by the relative cell responses. Such an index has a value of one when the cell responds only for one given orientation. A non-selective cell, giving equal responses to all orientations, has a selectivity index of zero.

Figure 8A illustrates the distribution of selectivity indices for luminance-defined, and Figure 8B for motion-defined gratings. Both distributions are broad, ranging from

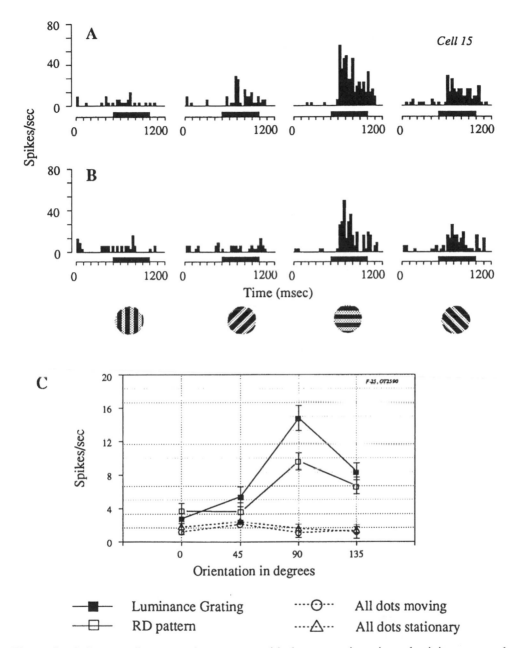

Figure 7 Cell responding to motion-contrast with the same orientation selectivity measured for luminance-contrast gratings. The ordinate in the plots is the cell's firing rate and the abscissa is the time in milliseconds. The horizontal thick line under each histogram indicates the duration of stimulus presentation. A. The cell responds well to the upwards drifting grating, less to upward and right drifting grating, and poorly to the other orientations. B. The response of the cell to gratings produced through motion-contrast. C. The orientation tuning plot for the two types of gratings, together with the responses for the control conditions. RD, random dot.

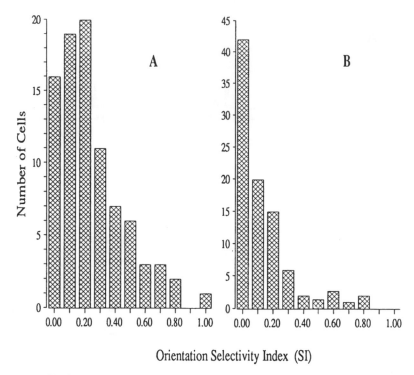

Figure 8 Distribution of cells responding to different types of gratings. A, luminance gratings; B, random dot gratings.

cells with no orientation selectivity to narrowly tuned cells. In this sample 38% of the cells showed significant orientation tuning for the luminance gratings, and about 20% of the cells responded selectively to the orientation of the motion-defined grating. Similar results have been reported recently for the inferior temporal cortex of the monkey (Sry *et al.*, 1993), where units were found to respond selectively to the two-dimensional (2D) shape of a visual stimulus defined by virtue of motion contrast.

Such experiments show that the temporal pathway utilizes motion information for analyzing shape. Area V4 receives input from both the P and the M channels (Ferrera *et al.*, 1993), and it has a number of cortical inputs from areas with high proportions of directional units, such as V3 or MT. Moreover, as indicated above, a limited number of neurons in this area show directional selectivity. Whether the response to relative motion is the result of local processing or a reflection of the afferent connections of this area remains to be seen.

6 DISRUPTION OF MOTION PROCESSING THROUGH LESIONS

The anatomical and physiological work reviewed in the previous sections show the existence of multiple, interconnected visual structures, which may be involved in the analysis of visual motion. Among the various visual structures, the M pathway and the extrastriate areas of the superior temporal sulcus are likely to play the most important role. It could be expected, therefore, that damage to the M pathway or to the STS areas results in severe deficits in all perceptual tasks that rely on motion analysis. The dramatic effects of selective cortical damage on motion perception in humans are in fact a strong hint as to what the consequence of ablation of such functionally specialized visual structures could be.

6.1   Clinical studies

A clinical report dating from the beginning of the century describes a patient with bilateral damage of the posterior extrastriate areas who could no longer perceive the movement of visual objects, but had the perceptual experience of a stimulus appearing in successive positions over time (Poetzl & Redlich, 1911). The patient's color and form vision, however, were unaffected.

That movement perception could be mediated by a specialized subsystem in the visual system was first proposed by Riddoch (1971), who noticed that the recovery of motion vision in patients blinded by gunshot wounds preceded that of pattern vision (cited by Zeki, 1991). Such observations, however, were entirely dismissed, mainly because of lack of anatomical and physiological evidence regarding functional organization within the visual system (see review by Zeki, 1991).

An unequivocal case of a patient suffering from 'motion blindness' was reported by Zihl and colleagues about ten years ago (Zihl et al., 1983). These authors describe in detail the perceptual deficits in motion perception occurring after bilateral damage to most of the parietal cortex, including areas 39, 40, 7. At the time of her testing, 19 months after her stroke, the patient displayed a number of perceptual deficits affecting her daily life, such as being unable to pour a full cup of coffee without having it overflow, or being frightened in the face of on-coming traffic, since a vehicle that would initially seem far away would be suddenly seen dangerously close. Detailed testing revealed normal temporal resolving power (normal critical flicker fusion values), unaffected spatial localization accuracy, saccadic eye movement accuracy, color vision, and acuity. Yet, motion perception was severely disturbed. Movement in depth was entirely abolished, and there was a nearly complete loss of motion vision for moving objects in the frontal plane for eccentricities greater than $15°$. For visual targets moving along the horizontal or vertical axes, within the central $15°$, movement perception was somewhat preserved, with fairly accurate velocity judgments.

This finding together with the physiological and anatomical demonstration of a network of visual areas, the neurons of which are particularly selective for motion attributes, raised an experimentally amenable question. What are the effects of experimentally induced lesions in the motion pathway?

6.2   The effects of selective magnocellular lesions

Disruption of either the P or the M channel was accomplished by systemic administration of the toxicant monomer, acrylamide, inducing severe damage to the P system

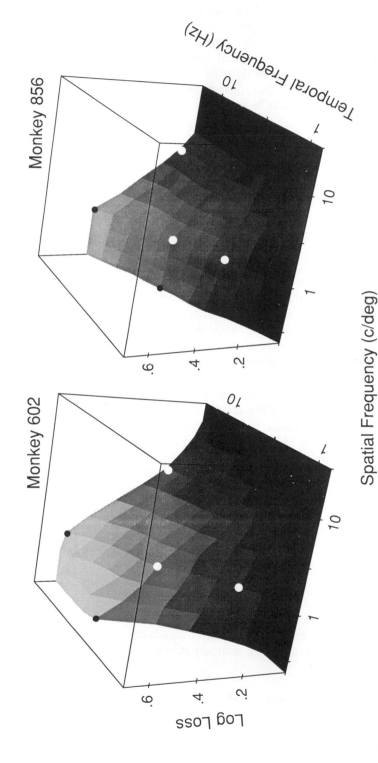

Figure 9 Loss of sensitivity for detecting drifting gratings produced by magnocellular lesions. Solid circles signify the measured values that were used for the surface interpolation. Reproduced with permission from Merigan *et al.* (1991a).

(Mulliken *et al.*, 1984), or by punctate injections of ibotenic acid in one dLGN sub-region. Ibotenic acid is a neurotoxin that is believed to destroy the cell bodies but leave the fibers of passage intact. Later research showed that this is to a large extent true, although considerable demyelination does initially occur due to an autoimmune response provoked by the neurotoxin (Coffey *et al.*, 1988). By recording neural activity with the same electrode-micropipette, ibotenic acid injection sites were restricted to desired visual field regions. These circumscribed lesions allowed direct comparison of lesioned and intact areas in the same experimental session (Merigan & Maunsell, 1990; Schiller *et al.*, 1990a, 1990b, Schiller & Logothetis, 1990; Merigan, 1991; Merigan *et al.*, 1991a, 1991b; Merigan & Maunsell, 1993).

Parvocellular lesions impaired high spatial frequency form vision, color vision, and high spatial frequency stereopsis, while leaving motion-related tasks unaffected. Magnocellular lesions, on the other hand, produced deficits in flicker and motion perception, but only for a certain spatiotemporal domain. Figure 9 shows the loss of contrast sensitivity for detecting drifting gratings resulting from lesions in the magnocellular layers. The ability of the monkeys to detect the gratings is affected for low spatial and high temporal frequencies (Schiller *et al.*, 1990a, 1990b; Merigan & Maunsell, 1990; Merigan *et al.*, 1991a). Increasing contrast, however, improved performance substantially, suggesting that the M pathway is more likely to be important for mediating the visibility of the gratings (Merigan et al., 1991a). The detectability of motion was found to depend also on the size of the stimulus (Figure 10A). Finally, the ability of the animals to discriminate the speed of random dot patterns has been found to be reduced but not eliminated by the magnocellular lesions (Figure 10). Thus, it appears that the motion system receives input from both pathways, the relative contribution of which is dependent on the spatiotemporal characteristics of the visual stimulus.

6.3 The effects of selective extrastriate lesions

The results reviewed here show that area MT, together with its satellite areas MST, FST and VIP, plays a predominant role in the analysis of visual motion. In fact, experiments showing that the sensitivity of MT neurons is comparable with the psychophysical performance in motion-discrimination tasks (Britten *et al.*, 1992; Salzman *et al.*, 1990), and experiments demonstrating that the activity of MT neurons during binocular motion rivalry reflects the perceived motion direction (Logothetis & Schall, 1989) suggest that single neuron activity in these areas may directly underlie motion perception.

However, initial attempts to examine deficits in motion perception following MT lesions in the monkey revealed a rather complicated picture. Newsome and Paré (1988) trained monkeys to discriminate upwards from downwards motion using dynamic random dot patterns. Since such patterns do not have figural elements, they preclude the detection of spatial displacements by simply remembering the position of a distinct contour (Lappin & Bell, 1976), thereby allowing isolation of the motion-sensing mechanism. The display consists of a set of dots coherently displaced (correlated motion signal), which is spatially dispersed among motion

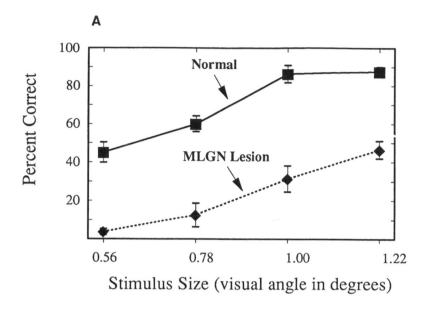

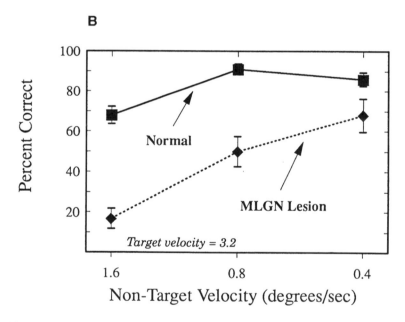

Figure 10 The effects of magnocellular lesions on motion detection (A), and velocity discrimination (B). Figure 10A also shows the effects of stimulus size in the detectability of motion. From Schiller *et al.* (1990b).

noise. The strength of the motion signal can be determined by the proportion of dots moving coherently in one direction.

Using this display Newsome and Paré tested monkeys before and after punctate lesions, made by injection of small amounts of ibotenic acid in electrophysiologically selected regions of area MT. When the MT-lesioned animals were tested within 24 hours after the injection, motion thresholds were found to be evaluated up to eight times those measured preoperatively, revealing a considerable dysfunction of the motion system. Contrast sensitivity was unaffected suggesting a motion-specific rather than a general visual impairment. The deficits vanished after a period of about two weeks. Similarly, the ability of the monkey to perform smooth pursuit eye movements, which have been shown to be elicited by the motion of the target's image across the retina, recovered within a short period of time, with only mild permanent deficits (Newsome et al., 1985).

Somewhat different results were obtained from a study examining the effects of restricted lesions in MT on the ability of the monkey to see relative motion, and 3D structure from motion. Andersen and colleagues (Andersen & Siegel, 1990; Siegel & Andersen, 1991) trained monkeys to detect shear motion in random dot displays. Shear motion is one type of relative motion in which the velocity of the moving dots, corresponding to one frame line, varies as a function of the line number, i.e. it varies perpendicular to the motion axis. This type of motion has been implicated in figure–ground segregation or in the extraction of depth through the kinetic depth effect (Rogers & Graham, 1979, 1982), and has been studied in detail in human psychophysics experiments. Psychophysical work has shown that macaque monkeys and humans have similar sensitivity profiles when tested for the detection of sinusoidally modulated shearing motion (Golomb et al., 1985). Testing monkeys with MT lesions revealed a dramatic increase in shear thresholds that was largely confined to the retinotopic locus of the lesion. Contrast thresholds were again unaffected. Similar to the previous studies, however, the impairment was transient with full recovery after a few days.

In contrast, severe deficits in the perception of 3D structure-from-motion lasted for at least 3 weeks (as long as the investigators tested the monkey). Structure from motion is an integrative motion process, requiring analysis of the velocity fields, and although depending on local motion signals, presumably requires appropriate pooling of all local directional responses. As such, it seems to be strongly affected even by partial MT lesions. Permanent deficits in the perception of structure-from-motion were reported recently for human patients suffering from occipital–parietal lesions. Thus, patients who had otherwise normal vision and could even perceive shape from motion (Vaina, 1989; Vaina et al., 1990), were unable to use motion information for the perception of the 3D shape of an object.

All experiments cited above indicate that area MT is an important part of the motion system. Nonetheless, the recovery of function observed in these experiments raises some questions. What is the mechanism mediating this recovery? Given that the pharmacological lesions were restricted within the lesioned area and seem to spare the white matter, why does recovery occur in only some tasks and not in others?

One possible mechanism for the recovery of function after chemical lesions is reorganization within area MT. Local reorganization of receptive fields after damage is not uncommon in cortex (Kaas et al., 1983; Merzenich et al., 1983, 1984), and recently it has also been shown for the visual system (Pettet & Gilbert, 1992). It would be reasonable to assume that reorganization could be more likely in areas having neurons with relatively large receptive fields like MT, and even more so for areas like MST with very little retinotopy and very large receptive field size. An obvious way to rule out the possibility of local reorganization is to lesion the entire area MT, or the entire region corresponding to the motion-related areas MT, MST and FST. Such experiments have been done with bilateral ibotenic acid injections at multiple sites along the STS, and with surgical removal of the entire caudal part of the posterior bank of STS by aspiration. Yet the results do not seem to be in total agreement.

Surgical bilateral removal of a region corresponding to areas MT, MST, and FST in the macaque permanently impaired the animal's ability to use a kinetic boundary for the discrimination of shape, while their ability to do so based on luminance cues was unaffected (Marcar & Cowey, 1992). Some of the animals were found unable to discriminate between a random dot display with zero correlation and an adjacent display in which a proportion of the dots oscillated coherently, even when the latter had no visual noise, i.e. all dots were oscillating coherently. Moreover, some other animals were unable to discriminate even the simplest motion of single dots (Cowey & Marcar, 1992). Different results were obtained in another study in which MT was ablated unilaterally by aspiration (Schiller, 1993). The animals in the latter study were tested in a number of visual tasks, such as contrast sensitivity, color vision, form vision, stereopsis, and flicker and motion perception. Lesioning MT yielded only mild to moderate deficits in motion and flicker perception, while shape from motion was unaffected. One possible explanation for the difference in these results is damage to the white matter underlying the posterior bank of STS, caused by the surgical manipulations. Histological examination of dLGN in these animals showed indeed retrograde cellular degeneration that is probably due to partial damage of the optic radiation.

A detailed and conclusive study was recently carried out by Pasternak and Merigan, who made multiple and extensive ibotenic acid injections in areas MT and MST (Pasternak & Merigan, 1993). They tested both local and global motion sensitivity using sinewave gratings of various spatiotemporal frequencies, and dynamic random dot patterns. In their random dot display each dot could take an independent 2D walk generated by a given spatial offset at each video frame (Williams & Sekuler, 1984). Displacement was usually held constant for all the dots, while direction varied stochastically in every frame. In such displays, when the range of the direction distribution is 360° no global motion is seen. Smaller ranges, however, yield the perception of a coherent flow along the mean of the direction distribution. The strength of the motion signal was determined by the distribution range, whereas noise levels were defined by varying the proportion of the dots that moved stochastically in any direction.

Using drifting luminance gratings, no permanent deficit was found in contrast sensitivity over a range of about four octaves, for either detection or direction discrimination. Thresholds for speed discrimination with drifting grating were elevated about two to four times for the animals that had extensive damage to both MT and MST. The monkeys could preoperatively discriminate 5–10% differences in the drift rate of the gratings over a broad range of contrasts. Postoperatively, the animals showed reduced speed sensitivity, but they could still discriminate speed differences of less than 40%.

The residual performance could be mediated by a functional, nondirectional mechanism underlying flicker discrimination. Nevertheless, such a mechanism has been shown to be less sensitive to temporal changes than the directional mechanism that processes speed information. Psychophysical work in humans and cats has shown that discrimination of flicker is less accurate than that of speed (McKee *et al.*, 1986; Pasternak, 1987), and the same seems to be true for at least one of the animals tested in this study. Moreover, flicker discrimination itself has been shown to depend on the directional mechanisms (Pasternak, 1987), since their elimination affects flicker discrimination as well (Schiller, 1993; Pasternak, 1987).

Testing global motion perception also revealed mild deficits when the random dot stimulus had a strong motion signal. The investigators varied both the range of the direction distribution and the noise level by changing the percentage of the dots that had 360° direction range. For zero noise levels, whereby the direction range of all dots was varied, postoperative testing for small differences in direction or speed showed only mild deficits. A permanent and severe deficit was only found for reduced signal-to-noise ratios.

Although the experiments discussed above yielded diverse and partly conflicting results, some conclusions can be drawn pertaining to the role of area MT in motion perception. For one, partial lesions at the extrastriate cortical level are unlikely to provide us with useful information as to the visual capacities mediated by the lesioned areas. Not only are the deficits transient, with very fast recovery, but even small sparings are evidently enough to support many visual functions at the threshold level. Large lesions seem to be effective in producing lasting deficits, particularly when neighboring areas subserving similar functions are included in the damaged area. However, even the entire ablation of the MT–MST area has small effects on primary visual motion tasks, such as direction or velocity discrimination, showing clearly that area MT is not all that we have to process motion information. Other areas must be involved, and an interesting question is whether such areas are normally serving different aspects of motion perception or are recruited after the disruption of processing within the superior temporal sulcus.

Finally, none of the lesions placed within STS causes the effects observed with human patients, suggesting some caution when interpreting human 'motion blindness' as a result of damage in the human homologue of area MT.

7 CONCLUDING REMARKS

Our knowledge of the organization of the visual system and of the transformations of its inputs at different levels of processing has greatly increased in the last thirty years. The notions of hierarchical and parallel processing helped us a great deal in the conceptualization of some of the problems that the visual system is called on to solve. Useful engineering concepts like 'channels' have been recruited to describe processing streams composed of neurons with related properties at different levels of the visual system. The 'motion channel' is one of those. It has been described as a pipeline starting at the retinal level through the P_{α} cell activity, and reaching the parietal lobe through specific laminae and areas in cortex. Many results presented in this review chapter refute this notion.

Information in the motion system seems to be flowing concurrently in both a 'vertical' and a 'horizontal' direction, within and among many interconnected subcortical and cortical visual areas. Lesions of the striate cortex show that extrageniculate

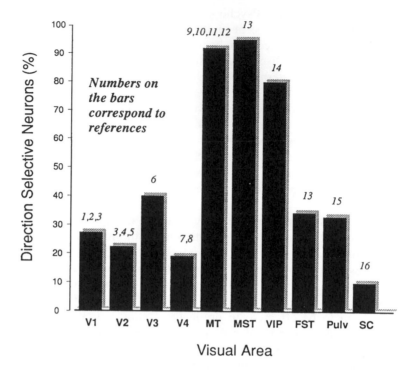

Figure 11 Proportion of direction-selective cells in various subcortical and cortical areas. Each column represents either a single value reported in one reference, or the mean of the values reported by different investigators. The numbers above each bar correspond to the following research studies: (1) De Valois *et al.*, 1982; (2) Schiller *et al.*, 1976; (3) Foster *et al.*, 1985; (4) Burkhalter & Van Essen, 1986; (5) DeYoe & Van Essen, 1985; (6) Felleman & Van Essen, 1987; (7) Desimone & Schein, 1987; (8) Ferrera *et al.*, 1993; (9) Albright, 1984; (10) Albright *et al.*, 1984; (11) Maunsell & Van Essen, 1983c; (12) Lagae *et al.*, 1993; (13) Desimone & Ungerleider, 1986; (14) Colby *et al.*, 1993; (15) Bender, 1982; (16) Goldberg & Wurtz, 1972.

pathways are capable of supporting motion processing in extrastriate cortex. Areas other than MT or MST seem to have a considerable proportion of neurons with highly directional responses (Figure 11), and a number of visual cortical areas must be taken into account for the generation of a uniform distribution of the neurons' preferred speeds (Figure 12).

Inputs to the motion system are unlikely to be of one simple sort. Whether luminance, chrominance or some second-order characteristic of the image, the motion system is likely to utilize a great variety of visual structures composing the *motion network*. Visual areas like those in the superior temporal sulcus may represent important functional links for some aspects of motion processing, but they are highly unlikely to be the *necessary* and *sufficient* neural substrate of motion perception as suggested by other investigators (Movshon & Newsome, 1992).

Area MT, in particular, seems to be an important part of the motion network. Damage to this area will cause some deficits, whether transient or permanent, that

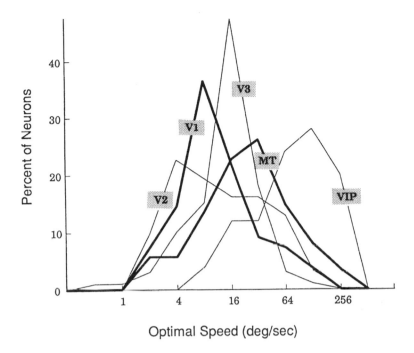

Figure 12. Distribution of speed preferences in different visual areas. V1 and V2 from Orban *et al.* (1986); V3 from Felleman and Van Essen (1987); MT from Maunsell and Van Essen (1983c), VIP from Colby *et al.* (1993).

are very specific to the motion system. Tasks requiring integrative activity such as pooling the responses of local motion detectors into some global representations, e.g. 3D surface descriptions or optic flow fields, appear to be primarily mediated by areas like MT or MST. Nonetheless, perceiving 3D structure from motion is currently the only tested capacity that is permanently and entirely lost, even with the application of small restricted MT lesions.

REFERENCES

Addams, R. (1964). An account of a peculiar optical phenomenon. In W. Dember (Ed.) *Visual Perception: The Nineteenth Century*, pp. 81–83. Wiley, New York.

Albright, T. D. (1983). Visual area MT and the processing of motion. *Dissertation Abstracts International*, **44**, 947.

Albright, T. D. (1984). Direction and orientation selectivity of neurons in visual area MT of the macaque. *J. Neurophysiol.*, **52**, 1106–1130.

Albright, T. D. (1992). Form-cue invariant motion processing in primate visual cortex. *Science*, **255**, 1141–1143.

Albright, T. D. & Desimone, R. (1987). Local precision of visuotopic organization in the middle temporal area (MT) of the macaque. *Exp. Brain Res.*, **65**, 582–592.

Albright, T. D., Desimone, R. & Gross, C. G. (1984). Columnar organization of directionally selective cells in visual area MT of the macaque. *J. Neurophysiol.*, **51**, 16–31.

Allman, J. M., Kaas, J. H., Lane, R. H. & Miezin, F. M. (1972). A representation of the visual field in the inferior nucleus of the pulvinar of the owl monkey (*Aotus trivargatus*). *Brain Res.*, **40**, 291–302.

Amthor, F. R., Oyster, C. W. & Takahashi, E. S. (1984). Morphology of ON–OFF direction-selective ganglion cells in the rabbit retina. *Brain Res.*, **298**, 187–190.

Amthor, F. R., Takahashi, E. S. & Oyster, C. W. (1989). Morphology of retinal ganglion cells with complex receptive fields. *J. Comp. Neurol.*, **280**, 97–121.

Andersen, R. A. & Siegel, R. M. (1990). Motion processing in the primate cortex. In G. M. Edelman, W. E. Gall & W. M. Cowan (Eds.) *Signal and Sense: Local and Global Order in Perceptual Maps*, pp. 163–184. John Wiley, New York.

Anstis, S. M. & Cavanagh, P. (1983). A minimum-motion technique for judging equiluminance. In J. D. Mollon & R. T. Sharpe (Eds.) *Color Vision: Physiology and Psychophysics*, pp. 155–166. Academic Press, New York.

Asanuma, C., Andersen, R. A. & Cowan, W. M. (1985). The thalamic relations of the caudal inferior parietal lobule and the lateral prefrontal cortex in monkeys: Divergent cortical projections from cell clusters in the medial pulvinar nucleus. *J. Comp. Neurol.*, **241**, 357–381.

Barlow, H. B. & Levick, W. R. (1965). The mechanism of directionally selective units in rabbit's retina. *J. Physiol.*, **178**, 477–504.

Bender, D. B. (1981). Retinotopic organization of macaque pulvinar. *J. Neurophysiol.*, **46**, 672–693.

Bender, D. B. (1982). Receptive-field properties of neurons in the macaque inferior pulvinar. *J. Neurophysiol.*, **48**, 1–17.

Bender, D. B. (1983). Visual activation of neurons in the primate pulvinar depends on cortex but not colliculus. *Brain Res.*, **279**, 258–261.

Benevento, L. A. & Fallon, J. H. (1975a). The projection of occipital cortex to the dorsal lateral geniculate nucleus in rhesus monkey (*Macaca mulatta*). *Exp. Neurol.*, **46**, 409–417.

Benevento, L. A. & Fallon, J. H. (1975b). The ascending projections of the superior colliculus in the rhesus monkey (*Macaca mulatta*). *J. Comp. Neurol.*, **160**, 339–362.

Benevento, L. A. & Miller, J. (1981). Visual responses of single neurons in the caudal lateral pulvinar of the macaque monkey. *J. Neurosci.*, **1**, 1268–1278.

Benevento, L. A. & Rezak, M. (1976). The cortical projections of the inferior pulvinar and adjacent lateral pulvinar in the rhesus monkey (*Macaca mulatta*): An autoradiographic study. *Brain Res.*, **108**, 1–24.

Blakemore, C. B. & Vital-Durand, F. (1986). Organization and postnatal development of the monkey's lateral geniculate nucleus. *J. Physiol. (Lond.)*, **380**, 453–491.

Boussaoud, D., Desimone, R. & Ungerleider, L. G. (1992). Subcortical connections of visual areas MST and FST in macaques. *Vis. Neurosci.*, **9**, 291–302.

Boycott, B. B. & Waessle, H. (1974). The morphological types of ganglion cells of the domestic cat's retina. *J. Physiol. (Lond.)*, **240**, 397–419.

Britten, K. H., Shadlen, M. N., Newsome, W. T. & Movshon, J. A. (1992). The analysis of visual motion: A comparison of neuronal and psychophysical performance. *J. Neurosci.*, **12**, 4745–4765.

Burkhalter, A. & Van Essen, D. C. (1986). Processing of color, form and disparity information in visual areas VP and V2 of ventral extrastriate cortex in the macaque monkey. *J. Neurosci.*, **6**, 2327–2351.

Campos-Ortega, J. A. & Hayhow, W. R. (1970). A new lamination pattern in the lateral geniculate nucleus of primates. *Brain Res.*, **20**, 335–339.

Campos-Ortega, J. A., Hayhow, W. R & Cluever, P. F. V. (1970). A note on the problem of retinal projections to the inferior pulvinar nucleus of primates. *Brain Res.*, **22**, 126–130.

Casagrande, V. A. & Debruyn, E. J. (1982). The Galago visual system: aspects of normal organization and developmental plasticity. In D. E. Haines (Ed.) *The Lesser Bushbaby (Galago) An Animal Model: Selected Topics*, pp. 138–168. CRC Press, Boca Raton, FL.

Cavanagh, P. & Anstis, S. (1991). The contribution of color to motion in normal and color-deficient observers. *Vision Res.*, **31**, 2109–2148.

Cavanagh, P. & Mather, G. (1989). Motion: The long and short of it. *Spatial Vision*, **4**, 103–129.

Cavanagh, P., Tyler, C. W. & Favreau, O. E. (1984). Perceived velocity of moving chromatic gratings. *J. Opt. Soc. Am. A*, **1**, 893–899.

Charles, E. R., Logothetis, N. K. & Schiller, P. H. (1993). Cue invariance in extra-striate visual cortex: responses of cells in areas MT and V4 to wavelength-, motion-, and luminance-defined stimuli (unpublished).

Chubb, C. & Sperling, G. (1988). Drift-balanced random stimuli: a general basis for studying non-Fourier motion perception. *J. Opt. Soc. Am. A*, **5**, 1986–2007.

Coffey, P. J., Perry, V. H., Allen, Y., Sinden, J. & Rawlins, J. N. P. (1988). Ibotenic acid induced demyelination in the central nervous system: a consequence of a local inflammatory response. *Neurosci. Lett.*, **84**, 178–184.

Colby, C. L., Gattass, R., Olson, C. R. & Gross, C. G. (1983). Cortical afferents to visual area PO in the macaque. *Soc. Neurosci. Abstr.*, **9**, 152.

Colby, C. L., Duhamel, J.-R. & Goldberg, M. E. (1993). Ventral intraparietal area of the macaque: Anatomic location and visual response properties. *J. Neurophysiol.*, **69**, 902–914.

Cole, G. R., Stromeyer, C. F. III & Kronauer, R. E. (1990). Visual interactions with luminance and chromatic stimuli. *J. Opt. Soc. Am. A*, **7**, 128–140.

Cowey, A. & Marcar, V. L. (1992). The effect of removing superior temporal cortical motion areas in the macaque monkey: I. Motion discrimination using simple dots. *Eur. J. Neurosci.*, **4**, 1219–1227.

Cragg, B. G. (1969). The topography of the afferent projections in the circumstriate-visual cortex of the monkey studied with the Nauta method. *Vision Res.*, **9**, 733–747.

Crook, J. M., Lange-Malecki, B., Lee, B. B. & Valberg, A. (1988). Visual resolution of macaque retinal ganglion cells. *J. Physiol. (Lond.)*, **392**, 193–211.

Cynader, M. S. & Berman, N. (1972). Receptive-field organization of monkey superior colliculus. *J. Neurophysiol.*, **35**, 187–201.

Davidson, R. M. & Bender, D. B. (1991). Selectivity for relative motion in the monkey superior colliculus. *J. Neurophysiol.*, **65**, 1115–1133.

De Monasterio, F. M. & Gouras, P. (1975). Functional properties of ganglion cells of the rhesus monkey retina. *J. Physiol. (Lond.)*, **251**, 167–195.

De Valois, R. L., Yund, E. W. & Hepler, N. (1982). The orientation and direction selectivity of cells in macaque visual cortex. *Vision Res.*, **22**, 531–544.

Derrington, A. M. & Lennie, P. (1984). Spatial and temporal contrast sensitivities of neurones in lateral geniculate nucleus of macaque. *J. Physiol.*, **251**, 167–195.

Derrington, A. M., Krauskopf, J. & Lennie, P. (1984). Chromatic mechanisms in lateral geniculate nucleus of macaque. *J. Physiol. (Lond.)*, **357**, 241–265.

Desimone, R. & Schein, S. J. (1987). Visual properties of neurons in area V4 of the macaque: sensitivity to stimulus form. *J. Neurophysiol.*, **57**, 835–867.

Desimone, R. & Ungerleider, L. G. (1986). Multiple visual areas in the caudal superior temporal sulcus of the macaque. *J. Comp. Neurol.*, **248**, 164–189.

Desimone, R., Schein, S. J., Moran, J. & Ungerleider, L. G. (1985). Contour, color, and shape analysis beyond the striate cortex. *Vision Res.*, **25**, 441–452.

De Yoe, E. A. & Van Essen, D. C. (1985). Segregation of efferent connections and receptive field properties in visual area V2 of the macaque. *Nature*, **317**, 58–61.

Diamond, I. T., Conley, M., Itoh, K. & Fitzpatrick, D. (1985). Laminar organization of geniculocortical projections in *Galago senegalensis* and *Aotus trivirgatus*. *J. Comp. Neurol.*, **242**, 584–610.

Douglas, R. J., Martin, K. A. C. & Nelson, J. C. (1993). The neurobiology of primate vision. *Baillière's Clin. Neurol.*, **2**, 191–225.

Dow, B. M. (1974). Functional classes of cells and their laminar distribution in monkey visual cortex. *J. Neurophysiol.*, **37**, 927–946.

Dubner, R. & Zeki, S. M. (1971). Response properties and receptive fields of cells in an anatomically defined region of the superior temporal sulcus in the monkey. *Brain Res.*, **35**, 528–532.

Duffy, C. J. & Wurtz, R. H. (1991a). Sensitivity of MST neurons to optic flow stimuli. I. A continuum of response selectivity to large-field stimuli. *J. Neurophysiol.*, **65**, 1329–1345.

Duffy, C. J. & Wurtz, R. H. (1991b). Sensitivity of MST neurons to optic flow stimuli. II. Mechanisms of response selectivity revealed by small-field stimuli. *J. Neurophysiol.*, **65**, 1346–1359.

Dursteler, M. R. & Wurtz, R. H. (1988). Pursuit and optokinetic deficits following chemical lesions of cortical areas mt and mst. *J. Neurophysiol.*, **60**, 940–965.

Eskew, R. T., Jr, Stromeyer, C. F. III & Kronauer, R. E. (1990). Cone-contrast comparison of luminance and chromatic sensitivities for movement, flicker, and flashes. *Technical Digest*, **148**.

Felleman, D. J. & Van Essen, D. C. (1987). Receptive field properties of neurons in area V3 of macaque monkey extrastriate cortex. *J. Neurophysiol.*, **57**, 889–920.

Felleman, D. J. & Van Essen, D. C. (1991). Distributed hierarchical processing in primate cerebral cortex. *Cerebral Cortex*, **1**, 1–47.

Felsten, G., Benevento, L. A. & Burman, D. (1983). Opponent-color responses in macaque extrageniculate visual pathways: The lateral pulvinar. *Brain Res.*, **288**, 363–367.

Ferrera, V. P., Nealey, T. A. & Maunsell, J. H. R. (1993). Responses in macaque visual area V4 following inactivation of the parvocellular and magnocellular pathways (in press).

Foster, K. H., Gaska, J. P., Nagler, M. & Pollen, D. A. (1985). Spatial and temporal frequency selectivity of neurons in visual cortical areas V1 and V2 of the macaque monkey. *J. Physiol.*, 331–364.

Fries, W. (1984). Cortical projections to the superior colliculus in the macaque monkey: A retrograde study using horseradish peroxidase. *J. Comp. Neurol.*, **230**, 55–76.

Fries, W. & Distel, H. (1983). Large layer VI neurons of monkey striate cortex (Maynert cells) project to the superior colliculus. *Proc. Roy. Soc. Lond. B*, **219**, 53–59.

Galletti, C., Battaglini, P. P. & Aicardi, G. (1988). 'Real-motion' cells in visual area V2 of behaving macaque monkeys. *Exp. Brain Res.*, **69**, 279–288.

Gattass, R. & Gross, C. G. (1981). Visual topography of striate projection zone (MT) in posterior superior temporal sulcus of the macaque. *J. Physiol.*, **46**, 621–637.

Gegenfurtner, K., Kiper, D. C., Beusmans, J. M. H., Carandini, M., Zaidi, Q. & Movshon, J. A. (1994). Chromatic properties of neurons in macaque MT. *Visual Neurosci.*, **11**, 455–461.

Giolli, R. A. & Tigges, J. (1970). The primary optic pathways and nuclei of primates. In C. R. Noback (Ed.) *Advances in Primatology*, pp. 29–54.

Goldberg, M. E. & Wurtz, R. H. (1972). Activity of superior colliculus in behaving monkey. I. Visual receptive fields of single neurons. *J. Neurophysiol.*, **35**, 542–559.

Golomb, B., Andersen, R. A., Nakayama, K., MacLeod, D. I. A. & Wong, A. (1985). Visual thresholds for shearing motion in monkey and man. *Vision Res.*, **25**, 813–820.

Gouras, P. (1969). Antidromic responses of orthodromically identified ganglion cells in monkey retina. *J. Physiol. (Lond.)*, **204**, 407–419.

Graham, J. (1982). Some topographical connections of the striate cortex with subcortical structures in *Macaca fascicularis*. *Exp. Brain Res.*, **47**, 1–14.

Gregory, R. L. (1966). *Eye and Brain*. McGraw-Hill, New York.

Gregory, R. L. (1977). Vision with isoluminant colour contrast: 1. A projection technique and observations. *Perception*, **6**, 113–119.

Harting, J. K., Hall, W. C., Diamond, I. T. & Martin, G. F. (1973). Anterograde degeneration study of the superior colliculus in *Tupaia glis*: Evidence for a subdivision between superficial and deep layers. *J. Comp. Neurol.*, **148**, 361–386.

Hawken, M. J., Parker, A. J. & Lund, J. S. (1988). Laminar organization and contrast sensitivity of direction-selective cells in the striate cortex of the Old World monkey. *J. Neurosci.*, **8**, 3541–3548.

Hendrickson, A. E., Wilson, M. E. & Toyne, M. J. (1970). The distribution of optic fibers in *Macaca mulatta*. *Brain Res.*, **23**, 425–427.

Hubel, D. H. & Livingstone, M. S. (1987). Segregation of form, color, and stereopsis in primate area 18. *J. Neurosci.*, **7**, 3378–3415.

Hubel, D. H. & Wiesel, T. N. (1968). Receptive fields and functional architecture of monkey striate cortex. *J. Physiol. (Lond.)*, **195**, 215–243.

Hubel, D. H. & Wiesel, T. N. (1972). Laminar and columnar distribution of geniculo-cortical fibers in the macaque monkey. *J. Comp. Neurol.*, **146**, 421–450.

Hubel, D. H., LeVay, S. & Wiesel, T. N. (1975). Mode of termination of retinotectal fibers in Macaque monkey: An autoradiographic study. *Brain Res.*, **96**, 25–40.

Itaya, S. K. & Van Hoesen, G. W. (1983). Retinal projections to the inferior and medial pulvinar nuclei in the old-world monkey. *Brain Res.*, **269**, 223–230.

Itoh, K., Conley, M. & Diamond, I. T. (1982). Retinal ganglion cell projections to individual layers of the lateral geniculate body in *Galago crassicaudatus*. *J. Comp. Neurol.*, **205**, 282–290.

Kaas, J. H., Huerta, M. F., Weber, J. T. & Harting, J. K. (1978). Patterns of retinal terminations and laminar organization of the lateral geniculate nucleus of primates. *J. Comp. Neurol.*, **182**, 517–554.

Kaas, J. H., Merzenich, M. M. & Killackey, H. P. (1983). The reorganization of somatosensory cortex following peripheral nerve damage in adult and developing mammals. *Ann. Rev. Neurosci.*, **6**, 325–356.

Kaplan, E. & Shapley, R. M. (1982). X and Y cells in the lateral geniculate nucleus of macaque monkeys. *J. Physiol.*, **330**, 125–143.

Kaplan, E. & Shapley, R. M. (1986). The primate retina contains two types of ganglion cells, with high and low contrast sensitivity. *Proc. Natl Acad. Sci. USA*, **83**, 2755–2757.

Kolb, H., Linberg, K. A. & Fisher, S. K. (1992). Neurons of the human retina: A Golgi study. *J. Comp. Neurol.*, **318**, 147–187.

Komatsu, H. & Wurtz, R. H. (1988). Relation of cortical areas mt and mst to pursuit eye movements I. Localization and visual properties of neurons. *J. Neurophysiol.*, **60**, 580–603.

Lagae, L., Raiguel, S. & Orban, G. A. (1993). Speed and direction selectivity of macaque middle temporal neurons. *J. Neurophysiol.*, **69**, 19–39.

Lane, R. H., Allman, J. M. & Kaas, J. H. (1971). Representation of the visual field in the superior colliculus of the grey squirrel (*Sciurus carolinensis*) and tree shrew (*Tupaia glis*). *Brain Res.*, **26**, 277–292.

Lappin, J. S. & Bell, H. H. (1976). The detection of coherence in moving random-dot patterns. *Vision Res.*, **16**, 161–168.

Lee, B. B., Creutzfeldt, O. D. & Elepfandt, A. (1979). The responses of magno- and parvocellular cells of the monkey's lateral geniculate body to moving stimuli. *Exp. Brain Res.*, **35**, 547–557.

Lee, B. B., Martin, P. R. & Valberg, A. (1989). Nonlinear summation of M- and L-cone inputs to phasic retinal ganglion cells of the macaque. *J. Neurosci.*, **9**, 1433–1442.

Lennie, P., Krauskopf, J. & Sclar, G. (1990). Chromatic mechanisms in striate cortex of macaque. *J. Neurosci.*, **10**, 649–669.

Leventhal, A. G., Rodieck, R. W. & Dreher, B. (1981). Retinal ganglion cell classes in old world monkey: morphology and central projections. *Science*, **213**, 1139–1142.

Livingstone, M. S. (1988). Art, illusion and the visual system. *Sci. Am.*, **258**, 78–85.

Livingstone, M. S. & Hubel, D. H. (1982). Thalamic inputs to cytochrome oxidase-rich regions in monkey visual cortex. *Proc. Natl Acad. Sci. USA*, **79**, 6098–6101.

Livingstone, M. S. & Hubel, D. H. (1988). Segregation of form, color, movement, and depth: anatomy, physiology, and perception. *Science*, **240**, 740–749.

Logothetis, N. K. (1991). Is movement perception color blind? *Curr. Biol.*, **1**, 298–300.

Logothetis, N. K. & Schall, J. D. (1989). Neuronal correlates of subjective visual perception. *Science*, **245**, 761–763.

Logothetis, N. K., Schiller, P. H., Charles, E. R. & Hurlbert, A. C. (1990). Perceptual deficits and the role of color-opponent and broad-band channels in vision. *Science*, **247**, 214–217.

Lund, J. S. (1973). Organization of neurons in the visual cortex, area 17, of the monkey (*Macaca mulatta*). *J. Comp. Neurol.*, **147**, 455–496.

Lund, J. S. & Boothe, R. G. (1975). Interlaminar connections and pyramidal neuron organization in the visual cortex, area 17, of the monkey (*Macaca mulatta*). *J. Comp. Neurol.*, **159**, 305–334.

Lund, J. S., Lund, R. D., Hendrickson, A. E., Bunt, A. H. & Fuchs, A. F. (1975). The origin of efferent pathways from the primary visual cortex, Area 17, of the macaque monkey as shown by retrograde transport of horseradish peroxidase. *J. Comp. Neurol.*, **164**, 287–304.

McKee, J., Nakayama, K. & Silverman, G. H. (1986). Presize velocity discrimination despite variations in temporal frequency and contrast. *Vision Res.*, **26**, 609–619.

Marcar, V. L. & Cowey, A. (1992). The effect of removing superior temporal cortical motion areas in the macaque monkey: II. Motion discrimination using random dot displays. *Eur. J. Neurosci.*, **4**, 1228–1238.

Marrocco, R. T. (1978). Conduction velocities of afferent input to superior colliculus in normal and dicorticate monkey. *Brain Res.*, **140**, 155–158.

Marrocco, R. T. & Li, R. H. (1977). Monkey superior colliculus: Properties of single cells and their afferent inputs. *J. Neurophysiol.*, **40**, 844–860.

Maunsell, J. H. R. & Newsome, W. T. (1987). Visual processing in monkey extrastriate cortex. *Ann. Rev. Neurosci.*, **10**, 363–401.

Maunsell, J. H. R. & Van Essen, D. C. (1983a). Functional properties of neurons in middle temporal visual area of the macaque monkey: II. Binocular interactions and sensitivity to binocular disparity. *J. Neurophysiol.*, **49**, 1148–1167.

Maunsell, J. H. R. & Van Essen, D. C. (1983b). The connections of the middle temporal visual area (MT) and their relationship to a cortical hierarchy in the macaque monkey. *J. Neurosci.*, **3**, 2563 2586.

Maunsell, J. H. R. & Van Essen, D. C. (1983c). Functional properties of neurons in middle temporal visual area of the macaque monkey: I. Selectivity for stimulus direction, speed, and orientation. *J. Neurophysiol.*, **49**, 1127–1147.

Maunsell, J. H. R., Nealey, T. A. & DePriest, D. D. (1990). Magnocellular and parvocellular contributions to responses in the middle temporal visual area (MT) of the macaque monkey. *J. Neurosci.*, **10**, 3323–3334.

Merigan, W. H. (1991). P and M pathway specialization in the macaque. In A. Valberg & B. B. Lee (Eds.) *From Pigments to Perception: Advances in Understanding Visual Processes*. Plenum Press, New York.

Merigan, W. H. & Maunsell, J. H. R. (1990). Macaque vision after magnocellular lateral geniculate lesions. *Vis. Neurosci.*, **5**, 347–352.

Merigan, W. H. & Maunsell, J. H. R. (1993). How parallel are the primate visual pathways. *Ann. Rev. Neurosci.*, **16**, 369–402.

Merigan, W. H., Byrne, C. E. & Maunsell, J. H. R. (1991a). Does primate motion perception depend on the magnocellular pathway? *J. Neurosci.*, **11**, 3422–3429.

Merigan, W. H., Katz, L. M. & Maunsell, J. H. R. (1991b). The effects of parvocellular lateral geniculate lesions on the acuity and contrast sensitivity of macaque monkeys. *J. Neurosci.*, **11**, 994–1001.

Merzenich, M. M., Kaas, J. H., Wall, J. T., Nelson, R. J., Sur, M. & Felleman, D. J. (1983). Topographic reorganization of somatosensory cortical areas 3B and 1 in adult monkeys following restricted deafferentation. *Neuroscience*, **8**, 33–55.

Merzenich, M. M., Nelson, R. J., Stryker, M. P., Cynader, M. S., Schoppmann, A. &

Zook, J. M. (1984). Somatosensory cortical map changes following digital amputation in adult monkeys. *J. Comp. Neurol.*, **224**, 591.

Montero, V. M. (1980). Patterns of connections from the striate cortex to cortical visual areas in superior temporal sulcus of macaque and middle temporal gyrus of owl monkey. *J. Comp. Neurol.*, **189**, 45.

Movshon, J. & Newsome, W. T. (1992). Neural foundations of visual motion perception. *Curr. Dev. Psychol. Sci.*, **1**, 36–39.

Mulliken, W. H., Jones, J. P. & Palmer, L. A. (1984). Receptive field properties and laminar distribution of X-like and Y-like simple cells in cat area 17. *J. Neurophysiol.*, **52**, 350–371.

Nakayama, K. (1985). Biological image motion processing: a review. *Vision Res.*, **25**, 625–660.

Newsome, W. T. & Paré, E. B. (1988). A selective impairment of motion perception following lesions of the middle temporal visual area (MT). *J. Neurosci.*, **8**, 2201–2211.

Newsome, W. T. & Wurtz, R. H. (1988). Probing visual cortical function with discrete chemical lesions. *Trends Neurosci.*, **11**, 394–400.

Newsome, W. T., Wurtz, R. H., Dursteler, M. R. & Mikami, A. (1985). Deficits in visual motion processing following ibotenic acid lesions of the middle temporal visual area of the macaque monkey. *J. Neurosci.*, **5**, 825–840.

Newsome, W. T., Wurtz, R. H. & Komatsu, H. (1988). Relation of cortical areas MT and MST to pursuit eye movements. II. Differentiation of retinal from extraretinal inputs. *J. Neurophysiol.*, **60**, 604–620.

Ogren, M. P. & Hendrickson, A. E. (1976). Pathways between striate cortex and subcortical regions in *Macaca mulatta* and *Siamiri sciureus*: Evidence for a reciprocal pulvinar connection. *Exp. Neurol.*, **53**, 780–800.

Ogren, M. P. & Hendrickson, A. E. (1979). Morphology and distribution of striate cortex terminals in the inferior and lateral subdivisions of the Macaca monkey pulvinar. *J. Comp. Neurol.*, **188**, 179–200.

Orban, G. A., Kennedy, H. & Bullier, J. (1986). Velocity sensitivity and direction selectivity of neurons in areas V1 and V2 of the monkey: influence of eccentricity. *J. Neurophysiol.*, **56**, 462–480.

Partlow, G. D., Colonnier, M. & Szabo, J. (1977). Thalamic projections of the superior colliculus in the rhesus monkey, macaca mulatta. *J. Comp. Neurol.*, **171**, 285–318.

Pasternak, T. (1987). Discrimination of differences in speed and flicker rate depends on directionally selective mechanisms. *Vision Res.*, **27**, 1881–1890.

Pasternak, T. & Merigan, W. H. (1993). Motion perception following lesions of the superior temporal sulcus in the monkey. *Cereb. Cortex* (in press).

Perry, V. H. & Cowey, A. (1984). Retinal ganglion cells that project to the superior colliculus and pretectum in the macaque monkey. *Neuroscience*, **12**, 1125–1137.

Perry, V. H., Oehler, R. & Cowey, A. (1984). Retinal ganglion cells that project to the dorsal lateral geniculate nucleus in the macaque monkey. *Neuroscience*, **12**, 1101–1123.

Petersen, S. E., Robinson, D. L. & Keys, W. (1985). Pulvinar nuclei of the behaving rhesus monkey: visual responses and their modulation. *J. Neurophysiol.*, **54**, 867–886.

Pettet, M. W. & Gilbert, C. D. (1992). Dynamic changes in receptive-field size in cat primary visual cortex. *Proc. Natl Acad. Sci. USA*, **89**, 8366–8370.

Poetzl, O. & Redlich, E. (1911). Demonstration eines Falles von bilateraler Affecktion beider Occipitallappen. *Wiener Klin. Wochenschr.*, **24**, 517–518.

Purpura, K., Kaplan, E. & Shapley, R. M. (1988). Background light and the contrast gain of primate P and M retinal ganglion cells. *Proc. Natl Acad. Sci. USA*, **85**, 4534–4537.

Ramachandran, V. S. & Gregory, R. L. (1978). Does color provide an input to human motion perception? *Nature*, **275**, 55–56.

Regan, D. (1986). Form from motion parallax and form from luminance contrast: Vernier discrimination. *Spatial Vision*, **4**, 305–318.

Regan, D. (1989). Orientation discrimination for objects defined by relative motion and objects defined by luminance contrast. *Vision Res.*, **29**, 1389–1400.

Regan, D. & Hamstra, S. (1991). Shape discrimination for motion-defined and contrast-defined form: squareness is special (in press).

Riddoch, G. (1971). Dissociation of visual perceptions due to occipital injuries with especial reference to appreciation of movement. *Brain*, **40**, 15–17.

Rogers, B. J. & Graham, M. (1979). Motion parallax as an independent cue for depth perception. *Perception and Psychophysics*, **8**, 125–134.

Rogers, B. J. & Graham, M. (1982). Similarities between motion parallax and stereopsis in human depth perception. *Vision Res.*, **27**, 261–270.

Roy, J.-P., Komatsu, H. & Wurtz, R. H. (1992). Disparity sensitivity of neurons in monkey extrastriate area MST. *J. Neurosci.*, **12**, 2478–2492.

Saito, H., Yukie, M., Tanaka, K., Hikosaka, K., Fukada, Y. & Iwai, E. (1986). Integration of direction signals of image motion in the superior temporal sulcus of the macaque monkey macaca-fuscata. Analysis of local and wide-field movements in the superior temporal visual areas of the macaque monkey macaca-fuscata. *J. Neurosci.*, **6**, 134–157.

Saito, H., Tanaka, K., Isono, H., Yasuda, M. & Mikami, A. (1989). Directionally selective response of cells in the middle temporal area MT of the macaque monkey to the movement of equiluminous opponent color stimuli. *Exp. Brain Res.*, **75**, 1–14.

Sakata, H., Shibutani, H. & Kawano, K. (1983). Functional propertics of visual tracking neurons in posterior parietal association cortex of the monkey. *J. Neurophysiol.*, **49**, 1364–1380.

Salzman, C. D., Britten, K. H. & Newsome, W. T. (1990). Cortical microstimulation influences perceptual judgements of motion direction. *Nature*, **346**, 174–177.

Schein, S. J., Marrocco, R. T. & De Monasterio, F. M. (1982). Is there a high concentration of color selective cells in area v-4 of monkey visual cortex? *J. Neurophysiol.*, **47**, 193–213.

Schiller, P. H. (1993). The effects of V4 and middle temporal (MT) area lesions on visual performance in the rhesus monkey. *Vis. Neurosci.*, **10**, 717–746.

Schiller, P. H. & Colby, C. L. (1983). The responses of single cells in the lateral geniculate nucleus of the rhesus monkey to color and luminance contrast. *Vision Res.*, **23**, 1631–1641.

Schiller, P. H. & Koerner, F. (1971). Discharge characteristics of single units in superior colliculus. *J. Neurophysiol.*, **34**, 920–936.

Schiller, P. H. & Logothetis, N. K. (1990). The color-opponent and broad-band channels of the primate visual system. *Trends Neurosci.*, **13**, 392–398.

Schiller, P. H. & Malpeli, J. G. (1977). Properties and tectal projections of the monkey retinal ganglion cells. *J. Neurophysiol.*, **40**, 428–445.

Schiller, P. H. & Malpeli, J. G. (1978). Functional specificity of lateral geniculate nucleus laminae of the rhesus monkey. *J. Neurophysiol.*, **41**, 788–797.

Schiller, P. H., Stryker, M. P., Cynader, M. S. & Berman, N. (1974). The response characteristics of single cells in the monkey superior colliculus following ablation or cooling of visual cortex. *J. Neurophysiol.*, **37**, 181–194.

Schiller, P. H., Finlay, B. L. & Volman, S. F. (1976). Quantitative studies of single-cell properties in monkey striate cortex. I. Spatiotemporal organization of receptive fields. *J. Neurophysiol.*, **39**, 1288–1319.

Schiller, P. H., Malpeli, J. G. & Schein, S. J. (1979). Composition of geniculo striate input to superior colliculus of the rhesus monkey. *J. Neurophysiol.*, **42**, 1124–1133.

Schiller, P. H., Logothetis, N. K. & Charles, E. R. (1990a). Functions of the color-opponent and broad-band channels of the visual system. *Nature*, **343**, 68–70.

Schiller, P. H., Logothetis, N. K. & Charles, E. R. (1990b). Role of the color-opponent and broad-band channels in vision. *Vis. Neurosci.*, **5**, 321–346.

Sclar, G., Maunsell, J. H. R. & Lennie, P. (1990). Coding of image contrast in central visual pathways of the macaque monkey. *Vision Res.*, **30**, 1–11.

Sekuler, R. & Ganz, L. (1963). After-effects of seen motion with a stabilized retinal image. *Science*, **139**, 419–420.

Shapley, R. (1990). Visual sensitivity and parallel retinocortical channels. *Ann. Rev. Psychol.*, **41**, 635–658.

Shapley, R. M., Kaplan, E. & Soodak, R. (1981). Spatial summation and contrast sensitivity of X and Y cells in the lateral geniculate nucleus of the macaque. *Nature*, **292**, 543–545.

Shipp, S. & Zeki, S. M. (1985). Segregation of pathways leading from area V2 to areas V4 and V5 of macaque monkey visual cortex. *Nature*, **315**, 322–325.

Shipp, S. & Zeki, S. M. (1989). The organization of connections between areas V5 and V1 in macaque monkey visual cortex. *Eur. J. Neurosci.*, **1**, 309–332.

Siegel, R. M. & Andersen, R. A. (1991). The Perception of structure from visual motion in monkey and man. *J. Cogn. Neurosci.*, **2**, 306–319.

Smith, V. C. & Pokorny, J. (1975). Spectral sensitivity of the foveal cone photopigments between 400 and 500 nm. *Vision Res.*, **15**, 161–172.

Snowden, R. J., Treue, S. & Andersen, R. A. (1992). The response of neurons in areas V1 and MT of the alert rhesus monkey to moving random dot patterns. *Exp. Brain Res.*, **88**, 389–400.

Sry, G., Vogels, R. & Orban, G. A. (1993). Cue-invariant shape selectivity of macaque inferior temporal neurons. *Science*, **260**, 995–997.

Standage, G. P. & Benevento, L. A. (1983). The organization of connections between the pulvinar and visual area MT in the macaque monkey. *Brain Res.*, **262**, 288–294.

Tanaka, K. & Saito, H. (1989). Analysis of motion of the visual field by direction, expansion/contraction, and rotation cells clustered in the dorsal part of the medial superior temporal area of the macaque monkey. *J. Neurophysiol.*, **62**, 626–641.

Tigges, M. & Tigges, J. (1970). The retinofugal fibers and their terminal nuclei in *Galago crassicaudatus* (primates). *J. Comp. Neurol.*, **138**, 87–102.

Trojanowski, J. Q. & Jacobson, S. L. (1975a). A combined horseradish peroxidase-autoradiographic investigation of reciprocal connections between superior temporal gyrus and pulvinar in squirrel monkey. *Brain Res.*, **85**, 347–353.

Trojanowski, J. Q. & Jacobson, S. L. (1975b). Peroxidase labeled subcortical afferents to pulvinar in rhesus monkey. *Brain Res.*, **97**, 144–150.

Trojanowski, J. Q. & Jacobson, S. L. (1977). The morphology and laminar distribution of cortico-pulvinar neurons in the Rhesus monkey. *Exp. Brain Res.*, **28**, 51–62.

Ungerleider, L. G. & Desimone, R. (1986). Cortical connections of visual area MT in the macaque. *J. Comp. Neurol.*, **248**, 190–222.

Ungerleider, L. G. & Mishkin, M. (1979). The striate projection zone in the superior temporal sulcus of *Macaca mulatta*: Location and topographic organization. *J. Comp. Neurol.*, **188**, 347–366.

Ungerleider, L. G., Galkin, T. W. & Mishkin, M. (1983). Visuotopic organization of projections from striate cortex to inferior and lateral pulvinar in rhesus monkey. *J. Comp. Neurol.*, **217**, 137–157.

Ungerleider, L. G., Desimone, R., Galkin, T. & Mishkin, M. (1984). Subcortical projections of area MT in the macaque. *J. Comp. Neurol.*, **223**, 368–386.

Vaina, L. M. (1989). Selective impairment of visual motion interpretation following lesions of the right occipito-parietal area in humans. *Biol. Cybernet.*, **61**, 347–359.

Vaina, L. M., Lemay, M., Bienfang, D. C., Choi, A. Y. & Nakayama, K. (1990). Intact 'biological motion' and 'structure from motion' perception in a patient with impaired motion mechanisms: A case study. *Vis. Neurosci.*, **5**, 353–369.

Van Essen, D. C., Maunsell, J. H. R. & Bixby, J. L. (1981). The middle temporal visual area in the macaque. myeloarchitecture, connections, functional properties and topographic organization. *J. Comp. Neurol.*, **199**, 293–326.

Van Essen, D. C., DeYoe, E. A., Olavarria, J. F., Knierim, J. J., Fox, J. M., Sagi, D. & Julesz, B. (1989). Neural responses to static and moving texture patterns in visual cortex of the macaque monkey. In D. Man-Kit Lam & C. D. Gilbert (Eds.) *Neural Mechanisms of Visual Perception*, pp. 137–154. Portofolio Publishing Company, The Woodlands, Texas.

Van Essen, D. C., DeYoe, E. A., Olavarria, J. F., Knierim, J. J., Fox, J. M., Sagi, D. & Julesz, B. (1991). Neural responses to static and moving texture patterns in visual cortex of the macaque monkey. In *Structure and Function of the Visual Cortex*, pp. 135–153.

Williams, D. W. & Sekuler, R. (1984). Coherent global motion percepts from stochastic local motions. *Vision Res.*, **24**, 55–62.

Wilson, M. E. & Toyne, M. J. (1970). Retino-tectal and cortico-tectal projections in *Macaca mulatta*. *Brain Res.*, **24**, 395–406.

Wohlgemuth, A. (1911). On the aftereffect of seen movement. *Br. J. Psychol. Monograph Suppl.*, **1**, 1–117.

Wong-Riley, M. T. T. (1979). Changes in the visual system of monocularly sutured or enucleated cats demonstrable with cytochrome oxidase histochemistry. *Brain Res.*, **171**, 11–28.

Wurtz, R. H. (1969). Visual receptive fields of striate cortex neurons in awake monkeys. *J. Neurophysiol.*, **32**, 727–742.

Wyatt, H. J. & Daw, N. W. (1975). Directionally sensitive ganglion cells in the rabbit retina: specificity for stimulus direction, size, and speed. *J. Neurophysiol.*, **38**, 613–626.

Zeki, S. M. (1969). Representation of central visual fields in prestriate cortex of the monkey. *Brain Res.*, **14**, 271–291.

Zeki, S. M. (1973). Colour coding in rhesus monkey prestriate cortex. *Brain Res.*, **53**, 422–427.

Zeki, S. M. (1974). Functional organization of a visual area in the posterior bank of the superior temporal sulcus of the rhesus monkey. *J. Physiol.*, **236**, 549–573.

Zeki, S. M. (1977). Colour coding in the superior temporal sulcus of rhesus monkey visual cortex. *Proc. R. Soc. Lond. B*, **197**, 195–223.

Zeki, S. M. (1978). Uniformity and diversity of structure and function in rhesus monkey prestriate visual cortex. *J. Physiol.*, **277**, 273–290.

Zeki, S. M. (1991). Cerebral akinetopsia (visual motion blindness). A review. *Brain*, **114**, 811–824.

Zihl, J., Von Cramon, D. & Mai, N. (1983). Selective disturbance of movement vision after bilateral brain damage. *Brain*, **106**, 313–340.

Part 4
Integration of Motion Signals

8

Models of Two-dimensional Motion Perception

Hugh R. Wilson
University of Chicago, USA

1 INTRODUCTION

It has long been known that many neurons in primary visual cortex (V1) are selective for the direction of motion of bars and edges (Hubel & Wiesel, 1962, 1968). Although one might suppose that these units are sufficient to determine the direction of motion of visual patterns, the problem turns out to be much more complicated than this. As a simple demonstration, consider the pattern of arrowheads in Figure 1. When viewed as a whole, these arrowheads are clearly seen to be moving directly to the right. If, however, one views a portion of the pattern through a circular hole or aperture in a piece of cardboard, one perceives diagonally upward motion for the upper edges of the arrowheads but diagonally downward motion for the lower edges. Thus, any neuron sampling a local region of a moving image via a spatially restricted receptive field can only detect motion perpendicular to the orientation of the local contours. Indeed, it is just this local contour motion to which V1 neurons respond (Movshon *et al.*, 1986).

If local motion detecting units can only detect the direction of motion perpendicular to local contours, the visual system must obviously employ some form of pooling to extract the direction of motion of more complex objects, such as the arrowheads in Figure 1. Direct psychophysical evidence for the existence of a pooling stage following extraction of contour motion has been provided by Welch (1989) and Derrington and Suero (1991). What sort of pooling operation would be appropriate? Adelson and Movshon (1982) have suggested a computational answer for rigid objects undergoing translation in the fronto-parallel plane. When an object composed of oriented contours undergoes rigid translation, the velocity of object motion is defined by a geometric construction. The underlying notion is a simple one: as only the velocity component perpendicular to a moving contour can be locally determined, the velocity component

VISUAL DETECTION OF MOTION
ISBN 0–12–651660–X

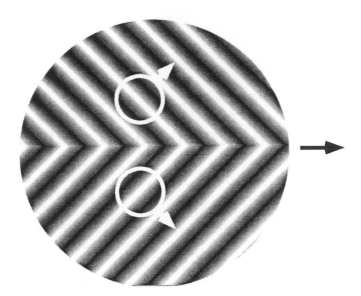

Figure 1 When this cosine arrowhead pattern translates to the right (black arrow), the direction of motion is correctly perceived. If, however, the pattern is occluded except for the small portion within one of the circular white apertures, then motion diagonally upwards is perceived in the top half but motion diagonally downwards in the lower half of the pattern (white arrows). This demonstrates that local motion information is inadequate to specify the direction of pattern motion (the 'aperture problem'). Thus, the visual system must contain a stage where local velocity estimates are pooled to compute pattern motion.

parallel to the moving contour is unknown. Thus, each local measurement of contour velocity merely imposes a constraint on the pattern velocity: it may have any vector component perpendicular to the direction of local contour motion. Adelson and Movshon (1982) pointed out that any pattern with just two contour orientations will permit a unique determination of the direction of pattern motion. As shown in Figure 2, all one need do is draw the two contour motion vectors from a common origin and then construct a perpendicular constraint line through the head of each vector. The direction of pattern motion will be given by the vector drawn from the origin to the unique point where these constraint lines intersect. This is known as the intersection of constraints (IOC) construction. For a rigidly translating object with many contour directions the IOC construction still provides a unique direction of pattern motion.

Granted that this construction is geometrically exact, one must next ask: does the visual system actually implement the IOC? As only two distinct contour orientations are required for the IOC construction, it will be convenient to focus discussion on patterns composed of two superimposed cosine gratings moving in different directions. Examples are shown in Figure 2 along with the IOC construction. These patterns are termed plaids and were first employed experimentally by Adelson and Movshon (1982). It will be useful to subdivide plaids into two distinct types. As shown in Figure 2, a Type I plaid has component grating motion vectors falling on opposite sides of the

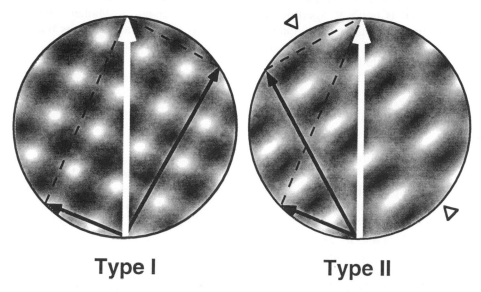

Type I   Type II

Figure 2 Two types of plaids, each consisting of two cosine gratings of identical spatial frequency but differing in orientation and direction of motion. Velocities of the component gratings are indicated by the black vectors, while the overall plaid velocity is given by the white vector. As each component grating only specifies the velocity component perpendicular to its bars, it is consistent with any velocity component orthogonal to this. Therefore, the velocity of plaid motion (white arrow) may be determined by drawing perpendiculars through the heads of the component vectors (dashed lines) and determining their unique point of intersection. This is the intersection of constraints (IOC) construction. In Type I plaids the two component vectors lie on opposite sides of the IOC resultant, while in Type II plaids both component vectors are on the same side of the IOC resultant. The triangles outside the Type II plaid mark a pattern node that will be detected by non-Fourier motion processes.

IOC resultant vector. In Type II plaids, on the other hand, both component motion vectors fall on one side of the IOC resultant.

The distinction between Type I and Type II plaids is useful for two reasons. First, there are a number of significant differences in the perception of Type II as opposed to Type I plaid motion (Ferrera & Wilson, 1987, 1990, 1991; Yo & Wilson, 1992). Second, Type II patterns always engender a large difference between the IOC direction and the vector sum of the component directions, while Type I patterns generally produce a much smaller difference between these two directions. Type II patterns therefore provide much more critical tests of visual system processing than do Type I patterns.

The goal of this chapter is the development of a neural model for the visual analysis of two-dimensional motion. This may be restated as a question: does the visual system implement the IOC construction, or does it embody a different computation that approximates the IOC in certain circumstances but not others? To answer this I will first summarize key data on the perceived motion of grating plaids, as these are among the simplest two-dimensional patterns with an unambiguous direction of motion.[1] It will next be necessary to introduce the concepts of Fourier and non-Fourier motion (Chubb & Sperling, 1988, 1989; Cavanagh & Mather, 1990) as a prelude to the development of a quantitative neural model that combines both types of motion. The neural model is next developed and extended to motion transparency. The chapter concludes with consideration of alternative motion models and ecological issues.

2 KEY DATA

The perceived direction of motion of Type II plaids differs from that of Type I plaids in several major ways. The perceived direction of Type I plaids generally agrees with the IOC direction in both fovea and periphery. The perceived direction of Type II plaids, however, is systematically biased away from the IOC and towards the component grating directions by at least 5–7° in the fovea (Ferrera & Wilson, 1990; Burke & Wenderoth, 1993) and by as much as 30° in peripheral vision (Yo & Wilson, 1992). Second, direction discrimination thresholds for Type I plaids average 0.5–1.0°, which is comparable to direction discrimination for single cosine gratings. Direction discrimination thresholds for Type II plaids are much larger, averaging about 6° (Ferrera & Wilson, 1990).

The systematic bias in the perceived direction of Type II plaids suggests that the visual system might be employing a strategy other than the IOC. More direct evidence for this comes from studies of the perceived direction of motion when the moving plaid is only presented briefly. As illustrated in Figure 3, Type II plaids are perceived to move in the vector sum of their component directions for 60 ms presentations, even when this differs by as much as 55° from the IOC direction (Yo & Wilson, 1992). These patterns only approach the IOC direction for durations of 150 ms or more. This important result has since been corroborated and extended to iso-luminant Type II gratings (Freedland & Banton, 1993) and to moving oblique line stimuli (Lorenceau *et al.*, 1993). Type I patterns were also tested, but for these patterns the IOC and vector sum directions roughly coincided, and no change in direction with duration was found. These results suggest that for durations up to about 60 ms the visual system computes the vector sum of the component grating directions.

Independent evidence for motion in the vector sum direction comes from an imaginative study by Mingolla *et al.* (1992). They employed an array of circular apertures, and within each a single bar moved at a fixed speed in a direction perpendicular to its orientation. When half the apertures contained bars moving at one velocity while the other half contained bars moving at a different velocity, subjects reported the appearance of rigid motion in the vector sum direction. This further challenges the IOC construction as a description of visual processing.

---

[1] It should be stressed that the conclusions of this chapter apply to a wide range of moving patterns, including random dot patterns. Plaid motion has been chosen as a focus because plaids have proven particularly useful for teasing apart the details of visual motion processing.

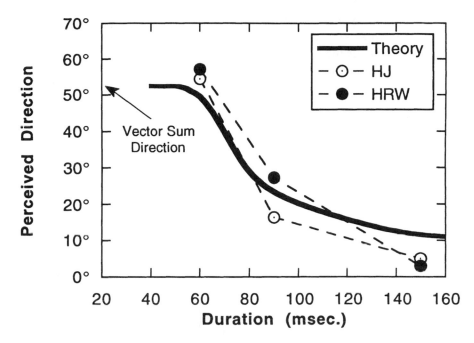

Figure 3   Perceived direction of a Type II plaid as a function of presentation duration. For both subjects, the pattern appeared to move in the vector sum of the component directions during 60 ms presentations (Yo & Wilson, 1992). This was true despite the fact that the vector sum direction differed by 55° from the IOC direction. Type II plaids are only perceived to move within 5° of the IOC direction after 150 ms. Solid curve shows the theoretical prediction of this perceived direction change.

3   FOURIER AND NON-FOURIER MOTION

Adelson and Movshon (1982) first suggested that the motion of two-dimensional patterns such as plaids was computed by a sequential two-stage process. Units in the first stage computed the directions of motion of the component gratings, and units in the second stage then combined these to extract the direction of pattern motion. Psychophysical evidence for this two-stage model has been provided by Welch (1989) and Derrington and Suero (1991). As the component gratings are the Fourier components of a plaid, units computing their motion may be described as Fourier motion units.

Several schemes have been proposed for the extraction of Fourier motion signals, but most represent variants or embellishments of the Reichardt (1961) correlation model.[2] As shown in Figure 3, the Reichardt correlator involves multiplication of two signals (one delayed by Δt) originating from spatially displaced receptive fields. A subtractive stage followed by a threshold yields two units that signal opposite

---

[2] Reichardt's 1961 paper contains one of the earliest and most powerful applications of sophisticated mathematical techniques to the simulation of visual information processing. This paper was an inspiration to me early in my own career.

directions of motion. A vertical cosine grating of contrast A moving to the right is described by the expression $\text{Acos}[2\pi\omega(x - vt)]$, where ω is spatial frequency in cycles per degree (cpd) and v is the velocity in deg/s. Trigonometric identities can be used to prove that the response R of a Reichardt unit will be:

$$R = S(A)^2 \sin(2\pi\omega v\Delta t) \sin(2\pi\omega\Delta x) \tag{1}$$

In this expression $S(A)$ is the sensitivity of the oriented receptive fields that provide inputs to the Reichardt correlator. As we shall see, contrast gain controls can be employed to minimize contrast dependent effects of the term $S(A)^2$.

A major alternative to the Reichardt correlator is the motion energy model devised

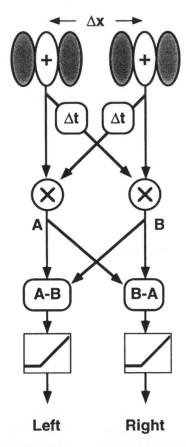

Figure 4 Scheme of a Reichardt (1961) motion correlator. The correlator employs responses of two oriented receptive fields (top, inhibitory zones in gray) separated spatially by Δx. A time delayed (Δt) copy of each response is multiplied by the current response of the other unit to produce signals A and B. These are then subtracted from each other to produce two difference signals: $(A - B)$ and $(B - A)$. Finally, neural thresholds only permit a response for signals greater than zero. Thus, this Reichardt network produces one unit signaling leftward motion and one signaling rightward motion.

by Adelson and Bergen (1985). This model starts with a pair of spatially displaced receptive fields and then incorporates a sequence of operations rather different from those of the Reichardt (1961) model. Recent experiments on complex cells in cat visual cortex support the motion energy formulation (Emerson *et al.*, 1992). If, however, responses of two motion energy units signaling opposite directions of motion are subtracted to form opponent direction units, the opponent responses can be shown to be identical to those of the Reichardt correlator. Accordingly, I shall use 'Reichardt unit' and 'motion energy unit' interchangeably on the assumption that an opponent motion stage occurs before the final computation of pattern direction.

Several laboratories have recently demonstrated the existence of a second type of motion that is invisible to simple motion energy units, and this has been termed non-Fourier motion (Badcock & Derrington, 1985, 1989; Chubb & Sperling, 1988, 1989; see Chapter 6 by Smith). Cavanagh and Mather (1990) have arrived at similar conclusions but have used the term 'second-order' motion. Typical one-dimensional stimuli that generate non-Fourier motion involve a stationary pattern (e.g. random dots or a cosine grating) that is contrast modulated by a moving envelope. One of the simplest non-Fourier motion stimuli is a stationary high frequency cosine with a low frequency moving contrast modulation (CM) envelope. This stimulus, introduced by Badcock and Derrington (1985, 1989), is defined by the equation:

$$CM = L_{mean}\left[\frac{1}{2} + \frac{1}{2}\cos(2\pi\omega_{CM}(x - vt))\right]\cos(2\pi\omega_{H}x) \qquad (2)$$

In this equation ω_H and ω_{CM} are the spatial frequencies of the high frequency carrier and the contrast modulation grating respectively, and v is the speed of the CM envelope.

Chubb and Sperling (1988, 1989) have shown that the visual analysis of non-Fourier motion involves either a squaring or full-wave rectifying nonlinearity. Figure 5 illustrates a simple sequence of operations that will extract non-Fourier motion. The stimulus is first processed by orientation selective filters similar to cortical simple cells, and recent work suggests that they may be rather broadly tuned for spatial frequency (Werkhoven *et al.*, 1993). This neural response is then full-wave rectified, an operation which can be accomplished easily by combining activity of on-center and off-center simple cells with the same orientation preference in a small region. The rectified neural activity is then processed by a second oriented filter at a lower spatial frequency, and the result provides input to a motion energy unit or Reichardt detector. The filter following rectification is chosen to be of lower spatial frequency tuning so as to eliminate components at $2\omega_H$ and above that result from rectification of the stationary high frequency grating.

The existence of both Fourier and non-Fourier motion processing in the visual system leads to a key question: are these two motion types processed in a single pathway or in parallel pathways. A clever experiment by Ledgeway and Smith (1993) has provided an answer. The stimulus in their experiments consisted of alternate frames of a 1.0 cpd luminance cosine and a 1.0 cpd contrast modulated random dot pattern. Between successive frames the spatial phase of the pattern was shifted by 90°. Thus, the stimulus sequence consisted of alternate Fourier and non-Fourier patterns with a 90° phase shift between each. Ledgeway and Smith (1993) reasoned that if

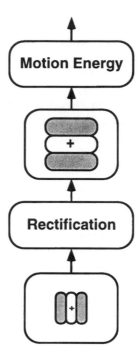

Figure 5   Minimal sequence of operations required to process non-Fourier motion. Following orientation selective filtering by an array of simple cells (bottom), there is a full wave rectification (or response squaring). In the cortex rectification is easily accomplished by pooling responses from on-center and off-center simple cells with the same preferred orientations and receptive field locations. This rectified image is then processed by a second oriented filter, which will generally be at a different orientation. To extract only non-Fourier components, this filter should be about an octave lower in spatial frequency than the first filter. Finally, motion energy is extracted using Reichardt units (see Figure 4).

Fourier and non-Fourier motion signals were both processed in a single channel, then smooth motion should result from this stimulus due to the 90° phase shift between frames. If, however, Fourier and non-Fourier motion are processed separately, then the phase shift within each motion pathway would be 180°, and the motion percept would be entirely ambiguous. Subjects all reported ambiguous motion with this stimulus. Furthermore, for phase shifts on the order of 120°, subjects perceived coherent motion in the opposite direction, as would be expected from the 240° phase shifts in separate Fourier and non-Fourier channels. These data provide strong support for independent, parallel processing of Fourier and non-Fourier motion signals. Further evidence for independent Fourier and non-Fourier pathways derives from the observation that the former generates optokinetic nystagmus, while the latter does not (Harris & Smith, 1992).

As the psychophysical evidence indicates that Fourier and non-Fourier motion signals are processed in parallel, one might expect there to be physiological evidence for such processing streams. The data are sparse at present, but a speculative scheme may be suggested. It has been reported that a population of neurons in primate area V2

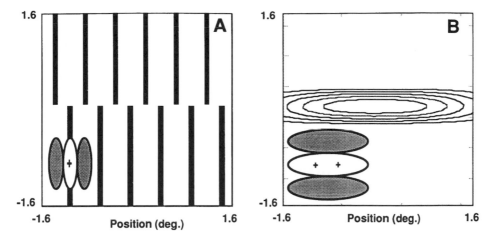

Figure 6 Extraction of a texture boundary ('illusory contour') by non-Fourier processing. In A the oriented first stage filters respond to the lines in the image but fail to respond along the contour. Following this filtering and rectification, the second stage filter shown in B accurately responds to the locus of the illusory contour. Neurons with these characteristics have been reported in V2 but not V1 (von der Heydt *et al.*, 1984; von der Heydt & Peterhans, 1989).

responds to the orientation of illusory contours like that shown in Figure 6, while cells in V1 lack this ability (von der Heydt *et al.*, 1984; von der Heydt & Peterhans, 1989). As shown in Figure 6, the sequence of oriented filtering, rectification, and second stage filtering at a lower spatial frequency will extract the locus of the illusory contour. However, this is exactly the processing sequence that is a precursor to the extraction of non-Fourier motion (see Figure 5). This leads naturally to the conjecture that V2 plays a major role in the extraction of non-Fourier motion (Wilson *et al.*, 1992). Further support for this conjecture is provided by the anatomical observation that V1 projects to the middle temporal motion area (MT) both directly and via V2 (Van Essen, 1985; Van Essen *et al.*, 1992). This is consistent with the V1 → MT pathway signaling Fourier motion and the V1 → V2 → MT pathway providing the additional processing machinery for parallel analysis of non-Fourier motion (Wilson *et al.*, 1992). This provides a plausible explanation for the existence of these parallel inputs to MT.

   Grosof *et al.* (1993) have recently claimed that a fair percentage of units in primate area V1 will in fact respond to illusory contours. This disagreement with the results of von der Heydt and Peterhans (1989) will certainly require further empirical research for its resolution. Regardless of the outcome, it must be emphasized that the accuracy of the neural motion model developed here to explain motion perception does not depend on the anatomical substrate of the non-Fourier pathway.

   One final point deserves mention here. The original studies of non-Fourier motion employed various CM stimuli like that described by equation 2 (Badcock & Derrington, 1985, 1989; Chubb & Sperling, 1988, 1989; Turano & Pantle, 1989; Pantle, 1992). As such patterns are certainly rare in nature, one might wonder whether non-Fourier motion has much relevance to the motion of rigid objects with myriad

Fourier components. A resolution is provided by the observation that the 'illusory contour' in Figure 6 is in fact a real feature of a rigid pattern: it is a texture boundary! Indeed, the non-linear processing required to extract illusory contours has been employed (with embellishments) by others to detect texture boundaries (Bergen & Landy, 1991; Graham, 1991; Graham *et al.*, 1992; Malik & Perona, 1990; Sutter *et al.*, 1989; Wilson & Richards, 1992). Thus, non-Fourier motion is perhaps best thought of as 'texture boundary motion' when considering rigid moving objects.

4   A MODEL FOR 2D MOTION

We have recently developed a quantitative neural model that combines Fourier and non-Fourier motion signals to determine the direction of two-dimensional motion (Wilson *et al.*, 1992). A schematic is shown in Figure 7. Based on the evidence cited above, the model incorporates parallel pathways that extract Fourier and non-Fourier motion signals. These are then brought together in a final processing stage that computes the vector sum of the Fourier and non-Fourier component motions. Studies of direction specific adaptation provide evidence that Fourier and non-Fourier motion signals are ultimately combined (Turano, 1991; Ledgeway, 1994). As will be seen, the vector sum of Fourier *and* non-Fourier components produces a result that is generally close to the IOC direction and is in good agreement with the psychophysics.

The vector sum computation may be accomplished by a neural network as follows. The motion components of the stimulus (both Fourier and non-Fourier) are first extracted by Reichardt or motion energy units. In two dimensions each of these component motion units will have a preferred direction of motion and a motion bandwidth. Preferred directions were assumed to be spaced every 15° for a total of 24 directions, and direction bandwidths were chosen to be ±22.5° as suggested by psychophysics (Phillips & Wilson, 1984). Spatial frequency bandwidths of 2.1 octaves were also measured psychophysically (Wilson *et al.*, 1983). None of these figures is critical as long as the spacing is no greater than the half bandwidth. In fact, a model with ±37.5° direction bandwidths and 30° spacing yields essentially identical results. As shown in equation 1, the response of these units contains a term $S(A)^2$ that is dependent on contrast. As the perceived direction of plaid motion depends relatively little on contrast (Stone *et al.*, 1990), gain controls are used to minimize this contrast term (Wilson *et al.*, 1992; Wilson & Kim, 1994b). Therefore each component unit will generate a response that depends on speed v through the term $\sin(2\pi\omega v\Delta t)$. For speeds that are not too high this is a monotonically increasing function of speed. Thus, each component motion unit can represent a vector quantity with its preferred direction and response level representing the components of the vector.

A network to compute the vector sum direction is shown in Figure 8. Each pattern or output unit in the network is stimulated by a weighted sum of the Fourier and non-Fourier component unit responses. The appropriate choice for a synaptic weighting function here is a cosine of the difference between the preferred direction of the component unit, θ_C, and that of the pattern unit, θ_P. Given this cosine weighting, it is possible to prove that the pattern unit with the maximum input

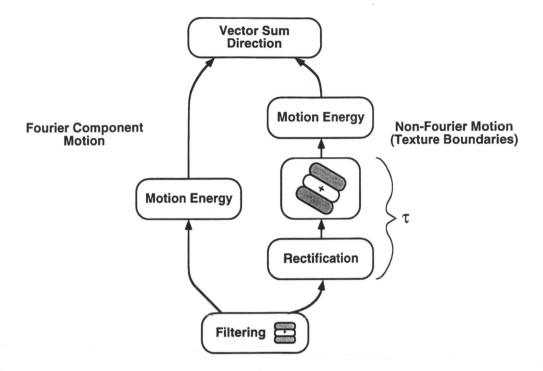

Figure 7 A neural model for two-dimensional motion perception (Wilson *et al.*, 1992). Following initial oriented filtering, two parallel motion processing pathways emerge. The one extracts the motion energy of Fourier pattern components (left), while the second incorporates additional rectification and filtering stages for the extraction of non-Fourier or texture boundary motion (right). These additional steps in non-Fourier processing require additional time τ. The Fourier and non-Fourier pathways then converge on a neural network that computes the vector sum direction (see Figure 8). Units at this final stage have properties similar to cells in MT, and the Fourier and non-Fourier pathways have been speculatively identified with the V1 → MT and V1 → V2 → MT projections respectively.

will have a preferred direction corresponding to the vector sum direction (Wilson *et al.*, 1992). That is:

$$\max_{\theta_P} \left[\sum_C R_C \cos(\theta_P - \theta_C) \right] = \text{vector sum direction} \qquad (3)$$

where R_C is the motion energy response of Fourier or non-Fourier component unit C. The maximum operation is easily implemented by the powerful recurrent inhibition in a 'winner take all' neural network (Wilson & Cowan, 1973; Feldman & Ballard, 1982). In this network each pattern unit strongly inhibits all other pattern units but does not inhibit itself. Due to the presence of neural thresholds the most strongly activated unit

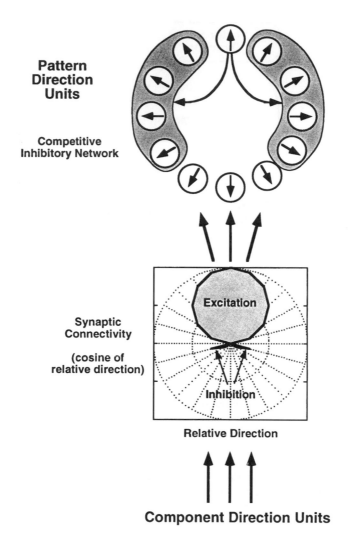

Figure 8 A neural network to compute the vector sum direction of combined Fourier and non-Fourier component motions (Wilson *et al.*, 1992). Each component unit provides input to a range of pattern units via a synaptic connectivity function. This function, plotted in polar coordinates in the center of the diagram, is a cosine of the relative direction between each pattern and component unit (see equation 3). To account for motion transparency, the negative or inhibitory lobe of this cosine connectivity function must be non-zero only for relative angles of ±105° and ±120°. Circled arrows at the top represent pattern units with different preferred directions of motion. (Preferred directions are separated by 30° here for simplicity, although the model embodies units spaced at 15° intervals.) Each pattern unit inhibits a range of other pattern units with preferred directions differing by up to ±120°. Thus, the pattern unit signaling upward motion would inhibit the range of units in the gray regions. As units signaling the opposite direction of motion are not inhibited, two directions of motion can be signaled in response to transparent stimuli. This competitive feedback network computes the vector sum of component directions.

will competitively shut off all other units.[3] Thus, a simple neural network that combines cosine synaptic weighting and competitive recurrent inhibition can readily compute the vector sum of the Fourier and non-Fourier component directions.

One minor modification to this cosine input weighting and competitive inhibitory scheme deserves mention here. In order to model both motion coherence and transparency it is necessary to restrict the range of both the input summation in equation 3 and the range of competitive inhibition to $\pm 120°$. If the full range of $\pm 180°$ is employed, the model will always generate a unique vector sum direction regardless of the stimulus. Restriction of the weighting and inhibition to $\pm 120°$ enables the model to generate two distinct vector sum directions under conditions where motion transparency is perceived. These issues will be taken up in detail later.

The network interactions are highly nonlinear due to the presence of neural thresholds, and therefore they must be expressed in terms of differential equations. If we denote the response of the pattern unit tuned to direction θ by P_θ, the appropriate equations for the 24 pattern units with preferred directions differing by $15°$ are:

$$\tau \frac{dP_\theta}{dt} = -P_\theta + S\left(\sum_C \cos(\theta - \theta_C)R_C - \sum_{k \neq \theta} \alpha_k P_k \right) \tag{4}$$

$$\text{where } S(x) = \begin{cases} 0 & \text{for } x \leq 0 \\ x & \text{for } x > 0 \end{cases}$$

In this equation the function $S(x)$ may be thought of as a function that generates a spike rate whenever the post-synaptic potential x is greater than zero. The first term in the argument of S represents the cosine weighted input of the component unit responses R_C, and the second term represents the recurrent inhibition from other pattern units with the connection strengths α_k. This set of coupled nonlinear equations was solved with the time constant $\tau = 10$ ms to generate model predictions (Wilson et al., 1992).

There is some evidence to suggest that the relevant neural circuitry is present in area MT, which is hypothesized to be the locus of the model neural network. Suppressive inhibition has been reported in MT between cells with different preferred directions of motion (Snowden et al., 1991). There is at present no direct evidence for the cosine synaptic weighting function. However, cells with different preferred directions of motion appear to be organized in an orderly columnar fashion in MT (Albright et al., 1984). Given this organization, cosine synaptic weighting is just a particular form of excitatory center and inhibitory surround connectivity, so cosine weighting would be easy to implement in MT.

The operation of the model is best elucidated by considering its response to the Type II pattern in Figure 2. Here the component gratings are moving at $30°$ and $70°$ relative to the vertically upward direction of pattern motion. This generates the pattern of Fourier component motion energy illustrated by the gray region in the polar

---

[3] In the Wilson et al. (1992) model the simple winner take all network was generalized to permit the most strongly activated unit and its nearest neighbors to survive the competitive inhibition. With three pattern units responding, the network can signal the vector sum direction to within $0.5°$ despite having units with preferred directions varying in $15°$ steps.

Component Responses

Pattern Unit Responses

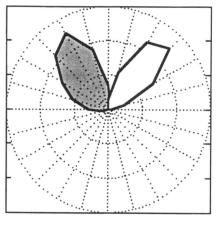
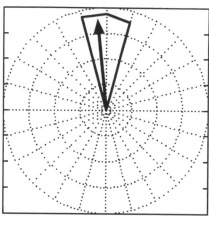

Direction

Direction

Figure 9   Responses of the motion model to the Type II plaid in Figure 2. The Fourier motion responses to the component gratings are plotted in gray on the left-hand polar plot, and the non-Fourier responses are plotted in white. Note that the non-Fourier responses tend to counter-balance the Fourier responses for Type II patterns. The neural network depicted in Figure 8 combines the Fourier and non-Fourier signals to generate the activity pattern shown in the right-hand polar plot. This activity predicts that the pattern will be perceived to deviate from the vertical IOC direction by about 6° (arrow), which is in agreement with psychophysics. Note that delay in the arrival of non-Fourier component signals will produce an initial response in the direction of the Fourier components, as is observed (Yo & Wilson, 1992; see Figure 3).

coordinate graph in Figure 9A. The non-Fourier pathway responds to motion of the nodes in the plaid, such as the one falling between the two small triangles in Figure 2. This is treated as a texture boundary (similar to the 'illusory contour' in Figure 6), and its motion generates non-Fourier signals in units tuned to the range of directions indicated by the white region in Figure 9A. Clearly, these units are signaling the direction of motion perpendicular to this boundary, and this is upwards and to the right. Given this pattern of Fourier and non-Fourier component motion energies, the final network with its cosine weighting and competitive inhibition generates the pattern of responses shown in Figure 9B. Interpolation among the three dominant pattern unit responses reveals that the vector sum direction deviates by about 6.0° from the IOC direction for this pattern, and that agrees with the perceived direction deviation for Type II plaids (Ferrera & Wilson, 1990).[4] Figure 10 shows theoretical predictions of perceived direction for both Type I and II plaids as a function of pattern

---

[4] As the modified winner take all network permits the maximally stimulated unit and its nearest neighbors to survive the competitive inhibition, parabolic interpolation was employed to extract the direction of motion. Whether such interpolation is actually employed by the visual system is irrelevant (although it is easy to implement); the important point is that highly accurate direction information is implicit in the firing rates of the three responding pattern units. This is a form of sparse population coding.

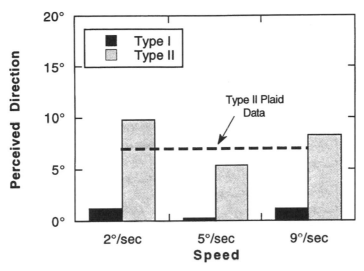

Figure 10 Predicted direction as a function of speed for Type I (black) and Type II (gray) plaids. In agreement with mean data (dashed line), Type II plaids are predicted to show a bias of about 6º away from the IOC direction, while Type I plaids all appear to move within 1º of the IOC direction (Ferrera & Wilson, 1990).

speed. Model responses are averages over a range of plaids with very different component directions but the same IOC resultant. In agreement with the psychophysics, Type I plaids are predicted to move within 1.0º of the 0º IOC direction, while Type II plaids show a systematic bias of about 7.0º towards the Fourier component directions (Ferrera & Wilson, 1990). Recently, biases in perceived direction for Type II plaids of 15° to 26° have been demonstrated by Burke and Wenderoth (1993). However, their largest biases were obtained with only a 10° difference between component directions, and recent calculations show that these larger biases are accurately predicted by the model in response to the appropriate stimuli (Wilson & Kim, 1994b).

It should be emphasized that the non-Fourier components play a role in the processing of both Type I and Type II plaids. Independent evidence for combination of Fourier and non-Fourier signals in plaid motion perception has been reported by Derrington *et al.* (1992). As is evident from Figure 9, however, the non-Fourier components in a Type II pattern signal a series of motion directions that counterbalance the Fourier components to produce motion near the IOC direction. It is this fact that allows the model to accurately predict two rather striking aspects of Type II motion. First, perceived direction of Type II plaids in the periphery (15° eccentricity in our experiments) is biased by about 30° away from the IOC direction and towards the direction of the Fourier components (Yo & Wilson, 1992). This is easily explained on the hypothesis that non-Fourier motion signals are weaker in peripheral than in foveal vision. Evidence for this hypothesis has been provided by Pantle (1992). The second observation is that Type II patterns appear to move in the vector sum direction during brief presentations (Yo & Wilson, 1992). As shown in Figure 3, this can deviate by up to 55° from the IOC direction. Reference to Figure 7 provides a theoretical explanation

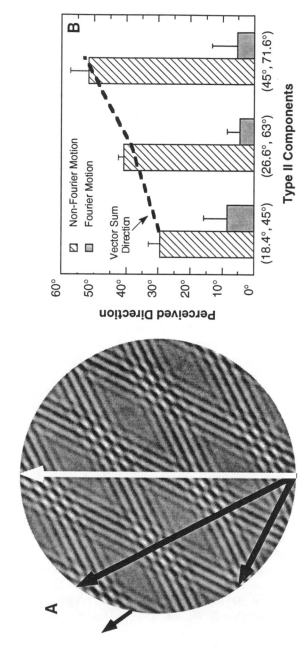

Figure 11 Perceived motion of Type II non-Fourier plaids. The plaid in A is composed of two CM gratings (see equation 2) with the CM envelopes moving in the directions shown by the two heavy black arrows (26.6°, 63.4°). The IOC direction for this non-Fourier, Type II plaid is shown by the white arrow, while the component vector sum direction is shown by the small black arrow on the circumference. Perceived directions for this and two other Type II plaids are plotted in B as averages for eight subjects (Wilson & Kim, 1994a). As shown by bars with diagonal markings, all subjects perceived non-Fourier, Type II plaids as moving in the vector sum direction (dashed line), and this held true even for very long presentations (minutes). In a control experiment, the high frequency carrier gratings were also moved, thereby introducing Fourier motion signals. The gray bars indicate that these patterns exhibited a small perceived bias comparable to that in Figure 10.

for this phenomenon. As the non-Fourier pathway requires additional rectification and second stage filtering processes, it should take some extra time τ before these signals arrive at the final network that computes the vector sum direction. For short durations, therefore, this network will compute the vector sum over only the non-Fourier components. The theoretical curve in Figure 3 shows model predictions on the assumption that the extra processing time $\tau = 60$ ms. For Type I plaids no such effect of presentation time is seen, because the vector sum of the Fourier components is close or identical to the IOC direction (Yo & Wilson, 1992).

Confirmation that a longer processing time is required for non-Fourier motion comes from a study by Derrington et al. (1993). They measured the stimulus duration required for accurate discrimination of motion direction. The results showed that direction of cosine gratings could be discriminated almost perfectly at 20 ms, while non-Fourier CM patterns necessitated about 200 ms for comparable performance (see also Chapter 6 by Smith). This is in reasonable agreement with the data in Figure 3.

The model in Figure 7 predicts that pattern motion is determined by computing the vector sum over Fourier and non-Fourier component motions, and this accurately predicts deviations from the IOC direction for Type II plaids under a variety of conditions. However, the model has a more dramatic implication: a plaid composed exclusively of non-Fourier components should always be perceived to move in the vector sum direction regardless of presentation duration and regardless of how far the vector sum deviates from the IOC direction. The most critical test of this prediction is performed by using Type II plaids, as these have the greatest difference between vector sum and IOC directions. Accordingly, we generated non-Fourier Type II plaids composed of two CM gratings (see equation 2) with different orientations and directions of movement (Wilson & Kim, 1994a). A typical non-Fourier Type II plaid and mean data for three subjects are plotted in Figure 11. For the three different non-Fourier plaids used, the perceived motion (hatched bars) was in the vector sum direction, even when this deviated by 53° from the IOC direction. This provides strong corroboration for the model in Figure 7 (Wilson et al., 1992). As a control, we reintroduced Fourier components by moving the high frequency carrier gratings at the same speeds as the CM envelopes. The solid bars show that these patterns appear to move with the 5–7° bias typical of Type II Fourier plaids (compare with Figure 10).

In an independent study Mingolla et al. (1992) examined the motion percept produced by an array of circular apertures. Half the apertures contained lines moving at one velocity, while the other half contained lines at a second velocity. Although not described in these terms by the authors, this is a stimulus with two Fourier components and no non-Fourier components. Subjects all perceived these patterns to move in the vector sum direction, which is in complete agreement with the Wilson et al. (1992) model.[5]

---

[5] The authors expressed their data in terms of the vector average. However, the vector average direction is identical to the vector sum direction.

5 SPEED PERCEPTION

The discussion thus far has focused on perceived motion direction and has ignored the issue of perceived speed. Thompson (1982; Stone & Thompson, 1992) has demonstrated that the perceived speed of one-dimensional patterns increases with increasing contrast. In addition, it has been proposed that a multiple temporal frequency channel model can explain speed perception (Thompson, 1983, 1984; Smith, 1987). This goes a long way towards solving the speed perception problem for one-dimensional patterns. However, additional issues arise when considering the speed of two-dimensional patterns. Suppose that a plaid is composed of components with directions of movement θ_i. If the true speed of plaid motion is v, then the component speeds are given by the expression $[v \cos(\theta_i)]$. How might the visual system extract v from these component signals?

   If the visual system had implemented the IOC construction, then perceived speed would equal the true pattern speed. Ferrera and Wilson (1991) tested this by comparing perceived plaid speed to that of a grating moving in the same direction. They found that the perceived speed of plaids was not accurately predicted by either the IOC speed or the speed of the faster Fourier component. Instead, the plaid speed seemed to be determined by the speed of the Fourier *or* non-Fourier component that moved in a direction closest to the computed direction of pattern motion. In particular, the perceived speed of Type II plaids was determined by the spatial frequency and perceived speed of their major non-Fourier component. This suggests that the visual system may first compute pattern direction using the scheme in Figure 7 and then estimate pattern speed by choosing the speed of the Fourier or non-Fourier component that is moving in a direction closest to the pattern as a whole. Corroborating evidence is provided by non-Fourier plaid motion. Although non-Fourier plaids move in the vector sum direction (see Figure 11), they are not perceived to move at the vector sum speed (Wilson & Kim, 1994a). Rather, the perceived speed is consistent with that of the faster non-Fourier component. Although these data are not definitive, they clearly rule out the IOC construction as a description of visual speed processing.

   Mingolla *et al.* (1992) proposed that the visual system computes the vector average of the component directions. As their study involved patterns with only two component motion directions, the data were consistent with the vector average hypothesis. However, theoretical considerations suggest that the vector average should not be a good predictor of perceived pattern speed. Consider any object moving at speed v that is delimited by a smoothly curved, closed contour. Simple examples would be a circle, a face, or an automobile moving in the fronto-parallel plane. As each of these objects contains boundary segments at all orientations, its component motion vectors will be described by the expression $[v \cos(\theta)]$, where θ varies from $\pm 90°$. As the projection of each of these onto the direction of pattern motion is $[v \cos(\theta) \cos(\theta)]$, the vector average speed is given by the formula:

$$\text{Vector average} = \frac{v \int_{-\pi/2}^{\pi/2} (\cos[\theta])^2 d\theta}{\int_{-\pi/2}^{\pi/2} d\theta} = \frac{v}{2} \qquad (5)$$

Thus, the vector average would underestimate the speed of these patterns by 50%! A discrete approximation assuming neural units spaced at 15° intervals and therefore averaging over ± 75° results in an underestimation of v by 45%. It seems unlikely that evolution would incorporate such a gross systematic error in speed processing.

The issue of two-dimensional speed perception remains not entirely resolved. However, a major candidate hypothesis is that pattern speed is determined by the speed of the Fourier or non-Fourier component moving in a direction closest to the pattern as a whole (Wilson & Kim, 1994a). For patterns bounded by continuous closed contours there will always be a component moving in the same direction as the whole pattern, and its speed will therefore be identical to the pattern speed. This may help to explain the observation that roughly half of the neurons in MT are tuned to component motion and not pattern motion (Movshon *et al.*, 1986; Rodman & Albright, 1989). It would be a simple matter for a winner take all network to select the largest of these. Furthermore, as this scheme would generate an accurate speed percept for the great majority of naturally occurring patterns (which are bounded by smooth contours), it seems plausible in evolutionary terms.

6 MOTION TRANSPARENCY

Adelson and Movshon (1982) noted that plaids composed of two gratings with very different spatial frequencies failed to move coherently. Rather, the two components

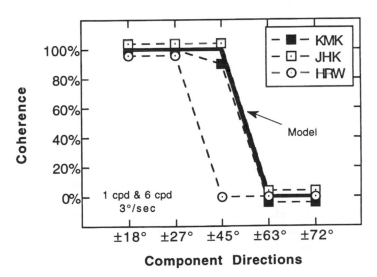

Figure 12 Coherent and transparent motion for plaids composed of 1 cpd and 6 cpd cosine gratings. As the relative angles between the component directions varied there was a sharp transition from 100% coherent motion to 100% transparency (0% coherence) near ±45° (Kim & Wilson, 1993). At 45° consistent inter-subject variations were observed. The prediction of a model with multiplicative facilitation across spatial scales is plotted as a heavy solid line. This facilitation was assumed only to affect pattern units tuned within ±30° of each other.

slid transparently across one another. Kim and Wilson (1993) examined plaid motion transparency in greater detail and found that the transition from coherent to transparent motion depended not only on the spatial frequency difference but also on the relative directions of component motion. In the main experiment plaids were produced from a 1.0 cpd and a 6.0 cpd cosine grating. As shown by the data in Figure 12, all subjects saw rigid, coherent motion when the angle between component motion directions was $\pm 18°$ or $\pm 27°$, and all saw only transparent motion with the larger angles of $\pm 63°$ and $\pm 72°$. At $\pm 45°$ consistent inter-subject differences were observed. Similar results were also obtained for plaids with spatial frequencies of 1.0 and 9.0 cpd and for a range of relative contrasts and speeds. The relative angle between component velocities is therefore a major determinant of motion transparency.

Stoner et al. (1990) examined the issue of motion transparency using the rectangular wave grating stimulus depicted in Figure 13A. When this pattern was moved vertically upwards, motion coherence and transparency depended on the relative luminance of the grating intersections. As shown by the data in Figure 13B, when these intersections were somewhat darker than the grating bars, transparency was the dominant percept, while for other luminances the motion was coherent. As the range of intersection luminances supporting motion transparency was similar to that for static multiplicative transparency of the two component gratings, Stoner et al. (1990) conjectured that motion transparency for these patterns was dependent upon a transparency interpretation of the static pattern. See Chapter 9 by Stoner and Albright for further discussion of this issue.

In light of the evidence for component directions as a major determinant of motion transparency with plaids (Figure 12), we wondered whether this variable might also play an important role in the Stoner et al. (1990) experiments. As they only moved their patterns vertically, the rectangular component gratings were moving at $\pm 68.2°$ relative to the vertical. At these angles plaids composed of different spatial frequencies always move transparently (see Figure 12). To test this idea, we simply moved the pattern in Figure 13A horizontally (Kim & Wilson, 1993), choosing the intersection luminance that had resulted in maximum perceived transparency. This is logically equivalent to reversing the direction of motion of one of the component rectangular gratings, and this manipulation results in component directions of $\pm 21.8°$ relative to the pattern direction. As predicted from the data in Figure 12, motion in this case was perceived to be 100% coherent by all subjects!

Given that the transparency results of Stoner et al. (1990) show a strong dependence on the angle between component directions, how is one to explain their original observation that transparency is also dependent on the relative luminance of the intersections? If an early visual nonlinearity (presumably retinal) computes the logarithm of pattern luminances, then the pattern in Figure 13A can be shown to possess neither Fourier nor non-Fourier motion components in the vertical direction when the intersection/bar luminance ratio is 0.4 (region of maximum transparent motion) (Kim & Wilson, 1993). As the luminance of the intersections is varied from this point, the strength of both Fourier and non-Fourier vertical components in these patterns increases. Relative strengths of these motion components are plotted by a solid line in Figure 13B, and they correlate rather well with the Stoner et al. (1990) data.

A simple hypothesis can provide a unified interpretation of these studies of motion coherence and transparency:

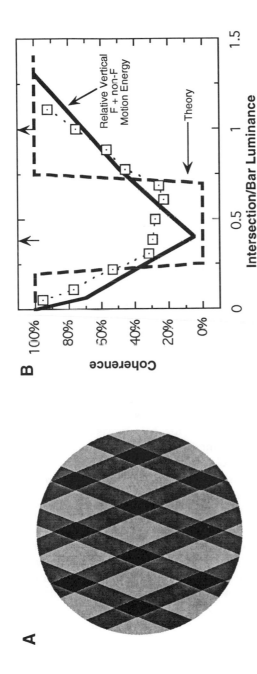

Figure 13 Transparency as a function of intersection luminance in rectangular wave plaids. Stoner *et al.* (1990) presented subjects with the pattern in A moving vertically upwards. Data were obtained as a function of the luminance of the diamond-shaped intersections relative to that of the dark bars. The average percentage of coherent motion reported by their subjects is plotted in B (squares). Coherence was lowest and transparent motion most common around the point where the intersections were at half the luminance of the bars. This agreed approximately with the range over which static pattern transparency could be validly inferred (range between arrows on abscissa). The heavy solid line plots the sum of Fourier and non-Fourier component motions in the vertical direction as a function of intersection luminance (assuming a retinal logarithmic transformation), and this is evidently highly correlated with the data. The heavy dashed line plots model coherence predictions based on the level of non-Fourier motion energy (Wilson & Kim, 1994b). If the patterns in A are moved to the right, the angle between the component directions is reduced from $\pm68.2°$ to $\pm22.8°$. In agreement with the data in Figure 12, the horizontal motion of these patterns was perceived to be 100% coherent by all subjects (Kim & Wilson, 1993).

> Consider all Fourier and non-Fourier component motion vectors in a stimulus. Now consider the maximum angle between each two adjacent vectors. If this angle is less than about 80° the motion will be coherent, but if it is greater than about 100° motion transparency will result.

This hypothesis fails to mention the range between 80° and 100° because the data in Figure 12 show that this range is fuzzy and subject to individual variations. It should also be noted that this hypothesis breaks down when one of the components is of much lower contrast than the others, and a full quantitative simulation must then be used for accurate prediction (Wilson & Kim, 1994b). This hypothesis is related to the proposal by Jasinschi *et al.* (1992) that transparency results when the component motion histogram is bimodal. However, it has been shown that bimodality is a necessary but not sufficient condition for transparency (Wilson & Kim, 1994b).

How does this hypothesis apply to the stimuli considered above? For a cosine plaid composed of two very different spatial frequencies, there will never be any non-Fourier components. This is because the first stage filter in the non-Fourier scheme will respond to at most one of the components but never both due to its spatial frequency tuning. Therefore, the second stage filter will have no texture boundaries to extract (see Figure 5). On the above hypothesis, therefore, motion coherence and transparency will be determined solely by the angle between Fourier components, and this agrees with the observations (Kim & Wilson, 1993; see Figure 12). In contrast, a cosine plaid with components of similar spatial frequency generates non-Fourier components moving in directions halfway between the components. Accordingly, the largest angle between adjacent components (Fourier plus non-Fourier) will be half that between the Fourier components alone, and this will almost always fall below 80°. These plaids will thus virtually always appear coherent.

The data of Stoner *et al.* (1990) are also explicable by this hypothesis. Consider vertical movement of the pattern in Figure 13A with the intersection luminance set to eliminate both Fourier and non-Fourier vertical motion components. Then the angle between adjacent components is 136.4°, and the hypothesis correctly predicts that transparency will result. If, however, the intersection luminance permits strong vertical motion components, then the relevant angle is 68.2°, and motion coherence is correctly predicted. Finally, if the pattern is moved horizontally the maximum angle among components can never be greater than 43.6° regardless of the strength of the non-Fourier components, and 100% rigid motion is correctly predicted to result.

How can this hypothesis be incorporated into the neural model developed above? As the model has been developed on a single spatial scale, i.e. with a single spatial frequency band for the first filtering stage, it is an easy matter to incorporate replicas of the circuitry on other scales. The data in Figure 12 can then be explained by interactions between units on different scales that signal pattern directions within ±30° of each other (Kim & Wilson, 1993). The appropriate coupling between scales is a multiplicative facilitation, which could easily be accomplished by modulatory synapses. As shown by the solid model curve in Figure 12, this scheme accurately predicts the data.

The form of motion transparency discovered by Stoner *et al.* (1990) requires a different set of model interactions, as processing of both rectangular grating components occurs in common on each spatial scale. To account for transparency on a single spatial scale, it is necessary to restrict the range of the cosine input weighting and the competitive inhibition to $\pm 120°$ (Wilson & Kim, 1994b). This simple restriction incorporates the above transparency hypothesis into the model. Consider the pattern of Fourier (gray) and non-Fourier (white) component unit responses in Figure 14A. This pattern of neural activity would be generated by the pattern in Figure 13A if the intersections were at the same luminance as the bars. The angle between adjacent lobes of activity is sufficiently small that the model, via cosine weighting and competitive feedback, combines them all to generate a single pattern response corresponding to coherent vertical motion (Figure 14B). Suppose, now, that the intersection luminance is reduced until the non-Fourier components vanish. This yields the pattern of Fourier component activity in Figure 14C. As the angle between these is now very large, the model generates the two pattern responses shown in Figure 14D, and thus transparent component motion is correctly predicted under the conditions where it was reported by Stoner $et\ al.$ (1990). As mentioned above, vanishing of the non-Fourier components at the appropriate intersection luminance requires logarithmic luminance compression before motion analysis. The heavy dashed line in Figure 13 plots the model predictions for coherence and transparency based on this analysis (Wilson & Kim, 1994b). The model correctly predicts transparency (0% coherence) in just the region where minimum coherence was reported by Stoner $et\ al.$ (1990). The fact that the model predicts such sharp transitions between coherence and transparency is due to its deterministic nature; more gradual transitions and a better fit to the data would result if a noise term had been included.

The model with this modification to predict transparency is an example of a cooperative non-linear network. As such it also displays hysteresis effects, which have been reported in motion perception (Williams $et\ al.$ 1986; Williams & Phillips, 1987). The cooperative simulation of motion transparency leads to an additional model prediction. Examination of Figure 14C and 14D reveals that the competitive inhibition in the pattern unit network has actually produced a repulsion between the directions of component motion. We therefore sought to measure motion repulsion between two sets of bars moving at $\pm 63°$ with respect to vertical (Kim & Wilson, 1994). One set of bars was maintained at 100% contrast, while the contrast of the second set was varied. As shown by the data in Figure 15, the perceived direction of the high contrast bars deviated systematically as the contrast of the second set increased. The heavy dashed line plots the motion repulsion predicted by the model, and it is in reasonable agreement with the data. Previous work has also demonstrated motion repulsion using random dot kinematograms (Marshak & Sekuler, 1979; Mather & Moulden, 1980; Snowden, 1989).

Welch and Bowne (1990) have shown that cosine gratings of the same spatial frequency can slide transparently when one of them is of much lower contrast than the other. The model can account for this on the assumption that the contrast threshold for the non-Fourier pathway is somewhat higher than that for the Fourier pathway, and there is some psychophysical evidence for this (Yo & Wilson, 1992). Given this, a pattern with one very low contrast component will generate mainly Fourier component responses, and these will yield transparency if the relative component directions are sufficiently different (Wilson & Kim, 1994b).

One further implication of motion transparency is worth mentioning. Emerson $et\ al.$ (1992) have demonstrated that direction selective complex cells in cat striate cortex

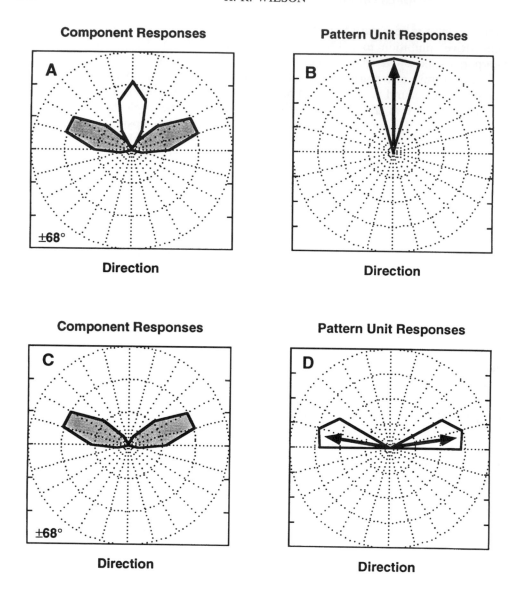

Figure 14 Model responses to the upward motion of patterns in Figure 13A. The component responses in A consist of Fourier responses to the dark bars (gray) plus a strong vertical non-Fourier response (white) generated for appropriate intersection luminances (e.g. 1.0 in Figure 13B). This Fourier and non-Fourier component activity produces the pattern unit responses in B, which correctly signal coherent upward motion (arrow). However, when the intersections are at a relative luminance (0.5 in Figure 13B) that eliminates the non-Fourier vertical motion signals, the component activity is exclusively Fourier, as shown by the gray regions in C. Under these conditions, the inhibitory network in Figure 8 generates the bimodal pattern unit responses in D. This corresponds to the percept of transparent motion that is empirically observed near this intersection luminance. Note that the two interpolated directions of motion indicated by the arrows in D are at a greater relative angle than the component inputs in C. This predicted repulsion is documented in Figure 15.

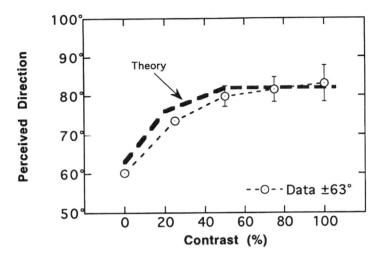

Figure 15 Repulsion of perceived directions during transparent motion. The stimulus consisted of two sets of bars moving at ±63° behind separate groups of adjacent apertures. Bars moving at 63° were always at 100% contrast, while those at − 63° were at the contrasts plotted on the abscissa. As the contrast of the second set of bars increased, the perceived direction of the high contrast bars was repelled from 63° to 82° (Kim & Wilson, 1994). Data are means for three subjects. The heavy dashed curve plots the model prediction (see Figure 14D).

are accurately described by a stage in the Adelson and Bergen (1985) model that occurs prior to motion opponency. At this stage units have a preferred direction of motion but also respond at a lower level to the opposite direction. If these units were to provide the direct input to pattern direction units, units responding weakly to the opposite direction would falsely signal the presence of motion transparency even for rigidly moving patterns. To avoid this, an opponent motion stage, which generates a response equivalent to the Reichardt (1961) model, must apparently occur before the computation of pattern motion. It is significant in this context that most neurons in MT show virtually no response to opposite direction motion, although neurons in V1 do (Albright *et al.*, 1984; Snowden *et al.*, 1991).

7 OTHER MOTION MODELS

The key aspects of the model presented here are summarized in Figure 7: parallel computation of Fourier and non-Fourier component motion energies is followed by a network that computes the vector sum direction for the two. Although this scheme is in agreement with the psychophysical data, it differs radically from a direct implementation of the IOC construction. Several alternative models have produced neural approximations to the IOC construction. Chief among these are the models of Hildreth (1984) and Heeger (1987). By approximating the IOC construction, however, both models should have difficulty predicting the bias in perceived direction for Type II patterns, particularly the 30–40° biases in the periphery (Yo & Wilson, 1992).

Both the Hildreth (1984) and Heeger (1987) models have difficulty in other areas. For brief durations all plaids appear to move in the vector sum direction (Yo & Wilson, 1992). Neither model has yet been shown to predict this or the change in perceived direction over time (see Figure 3; also, Freedland & Banton, 1993; Lorenceau *et al.*, 1993). Furthermore, both models were developed before the importance of non-Fourier motion was recognized. Thus, it is understandable that both fail to incorporate non-Fourier motion signals. However, current physiological evidence clearly indicates that the great majority of MT neurons respond to both Fourier and non-Fourier motion (Albright, 1992).

Perrone (1990) developed a model for pattern motion that employed cosine weighting of component inputs followed by maximum detection. However, that model differed from the present one in only including Fourier component vectors. Exclusion of non-Fourier vectors in this model suggests that it is computing the vector sum of Fourier component directions, and this disagrees with the psychophysics for Type II Fourier plaids, except for brief presentations (Figure 3).

Another class of models are those based on regularization algorithms (Poggio *et al.*, 1985). The deep insight here is due to Marr (1982): many visual problems in stereo as well as motion are not adequately constrained, and the data therefore admit of multiple interpretations. The motion of plaids with cosine components of radically different spatial frequency is a case in point: the visual stimulus is always consistent with either rigid motion or transparent sliding of the two components. Yet the data show that the visual system resolves this ambiguity in different ways for different component directions (see Figure 12). Thus, the visual system must have evolved constraints in the form of network connectivity to solve these problems in a biologically relevant way, an observation with interesting philosophical implications (Wilson, 1991). Beyond this, however, regularization theory is of limited utility. It prescribes minimization of a quantity subject to constraints, but it dictates neither what quantity is to be minimized nor what additional constraints should be imposed. Furthermore, regularization generally involves maximizing the smoothness of solutions over space. Thus, regularization theory performs poorly with motion transparency and the detection of motion boundaries (e.g. Yuille & Grzywacz, 1988). For example, the model of Grzywacz and Yuille (1990) incorrectly predicts that all of the plaids in Figure 12 should move coherently. Finally, regularization theory proper says nothing about the time evolution of the system response, and thus it does not predict either changes in perceived direction with time or motion hysteresis.[6]

Sereno (1993) has adopted a network learning approach to modeling motion processing. Her simulation involved arrays of V1 and MT units covering all directions in 15° intervals and a range of speeds in octave intervals. This network satisfactorily learned a set of connections from the V1 to the MT units that approximates the IOC construction. It also correctly predicts a number of psychophysical

---

[6] Those who employ regularization theory frequently comment that the dynamics are part of the 'neural implementation'. From a mathematical perspective, however, regularization theory amounts to finding a Liapunov function that has a minimum at the desired point in the solution space. However, a given Liapunov function is generally consistent with an infinite number of different dynamical systems that all approach the same final state. Hence, the function chosen for minimization contains no information about system dynamics.

phenomena, including transparency for patterns with widely separated component directions. However, this model also fails to deal with non-Fourier motion. Thus, it cannot explain data on temporal changes in perceived direction of Type II plaids (Figure 3). It would be interesting to see whether Sereno's (1993) network learning approach would generate a different solution if it were trained with plaids containing both Fourier and non-Fourier components.

8   DISCUSSION AND ECOLOGICAL CONSIDERATIONS

This chapter has shown that the model schematized in Figure 7 can accurately predict a wide range of psychophysical data. The model first computes Fourier and non-Fourier component motions in parallel and then combines the two in a neural network that computes the vector sum of these directions (Wilson et al., 1992). Definitive evidence for parallel Fourier and non-Fourier pathways has been presented by Ledgeway and Smith (1993), while Derrington et al. (1992) have produced independent evidence that Fourier and non-Fourier motion signals are combined in plaid processing. Visual computation of the vector sum direction is supported by the discovery that all plaids move in the vector sum direction for brief presentations (Yo & Wilson, 1992) and by the observation that non-Fourier Type II plaids move in the vector sum direction (Wilson & Kim, 1994a). Furthermore, Mingolla et al. (1992) have shown that arrays containing only Fourier component motions are also perceived to move in the vector sum direction.

There is anatomical evidence for parallel motion pathways with subsequent convergence: area MT receives its two major inputs from V1 and V2 (Van Essen, 1985; Krubitzer & Kaas, 1990; Van Essen et al., 1992). As the major input to V2 comes from V1, this second pathway to MT obviously involves additional neural processing. As additional processing is necessary for the extraction of non-Fourier motion, it is natural to suppose that the V1 → V2 → MT pathway carries the non-Fourier motion signal in primates (Wilson et al., 1992). This is consistent with physiological reports that V2 neurons (but not V1) respond to illusory contours like those in Figure 6 (von der Heydt et al., 1984; von der Heydt & Peterhans, 1989). As Grosof et al. (1993) have reported that a percentage of V1 neurons will respond to certain illusory contours, this speculative identification of V2 with non-Fourier processing may require modification. Finally, Albright (1992) has reported that 87% of MT neurons are direction selective for both Fourier and non-Fourier motion. This is the expected result if the model vector sum network is assumed to represent processing in MT. Indeed, the model pattern units respond to the two-dimensional motion of both gratings and plaids in the same manner as MT pattern units (Movshon et al., 1986; Wilson et al., 1992). This is evident from a comparison of the Fourier responses to the component gratings in Figure 14A (gray) with the pattern unit responses to the direction of plaid motion in 14B.

The tentative identification of the final model processing stage with MT leads to a number of physiological predictions. One of the simplest to test is that MT neurons will prefer motion in the vector sum direction when stimulated with non-Fourier Type II plaids, but will shift their preference to near the IOC resultant when stimulated with ordinary plaids (see Figure 11). A second prediction is that MT neurons should shift their direction preference towards the vector sum direction for brief presentations of Type II plaids (see Figure 3).

In concluding this chapter, it is worth considering some of the ecological and evolutionary implications of these ideas as well as some unresolved issues. Key among the unresolved issues is the size of spatial integration areas for motion and the extraction of motion boundaries. The discussion to this point has assumed that the stimulus was uniform throughout the relevant region of visual space. However, this is clearly not typical: a rigidly moving object is viewed against a (usually) stationary background, and other objects with different directions of motion may be present nearby. Clearly, therefore, the integration of Fourier and non-Fourier component motion directions implemented by the model in Figure 7 must operate over a restricted range of visual space. Physiological data suggest that receptive fields in MT are about ten times larger than those in V1, so substantial pooling must occur. What is really needed, however, is a psychophysical estimate of these pooling regions in human vision. Nawrot and Sekuler (1990) have provided a promising approach to this problem. They analyzed perceived motion in spatially adjacent strips of a random dot pattern and found that pooling occurred for strip widths up to about 1.0°. For wider strips, motion repulsion akin to that documented in Figure 15 occurred. It will be useful to extend this type of paradigm to the motion of adjacent plaid patterns. The issue of spatial pooling of motion signals is considered in more depth in Chapter 10 by Williams and Brannan.

In addition to the neglect of spatial pooling areas, this chapter has not considered higher aspects of motion processing, such as computation of rotary motion or expanding motion, etc. (see Chapter 11 by M. Harris). However, many of these computations might be accomplished by appropriate spatial combinations of the two-dimensional motion networks discussed here. Indeed, there is some evidence that these higher order motion computations are carried out in area MST, which receives major input from MT (Saito et al., 1986).

Consideration of spatial pooling areas for motion also suggests an insight into motion transparency. Motion transparency is a relatively infrequent phenomenon in the natural world, although it is easy to create in the laboratory. One might therefore ask: why has local motion integration evolved to permit transparent solutions? A speculative answer is provided by the observation that moderately large motion integration areas will frequently overlap boundaries between two very different motion regimes. Therefore, when a local processing network signals two different directions of motion, as in Figure 14C–D, the primary ecological significance may be that it is signaling the locus of a motion boundary. Indeed, it is possible to construct neural networks that use the pattern unit responses to signal the presence or absence of a motion boundary in a region (Wilson & Kim, 1994b). Furthermore, these networks explicitly signal motion coherence and transparency.

The dependence of motion transparency on angular differences in component directions also has a plausible ecological interpretation. When the inputs to a pattern unit fall within a rather small range of directions, it is likely that all result from the motion of a single object. For example, running animals have a unique motion direction despite local variations resulting from the bodily plasticity of running. On the other hand, a dramatically bimodal pattern of local motion vectors would most likely signify that the unit was pooling across a motion boundary. Thus, the perception of motion transparency may be largely an epiphenomenon of motion boundary extraction. Further psychophysical work is clearly needed to explore these ideas.

If the direction of rigid translational motion is exactly given by the IOC construction, why might the visual system have evolved a vector sum computation instead? A plausible answer readily presents itself. In the natural world moving objects are typically bounded by smoothly curved, closed contours. These objects therefore contain all Fourier motion components within $\pm 90°$ of the direction of object motion, and the vector sum of these directions is exactly the direction of object motion. The same is true for any bilaterally symmetric object, such as an arrow: the vector sum of the component motion directions correctly predicts the direction of flight. Thus, computation of the vector sum direction is generally accurate in the evolutionary environment. Perhaps related to this is the observation that the vector sum direction is so easy for a neural network to compute: all that is needed is a cosine synaptic weighting function on the inputs followed by a competitive inhibitory network (Wilson et al., 1992). Indeed, current evidence indicates that the cortex has also evolved vector sum computations for directing motor movements (Georgopoulos et al., 1986; Georgopoulos et al., 1993), for integrating receptive field location with eye and head position signals (Andersen et al., 1993), and for face recognition (Young & Yamane, 1992). Thus, networks implementing a vector sum computation seem to be common in the cortex.

As noted above, a key aspect of the model is parallel processing of Fourier and non-Fourier motion followed by their combination. This leads to an interesting speculation. The Fourier pathway, which is simpler, might have been the first to evolve. This pathway alone would provide accurate two-dimensional motion information for the majority of real world objects. Subsequent evolution of a parallel non-Fourier pathway would then have simply required incorporating additional signals within a preestablished V1 → MT network. This addition of non-Fourier component motion signals would have improved and extended the range of motion processing accuracy.

The model illustrated in Figure 7 is certainly incomplete, and many details will doubtless turn out to be incorrect. For example, the model fails to suggest a role for the known neural feedback from MT to both V1 and V2 (Van Essen, 1985; Van Essen et al., 1992). The major insight is that combination of Fourier and non-Fourier component motion signals is crucial for visual analysis of virtually all two-dimensional motion. Within this context the model provides a coherent explanation for a wide range of psychophysical data and has accurately predicted several new phenomena (Wilson & Mast, 1993; Wilson & Kim, 1994a).

ACKNOWLEDGMENT

Preparation of this chapter was supported in part by NIH grant EY02158 to the author.

REFERENCES

Adelson, E. H. & Bergen, J. R. (1985). Spatiotemporal energy models for the perception of motion. *J. Opt. Soc. Am. A*, **2**, 284–299.
Adelson, E. H. & Movshon, J. A. (1982). Phenomenal coherence of moving visual patterns. *Nature*, **300**, 523–525.

Albright, T. D. (1992). Form-cue invariant motion processing in primate visual cortex. *Science*, **255**, 1141–1143.

Albright, T. D., Desimone, R. & Gross, C. G. (1984). Columnar organization of directionally selective cells in visual area MT of the macaque. *J. Neurophys.*, **51**, 16–31.

Andersen, R. A., Snyder, L. H., Li, C.-S. & Stricanne, B. (1993). Coordinate transformations in the representation of spatial information. *Curr. Opinion Neurobiol.*, **3**, 171–176.

Badcock, D. R. & Derrington, A. M. (1985). Detecting the displacement of periodic patterns. *Vision Res.*, **25**, 1253–1258.

Badcock, D. R. & Derrington, A. M. (1989). Detecting the displacements of spatial beats: no role for distortion products. *Vision Res.*, **29**, 731–739.

Bergen, J. R. & Landy, M. S. (1991). Computational modeling of visual texture segregation. In M. Landy & J. A. Movshon (Eds.) *Computational Models of Visual Processing*. MIT Press, Cambridge, MA.

Burke, D. & Wenderoth, P. (1993). The effect of interactions between one-dimensional component gratings on two-dimensional motion perception. *Vision Res.*, **33**, 343–350.

Cavanagh, P. & Mather, G. (1990). Motion: the long and short of it. *Spatial Vision*, **4**, 103–129.

Chubb, C. & Sperling, G (1988). Drift-balanced random stimuli: a general basis for studying non-Fourier motion perception. *J. Opt. Soc. Am. A*, **5**, 1986–2007.

Chubb, C. & Sperling, G. (1989). Two motion perception mechanisms revealed through distance-driven reversal of apparent motion. *Proc. Natl Acad. Sci. USA*, **86**, 2985–2989.

Derrington, A. & Suero, M. (1991). Motion of complex patterns is computed from the perceived motions of their components. *Vision Res.*, **31**, 139–149.

Derrington, A. M., Badcock, D. R. & Holroyd, S. A. (1992). Analysis of the motion of 2-dimensional patterns: evidence for a second-order process. *Vision Res.*, **32**, 699–707.

Derrington, A. M., Badcock, D. R. & Henning, G. B. (1993). Discriminating the direction of second-order motion at short stimulus durations. *Vision Res.*, **33**, 1785–1794.

Emerson, R. C., Bergen, J. R. & Adelson, E. H. (1992). Directionally selective complex cells and the computation of motion energy in cat visual cortex. *Vision Res.*, **32**, 203–218.

Feldman, J. A. & Ballard, D. H. (1982). Connectionist models and their properties. *Cogn. Sci.*, **6**, 205–254.

Ferrera, V. P. & Wilson, H. R. (1987). Direction specific masking and the analysis of motion in two dimensions. *Vision Res.*, **27**, 1783–1796.

Ferrera, V. P. & Wilson, H. R. (1990). Perceived direction of moving two-dimensional patterns. *Vision Res.*, **30**, 273–287.

Ferrera, V. P. & Wilson, H. R. (1991). Perceived speed of moving two-dimensional patterns. *Vision Res.*, **31**, 877–893.

Freedland, R. L. & Banton, T. (1993). Type II near-isoluminant plaids: shifts in perceived direction at brief durations. *ARVO Abstract Issue, Invest. Ophthalmol. Vis. Sci.*, **34**, 1031.

Georgopoulos, A. P., Schwartz, A. B. & Kettner, R. E. (1986). Neuronal population coding of movement direction. *Science*, **233**, 1416–1419.

Graham, N. (1991). Complex channels, early local nonlinearities, and normalization in texture segregation. In M. Landy & J. A. Movshon (Eds.) *Computational Models of Visual Processing*. MIT Press, Cambridge, MA.

Graham, N., Beck, J. & Sutter, A. (1992). Nonlinear processes in spatial frequency channel models of perceived texture segregation: effects of sign and amount of contrast. *Vision Res.*, **32**, 719–743.

Grosof, D. H., Shapley, R. M. & Hawken, M. J. (1993). Macaque V1 neurons can signal 'illusory' contours. *Nature*, **365**, 550–552.

Grzywacz, N. M. & Yuille, A. L. (1990). A model for the estimate of local image velocity by cells in the visual cortex. *Proc. R. Soc. Lond. B*, **239**, 129–161.

Harris, L. R. & Smith, A. T. (1992). Motion defined exclusively by second-order characteristics does not evoke optokinetic nystagmus. *Vis. Neurosci.*, **9**, 565–570.

Heeger, D. H. (1987). Model for the extraction of image flow. *J. Opt. Soc. Am. A*, **4**, 1455–1471.

Hildreth, E. C. (1984). *The Measurement of Visual Motion*. MIT Press, Cambridge, Mass.

Hubel, D. H. & Wiesel, T. N. (1962). Receptive fields, binocular interaction, and functional architecture in the cat's striate cortex. *J. Physiol.*, **160**, 106–154.

Hubel, D. H. & Wiesel, T. N. (1968). Receptive fields and functional architecture of monkey striate cortex. *J. Physiol.*, **195**, 215–243.

Jasinschi, R., Rosenfeld, A. & Sumi, K. (1992). Perceptual motion transparency: the role of geometrical information. *J. Opt. Soc. Am. A*, **9**, 1865–1879.

Kim, J. & Wilson, H. R. (1993). Dependence of plaid motion coherence on component grating directions. *Vision Res.*, **33**, 2479–2489.

Kim, J. & Wilson, H. R. (1994). Direction repulsion between components in motion transparency. *Vision Res.* (in press).

Krubitzer, L. A. & Kaas, J. H. (1990). Cortical connections of MT in four species of primates: areal, modular, and retinotopic patterns. *Vis. Neurosci.*, **5**, 165–204.

Ledgeway, T. (1994). Adaptation to second-order motion results in a motion after-effect for directionally ambiguous test stimuli. Submitted for publication.

Ledgeway, T. and Smith, A. T. (1993). Separate mechanisms for the detection of first and second order motion in human vision. *ARVO Abstract Issue, Invest. Ophthalmol. Vis. Sci.*, **34**, 1363.

Lorenceau, J., Shiffrar, M., Wells, N. & Castet, E. (1993). Different motion sensitive units are involved in recovering the direction of moving lines. *Vision Res.*, **33**, 1207–1217.

Malik, J. & Perona, P. (1990). Preattentive texture discrimination with early vision mechanisms. *J. Opt. Soc. Am. A*, 923–932.

Marr, D. (1982). *Vision: A Computational Investigation into the Human Representation and Processing of Visual Information*. W. H. Freeman, San Francisco.

Marshak, W. & Sekuler, R. (1979). Mutual repulsion between moving visual targets. *Science*, **205**, 1399–1401.

Mather, G. & Moulden, B. (1980). A simultaneous shift in apparent direction: further evidence for a 'distribution shift' model of direction coding. *Q. J. Exp. Psych.*, **32**, 325–333.

Mingolla, E., Todd, J. T. & Norman, J. F. (1992). The perception of globally coherent motion. *Vision Res.*, **32**, 1015–1031.

Movshon, J. A., Adelson, E. H., Gizzi, M. S. & Newsome, W. T. (1986). The analysis of moving visual patterns. In C. Chagas, R. Gattass & C. Gross (Eds.) *Pattern Recognition Mechanisms*, pp. 117–151. Springer-Verlag, New York.

Nawrot, M. & Sekuler, R. (1990). Assimilation and contrast in motion perception: explorations in cooperativity. *Vision Res.*, **30**, 1439–1451.

Pantle, A. (1992). Immobility of some second-order stimuli in human peripheral vision. *J. Opt. Soc. Am. A*, **9**, 863–867.

Perrone, J. A. (1990). Simple technique for optical flow estimation. *J. Opt. Soc. Am. A*, **7**, 264–278.

Phillips, G. C. & Wilson, H. R. (1984). Orientation bandwidths of spatial mechanisms measured by masking. *J. Opt. Soc. Am. A*, **1**, 226–232.

Poggio, T., Torre, V. & Koch, C. (1985). Computational vision and regularization theory. *Nature*, **317**, 314–319.

Reichardt, W. (1961). Autocorrelation, a principle for the evaluation of sensory information by the central nervous system. In W. A. Rosenblith (Ed.) *Sensory Communication*, pp. 303–317. Wiley, New York.

Rodman, H. R. & Albright, T. D. (1989). Single-unit analysis of pattern-motion selective properties in the middle temporal visual area (MT). *Exp. Brain Res.*, **75**, 53–64.

Saito, H., Yukie, M., Tanaka, K., Hikosaka, K., Fukada, Y. & Iwai, E. (1986). Integration of direction signals of image motion in the superior temporal sulcus of the macaque monkey. *J. Neurosci.*, **6**, 145–157.

Sereno, M. E. (1993). *Neural Computation of Pattern Motion*. MIT Press, Cambridge, MA.

Smith, A. T. (1987). Velocity perception and discrimination: relation to temporal mechanisms. *Vision Res.*, **27**, 1491–1500.

Snowden, R. J. (1989). Motions in orthogonal directions are mutually suppressive. *J. Opt. Soc. Am. A*, **6**, 1096–1101.

Snowden, R. J., Treue, S., Erickson, R. G. & Andersen, R. A. (1991). The response of area MT and V1 neurons to transparent motion. *J. Neurosci.*, **11**, 2768–2785.

Stone, L. S. & Thompson, P. (1992). Human speed perception is contrast dependent. *Vision Res.*, **32**, 1535–1549.

Stone, L. S., Watson, A. B. & Mulligan, J. B. (1990). Effect of contrast on the perceived direction of a moving plaid. *Vision Res.*, **30**, 1049–1067.

Stoner, G. R., Albright, T. D. & Ramachandran, V. S. (1990). Transparency and coherence in human motion perception. *Nature*, **344**, 153–155.

Sutter, A., Beck, J. & Graham, N. (1989). Contrast and spatial variables in texture segregation: testing a simple spatial frequency channels model. *Percept. Psychophys.*, **46**, 312–332.

Thompson, P. (1982). Perceived rate of movement depends on contrast. *Vision Res.*, **22**, 377–380.

Thompson, P. (1983). Discrimination of moving gratings at and above detection threshold. *Vision Res.*, **23**, 1533–1538.

Thompson, P. (1984). The coding of velocity of movement in the human visual system. *Vision Res.*, **24**, 41–45.

Turano, K. (1991). Evidence for a common motion mechanism of luminance and contrast modulated patterns: selective adaptation. *Perception*, **20**, 455–466.

Turano, K. & Pantle, A. (1989). On the mechanism that encodes the movement of contrast variations: velocity discrimination. *Vision Res.*, **29**, 207–221.

Van Essen, D. C. (1985). Functional organization of primate visual cortex. In A. Peters & E. G. Jones (Eds.) *Cerebral Cortex*, Plenum, New York, pp. 259–329.

Van Essen, D. C., Anderson, C. H. & Felleman, D. J. (1992). Information processing in the primate visual system: an integrated systems perspective. *Science*, **255**, 419–423.

von der Heydt, R. & Peterhans, E. (1989). Mechanisms of contour perception in monkey visual cortex. I. Lines of pattern discontinuity. *J. Neurosci.*, **9**, 1731–1748.

von der Heydt, R., Peterhans, E. & Baumgartner, G. (1984). Illusory contours and cortical neuron responses. *Science*, **224**, 1260–1262.

Welch, L. (1989). The perception of moving plaids reveals two motion processing stages. *Nature*, **337**, 734–736.

Welch, L. & Browne, S. F. (1990). Coherence determines speed discrimination. *Perception*, **19**, 425–435.

Werkhoven, P., Sperling, G. & Chubb, C. (1993). The dimensionality of texture-defined motion: a single channel theory. *Vision Res.*, **33**, 463–485.

Williams, D. & Phillips, G. (1987). Cooperative phenomena in the perception of motion direction. *J. Opt. Soc. Am. A*, **4**, 878–885.

Williams, D., Phillips, G. & Sekuler, R. (1986). Hysteresis in the perception of motion direction as evidence for neural cooperativity. *Nature*, **324**, 253–255.

Wilson, H. R. (1991). Shadows on the cave wall: philosophy and visual science. *Phil. Psychol.*, **4**, 65–78.

Wilson, H. R. & Cowan, J. D. (1973). A mathematical theory of the functional dynamics of cortical and thalamic nervous tissue. *Kybernetik*, **13**, 55–80.

Wilson, H. R. & Kim, J. (1994a). Perceived motion in the vector sum direction. *Vision Res.* **34**, 1835–1842.

Wilson, H. R. & Kim, J. (1994b). A model of motion coherence and transparency. *Vis. Neurosci.* (in press).

Wilson, H. R. & Mast, R. (1993). Illusory motion of texture boundaries. *Vision Res.*, **33**, 1437–1446.

Wilson, H. R. & Richards, W. A. (1992). Curvature and separation discrimination at texture boundaries. *J. Opt. Soc. Am. A*, **9**, 1653–1662.

Wilson, H. R., McFarlane, D. K. & Phillips, G. C. (1983). Spatial frequency tuning of orientation selective units estimated by oblique masking. *Vision Res.*, **23**, 873–882.

Wilson, H. R., Ferrara, V. P. & Yo, C. (1992). Psychophysically motivated model for two-dimensional motion perception. *Vis. Neurosci.*, **9**, 79–97.

Yo, C. & Wilson, H. R. (1992). Perceived direction of moving two-dimensional patterns depends on duration, contrast, and eccentricity. *Vision Res.*, **32**, 135–147.

Young, M. P. & Yamane, S. (1992). Sparse population coding of faces in the inferotemporal cortex. *Science*, **256**, 1327–1331.

Yuille, A. L. & Grzywacz, N. M. (1988). A computational theory for the perception of coherent visual motion. *Nature*, **333**, 71–74.

9

Visual Motion Integration: A Neurophysiological and Psychophysical Perspective

Gene R. Stoner and Thomas D. Albright

The Salk Institute for Biological Studies, La Jolla, USA

1 INTRODUCTION

The light reflected from different moving objects frequently falls onto the same or adjacent region(s) of the retina. Despite this anonymous mixing of light, the primate visual system effortlessly groups only those image motions that arise from a common physical motion. The previous chapter by Wilson detailed the computational problems inherent in this remarkable ability and reviewed candidate solutions to those problems. In this chapter we examine evidence bearing on the plausibility of these different solutions.

Two fundamental questions will concern us. First, '*How* are the motions encoded by individual detectors "integrated" to represent a moving object?' Second, '*Which* signals should be integrated to represent a moving object?' The moving plaid pattern stimulus (Figure 1) (De Valois *et al.*, 1979a; Adelson & Movshon, 1982) captures, in a vastly simplified form, key features of the issues raised by more complex natural stimuli, and hence permits a systematic investigation of these two important questions. This stimulus has accordingly emerged as a central experimental tool and studies that have employed it figure prominently in our review. Simply stated, the 'how' of the integration process is approachable by considering the perception of coherently moving plaid patterns, whereas the selectivity (the 'which') of that process can be revealed by exploring the conditions under which a percept of non-coherent motion is obtained.

VISUAL DETECTION OF MOTION
ISBN 0–12–651660–X

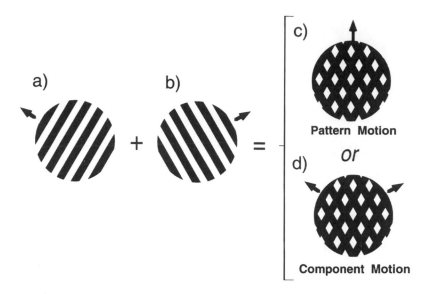

Figure 1   Moving plaid patterns are produced by superimposition of two drifting periodic gratings (a and b). The resultant percept is either that of a coherently moving two-dimensional pattern (c) or two one-dimensional gratings sliding past one another (d), depending on a variety of stimulus parameters.

2   COMPUTATIONAL ISSUES

2.1   Primary Visual Cortex and the Ambiguity of One-dimensional Measurements

Before considering approaches to the problem of motion signal integration, we must understand why such integration is necessary. As described in Chapter 3 by Snowden, the earliest stage at which motion detection is expressed in the primate visual system is at the level of single neurons in primary visual cortex (V1). Information provided by these early motion detectors is limited by the 'aperture problem'. This problem derives its name from the observation that, when viewed through an aperture, the direction of movement of a single one-dimensional (1D) image contour (such as the edge of an object) is ambiguous with respect to the two-dimensional (2D) motion of the object of which it is part (Figure 2A). This ambiguity arises because no image variation exists parallel to the axis of orientation of the contour; hence, movement along that axis is invisible. This directional uncertainty afflicts *any* motion detector (whole observer, neuron, or machine) having a field of view limited to local 1D boundaries of 2D image features.

   The information provided by early neuronal motion detectors may thus be limited by the spatial structure of the image feature present in the neuronal receptive field. A potentially more serious aperture problem, however, is attributable to the structure of the receptive field itself. The majority of directionally selective neurons in area V1

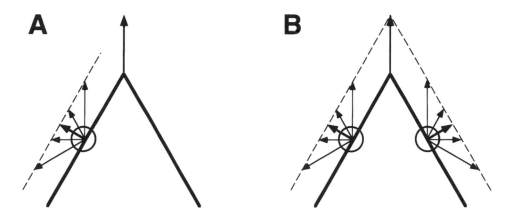

Figure 2 A. The aperture problem. When examined locally (as implied by circle), the apparent motion of a 1D feature embedded in 2D pattern is inherently ambiguous. Motion detectors possessing oriented receptive fields are sensitive only to a single axis of image variation; hence their signals are ambiguous even when multiple 1D features are spatially superimposed. The motion signals provided by such oriented motion detectors are physically consistent with a family of 2D image motions, as suggested by the vectors that terminate on the dotted constraint line shown. The relationship between the 1D measurement and the family of possible 2D motions is given by:

$$\text{speed}_{2D} = \text{speed}_{1D}/\text{cosine}(\alpha_{1\Delta} - \alpha_{2\Delta})$$

where Speed_{2D} and Speed_{1D} are the speeds of the possible 2D and measured 1D motions respectively and $(\alpha_{1\Delta} - \alpha_{2\Delta})$ is the directional difference between those motions.
B. Intersection of constraints solution to the aperture problem. Each solitary measurement of motion defines an ambiguous constraint line; the intersection of two constraints (IOC) yields the 2D velocity consistent with both measurements.

possess 1D spatial sensitivity profiles (e.g. Hubel & Wiesel, 1968). In consequence, they respond to only a single spatial dimension of image variation even if stimulated with 2D image features. The motion information carried by such neurons is wholly ambiguous and physically consistent with a family of 2D image motions (Figure 2A); the 'true' 2D motion cannot be divined from the activity of any single neuron of this type.

2.2 The middle temporal visual area (area MT) and the Disambiguation of 2D Motion

Unlike area V1 neurons, a sub-population of neurons in cortical visual area MT (see Chapter 3 by Snowden and Chapter 7 by Logothetis; Albright, 1993) have been shown to provide unambiguous 2D velocity information (Movshon *et al.*, 1985; Rodman & Albright, 1989). The problem of how area MT neurons achieve this property remains unresolved. The neurophysiological data have led many to propose a two-stage model of 2D neuronal motion processing in which extraction of 1D components is followed by integration of those measurements. Others have suggested that the motion system might directly 'track' the motion of composite image features that result from the superimposition of multiple 1D image features. Candidate features include second-order image variation and 2D luminance variation such as the 'blobs' characteristic of plaids. Indeed, considerable light has been shed on the problem by numerous modeling efforts (e.g. Marr & Ullman, 1981; Adelson & Movshon, 1982; Albright, 1984; M. I. Sereno, 1989; M. E. Sereno, 1993; Bülthoff *et al.*, 1989; Wang *et al.*, 1989; Heeger, 1987; Grzywacz & Yuille, 1991; Wilson *et al.*, 1992; Hildreth, 1984; Nowlan & Sejnowski, 1993; Perrone, 1990; Yuille & Grzywacz, 1988; Marshall, 1988; Sporns *et al.*, 1989; Chou & Adelson, 1993). Some of the distinguishing features of these models are outlined below.

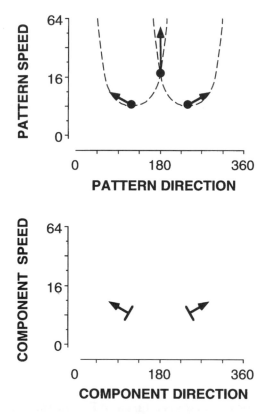

Figure 3 Neural projective fields between component and pattern neurons that embody the cosine velocity relationship given in Figure 2. Each coordinate in these rectilinear speed/ direction plots corresponds to a component or pattern neuron (or column of neurons with common velocity tuning) having a particular speed and direction preference. The constraint lines of Figure 2 become the U-shaped curves in this rectilinear space. The two arrows within each component plot (*bottom*) correspond to the preferred speed and direction preferences of the maximally activated component neurons. The arrows centered at 180° in the pattern plots (*top*) lie at the intersection of the projective fields of those neurons and thus define IOC solutions.

2.3 Integration Schemes

2.3.1 Intersection of constraints

One straightforward computational approach is suggested by the fact that 2D pattern motion can be uniquely determined from two or more 1D motion measurements (Figure 2B) (Fennema & Thompson, 1979; Adelson & Movshon, 1982). It has been suggested (e.g. Albright, 1984) that the constraint lines of this intersections of constraints (IOC) solution might be translated directly into the neural 'projective fields' (Lehky & Sejnowski, 1988) existing between 1D (V1/MT) and 2D (MT) motion processing stages. More specifically, 1D motion detectors might project to 2D velocity tuned neurons possessing a range of direction/orientation preferences that span 180° (as dictated by the cosine relationship given in Figure 2B). As illustrated in Figure 3, the projection patterns from two neurons with IOC projective fields would intersect in a manner formally equivalent to the geometrical IOC solution.

2.3.2 Spatio-temporal integration

IOC models of the type described above require convergence of information from neurons tuned to both direction and speed. A potential difficulty arises, however, because V1 neurons, as well as many MT neurons (Newsome et al., 1983) are tuned to temporal frequency rather than speed per se (manifest by a dependence of preferred speed on the spatial frequency profile of a moving image (e.g. Foster et al., 1985; Movshon, 1988). 'Spatio-temporal integration' models incorporate this additional dimension of ambiguity (e.g. Heeger, 1987; Grzywacz & Yuille, 1991).

According to spatio-temporal integration models, second-stage integration neurons receive converging inputs from a subset of first-stage detectors. The tuning of those detectors falls on the spatio-temporal energy plane that defines the velocity preference of the associated second-stage neuron. One important consequence of this convergence is that spatio-temporal integration models, like IOC models, possess extensive 'cross-directional' connections between primary motion detectors and second-stage neurons.

2.3.3 Integration schemes lacking cross-directional interconnectivity

Several integration schemes have been put forth that lack the cross-directional interconnectivity of IOC and spatio-temporal integration models. Such models typically vary the weights between component and pattern 'neurons' as a function of the cosine of the directional difference between those units. Although cosine-weighting alone yields a vector-sum estimate of 2D pattern motion, which will not always be in agreement with the veridical (IOC) solution, this limitation can be corrected by incorporating other types of image information. Perrone (1990), for example, has proposed a cosine-weighted model that utilizes non-oriented first-order motion detectors. Notably, this model yields a near-veridical solution only if 2D features are sufficiently prevalent in the moving image. Another approach has been taken by Wang et al. (1989) who claim that their cosine-weighted model converges upon an

IOC solution as a result of 'smoothing' operations[1] (implemented via lateral connections) between velocity measurements representing spatially adjacent locations in the visual image. Finally, an innovative cosine-weighted model proposed by Wilson et al. (1992) incorporates a supplementary second-order/non-Fourier pathway (see Chapter 8 by Wilson). Because many 2D patterns possess second-order components that move in the same direction as the whole pattern, this model's solution is frequently accurate.

Although differing from one another in terms of their input types, these non-IOC schemes are distinguished uniformly from IOC and spatio-temporal integration models by the fact that their first-order motion detectors provide significant excitatory input only to those pattern neurons having similar directional preferences.

2.3.4 Formal vs. neural network solutions

In order to reproduce the formal solution, IOC models would have to precisely identify the velocity corresponding to the intersection of the activated projection fields (center arrow in Figure 3). Mechanisms that could achieve this include a logical ANDing of the inputs (see Movshon et al., 1985) and/or some variant of a 'winner-take-all' mechanism in which maximally activated neurons suppress those receiving less input (e.g. Bültoff et al., 1989). It is probably not realistic to assume, however, that the activity of pattern neurons receiving non-maximal input (i.e. those not lying exactly at the intersection) will have no effect on the 2D motion computation. An example of a more plausible approach comes from Sereno (1993) who derives pattern velocity from her IOC-style model by taking a weighted average over the population of active pattern neurons. The resulting solution deviates predictably from formal IOC solutions for many stimulus conditions. This evidences the critical distinction existing between formal and neural network IOC solutions – a point that will be highly relevant when we later weigh the psychophysical evidence for an IOC-type integration scheme.

3 INVESTIGATING THE MOTION INTEGRATION PROCESS: HOW ARE MOTION SIGNALS COMBINED?

3.1 Psychophysical Studies of Motion Integration

3.1.1 Evidence for component motion extraction

Many attempts have been made to test, using psychophysical methods, the hypothesis that 2D motion coherence requires integration of 1D motion measurements. Movshon and colleagues (1985) reasoned that if this hypothesis were valid, it should be possible to interfere with motion coherence by masking 1D motion signals with 1D dynamic visual noise. Using moving plaid patterns as test stimuli, they found support for their

---

[1] Smoothing operations are common to many motion models, including: Hildreth (1984), Anandan and Weiss (1985), Horn and Schunck (1981), Nagel and Enkelmann (1986), Yuille and Grzywacz (1988).

hypothesis in the observation that motion coherence was least likely to occur when the noise was oriented parallel to one of the component gratings (i.e. perpendicular to the direction of component motion).

Others have approached this issue by taking advantage of the fact that there are conditions under which the visual motion system mis-estimates the speed of moving gratings. For instance, it has been shown that the apparent speed of a grating depends upon its spatial frequency (e.g. Diener *et al.*, 1976; Campbell & Maffei, 1980; Smith & Edgar, 1990). Accordingly, when a plaid is composed of components having different spatial frequencies, the perceived direction of the plaid is consistent with an IOC integration of the perceived – not actual – speeds of the gratings when presented alone (Smith & Edgar, 1990). Similarly, the reduction of perceived speed that accompanies a decline in luminance contrast of a single moving grating (Thompson, 1982), has been shown by Stone *et al.* (1990) to alter perceived pattern direction in a manner consistent with an IOC utilization of that reduced speed. In addition, Derrington and Suero (1991) adapted subjects to a grating that moved in the same direction as one of the components of a subsequently presented plaid pattern. This manipulation modifies the perceived direction of the plaid as predicted by an IOC solution and the adaptation-induced reduction of perceived component speed.

In a study with similar implications, Burke and Wenderoth (1993) hypothesized and confirmed that the phenomenon of perceptual motion 'repulsion' (Marshak & Sekuler, 1979) occurs when two moving gratings are superimposed to form plaids. Furthermore, based on their findings, they argued that the non-veridical directional estimates associated with 'Type II' plaids (see below; Kim & Wilson, 1993; Wilson *et al.*, 1992; Wilson, Chapter 8, this volume) could be accounted for by (1) the fact that repulsion causes the apparent motions of the gratings to be quite different from their physical motions, and (2) allowing that the IOC process operates upon the apparent rather than actual component motions.

Finally, using a different strategy, Welch (1989) found that the ability to accurately judge velocity differences in moving plaids was a function of the velocity of the component gratings rather than the velocity of the plaid itself. This counter-intuitive result strongly suggests that the perception of coherent pattern motion relies upon preliminary extraction of components.

In sum, the results of these various psychophysical studies constitute compelling collective evidence for an IOC-type integration scheme operating on imperfectly (but predictably) extracted 1D component motions. Despite this evidence, there are some indications that image features other than 1D oriented Fourier components might contribute to the perception of moving 2D patterns. This evidence is discussed next.

3.1.2   *Evidence for extraction of composite features*

Derrington and Badcock (1992) found that the direction of a plaid's motion could be discriminated under conditions in which the movements of the component gratings (presented individually) could not. Since the motions of the components were apparently invisible, they proposed that pattern motion detection must have been accomplished by extraction of composite features, such as the lattice of 2D contrast variations ('blobs') characteristic of plaids.

Further evidence for a contribution from blob motion detectors comes from Burke

and Wenderoth (1993) who, by comparing the motion aftereffects elicited by simul-
taneously- versus alternately-presented plaid components, concluded that pattern
motion perception utilizes *both* an IOC algorithm and blob tracking mechanism.
They, furthermore, concluded that the IOC mechanism is binocular (consistent with
its presumed extrastriate substrate) whereas the blob mechanism is located at a site
prior to binocular convergence.

It is now well established that mechanisms exist for detection of second-order or
'non-Fourier' motion, as defined by cues such as flicker or contrast amplitude
modulation (e.g. Cavanagh & Mather, 1990; Chubb & Sperling, 1988; Albright,
1992; Zhou & Baker, 1993; see Chapter 6 by Smith and Chapter 7 by Logothetis).
In a clever set of experiments, Derrington *et al.* (1992) explored the contribution of
such mechanisms to the processing of conventional (luminance-defined) plaid patterns. When the component gratings of a moving plaid were each displaced by 3/8 of
their periods, the perceived motion of the plaid was in the direction *opposite* to that
predicted by an IOC integration of the motions perceived when the individual gratings
were presented alone. The reported direction was, however, consistent with the motion
of second-order image components. Accordingly, Derrington *et al.* proposed that
second-order mechanisms can play a critical role in the disambiguation of 2D moving
patterns.

3.1.3 Motion signal integration: IOC versus vector sum

Vector-summation of 1D component motions has been offered as an alternative to IOC
integration schemes (see section 2). In practice, a vector-sum integration can be
achieved by models that utilize cosine-weighted projections from 1D to 2D stages
(Wilson *et al.*, 1992). Simple vector summation of conventional 1D signals generally
yields a non-veridical 2D motion, and hence cosine-weighted projection schemes
would appear insufficient. As proposed by Wilson and colleagues, however, vector
summation models may be salvageable by incorporating inputs conveying second-
order as well as conventional Fourier motion signals. Under many stimulus conditions,
such a scheme yields a solution that approximates the veridical or IOC solution.

Experimental evidence that favors this hybrid vector-sum scheme over IOC-style
models comes mainly from use of 'Type II' plaids (Wilson *et al.*, 1992; Ferrera &
Wilson, 1987, 1990; Ferrera & Maunsell, 1991; Kim & Wilson, 1993), which yield
dissimilar predictions from vector-sum and IOC determinations of velocity. Subjects
viewing such plaids report motion in a direction that matches the vector sum of
Fourier and second-order components. This perceived direction often deviates substantially from that predicted by geometrically derived IOC velocity determinations.
In addition, Yo and Wilson (1992) have noted that stimulus duration influences
perceived direction in a predictable fashion: when presented for brief durations
(< 150 ms), Type II plaids appear to move in the direction of the vector sum of the
Fourier components alone. This latter finding falls in favor of independent Fourier and
non-Fourier pathways and suggests a temporal delay in the non-Fourier pathway, such
that early perceptual judgments rely solely on the Fourier components. Finally, Wilson
and Kim (1993) have shown that Type II plaids composed of either (1) two non-
Fourier components, or (2) one non-Fourier and one Fourier component, appear to

move in the direction predicted from the vector sum of those components. Although these collective results are consistent with the Fourier/non-Fourier vector-sum integration model advocated by Wilson and colleagues, these findings also support alternative interpretations.

Of primary importance is the already-discussed fact that neural networks possessing IOC projection patterns can deviate from geometric IOC solutions. The finding that visual perception deviates from geometrically derived IOC solutions cannot, therefore, be interpreted as ruling out IOC-style models. Indeed, as illustrated in Figure 4, the pattern of activation resulting from stimulation with Type II patterns suggests that IOC models might well yield solutions consistent with the observations of Wilson and colleagues.

A second point of concern arises from the aforementioned experiments of Burke and Wenderoth (1993) who point out that the systematic velocity judgment errors observed for Type II plaids are fully consistent with the IOC solution if it is, 'formulated in terms of perceived or extracted component directions'.

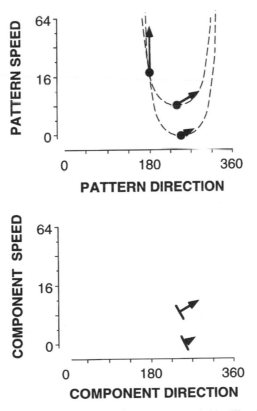

Figure 4 The IOC projective fields activated by 'Type II' plaids. The direction preferences of the pattern neurons receiving inputs from only one projective field lie mostly to one side of the neuron(s) receiving inputs from both projective fields. The contribution of these less activated neurons could account for the non-veridical percepts peculiar to Type II plaids.

Thirdly, the reported deviations from the IOC prediction that occur for Type II plaids may arise from an unconfirmed failure of the coherence mechanism. Welch and Bowne (1990) have shown that subjects reliably judge plaid velocities only when the components move coherently; when coherence fails these judgments are 'contaminated' by the component motions. Subjects may be less likely to report a coherent motion percept when viewing Type II plaids than they are when viewing Type I plaids (Stoner & Albright, 1993a; but see Kim & Wilson, 1993). Under Type II viewing conditions in which perceptual coherence is not monitored, therefore, the possibility exists that perceptual reports are based partly on the motions of the components.

Finally, the temporal progression in perceived direction observed for Type II plaids may reflect the evolution of competitive interactions amongst pattern neurons rather than the tardy contribution of a non-Fourier pathway. Mutual inhibition between pattern neurons may, after a small delay, suppress pattern neurons receiving non-maximal input. Since, for Type II plaids, those less activated neurons encode directions more similar to that of the individual components than that defined by the IOC (as noted above – see Figure 4), initial direction judgments should be biased more heavily toward component velocities than later judgments. This too is consistent with the observations of Wilson and colleagues.

3.2   Neurophysiological Studies of Motion Integration

We have already alluded to the existence of two types of directionally selective neurons – those that carry 1D motion signals and those that extract 2D 'pattern motion'. A review of the means by which the neurophysiologist differentiates between these two response types is the goal of the next few sections of this chapter.

3.2.1   Orientation-independent motion detection

In seminal neurophysiological experiments, Movshon and colleagues (Movshon *et al.*, 1985) sought a method to identify neurons possessing directional tuning that is (like perception) invariant over changes in the orientational composition of a moving stimulus. They proposed a test based on a comparison of directional tuning curves resulting from stimulation with (1) individual oriented gratings vs. (2) plaid patterns formed by superimposition of two such gratings (Figure 1). This test is grounded on the directional dissociation existing between the motions of the oriented components that compose a plaid pattern (each of which appears to move orthogonal to its orientation) and the composite motion of the plaid itself. If a neuron responded only to a plaid pattern's 1D components, this test would identify it as a 'component motion' detector. Conversely, neurons exhibiting selectivity for the composite motion of the plaid pattern would be classified as 'pattern motion' detectors.

3.2.2   Component and pattern response predictions

The first step in the identification of these two idealized neuronal types is the generation of component and pattern predictions based on directional tuning for individual gratings. A neuron responding only to the motions of a plaid pattern's

PREDICTIONS

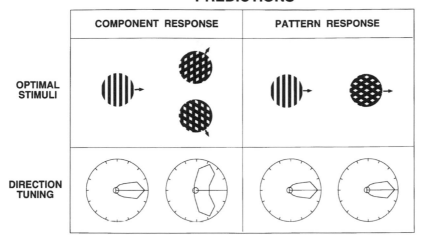

Figure 5 *Left*: optimal stimuli (*top*) and predicted direction tuning curves (*bottom*) for a hypothetical component-type neuron. Direction tuning is plotted on polar axes in which radius represents neuronal response magnitude and polar angle represents direction of stimulus motion. The small circle in the center of each plot indicates the baseline firing rate. The response to the single grating is best for rightward motion. Direction tuning for the plaid pattern is predicted from an appropriate linear sum of responses to the single grating components (see text). The predicted optimal plaid stimuli and the plaid direction tuning reflect sensitivity to both of the oriented components. *Right*: optimal stimuli (*top*) and direction tuning curves (*bottom*) for a hypothetical pattern-type neuron. The predicted plaid tuning curve is identical to that of the single grating, indicating sensitivity of the neuron to the direction of coherent pattern motion. Adapted from Rodman and Albright (1989).

component gratings should yield a neuronal response that is the sum of the responses to the same component gratings when presented individually. Hence, the component prediction is:

$$R(\text{Plaid}_\alpha) = R(\text{Grating}_{\theta1}) + R(\text{Grating}_{\theta2}) \tag{1}$$

where α is the direction of plaid motion and $\theta1$ and $\theta2$ refer to the individual component directions.

The pattern prediction, on the other hand, assumes that the directional tuning curve for a plaid pattern should be identical to that obtained using a single oriented component:

$$R(\text{Plaid}_\alpha) = R(\text{Grating}_\alpha) \tag{2}$$

A graphical depiction of the means by which these predictions are constructed is shown in Figure 5. To precisely quantify the degree of correspondence between tuning curves and the component and pattern predictions, Movshon and colleagues proposed the use of a similarity metric derived as a partial correlation:

$$R_p = (r_p - r_c r_{pc})/[(1 - r_c^2)(1 - r_{pc}^2)]^{0.5} \tag{3}$$

where:

R_p = partial correlation for pattern prediction
r_c = raw correlation of plaid direction turning data with component prediction
r_p = raw correlation of plaid direction turning data with pattern prediction
r_{pc} = correlation of the two predictions

Similarly, the component partial correlation R_c is computed by exchanging r_p and r_c.

Note that as the similarity of the component and pattern predictions increases, the discriminating power of this classification scheme decreases. Factors affecting the similarity of these predictions include both the directional tuning of the neuron and the angular difference between the component and pattern motions. Because the ability to classify a neuron using this metric thus depends as much on the physical characteristics of the visual stimulus as it does on the intrinsic properties of the neuron, those classifications should *not* be considered definitive. Another point with important consequences for distinguishing component and pattern type behavior is borne from the observation that some directionally selective neurons (in MT and elsewhere) possess multi-peaked direction tuning curves for individual components (e.g. Albright, 1984; De Valois *et al.*, 1982; Felleman and Van Essen, 1987). In consequence, component and pattern predictions do not always have the idealized unilobed and bilobed forms shown in Figure 5.

These inherent subtleties in the classification of component and pattern neurons require that care be taken in dubbing either real or model neurons as pattern-type. To our knowledge, model 'pattern-type' neurons have yet to be subjected to the Movshon *et al.* classification method. Full evaluation of the neurobiological relevance of these models requires that modelers utilize the criteria adopted by the neurophysiologist.

3.2.3 Component and pattern responses

In their search for pattern-type behavior in the primate visual system, Movshon *et al.* (1985) first examined motion sensitive neurons in area V1 of macaques. Consistent with the known orientation tuning of these cells, they were found to signal only the motions of the 1D components embedded in the plaid patterns (see also De Valois *et al.*, 1979). As such, their behavior was consistent with the component prediction (Figure 5). When directionally selective MT neurons were tested with plaid patterns (Movshon *et al.*, 1985; Rodman & Albright, 1989), many exhibited component sensitivity similar to that seen in V1 (Figure 6a). Some MT neurons, however, clearly exhibited pattern-type responses (Figure 6b).

The behavior of a representative sample of MT neurons studied under these conditions (Rodman & Albright, 1989) is summarized in Figure 7 by cross-plotting the values of the partial correlations R_c and P_p computed for each cell in the sample. These values form a nearly continuous distribution across the +/+ quadrant of this pattern/component correlation space. A subset of the sample possessed high $R_c : R_p$ ratios (~ 33%) and were classified as component type on these grounds. Conversely, those cells possessing low $R_c : R_p$ ratios (~ 30%) were classified as pattern type.

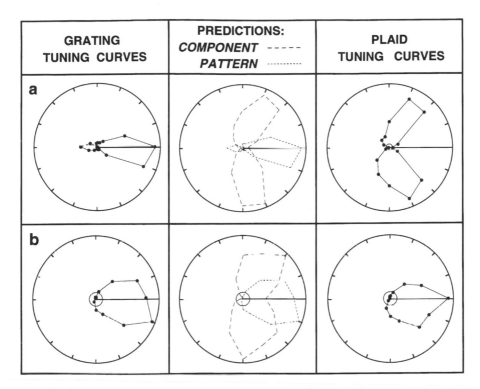

GRATING TUNING CURVES	PREDICTIONS: *COMPONENT* ----- *PATTERN* ··········	PLAID TUNING CURVES
a		
b		

Figure 6 Data from two neurons representing 'component' (*row a*) and 'pattern' (*row b*) stages of motion processing in cortical visual area MT of the rhesus monkey. Directional tuning curves were acquired using a drifting sine-wave grating (*first column*) or a perceptually coherent plaid pattern (*third column*). Responses elicited by each stimulus type, moving in each of 16 different directions of motion, are plotted in a polar format (see legend to Figure 5). Both cells exhibited a single peak in the grating tuning curve. From these curves, responses to the moving plaid pattern were predicted in accordance with either component or pattern assumptions (*second column*). By definition, the behavior of the component neuron in response to the moving plaid pattern conforms to the component prediction, while that of the pattern neuron conforms to the pattern prediction. Adapted from Rodman and Albright (1989).

Similar proportions of pattern and component type cells have been seen in other studies (Movshon *et al.*, 1985; Stoner & Albright, 1992a). Because pattern and component predictions lack complete independence (for reasons cited above), it is not surprising that a substantial fraction of MT neurons prove unclassifiable by these means. This inability to classify is intrinsic to the classification scheme and, hence, it is premature to dismiss 'unclassifiable' neurons as neither pattern nor component type.

These data have led to the assumption that pattern type behavior first evolves in area MT (e.g. Movshon *et al.*, 1985; Stoner & Albright, 1992a). Some recent reports, however, suggest other possibilities. Levitt and Movshon (1993), for example, have reported the existence of a small population of neurons exhibiting pattern-type behavior in area V2, an area that gives rise to a strong input to MT (Maunsell &

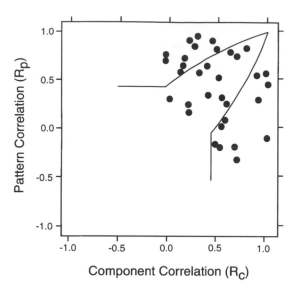

Figure 7   Pattern/component correlation space (adapted from Movshon *et al.*, 1985) used to indicate classifications of component- and pattern-type selectivity. Each data point indicates the degree to which the direction tuning for an MT neuron is correlated with pattern and component response predictions (see text). This sample of MT neurons forms a nearly continuous distribution in the +/+ quadrant of this space. Cells possessing high $R_c : R_p$ ratios are classified as component type, whereas cells with low $R_c : R_p$ ratios are classified as pattern type. For statistical reasons (see text), cells with intermediate $R_c : R_p$ ratios cannot be positively assigned using this metric. Adapted from Rodman and Albright (1989).

Van Essen, 1983a; Ungerleider & Desimone, 1986). In addition, Ferrera and Maunsell (1991) have reported that some V1 neurons exhibit pattern type behavior if visual stimuli are scaled appropriately. A more complete understanding of the relative contributions of these early visual areas to the general problem of motion signal integration must await further studies of neuronal response properties and patterns of anatomical connectivity.

3.2.4   *Non-oriented and second-order motion detectors*

Despite psychophysical evidence suggesting that pattern motion perception may rely partly upon the detection of 'composite' image features such as 2D contrast blobs and oriented second-order features, where this detection takes place is uncertain. Importantly, for conventional plaid patterns, both types of composite features (i.e. blobs and second-order components) move in the pattern direction; hence motion detectors sensitive to either type of feature should exhibit pattern-type responsivity. Possible sources of *non-oriented* inputs include both sub-cortical (tectofugal) structures, such as the pulvinar, and cortical neurons. Neuronal responses to *second-order* or 'non-Fourier' moving stimuli have been demonstrated in both V1 and MT (Albright, 1992; Zhou & Baker, 1993; see Chapter 6 by Smith), but the relationship between that type of responsivity and pattern-type directional tuning remains unexplored. Further identification of the sources of non-oriented and second-order input types is an exciting challenge for future research.

3.2.5 Orientation tuning and pattern motion selectivity

Albright (1984) examined the tuning of MT neurons to stationary oriented stimuli and found that most neurons (60%; termed 'Type I') responded best when the orientation of the stimulus was perpendicular to the preferred direction of motion (as is also the case for V1 neurons). Unexpectedly, a subset of MT neurons (30%; 'Type II') were found to possess orientation preferences for stationary stimuli that were *parallel* to their preferred directions of motion. It was noted that the seemingly anomalous behavior of these Type II neurons might be functionally identified with pattern motion detection. This hypothesis follows naturally from the physical properties of moving 2D patterns: there exists a cosine relationship between the speed at which a component moves and its direction of motion relative to the motion of the pattern it is a part of (i.e. $\text{speed}_{1D} = \text{speed}_{2D} * \text{cosine}(\alpha_{1\Delta} - \alpha_{2\Delta})$; see Figure 2). Pattern type neurons were thus predicted to exhibit Type II behavior because the only direction of pattern motion consistent with an apparently stationary component is one that is parallel to the component's orientation (Albright, 1984). This predicted identity between Type II and pattern motion neurons in area MT was confirmed in a subsequent neurophysiological study (Rodman & Albright, 1989).

The ability to account for Type II orientation tuning is thus an important criterion by which one can evaluate models purporting to explain pattern motion detection. IOC and spatio-temporal integration solutions meet this criterion owing to their 'cross-directional' interconnectivity patterns. Although such models are not unique in this respect (see Albright, 1984), non-IOC projection schemes incorporating oriented inputs would not be expected to exhibit Type II properties. Proponents of such schemes are thus obliged to explain these properties by appeal to other mechanisms.

3.2.6 Do component neurons form the primary input to pattern neurons?

Movshon *et al.* (1985) reported that component neurons in area MT were more commonly found in input laminae (layers 4 and 6), whereas pattern neurons were usually located in laminae that give rise to the principal outputs (layers 2, 3 and 5). Movshon and Newsome (1984) have, furthermore, provided evidence indicating that the input from V1 to MT is largely from directionally selective, orientation tuned neurons. Taken together, these observations suggest a hierarchical scheme in which V1 component neurons feed into MT component neurons. The latter may project, in turn, locally to pattern neurons. The supporting neurobiological data are, however, rather indirect and it is important to keep in mind that the common assumption of a hierarchical scheme, in which component neurons feed into pattern type neurons, has yet to be fully substantiated.

3.3   Summary: Motion Integration

We have reviewed evidence bearing on both the nature of elemental motion detectors and the means by which the information these detectors provide is combined to create a percept of coherent motion. Most evidence suggests that information about the motions of oriented Fourier components constitutes the major input to the integration mechanism. Component neurons in areas V1 and MT are clear carriers of such information. Recent psychophysical evidence also suggests that, under some stimulus conditions, determination of 2D pattern velocity also relies upon detection of the motion of non-oriented image features and/or oriented second-order features. The identity of the neuronal conveyers of these latter two types of information is currently unknown. Moreover, although there are plentiful psychophysical data bearing on the issue of how component and pattern neurons are connected, much of these data are open to varied interpretations. It is hoped that future studies utilizing anatomical tracers and/or cross-correlational analyses of functional connectivity will provide the knowledge needed to settle this debate.

4   INVESTIGATING THE SELECTIVITY OF MOTION INTEGRATION: *WHICH* MOTION SIGNALS ARE COMBINED?

In this section we review experiments that illuminate the primate visual system's ability to combine only those retinal image motions that arise from a common physical motion.

4.1   Psychophysical Studies of Selective Integration

Experimental strategies in this genre can be roughly grouped into two classes. The first class has been concerned with the effect of *component similarity* on the perceptual coherence of moving plaid patterns. It has been shown, for example, that perceptual coherence becomes less likely if component gratings differ significantly along a particular stimulus dimension, such as spatial frequency, luminance contrast, or color. The interpretation of these results has been largely mechanistic – built upon channel-specific integration mechanisms – and we will first consider them in that guise.

The second class of studies has been motivated by a more functional interpretation of the integration process. Specifically, it has been proposed that the primate visual system's demonstrated ability to reconstruct object motion from the motions projected onto the retinae implies tacit knowledge of the 'rules' by which 2D retinal images are formed from their real-world 3D counterparts (see Stoner & Albright, 1993b). Support for this idea comes from studies that have deliberately introduced *image segmentation cues* – 2D retinal cues derived from the rules of image formation and indicative of object boundaries and depth relationships – into moving images. Although these cues (such as luminance or binocular disparity relationships indicative of occlusion or transparency) are known to play a critical role in perception of static surface relationships and object recognition, acceptance of their role in motion processing has been slow.

4.1.1   Channel-specific motion integration?

The role that component similarity plays in motion signal integration was first explored by Adelson and Movshon (1982), who found that the likelihood of perceptual coherence for moving plaid patterns decreases when component gratings differ significantly along the dimensions of: (1) direction of motion, (2) spatial frequency, or (3) luminance contrast. Subsequently, Adelson and Movshon (1984) reported that component gratings having different binocular disparities (thus appearing to lie in different depth planes) are also less likely to cohere. Finally, the presence of chromatic similarity between a plaid pattern's components influences both perceptual motion coherence (Kooi et al., 1989; Krauskopf & Farell, 1990) and oculomotor tracking (Dobkins et al., 1992).

A failure of coherence in the presence of dissimilar components suggests that the component motions are processed by independent neural 'channels'. For example, as a straightforward explanation of the role of spatial frequency in motion coherency, Movshon et al. (1985) proposed that integration is preceded by band-pass filtering of the image into a set of spatial frequency channels, each of which processes a narrow and non-overlapping range of frequencies. A percept of coherent pattern motion was proposed to occur only when the components activate the same channel.

Channel-specific integration schemes can also be offered to account for the other component similarity effects that have been observed. As originally conceived, channel-specific motion integration schemes require that segregated channels exist at the level at which motion integration is thought to occur. The plausibility of this type of scheme thus depends, in part, on independent evidence for channels within area MT. Relevant data come from both neurophysiological and psychophysical sources.

4.1.1.1   Evidence for channels within area MT
The plausibility of channel-based explanations for selective integration effects depends heavily on the parameter in question. Neurophysiological studies have found that MT neurons exhibit some degree of differential selectivity for spatial frequency (Newsome et al., 1983), although it is quite unclear whether the specificity of these neuronal channels is sufficient to account for the similarity effects on perceptual motion coherence. Furthermore, although there exists considerable psychophysical evidence for spatial frequency channels in the primate visual system (e.g. Campbell & Robson, 1968), their role in the segregation of motion signals is unclear. It is known, for example, that low spatial frequencies can 'capture' the motion of high spatial frequencies (Ramachandran et al., 1983). Also of direct relevance are experiments by Smith (1992), which demonstrate that, under some conditions, perceptual motion coherence can occur when component spatial frequencies differ by as much as four octaves. These observations are clearly incompatible with *complete* segregation of spatial frequency, as idealized in the original channel-specific coherence model (Movshon et al., 1985).

Evidence for luminance contrast channels is non-existent. Thus, as applied to luminance contrast, channel-specific integration is a highly implausible explanation for the observed component similarity effects (and was not advanced by Movshon and colleagues).

Both neurophysiological (e.g. Maunsell & Van Essen, 1983b) and psychophysical (e.g. Richards, 1970) evidence can be interpreted as supporting the existence of the requisite binocular disparity channels. Other experiments (reviewed below), however, have shown that the influence of disparity on motion coherence is contingent upon stimulus properties unrelated to disparity per se. Thus, as we shall see, channel mechanisms alone cannot explain the role of binocular disparity in the selective integration of motion signals.

The role of color in motion coherence is particularly intriguing because there exists considerable psychophysical (e.g. Hurvich & Jameson, 1957; Krauskopf et al., 1982) and neurophysiological evidence (see Lennie & D'Zmura, 1988 for review) for chromatic channels. However, although the requisite channels may certainly be said to exist in the lateral geniculate nucleus (LGN), their presence in V1 is controversial (e.g. Lennie et al., 1991) and they are clearly absent from MT. A possible means by which channels remote from the motion integration stage might influence perceptual motion coherence is considered below.

4.1.1.2   Second-order channel mechanisms

Although not considered by Movshon and colleagues, the channel idea can be extended to include within-channel detection of second-order image features. This speculative extension of the channel idea is intriguing because it offers the opportunity to explain similarity effects without the need for segregated channels to be present at the motion integration stage. To appreciate this, suppose that color tuned neurons in the LGN subjected their inputs to a non-linearity. When stimulated with moving plaid patterns, this non-linearity would introduce a second-order image component (oriented perpendicular to the plaid motion) only if both gratings activated the same channel. The neurons receiving input from these color channels would inherit sensitivity to this second-order component. Importantly, these higher-order neurons would retain this sensitivity even if multiple color channels converged upon them and those neurons thereby lost selectivity to color variation. The motion of this second-order component could be detected by still higher-order neurons that provide input into the integration stage. In consequence, an additional pattern motion signal would be present at the integration stage only when both components activate the same color channel. Although the need for this sort of explanation is clearest in the case of color, similarity effects along other stimulus dimensions might also plausibly arise from channel-specific non-linearities (see Chapter 8 by Wilson for a related proposal regarding the role of spatial frequency in motion coherency).

4.1.2   Feature classification and selective motion integration

The plaid stimuli most commonly used in the experiments cited up to this point were created by additive superimposition of the luminances of two component gratings (e.g. Movshon et al., 1985). Light reflected from overlapping real-world objects, however, is rarely combined in an additive fashion in the formation of the retinal image. More typically, when one object passes in front of another it occludes, at least partially, the

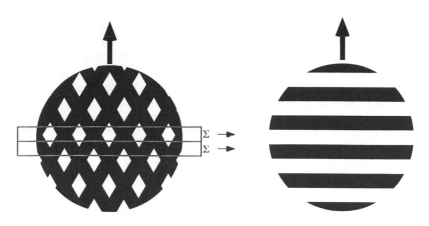

Figure 8 Summing image intensity along the axis oriented perpendicular to the direction of plaid pattern motion reveals the net luminance variation, or Fourier energy, moving in the pattern direction. This 'phantom' grating has zero contrast when the luminances of the two component gratings are simply added; under such conditions it should not contribute to motion coherency. When component gratings are configured to resemble one transparent or occlusive grating overlying another, however, a high-contrast image component is produced which moves in the pattern direction. It has been suggested that the visual system might utilize a log transformation of the image to reduce the influence of this spurious image motion (see main text for details).

distant object. At the point of overlap, the resultant luminance may be exclusively that of the proximal object. A consequence of this non-additive mixing of light is the presence of image components that move in a direction consistent with the composite motion of the occluder and the occluded (Figure 8). The resultant motion signal is a product, however, *not of a single object's motion*, but of two objects interacting incidently via image formation. For most behavioral tasks the composite motion so defined has little or no relevance. We next examine evidence suggesting that the visual motion system selectively *avoids* integration of image motions that are only spuriously associated with one another.

4.1.2.1 Occlusion as a basis for feature classification

Consider, once again, a simple 1D image component moving within an aperture (e.g. Figure 2A). At the boundary between the grating and the aperture, a series of line terminators exists. Each terminator is, by definition, a 2D contrast modulation and the motion of each is – unlike the 1D component itself – unambiguous in two dimensions. Terminators might, therefore, prove useful in disambiguating the motion of the associated component. When the aperture is circular, however, the average motion of the terminators is in a direction parallel to the motion of the 1D component, thus providing no additional information. By contrast, when the same component is viewed through a rectangular aperture, the average direction of terminator motion is skewed toward the long axis of the aperture. When one views such a stimulus, the component typically appears to move along this axis. This phenomenon was studied extensively

by Wallach (1935) and is commonly known as the barber-pole illusion. This effect of
aperture shape on motion perception may be accounted for by a 'smoothing' operation, in which motion signals arising from the terminators are integrated with those
from the interior of the grating (e.g. Bülthoff *et al.*, 1989; Wang *et al.*, 1989). Such
proposals, however, ignore the fact that terminators in natural images are frequently not
a real, or 'intrinsic', property of the associated contour, but rather may be an 'extrinsic'
property – a consequence of two surfaces overlapping in the retinal image, rather than a
physical property of the surfaces themselves. In the extrinsic case, terminator motion is
irrelevant and should be disregarded by the motion integration process.

Shimojo *et al.* (1989) devised a variant of the barber-pole illusion to test the idea
that the motion system is sensitive to the rules of image formation and selectively
disregards some image features in accordance with their figural interpretation. In
particular, Shimojo *et al.* predicted that when terminators are seen to be an intrinsic
property of the grating they will influence its perceived motion. Conversely, terminators interpreted as extrinsic, by consequence of perceived occlusion, should have no
such influence. Figural interpretation of the terminators in the barber-pole display was
biased using binocular disparity to manipulate perceived depth ordering of the grating
and the rectangular aperture (Figure 9, center and right panels). As hypothesized, the
barber-pole illusion was no longer perceived when the grating was seen to lie behind
the aperture. Instead, the grating appeared to move in a direction perpendicular to its
orientation, apparently uninfluenced by terminator motion.

4.1.2.2 Perceptual transparency as a basis for feature classification

Complete occlusion is not the only type of projective interaction that can occur
between elements in a visual scene. Interactions between surfaces that are *partially*

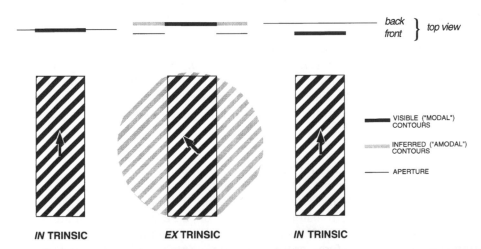

Figure 9 Intrinsic and extrinsic figural properties can be used to explain the barber-pole
illusion (Wallach, 1935). The original illusion is shown schematically at *left*. Viewed through a
rectangular aperture, the grating appears to move along the long axis of the aperture (*bold
arrow*), not perpendicular to grating orientation. As shown by Shimojo *et al.* (1989), depth
ordering (*center and right*) affects figural interpretation (intrinsic vs. extrinsic) of line terminators in the moving grating. Inferred terminator motion, in turn, influences perceived motion
of the grating. See text for details. From Stoner and Albright (1993b).

occluding or transparent also yield characteristic patterns of luminance in the retinal image. An appreciation of these luminance relationships permits inference about the spatial relationships between objects in the visual scene and the classification of image features. Consider an object composed of many small occluders with incomplete coverage (e.g. a wire mesh or a bush). When another object passes behind this perforated object, light reflected from the surface of the more distant object is only partially transmitted and it mixes with the light reflected off of the foreground object. If the individual occluders of the foreground object are too small to be resolved, the light mixture at the point of overlap appears uniform – being equal to the sum of the attenuated background luminance and that of the foreground surface. As a result, the foreground object is *perceptually transparent*;[2] overlapping foreground and background objects are simultaneously perceived.

It is also useful to consider a special case of perceptual transparency in which the foreground object reflects no light of its own. Under these conditions, the foreground object multiplicatively attenuates[3] (by a transmittance factor approximated by the area ratio of perforations to occluders) the light reflected off of the more distant surface. From a functional perspective, this unique multiplicative luminance relationship is particularly interesting because it is compatible with the transparent 'surface' being a shadow.

These luminance relationships and related figural cues are captured by the 'laws of perceptual transparency' and are known to be crucial for image segmentation and feature classification (Metelli, 1974; Beck *et al.*, 1984; Kersten, 1991; Stoner & Albright, 1993b). Patterns of retinal image luminance falling between the extremes of occlusion (Figure 10, *right*) and multiplicative (shadow-like) transparency (Figure 10, *left*), can arise from optical projection of spatially overlapping objects. Accordingly, luminance variations occurring at the region where two surfaces overlap (region A in Figure 10) define extrinsic features. Because luminance variations that lie outside of this occlusion–transparency range cannot be accounted for by object inter-relationships, such variations define intrinsic features, such as differences in surface reflectance. The figural relationships implied by these feature classifications should gate the selective integration of visual motion signals.

4.1.2.3   Feature classification and plaids

The generality of this feature classification hypothesis can be tested using plaid patterns constructed with reference to the laws of perceptual transparency. When the luminance variations in a plaid pattern are made to be physically consistent with one transparent or occluding grating overlying another, the points of overlap

---

[2] Contrary to the common usage of the term 'transparent', which characterizes materials (such as clean glass or air) that permit unattenuated viewing of distant objects, 'perceptually transparent' is used here to refer to surfaces that allow only partial transmission of light from more distant surfaces and are themselves visible (a property that is sometimes termed *translucency*).

[3] Multiplicative attenuation means that the resultant image luminance is simply the product of the unobscured background luminance and the transmittance of the foreground surface. Because no foreground luminance contributes to the equation, the contrast of the attenuated region is equal to the contrast of the unobscured background – only the mean luminance differs.

"PURE" TRANSPARENCY   "PURE" OCCLUSION

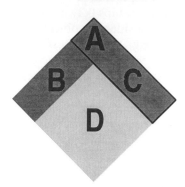

Figure 10 The procedures for creating perceptually transparent plaid patterns are derived from the physics of transparency (Metelli, 1974; Beck *et al.*, 1984). Simply put, luminance ratios within the pattern must be physically consistent with the transmittance of light from a far surface *through* a near surface. Appropriate luminance ratios convey a sense of depth ordering (and hence image segmentation) in a pattern devoid of other depth cues. The zone of perceptual transparency is defined by two extremes – 'pure' transparency (left) and 'pure' occlusion (right), illustrated at *bottom left* and *bottom right*, respectively. Pure transparency occurs when the near, transparent surface A–C reflects no light but transmits light from the surface behind it. Pure occlusion occurs when the near surface A–C reflects light but transmits no light from the surface behind it. See text for details. From Stoner and Albright (1993b).

may be classified as extrinsic features. Their motions should, accordingly, not contribute to perceptual motion coherence. Alternatively, when the plaid luminance variations are not consistent with transparency or occlusion, they may be interpreted as intrinsic features of a 2D pattern and should facilitate motion coherence.

Following this logic, Stoner *et al.* (1990) configured plaid patterns such that they were either consistent or inconsistent with transparency/occlusion (Figure 11). The results of that study appear in Figure 12. As predicted, when the luminance relationships within the plaid pattern were adjusted so that the component gratings appeared transparent or occlusive, non-coherent motion was usually reported. On the other hand, when the luminance configuration was incompatible with transparency/occlusion, subjects were much more likely to report coherent motion.

4.1.2.4 Possible mechanisms: non-linearities?

Although the results of Stoner *et al.* (1990) are congruous with our proposal that motion signal integration is gated by perceptual transparency and occlusion, they reveal little about the mechanism. One commonly considered possibility (e.g. Stoner & Albright, 1992b; Trueswell & Hayhoe, 1993; Mulligan, 1993; Noest & van den Berg, 1993; Plummer & Ramachandran, 1993) follows from the finding that the visual system appears to subject the retinal image to an early compressive non-linearity (see MacLeod, 1978; MacLeod *et al.*, 1992). This proposal can be understood by recalling that plaid patterns constructed using rules of perceptual transparency or occlusion

Figure 11 Each plaid can be viewed as a tessellated image composed of four distinct, repeating subregions, identified as A, B, C and D. Region D is normally seen as background (owing to its relatively large size). Regions B and C are seen as narrow overlapping surfaces and the remaining region A as their intersection. Perceptual transparency was manipulated in these experiments (Stoner *et al.*, 1990; Stoner & Albright, 1992a) by setting the luminance of region A to one of several values, while the luminances of B, C and D were held constant. Adapted from Stoner and Albright (1993b).

typically contain (unlike conventional plaids formed by linear summation of component luminances) 'phantom' Fourier components that move in the pattern direction (Figure 8). If these visual images were subjected to an early logarithmic signal transformation, phantom Fourier components would be eliminated from the neuronal representation of transparent plaids created by multiplicative rules (i.e. shadow-like plaids). Conversely, this type of non-linearity would *introduce* image components, moving in the pattern direction, to plaids not formed by multiplicative rules. This simple mechanism leads to the prediction that motion coherence should be minimal for transparent plaids created by multiplicative rules and roughly accounts for the observations of Stoner *et al.* (1990). The adequacy of this mechanism is called into question, however, when one considers that perceptual transparency is influenced by a variety of segmentation cues (e.g. figural continuity, binocular disparity, etc.) beyond the scope of point-wise non-linearities (Metelli, 1974). Inasmuch as perceptual motion coherence is dependent upon the same subtle cues that influence perceptual transparency, a signal compression mechanism of the sort proposed cannot be the sole explanation for the effects of transparency on motion integration. We turn now to studies that address this issue directly.

4.1.2.5   Foreground/background assignment and transparency

The perception of transparency is inherently dependent upon the visual system's ability to interpret foreground and background relationships in a visual scene. This point can be understood by referring to the left panel of Figure 10. The perception of transparency for surface (A–C) is supported, in part, by the fact that the luminance contrast between regions A and C does not exceed that between B and D. This constraint is captured by one of the corollaries of transparency: transparent surfaces normally dilute rather than enhance the contrast of background surfaces seen through

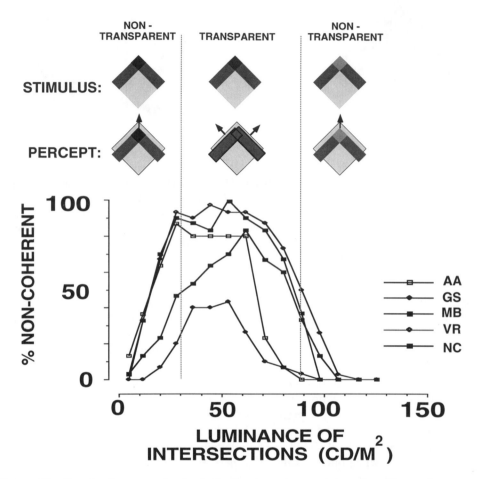

Figure 12 Results from psychophysical experiments examining the effects of perceptual transparency on motion coherency. Probability of the component motion percept is plotted as a function of the intersection luminance for appropriately configured plaid patterns (see Figures 9 and 10). Both gratings were of the same spatial frequency (1.75 cycles per degree). On each trial the individual gratings were moved at an angle of 135° relative to one another at a speed of 3°/s, resulting in a pattern speed of 8°/s. Pattern direction was either up or down, and varied on a random schedule. Intersection luminance was varied in equal steps, such that it was either compatible or incompatible with transparency. The 'transparency zone' extends from pure (multiplicative) transparency (35 cd/m^2) up to the point of occlusion (90 cd/m^2). A percept of non-coherent component motion is most likely within a region roughly centered on the transparency zone. Each data point represents the mean of 30 trials for each intersection luminance value. Data are shown for five subjects. Adapted from Stoner *et al.* (1990).

them. The visual system must apply this rule to gain a percept of transparency. To do so, however, the system must 'know' which portion of the image corresponds to the unobscured background and which corresponds to the portion viewed through the transparent foreground surface. It follows that pictorial cues that influence foreground/background interpretation should have marked effects on the perception of transparency. The latter should, in turn, have similarly marked effects on perceptual motion coherence – effects that cannot be accounted for solely by a simple signal compression mechanism.

In order to test whether foreground/background interpretation (F/B) influences motion coherency, pictorial cues that affect F/B may be employed in the construction of plaid patterns. Stoner and Albright (1992b) devised two different methods of doing so (Figure 13). Human subjects viewed plaid patterns for which F/B was manipulated by these means and reported whether they saw coherent or non-coherent motion. As was the case for the original transparency experiments described above (Stoner *et al.*, 1990), only the luminance of one sub-region (region A in Figure 13) in the plaid pattern was varied. F/B was manipulated such that region A was perceived to be either (1) the *intersection* of the two gratings (as was always the case in the previous study), or (2) the unobstructed background. Both perceptual transparency (for static plaids) and coherence (for moving plaids) depended dramatically upon F/B, as manipulated by either technique. This influence of F/B, in the absence of changes in the basic luminance configuration, disallows fixed non-linear transformation as the explanation for either perceptual transparency or the effects of transparency on motion coherency. It supports our assertion that the motion system has access to image segmentation processes with tacit knowledge of the rules governing retinal image formation from natural scenes. We shall see further evidence of this in the studies reviewed below.

4.1.2.6 Transparency, occlusion, binocular disparity, and motion parallax

Transparency and occlusion, as defined by luminance cues, imply a depth ordering of surfaces in the visual scene. As such, these luminance cues are normally in agreement with stereoscopic and motion parallax cues for depth relationships. We have seen that some of these cues independently influence motion signal integration in a manner predictable from depth ordering (Stoner *et al.*, 1990; Adelson & Movshon, 1984). In view of the marked dependence of luminance-defined perceptual transparency/occlusion and motion coherence on the presence of consistent pictorial cues for depth relationships (Stoner & Albright, 1992b), we might also expect these phenomena to be sensitive to the presence of consistent stereoscopic and motion parallax cues. Specifically, we should expect that motion signal integration depends not simply upon the presence of either luminance, binocular disparity, or motion parallax cues for transparency/occlusion, but rather upon whether the depth ordering implied by each cue *agrees* with that implied by the others.

Trueswell and Hayhoe (1993) tested this hypothesis by exploiting the fact that particular luminance configurations dictate not only whether transparency is likely to be perceived but also which of the two overlapping surfaces will appear closer. The depth relationships resulting from this luminance cue were pitted against binocular disparity cues for depth. It was found that perceptual motion coherence was least likely when the two cues mutually supported segmentation of the gratings into separate surfaces. This result further highlights the inadequacy of a simple non-linear

LUMINANCE CONTRAST VIEWED THROUGH
TRANSPARENT FOREGROUND SURFACE [*AC*] SHOULD BE
LESS THAN THAT OF BACKGROUND [*BD*]

**PERCEPTUALLY
TRANSPARENT**   **PERCEPTUALLY
NON-TRANSPARENT**

FOREGROUND CONTRAST (*A/C*)
IS LESS THAN
BACKGROUNDCONTRAST (*B/D*)

FOREGROUND CONTRAST (*B/D*)
IS GREATER THAN
BACKGROUND CONTRAST (*A/C*)

Figure 13   Schematic illustration of two methods used by Stoner and Albright (1992b) to manipulate the perception of foreground and background in plaid patterns. The plaids were tessellated versions of these basic patterns (see Figure 10). One method, depicted in the *top row*, involved manipulating the relative sizes of the plaid sub-regions. The larger regions (region D on the *left* and region A on the *right*) are usually seen as background (Petter 1956, cited in Vallortigara & Bressan, 1991). A second method, shown in the *bottom row*, was to place a static checkerboard in the putative background region. The plaid motion progressively occluded/disoccluded this pattern, causing the textured region (region D on the *left* and region A on the *right*) to be seen as background (note that the 'occlusion' of this checkerboard by the putatively transparent gratings is physically consistent with the contrast reduction associated with transparent surfaces or the blurring common with translucent surfaces such as smoked glass). Both methods reliably influence perceptual assignment of foreground/ background, while leaving the space-averaged luminance of the four regions constant. The reversal of foreground/background assignment, in turn, had a profound effect on both perceptual transparency and motion coherency judgments by human observers. From Stoner and Albright (1992b).

signal transformation as an explanation for the effects of transparency on motion signal integration and it points to the limitations of the channel model as explained for the corresponding effects of binocular disparity.

A similar experimental strategy (conceptually similar, but not involving plaid patterns) was adopted by Kersten *et al*. (1992) in a study pitting the depth ordering implied by luminance cues for transparency/occlusion cues against that implied by motion parallax (Wallach & O'Connell, 1953). Two overlapping rectangles were rocked back and forth in a manner consistent with one rectangle being in front of another. When luminance cues were consistent with this implied depth relationship, subjects reported a percept of two overlapping rectangles rocking independently back and forth in depth – the region of overlap was apparently classified as extrinsic and its motion 'ignored'. When the two cues conflicted, however, neither was sufficient to support segregation, the region of overlap was interpreted as an intrinsic feature, and subjects reported a strong percept of non-rigid motion. The interdependency of these various stimulus parameters demonstrates, once again, that the visual motion system has access to image segmentation processes of considerable sophistication. This result is also particularly interesting because it reveals that (1) similar selective integration rules hold for motion in the third dimension, and (2) the process is recursive, in that motion signals are themselves among the potential cues for depth ordering and hence cooperate with other cues in their own integration.

4.1.2.7 Segmentation cues acting at a distance: influence of aperture configuration

The barber pole experiments of Shimojo *et al*. (1989), reviewed above, demonstrate the powerful influence of terminator classification on motion processing. The significance of those observations was extended by Kooi *et al*. (1992), who explored the role of aperture configuration on motion coherency in plaid patterns. The conventional means of displaying plaid patterns involves viewing them through a circular aperture. In one set of experiments, however, Kooi *et al*. constructed plaid stimuli such that, at the margins of the display, each grating moved within its own rectangular aperture (oriented so that one axis was parallel to the grating). Twin barber-pole illusions were obtained and the two gratings were seen to move in different directions. These independent motions propagated into the interior of the display, resulting in perceptual non-coherency. This phenomenon may reflect the visual system's attempt to minimize the number of distinct motions – perceptual non-coherency, in this case, requires only two global motions whereas a coherent percept would require three (one in the interior and two at the margins).

In another experiment, Kooi *et al*. (1992) examined the influence of stereoscopic cues that provide for depth ordering of the aperture and plaid pattern. Plaids were viewed through a square aperture aligned such that the two axes were parallel to the directions of the two moving components. A percept of motion coherence was most likely when the viewing aperture was placed in front of the moving plaid. This finding is entirely consistent with the observations of Shimojo *et al*. (1989) and can be interpreted in the same vein: when the terminators are perceived to be extrinsic – a consequence of occlusion by the aperture – they are ignored. These results are particularly important because they emphasize the ability of cues for feature classification to 'act at a distance', governing the integration of motion signals at locations in the image where segmentation cues are either absent or ambiguous. Models of motion signal integration must provide for the influence of such non-local information.

4.1.2.8   Contrast and spatial frequency revisited: cues for depth ordering/segmentation?

We have explored several lines of evidence supporting our proposal that image segmentation cues control the selective integration of visual motion signals. It is worth considering whether the original well-documented effects of contrast and spatial frequency on motion coherence (Adelson & Moushon, 1982) can also be interpreted in this functional context. A relationship between contrast and image segmentation is suggested by the observation that objects that contrast more with background are perceived to be closer than those having lesser contrast (Egusa, 1982). Taking note of this relationship, Vallortigara and Bressan (1991) tested the hypothesis that perceptual motion coherence is influenced by contrast because contrast is a cue for relative depth. They did so by examining the interdependence of contrast and luminance-defined occlusion as cues for depth ordering. If contrast acts as a segmentation cue in this context, its effects should be manifest only when they are in agreement with luminance cues for occlusion (analogous to the interdependence of stereoscopic and luminance cues, described above). Plaid patterns were constructed such that the two cues were either consistent (high contrast grating placed in front of low contrast grating, via luminance-defined occlusion) or inconsistent (high contrast grating placed behind low contrast grating). In line with predictions, perceptual motion coherence was found to be minimal when contrast and luminance cues were in agreement, supporting the possibility that contrast acts as a segmentation cue and casting doubt on the adequacy of simple channel-based schemes for selective integration.

A possible role for spatial frequency in image segmentation is suggested by the fact that relative size is also a cue for relative depth; the larger of two superimposed figures generally appears to lie in front of the other (Vallortigara & Bressan, 1991; Petter, 1956; cited in Vallortigara & Bressan, 1991; see also Kanizsa, 1979). The size of each 'bar' of a periodic grating is a direct function of spatial frequency. This suggests that frequency-dependent depth segregation of plaid components might account for differential perceptual motion coherence. Vallortigara and Bressan (1991) tested this hypothesis by varying the relative size difference between the bars of two superimposed square-wave gratings. This manipulation alone had marked effects on perceptual coherence; an observation that is scarcely surprising since the manipulation also alters the relative spatial frequency content of the two gratings. This relative size/frequency dependence could be destroyed, however, by simply occluding each point of component intersection with a gray disc. This result was interpreted as indicating that, when the intersections were unavailable as a source of segmentation information, relative size/frequency was unimportant.

While both sets of results are suggestive, it remains to be seen whether these contrast and size manipulations alter luminance gradients (i.e. Fourier components) such that perceptual motion coherence could be explained by simpler mechanisms. In addition, there exists no independent evidence regarding the role of spatial frequency in the depth ordering of conventional sine-wave plaids. Despite these qualifications, in the light of mounting evidence for the sensitivity of the motion system to the rules of

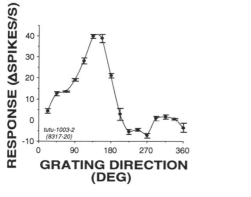

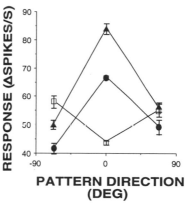

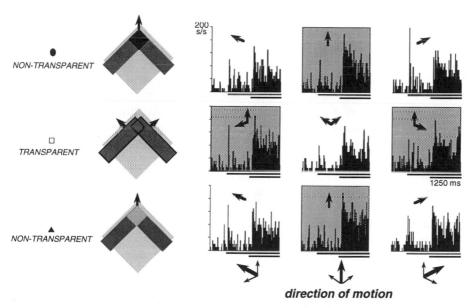

direction of motion

Figure 14   Neural correlates of perceptual motion signal integration. Differential responses of an MT pattern-type neuron to coherent vs. non-coherent plaid patterns. Directional tuning for a single drifting grating is plotted at *top left*. Responses to coherent and non-coherent plaids are plotted at *top right*. When stimulated with coherent plaid patterns, response was maximal when the pattern moved in the neuron's preferred direction (0° in graph at *top right*; highlighted histograms in *top and bottom rows* of lower figure). However, when stimulated with non-coherent plaid patterns, responses were maximal when either component moved in the preferred direction (±67.5° in graph at *top right*; highlighted histograms in *center row* of lower figure). Adapted from Stoner and Albright (1992a).

retinal image formation, it seems plausible that contrast and spatial frequency exert their effects as cues for segmentation rather than strictly as channel parameters.

4.2 Neurophysiological Studies of Selective Integration

We have reviewed numerous psychophysical experiments bearing on the selective nature of motion signal integration and we have attempted to interpret the results of these experiments in the context of a coherent theoretical framework. Ultimately, however, a complete understanding of this process will depend upon an appreciation of the neural elements and events that underlie it. As discussed in the first half of our chapter, the weight of neurobiological evidence points to extrastriate visual area MT as the locus of integration. Only recently, however, has progress been made toward identifying neural events that are responsible for *selective* integration.

To address this issue directly, Stoner & Albright (1992a) examined whether the activity of MT neurons representing the integration stage could be modified by factors known to influence perceptual motion coherence. As in their earlier psychophysical experiments (Stoner *et al.*, 1990), perceptual motion coherence was manipulated by altering the stimulus conditions such that they were either consistent or inconsistent with transparency. Directional selectivity of MT neurons was then assessed using each of three different plaid configurations. Two of these stimuli elicited a percept of coherent pattern motion, while the third elicited a percept of independently moving components. Data obtained from a typical neuron are shown in Figure 14. When presented with either of the perceptually coherent plaids, this cell responded more strongly when the pattern moved in the preferred direction than when either of the components moved in this same direction. This type of directional tuning is characteristic of pattern neurons (Movshon *et al.*, 1985; Rodman & Albright, 1989). When presented with the perceptually non-coherent plaid, however, this neuron's behavior underwent a striking transformation: the pattern response dropped substantially and component responses became elevated. The resultant bilobed directional tuning curve is recognizable as the characteristic signature of a component neuron (see Figure 6). Thus, as was the case for the majority of neurons studied, this cell became more sensitive to component motion (and less sensitive to pattern motion) when the stimulus was configured to render component motion as the dominant percept.

These results demonstrate that visual stimulus conditions capable of affecting image segmentation and, hence perceptual motion coherence, also cause systematic differences in the directional tuning of those neurons thought to underlie motion integration. Although it remains to be seen whether this neuronal–perceptual correlation accompanies the full range of stimulus conditions known to influence perceptual coherence, these new results sharply refine our understanding of the neural mechanisms underlying selective motion integration.

4.3 How might Image Segmentation Cues Interact with Motion Signals?

We have considered abundant evidence for the selective nature of motion signal integration and we have argued that many, if not all, of these phenomena can be explained as a predictable consequence of image segmentation. This functional context, however alluring it may be, does not address the critical issue of mechanism. As we have seen, however, many of the experiments reviewed above have led some investigators to champion specific mechanisms for selective integration. We find it useful to group candidate mechanisms into two general classes. The first consists of those in which segmentation and motion integration go hand-in-hand, in the sense that it is the integration mechanism itself that dictates criteria for selectivity. Channel-specific integration schemes and those that depend upon non-linear signal transformations fall into this class. Although, as we have seen, such mechanisms can plausibly account for some specific perceptual phenomena, they cannot easily explain many other examples of selective integration (e.g. the influence of foreground/background interpretation or the conjoint influence of multiple depth stratification cues).

The alternative class of mechanisms consists of those in which the segmentation process is remote from the stage at which motion signal integration occurs. According to such schemes, once the visual scene is parsed into objects, appropriate signals are relayed to the integration stage such that motion signals are bound together in an object-specific manner. There are numerous advantages of a system of this sort, including the fact that it allows gating by a generalized (or cue-independent) segmentation signal, and the fact that, once extracted, segmentation signals could be used for other computations (e.g. object recognition) in addition to those motion-related. The obvious limitation of such proposals, however, is that they side step the problem of segmentation itself – it is simply assumed that segmentation occurs and that the product is made available to the motion system. If we adopt a willingness to defer a solution to this problem-of-problems in vision, we can consider more specifically the means by which segmentation signals might interact with motion signals at the integration stage.

The influence of segmentation cues could be effected by a variety of different mechanisms. One possibility, alluded to above, is that early motion signals are 'tagged' as either intrinsic or extrinsic (on the basis of segmentation signals) and only intrinsic signals are admitted to the integration stage (Figure 15). Accordingly, the probability of coherence would be influenced by whether this additional signal was present. Another possibility, not exclusive of the first, is that mechanisms sensitive to locally present segmentation cues (e.g. local luminance relationships seen at regions of object overlap in conjunction with disparity), are capable of 'linking together' more global image primitives (e.g. the regions of non-overlap) by promoting selective synchronization of activity among component neurons. Motion signal integration might then be achieved at the pattern neuron stage by close temporal synchrony between active component inputs. Indeed, in light of the ease with which motion signal integration can be manipulated, this would seem to be an ideal experimental paradigm for testing the hypothesis that synchronized firing 'binds' locally detected features into objects (Gray et al., 1989; Crick & Koch, 1990).

What might be the source of the segmentation information that gates motion signal integration? Psychophysical evidence points to an early representation of surface relationships (Ramachandran & Cavanagh, 1985; Nakayama & Shimojo, 1990; Shimojo et al., 1989). Neurophysiological evidence corroborates this hypothesis; area V2, an early stage in the visual hierarchy, contains neurons that are orientation and disparity tuned (DeYoe & Van Essen, 1985) as well as those that are responsive to subjective contours (von der Heydt et al., 1984; von der Heydt & Peterhans, 1989).

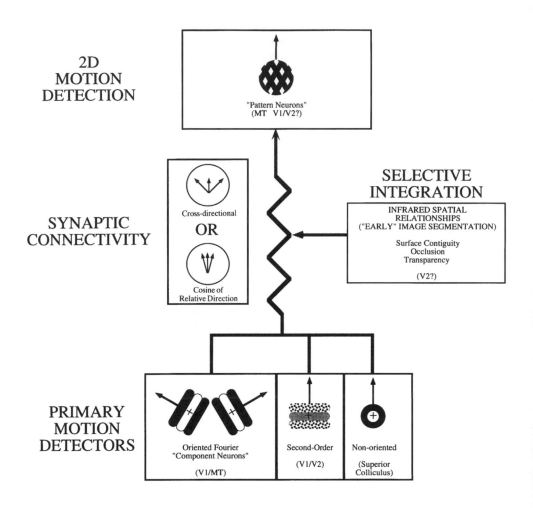

2D
MOTION
DETECTION

"Pattern Neurons"
(MT V1/V2?)

SELECTIVE
INTEGRATION

INFRARED SPATIAL
RELATIONSHIPS
("EARLY" IMAGE SEGMENTATION)

Surface Contiguity
Occlusion
Transparency

(V2?)

SYNAPTIC
CONNECTIVITY

Cross-directional

OR

Cosine of
Relative Direction

PRIMARY
MOTION
DETECTORS

Oriented Fourier
"Component Neurons"

(V1/MT)

Second-Order

(V1/V2)

Non-oriented

(Superior
Colliculus)

MOVING
IMAGE

Thus, area V2 may possess the means to differentiate between intrinsic and extrinsic image features. Projections between V2 thick stripes and MT are substantial (e.g. DeYoe & Van Essen, 1985) and may provide a route by which segmentation cues interact with motion signals. It should be stressed, however, that any such proposed function for these connections is quite speculative. More precise determination of the neural connections and events responsible for imparting image segmentation information to the motion system is an exciting challenge for future research.

5   CONCLUDING REMARKS: THE FRONTIER

Unlike many other issues faced in efforts to understand primate vision, the integration of motion signals constitutes a well-defined computational problem, to which we can link a specific neural substrate, and upon which visually-guided behavior critically depends. In that sense, it represents a model system for exploring the relationships among neuronal phenomena, perception, and behavior. In reviewing essential aspects of motion integration we have attempted to cast them in this broad functional context. It should be apparent, however, that there are many unanswered questions. Still unresolved are fundamental issues pertaining to the specific patterns of neuronal connections that yield pattern motion properties, and the routes by which different types of information (motion and segmentation) reach the integration stage. Knowledge of this sort is clearly needed to resolve controversies regarding the form of the integration mechanism and the means through which it achieves selectivity. Thus, as is often the case, the limiting step is formidable empiricism and the ball falls soundly in the court of the neurobiologist.

ACKNOWLEDGMENTS

Preparation of this chapter was supported in part by a grant from the National Institute of Mental Health (MH50706) and a Development Award from the McKnight

---

Figure 15   Schematic diagram illustrating hypothesized motion processing stages and neural connectivity patterns that could account for selective integration of motion signals in the primate visual system. The oriented components of a moving 2D retinal stimulus (a plaid pattern, in this example) are known to give rise to activity in 1D (oriented or 'component type') motion detectors at the first processing motion stage (V1 and MT). The same stimulus may also elicit activity amongst populations of second-order/non-Fourier and non-oriented motion detectors, both of which could indicate the movement of image features in the pattern direction. Different classes of motion integration models postulate different patterns of synaptic connectivity between primary motion detectors and pattern-type neurons. IOC and spatio-temporal models posit synaptic connectivity that spans directional preferences of 180°. Conversely, some modelers argue for synaptic connectivity in which primary motion detectors project mainly to 2D motion detectors having near-identical directional selectivity. The processing of segmentation cues (perhaps by area V2) allows image features (and corresponding motion signals) to be classified as intrinsic or extrinsic and ensures that only motion signals arising from a common physical motion are integrated.

Endowment Fund for Neuroscience (TDA). Gene Stoner was partially supported by a Research Fellowship from the McDonnell-Pew Center for Cognitive Neuroscience at San Diego.

REFERENCES

Adelson, E. H. & Movshon, J. A. (1982). Phenomenal coherence of moving visual patterns. *Nature*, **300**, 523–525.

Adelson, E. H. & Movshon, J. A. (1984). Binocular disparity and the computation of two-dimensional motion. *J. Opt. Soc. Am. A*, **1**, 1266.

Albright, T. D. (1984). Direction and orientation selectivity of neurons in visual area MT of the macaque. *J. Neurophys.*, **52**, 1106–1130.

Albright, T. D. (1992). Form-cue invariant motion processing in primate visual cortex. *Science*, **255**, 1141–1143.

Anandan, P. & Weiss, R. (1985). Introducing a smoothness constraint in a matching approach for the computation of optical flow fields. *Proceedings of the IEEE Workshop on Computer Vision: Representation and Control*. Bellaire, Michigan, October, pp. 186–194.

Beck, J., Prazdny, K. & Ivry, R. (1984). The perception of transparency with achromatic colors. *Percept. Psychophys.*, **35**, 407–422.

Bülthoff, H., Little, J. & Poggio, T. (1989). A parallel algorithm for real-time computation of optical flow. *Nature*, **337**, 549–552.

Burke, D. & Wenderoth, P. (1993). The effect of interactions between one-dimensional component gratings on two-dimensional motion perception. *Vision Res.*, **33**, 343–350.

Campbell, F. W. & Maffei, L. (1980). The influence of spatial frequency and contrast on the perception of moving patterns. *Vision Res.*, **21**, 713–721.

Campbell, F. W. & Robson, J. G. (1968). Application of Fourier analysis to the visibility of gratings. *J. Physiol. (Lond.)*, **197**, 551–566.

Cavanagh, P. & Mather, G. (1990). Motion: the long and short of it. *Spatial Vision*, **4**, 103–129.

Chou, G. & Adelson, E. (1993). A computational model for the integration of form and motion. *Invest. Ophthalmol. Vis. Sci.*, **34**, 1029.

Chubb, C. & Sperling, G. (1988). Drift-balanced stimuli: A general basis for studying non-Fourier motion perception. *J. Opt. Soc. Am. A*, **5**, 1986–2006.

Crick, F. & Koch, C. (1990). Towards a neurobiological theory of consciousness. *Sem. Neurosci.*, **2**, 263–275.

Derrington, A. M. & Badcock, D. R. (1992). Two-stage analysis of the motion of 2-dimensional patterns, what is the first stage? *Vision Res.*, **32**, 691–698.

Derrington, A. M. & Suero, M. (1992). Motion of complex patterns is computed from the perceived motions of their components. *Vision Res.*, **31**, 139–149.

Derrington, A. M., Badcock, D. R. & Holroyd, S. A. (1992). Analysis of the motion of 2-dimensional patterns: evidence for a second-order process. *Vision Res.*, **32**, 699–707.

De Valois, K. K., De Valois, R. L. & Yund, E. W. (1979). Responses of striate cortex cells to grating and checkerboard patterns. *J. Physiol. (Lond.)*, **291**, 483–505.

De Valois, R. L., Yund, E. W. & Hepler, N. (1982). The orientation and direction selectivity of cells in macaque visual cortex. *Vision Res.*, **22**, 531–544.

DeYoe, E. A. & Van Essen, D. C. (1985). Segregation of efferent connections and receptive field properties in visual area V2 of the macaque. *Nature*, **317**, 58–61.

Diener, H. C., Wist, E. R., Dichgans, J. & Brandt, T. H. (1976). The spatial frequency effect on perceived velocity. *Vision Res.*, **16**, 169–176.

Dobkins, K. R., Stoner, G. R. & Albright, T. D. (1992). Oculomotor responses to perceptually coherent and non-coherent plaids. *Soc. Neurosci. Abstr.*, **18**, 1034.

Egusa, H. (1982). Effect of brightness on perceived distance as a figure-ground phenomenon. *Perception*, **11**, 671–676.

Felleman, D. J. & Van Essen, D. C. (1987). Receptive field properties of neurons in area V3 of macaque monkey extrastriate cortex. *J. Neurophysiol.*, **57**, 889–920.

Fennema, C. L. & Thompson, W. B. (1979). Velocity determination in scenes containing several moving objects. *Computer Graphics & Image Processing*, **9**, 301–315.

Ferrera, V. P. & Maunsell, J. H. R. (1991). Responses of single units in macaque V1 to moving patterns. *Soc. Neurosci. Abstr.*, **17**, 177.

Ferrera, V. P. & Wilson, H. R. (1987). Direction specific masking and the analysis of motion in two dimensions. *Vision Res.*, **27**, 1783–1796.

Ferrera, V. P. & Wilson, H. R. (1990). Perceived direction of moving two-dimensional patterns. *Vision Res.*, **30**, 273–287.

Ferrera, V. P. & Wilson, H. R. (1991). Perceived speed of moving two-dimensional patterns. *Vision Res.*, **31**, 877–893.

Foster, K. H., Gaska, J. P., Nagler, M. & Pollen, D. A. (1985). Spatial and temporal frequency selectivity of neurons in visual cortical areas V1 and V2 of the macaque monkey. *J. Physiol. (Lond.)*, **365**, 331–363.

Gray, C. M., Koenig, P., Engel, A. K. & Singer, W. (1989). Oscillatory responses in cat visual cortex exhibit inter-columnar synchronization which reflects global stimulus properties. *Nature*, **338**, 334–337.

Grzywacz, N. M. & Yuille, A. L. (1991). Theories for the visual perception of local velocity and coherent motion. In M. S. Landy & J. A. Movshon (Eds.) *Computational Models of Visual Processing*, pp. 231–252. MIT Press, Cambridge, MA.

Heeger, D. J. (1987). Model for the extraction of image flow. *J. Op. Soc. Am. A*, **4**, 1455–1471.

Hildreth, E. C. (1984). *The Measurement of Visual Motion*, MIT Press, Cambridge, MA.

Horn, B. K. P. (1986). *Robot Vision*. MIT Press, Cambridge, MA.

Horn, B. K. P. & Schunck, B. G. (1981). Determining optic flow. *Artificial Intelligence*, **17**, 185–203.

Hubel, D. H. & Wiesel, T. N. (1968). Receptive fields and functional architecture of monkey striate cortex. *J. Physiol.*, **195**, 215–243.

Hurvich, L. M. & Jameson, D. (1957). An opponent-process theory of color vision. *Psychol. Rev.*, **64**, 384–390; 397–404.

Kanizsa, G. (1979). *Organization in Vision: Essays on Gestalt Perception*. Praeger Publishers, New York.

Kersten, D. (1991). Transparency and the cooperative computation of scene attributes.

In M. S. Landy & J. A. Movshon (Eds.) *Computational Models of Visual Processing*, pp. 209–228, MIT Press, Cambridge, MA.

Kersten, D., Bülthoff, H. H., Schwartz, B. L. & Kurtz, K. J. (1992). Interaction between transparency and structure from motion. *Neural Computation*, **4**, 573–589.

Kim, J. & Wilson, H. R. (1993). Dependence of plaid motion coherence on component grating directions. *Vision Res.*, **33**, 2479–2489.

Kooi, F. L., De Valois, K. K., Grosof, D. H. & Switkes, E. (1989). Coherence properties of colored moving plaids. *Invest. Ophthalmol. Vis. Sci.*, **30**, 389.

Kooi, F. L., De Valois, K. K. & Switkes, E. (1992). Higher order factors influencing the perception of sliding and coherence of a plaid. *Perception*, **21**, 583–598.

Krauskopf, J. & Farell, B. (1990). Influence of colour on the perception of coherent motion. *Nature*, **348**, 328–331.

Krauskopf, J., Williams, D. R. & Heeley, D. W. (1982). Cardinal directions of color space. *Vision Res.*, **22**, 1123–1131.

Lehky, S. R. & Sejnowski, T. J. (1988). Network model of shape-from-shading: Neural function arises from both receptive and projective fields. *Nature*, **333**, 452–454.

Lennie, P. & D'Zmura, M. (1988). Mechanisms of color vision. *Crit. Rev. Neurobiol.*, **3**, 649–669.

Lennie, P., Krauskopf, J. & Sclar, G. (1990). Chromatic mechanisms in striate cortex of macaque. *J. Neurosci.*, **10**, 649–669.

Levitt, J. & Movshon, J. A. (1993). Personal communication.

MacLeod, D. I. A. (1978). Visual sensitivity. *Ann. Rev. Psychol.*, **29**, 613–645.

MacLeod, D. I. A., Williams, D. R. & Makous, W. (1992). A visual non-linearity fed by single cones. *Vision Res.*, **32**, 347–363.

Marr, D. C. & Ullman, S. (1981). Directional selectivity and its use in early visual processing. *Proc. R. Soc. Lond. (Biol.).*, **211**, 151–180.

Marshak, W. & Sekuler, R. (1979). Mutual repulsion between moving visual targets. *Science*, **205**, 1399–1401.

Marshall, J. A. (1988). Neuronal networks for computational vision: Motion segmentation and stereo fusion. Unpublished doctoral dissertation.

Maunsell, J. H. R. & Van Essen, D. C. (1983a). The connections of the middle temporal visual area (MT) and their relationship to a cortical hierarchy in the macaque monkey. *J. Neurosci.*, **3**, 2563–2586.

Maunsell, J. H. R. & Van Essen, D. C. (1983b). Functional properties of neurons in middle temporal visual area of the macaque monkey. II. Binocular interactions and sensitivity to binocular disparity. *J. Neurophysiol.*, **49**, 1148–1167.

Metelli, F. (1974). The perception of transparency. *Sci. Am.*, **230**, 91–95.

Movshon, J. A. (1988). Spatio-temporal tuning and speed sensitivity in macaque visual cortical neurons. *Invest. Ophthalmol. Vis. Sci. Suppl.*, **29**, 327.

Movshon, J. A. & Newsome, W. T. (1984). Functional characteristics of striate cortical neurons projecting to MT in the macaque. *Soc. Neurosci. Abstr.*, **10**, 933.

Movshon, J. A., Adelson, E. A., Gizzi, M. & Newsome, W. T. (1985). The analysis of moving visual patterns. In C. Chagas, R. Gattass & C. G. Gross (Eds.) *Study Group on Pattern Recognition Mechanisms*, pp. 117–151. Pontifica Academia Scientiarum: Vatican City.

Mulligan, J. B. (1993). Non-linear combination rules and the perception of visual motion transparency. *Vision Res.*, **33**, 2021–2030.

Nagel, H. H. & Enkelmann, W. (1986). An investigation of smoothness constraints for the estimation of displacement vector fields for sequences. *IEEE Transactions on Pattern and Analysis Machine Intelligence PAMI*, **8**, 565–593.

Nakayama, K. & Shimojo, S. (1990). Toward a neural understanding of visual surface representation. *The Brain: Cold Spring Harbor Symposia on Quantitative Biology*, **55**, 911–924.

Newsome, W. T., Gizzi, M. S. & Movshon, J. A. (1983). Spatial and temporal properties of neurons in macaque MT. *Invest. Ophthalmol. Vis. Sci. Suppl.*, **24**, 106.

Noest, A. J. & van den Berg, A. V. (1993). The role of early mechanisms in motion transparency and coherence. *Spatial Vision*, **7**(2), 125–147.

Nowlan, S. J. & Sejnowski, T. J. (1994). Filter selection model of motion processing in area MT of primates. *J. Neurosci.* (in press).

Perrone, J. A. (1990). A simple technique for optic flow estimation. *J. Opt. Soc. Am. A*, **7**, 264–278.

Petter, G. (1956). Nuove ricerche sperimentali sulla totalizzazione percettiva. *Rivista di psicologia*, **50**, 213–227.

Plummer, D. J. & Ramachandran, V. S. (1993). Perception of transparency in stationary and moving images. *Spatial Vision*, **7**(2), 113–123.

Ramachandran, V. S. & Cavanagh, P. (1985). Subjective contours capture stereopsis. *Nature*, **317**, 527–530.

Ramachandran, V. S., Ginsburg, A. P. & Anstis, S. M. (1983). Low spatial frequencies dominate apparent motion. *Perception*, **12**, 457–462.

Richards, W. A. (1970). Stereopsis and stereoblindness. *Exp. Brain Res.*, **10**, 380–388.

Rodman, H. R. & Albright, T. D. (1989). Single-unit analysis of pattern-motion selective properties in the middle temporal visual area (MT). *Exp. Brain Res.*, **75**, 53–64.

Sereno, M. E. (1993). *Neural Computation of Pattern Motion*. MIT Press, Cambridge, MA.

Sereno, M. I. (1989). Learning the solution to the aperture problem for pattern motion with a Hebb rule. *Adv. Neural Info. Proc. Sys.* 2, Morgan-Kaufman.

Shimojo, S., Silverman, G. H. & Nakayama, K. (1989). Occlusion and the solution to the aperture problem for motion. *Vision Res.*, **29**, 619–626.

Smith, A. T. (1992). Coherence of plaids comprising components of disparate spatial frequencies. *Vision Res.*, **32**, 1467–1474.

Smith, A. T. & Edgar, G. K. (1990). The influence of spatial frequency on perceived temporal frequency and perceived speed. *Vision Res.*, **30**, 1467–1474.

Sporns, O., Gally, J. A., Reeke, G. N. Jr & Edelman, G. M. (1989). Reentrant signaling among simulated neuronal groups leads to coherency in their oscillatory activity. *Proc. Natl Acad. Sci. USA*, **86**(18), 7265–7269.

Stone, L. S., Watson, A. B. & Mulligan, J. B. (1990). Effect of contrast on the perceived direction of a moving plaid. *Vision Res.*, **30**(7), 1049–1067.

Stoner, G. R. & Albright, T. D. (1992a). Neural correlates of perceptual motion coherence. *Nature*, **358**, 412–414.

Stoner, G. R. & Albright, T. D. (1992b). The influence of foreground/background assignment on transparency and motion coherence in plaid patterns. *Invest. Opthalmol. Vis. Sci. Suppl.*, **33**, 1050.

Stoner, G. R. & Albright, T. D. (1993a). Unpublished observations.

Stoner, G. R. & Albright, T. D. (1993b). Image segmentation cues in motion processing: Implications for modularity in vision. *J. Cogn. Neurosci.*, **5**(2), 129–149.

Stoner, G. R., Albright, T. D. & Ramachandran, V. S. (1990). Transparency and coherence in human motion perception. *Nature*, **344**, 153–155.

Thompson, P. (1982). Perceived rate of movement depends on contrast. *Vision Res.*, **22**, 377–380.

Trueswell, J. C. & Hayhoe, M. M. (1993). Surface segmentation mechanisms and motion perception. *Vision Res.*, **33**, 313–328.

Ungerleider, L. G. & Desimone, R. (1986). Cortical connections of visual area MT in the macaque. *J. Comp. Neurol.*, **248**, 190–222.

Vallortigara, G. & Bressan, P. (1991). Occlusion and the perception of coherent motion. *Vision Res.*, **31**, 1967–1978.

von der Heydt, R. & Peterhans, E. (1989). Mechanisms of contour perception in monkey visual cortex. I. Lines of pattern discontinuity. *J. Neurosci.*, **9**, 1731–1748.

von der Heydt, R., Peterhans, E. & Baumgartner, G. (1984). Illusory contours and cortical neuron responses. *Science*, **224**, 1260–1262.

Wallach, H. (1935). Uber visuell wahrgenommenr Bewegungsrichtung. *Psychol. Forsch.*, **20**, 325–380.

Wallach, H. & O'Connell, D. N. (1953). The kinetic depth effect. *J. Exp. Psychol.*, **45**, 205–217.

Wang, H. T., Mathur, B. & Koch, C. (1989). Computing optical flow in the primate visual system. *Neural Computation*, **1**, 92–103.

Welch, L. (1989). The perception of moving plaids reveals two motion-processing stages. *Nature*, **337**, 734–736.

Welch, L. & Bowne, S. F. (1990). Coherence determines speed discrimination. *Perception*, **19**(4), 425–435.

Wilson, H. R. & Kim, J. (1994). Perceived motion in the vector sum direction. Submitted for publication.

Wilson, H. R., Ferrera, V. P. & Yo, C. (1992). A psychophysically motivated model for two-dimensional motion perception. *Vis. Neurosci.*, **9**, 79–97.

Yo, C. & Wilson, H. R. (1992). Perceived direction of moving two-dimensional patterns depends on duration, contrast and eccentricity. *Vision Res.*, **32**, 135–147.

Yuille, A. L. & Grzywacz, N. M. (1988). A computational theory for the perception of coherent visual motion. *Nature*, **333**, 71–74.

Zhou, Y. X. & Baker, N. K. (1993). A processing stream in mammalian visual cortex neurons for non-Fourier responses. *Science*, **261**, 98–101.

10

Spatial Integration of Local Motion Signals

Douglas W. Williams[1] and Julie R. Brannan[2]
[1] *Rockefeller University, New York, and* [2] *State University of New York, USA*

1 INTRODUCTION

For the most part, the initial stages of visual processing are localized in space. For both color vision and spatial vision, processing is carried out by means of localized receptive fields. There is evidence that this also holds true for motion perception. Spatial receptive fields only respond to motion over a limited distance. Additionally, for moving patterns linear contrast sensitivity summation is limited to less than one degree (Anderson & Burr, 1987). In essence, every visual scene is fragmented into many different local entities. In order for perception of the entire scene to occur, these individual local responses must be integrated into a global construct.

The global percept that results from the spatial integration of local motion signals is not always simply the aggregate collection of the local motions. In particular, the combination of several different motion vectors can produce a percept of coherent motion in a single direction. Adelson and Movshon (1982), for example, found that two sinusoidal gratings of similar spatial frequencies which move in different directions may in fact appear to cohere into a single moving checkerboard-like pattern. Additionally, Levinson *et al.* (1982) demonstrated that two spatially interspersed random dot patterns moving in orthogonal directions can generate a percept of motion representing the mean of the two directions (if contrast is near threshold). Early Gestaltists were aware that the motion of a feature depends on the behavior of features nearby, and proposed the law of shared common fate (Koffka, 1935). This law states that features in motion tend to be perceived as moving coherently together.

2 DETERMINANTS OF LOCAL MOTION SIGNALS

When two moving sinusoidal gratings of different orientations, each of which moves perpendicularly to its orientation, cohere into a plaid or checkerboard-like pattern, the perceived motion of the plaid is not always the vector sum of the sinusoidal components. If the direction of motion *is* the vector sum of the components, the coherence is denoted Type I; otherwise, it is Type II. To account for the two types of patterns, Adelson and Movshon (1982) proposed a computational scheme known as the intersection of constraints (IOC) solution. For this theory, the determination of global coherence is a two-stage process. In the first stage, the direction and speed of the Fourier components are extracted by oriented linear spatial filters (i.e. the speed and direction of the sinusoidal components). Then in the second stage, the direction of global coherent motion is computationally represented in velocity space by the intersection of vectors perpendicular to the velocity vectors of the Fourier components (see Chapter 9 by Stoner and Albright).

The two types of patterns exhibit many different properties, suggesting that they are not generated by the same process. For example, Type I patterns produce significantly stronger masking effects than Type II patterns (Ferrera & Wilson, 1987). Direction discrimination thresholds are found to be much poorer for the Type II plaids (5.0–7.0 degrees) than for Type I (0.5–1.0 degrees). In the periphery, the perceived direction of Type II patterns differs by as much as 40 degrees from what the IOC would predict (Yo & Wilson, 1992). Finally, for brief presentations (60 ms) in foveal viewing, Type II patterns are perceived as moving in the vector sum direction (Yo & Wilson, 1992). These findings argue that the IOC does not hold.

Studies have revealed that stimulus motion which would be undetected by a Fourier-based motion pathway can contribute to motion perception (Chubb & Sperling, 1988). In the plaid patterns, luminance 'blobs' are formed by the overlap of the two component gratings. For both Type I and Type II patterns these blobs move in the direction of the resultant plaid pattern. The motion of these luminance blobs would be transparent to a Fourier-based motion system. Derrington *et al.* (1992) found evidence that the perception of the plaid motion depends on both the Fourier and non-Fourier motion components.

Wilson *et al.* (1992) have proposed a different model for plaid motion perception in which both a Fourier pathway and a non-Fourier pathway provide inputs to a final processing stage. There is evidence that this final stage occurs in the middle-temporal cortex (MT). For example, many neurons in MT respond to the direction of plaid motion rather than to the actual motion of the component gratings (Movshon *et al.*, 1986). MT receives input directly from primary visual cortex (V1) but also receives input routed from V1 to secondary cortex (V2) (Van Essen, 1985). The model of Wilson *et al.* (1992) suggests that motion energy units in V1 extract the direction of motion of the component gratings (Fourier) in the plaid, while V2 neurons extract information about the motion of texture boundaries (non-Fourier).

The model of Wilson *et al.* (1992) does not address the issue of the spatial integration regions of the local motion components. Because of the periodic nature of sinusoidal gratings, stimuli composed of two sinusoidal gratings moving in different directions provide little opportunity to segregate different local motion information in order to examine possible spatial interactions.

3 GLOBAL COHERENT MOTION FOR STOCHASTIC LOCAL NETWORKS

To explore the issue of global percepts arising from disparate local information, moving random dot kinematograms are an ideal stimulus. Many different rules may be used to generate such stimuli, resulting in as many different kinematograms. For example, large subsets of the dots often move in one direction. These stimuli would not suit our purposes, however: multiple motion vectors in one direction are redundant with the local information and thus obscure the contribution of the motion of individual dots to the general percept. Therefore, Williams and Sekuler (1984) developed a kinematogram in which the direction of motion of each dot is individually defined. Since these stimuli are crucial to the discussion of global motion, more detail will be given than would ordinarily appear in a book chapter. In these kinematograms, dots are initially distributed randomly over the display area. Then each dot takes an independent two-dimensional random walk. All dots move the same distance from frame to frame, but each dot's direction of displacement from one frame to the next is independent of the directions in which the other dots move. Additionally, the direction in which each dot travels from one frame to the next is independent of the direction of its own previous displacements. Possible directions in which all dots move are chosen from the same uniform probability distribution.

Using these stimuli, Williams and Sekuler (1984) showed that if the range of the distribution of possible directions extends over the entire display area (360 degrees) only local, random movement of the dots is perceived. This effect resembles the pattern of a detuned television set. However, if the range of possible directions is less than 180 degrees, the pattern could appear to flow in one direction which represented the mean of the individual dots' directions. This was true even though the individual movement of the dots could still be perceived. This effect can be visualized by imagining snowflakes in a storm which, although perturbed individually by local wind currents, still appear to drift together downward in one direction.

Williams and Sekuler (1984) systematically explored the probability of seeing a global percept of motion in one direction from local motion vectors. They varied the range of the distribution of vectors, measuring the probability of seeing unidirectional flow in a direction representing the mean of local movements. Additionally, they examined how this perceived coherence of motion in one direction varied with such various parameters as spatial frequency, step size of the random walk, density of dots in each display, and duration of the movement.

These stimuli provide two important pieces of information regarding how the global coherent percept results from the local motion vectors. First, there is temporal summation. The strength of the global percept increases in a nonlinear manner with stimulus durations. The probability of seeing unidirectional flow increased with the number of frames, up to 11 frames, but perceptibility was not augmented by additional frames. This result then leads to the issue of whether the percept of coherent motion is dependent on the overall set of directions present from frame to frame, or whether it is dependent on the particular path each dot took over time. Williams and Sekuler (1984) answered this question by measuring coherent motion under two conditions. The first condition used stimulus patterns comprised of two sets of spatially interspersed random dots. For one set of dots (the 'noise' set), the distribution of directions was uniform over all 360 degrees of directions. In the other set (the 'signal' set), dots moved only upward. The set assignments of the dots remained the same over all frames presented; that is, some dots moved upward in every frame while other dots moved randomly in every frame. The second condition was identical to the first except that in each frame the dots constituting the signal set and those constituting the noise set were chosen independently from the assignments in the past frame. Although the proportion of dots constituting the signal remained constant in all frames, there were not two separate sets of dots ('signal' and 'noise') as there were in the first condition. Interestingly, there was no significant difference between the results from the two conditions.

Thus the global percept depends only on the set of directions from frame to frame, and not on the particular path taken by each dot over time. These results are consistent with the idea that the directions of the individual steps are independently detected, then pooled over time and space to generate the perception of coherent global flow.

The second interesting finding from these experiments is that they place constraints on the pooling process. In particular, although for the stimuli discussed to this point the direction of the global coherent motion is along the mean of the directions of the local random motion, a computational scheme of simple vector addition will not suffice to account for the pooling processes. The stimulus defined by a uniform distribution of directions with a range of 180 degrees generates a percept of unidirectional flow in the direction of the mean on nearly 100% of the trials. If directions within 20 degrees of the center (the mean) are removed from the distribution, then the frequency of seeing coherent flow along the mean is reduced to 50%. The mean of the distribution is unchanged by removing these directions. Furthermore, 98% of the dots will still have a vector component of motion in the direction of the mean. This suggests that more than simple vector addition is involved in the spatial pooling of the local responses. Spatial integrations must involve the pooling of responses of direction-selective mechanisms that are tuned to the mean direction of the distributions.

4   COOPERATIVITY FOR COHERENT GLOBAL MOTION

Within the framework of motion, correspondence is defined as the process by which we identify elements in different views as representing the same object at different times. In this way we can maintain a perceptual identity of objects in motion. Random dot kinematograms illustrate this correspondence problem. When a set of dots is displaced, there is no inherent information which determines which dot in the initial position matches which dot after the displacement. One might guess that a dot is matched with the next dot; that is, it is perceived to move to the nearest displaced dot. This is not always the case (Ullman, 1979).

There is evidence that the solution to this correspondence problem requires 'cooperative' interactions. If the local elements of a parallel network are extensively interconnected and are allowed to interact, then global behavior can be generated that would not occur if the mechanisms were isolated from each other. According to the Gestalt school of thought, the act of perception involves far more than a simple assimilation of individual sensations. In this sense, cooperativity complements well the Gestalt concept that the whole is more than the sum of its parts.

Cooperative parallel networks which are extensively interconnected are capable of exhibiting three properties: multi-stable states, order–disorder transitions, and 'hysteresis'. The solution to the motion correspondence problem requires parallel cooperative interactions, as demonstrated by a phenomenon called 'pulling'. If each element in a kinematogram is independent, then the action of a few elements could not influence the others. However, if interactions are occurring, the behavior of a few elements could change the overall state of the stimulus. This pulling has been demonstrated by Chang and Julesz (1984). In their stimulus, the motion for each dot in the kinematogram is ambiguous. Each dot had a potential matching dot after each displacement to the left and to the right. Dots should be perceived as moving to the left or the right with equal probability. Chang and Julesz (1984) demonstrated that with as few as 4% of the dots biased to move in one direction, all dots were perceived as moving in that direction. This cooperative property in motion would help to maintain a stable motion percept in the presence of noise.

Hysteresis is a form of memory in which a system, having reached a stable state, shows resistance to additional change. The consequence of this behavior is that the history of stimulation affects the system's response. To search for hysteresis in the generation of global coherent motion from stochastic local motion, one measures the transition points marking the change from global coherent motion to local random motion and vice versa. This is done by gradually changing the directional content of the stimulus between the two extremes of a uniform distribution with a range of 180 degrees or less and another with a range of 360 degrees. If the directional content of the stimulus for which these transitions occur is dependent upon whether the change is from global to local motion or vice versa, it is indicative of hysteresis.

Evidence has been found for the presence of hysteresis in the global coherent motion percept (Williams *et al.*, 1986; Williams & Phillips, 1987). The stimuli used for these experiments were also random dot kinematograms, in which each dot took an independent two-dimensional random walk of constant step size. The direction of motion for each dot was taken from either one of two uniform distributions, or from a mixture of the two distributions. The mean of the distribution of movement was always in an upward direction. When the dots drew their movements from a uniform distribution of 360 degrees (referred to as 'noise'), only the local random movement of the dots was perceived. When all the dots drew their position from a distribution of 180 degrees or less (referred to as 'signal'), however, the dots appeared to move together in the direction of the mean of the distribution. This was true despite the fact that individual movements of the dots could still be perceived. When dots drew their position from a combination of the signal and noise distributions, each dot's displacement came randomly from either distribution. What was perceived in this case depended upon the relative proportion of signal versus noise.

Two different types of trials were run in random order. In the first case, the direction of motion for each dot was chosen initially from the signal distribution, producing the perception of upward flow. After a random interval of up to 12 seconds, the proportion of dots drawing directions from the signal distribution was then progressively decreased while the proportion of dots choosing from the noise distribution was increased. This process continued until the observer responded that the kinematogram had changed its movement from upward flow to local random motion. After this response, the proportion continued to decrease for another random interval of up to

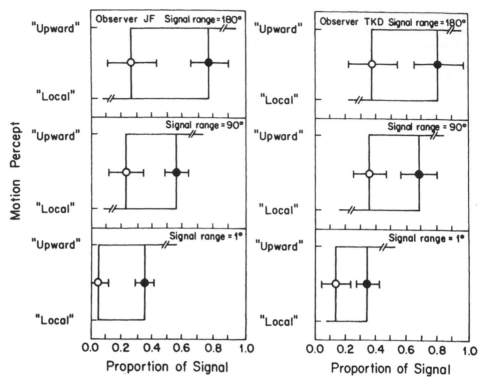

Figure 1 Perceptual transitions measured under two different histories of stimulus exposure for three signal distributions. The results for both observers JF and TKD were obtained for signal distributions whose ranges were 180 degrees, top; 90 degrees, middle; and 1 degree, bottom. Measurements were collected using a step size of 0.9 degree. Data points show the proportion of 'signal' dots required for a perceptual transition from local random motion to global upward flow (closed circles) and for perceptual transition from global upwards flow to random local motion (open circles). Error bars represent one standard deviation (100 measurements). In each panel the separation between transition points measured with the different exposure histories is an index of hysteresis. Note the narrowing and the leftward shift of the profiles with decreasing and signal range. Reproduced with permission from Williams and Phillips (1987).

six seconds. At this point, the process was reversed, and the proportion of signal was increased until the observer responded that the upward flow had reappeared. After this response, the trial ended. In the second type of trial, the sequence of stimuli was reversed: initially all dots chose direction from the noise distribution (resulting in the perception of local random motion) and the point of transition to a perception of upward flow and back to local random motion was measured. These two types of trials produced different histories of directional exposure, revealing any perceptual biases dependent upon the history of stimulation.

The experiment was repeated using two narrower ranges of signal distribution: 90 degrees and one degree. Each of the three distributions generated a different history of exposure. One would expect that as the distribution narrows, the signal would become more effective in the stimulation of directionally selective visual elements that are tuned to upward motion. As a result, a transition between perceived states might require fewer dots.

The results from two observers are shown in Figure 1. Each panel in the figure represents data from a different range of signal. Note that the transitions differ for the signal ranges of 180 degrees, 90 degrees, and one degree. Interestingly, the transition from local random movement to upward flow requires a larger number of dots sampled from the signal distribution than does the transition from upward flow to local random motion. These two types of transition occur at significantly different ratios of signal to noise for all three signal distributions (p < 0.005).

A number of alternate explanations must be considered and rejected before the results can be solely attributed to hysteresis. For example, the use of a fixation point did not affect the results, suggesting that eye movements played little if any role in the effect. Also, if the motion aftereffect (or waterfall illusion) was responsible, it would have facilitated rather than retarded the transition from upward motion to local random noise. Finally, given the slow time course for changing the signal proportion, reaction time could not provide an explanation.

To explore how spatial parameters might affect hysteresis, a 180 degree distribution was used under four additional conditions: (1) with a four-fold decrease in the spatial density of the kinematogram's dots; (2) with a four-fold decrease in the area of the display; (3) with a nine-fold decrease in step size; and (4) with the display shifted horizontally into the periphery so that the nearest dot was four degrees from fixation. The results from all four conditions did not differ significantly from the original findings (Williams & Phillips, 1987), although it is probable that extreme changes in the variables would affect hysteresis. However, due to these data and to other evidence that spatial variables have little effect on the dot interactions responsible for motion in kinematograms (Baker & Braddick, 1982; Williams & Sekuler, 1984), the development of a model network to account for these results need not treat space explicitly. This hysteresis could be accounted for by cooperative, nonlinear excitatory and inhibitory interactions among direction-selective motion mechanisms.

The cooperative model proposed to account for coherent global motion (Williams & Phillips, 1987) comprises a set of direction-selective mechanisms which cover all 360 degrees of motion direction, with each mechanism having a Gaussian profile for directional selectivity. Based on previous results (Williams et al., 1984), the half-amplitude half-bandwidth of each mechanism's Gaussian sensitivity profile was set to 30 degrees. The model's mathematical formulation is a modification of a cooperative neural network model previously proposed by Wilson and Cowan (1973). The model assumes nonlinear excitatory interactions among mechanisms which are sensitive to similar directions of motion, and nonlinear inhibition among mechanisms sensitive to different directions. This cooperative system's dynamic response can be represented by a pair of coupled differential equations. For the excitatory activity, E_i, in direction channel i:

$$dE_i/dt = - E_i + (1 - E_i) S (P_i + \Sigma a_{ij}E_j - \Sigma b_{ij}I_j) \qquad (1)$$

where S is a nonlinear function of sigmoidal shape, P_i is the external input to channel i, and a_{ij} and b_{ij} are the excitatory and inhibitory weights, respectively, of channel j with respect to channel i. A similar equation gives the inhibitory activity I_i in channel i:

$$dI_i/dt = - I_i + (1 - I_i) \, S \, (\Sigma c_{ij} E_j - \Sigma d_{ij} I_j) \qquad (2)$$

where S is again the nonlinear function of sigmoidal shape, and c_{ij} and d_{ij} are the excitatory and inhibitory weights, respectively, of channel j with respect to channel i. The functional form of the nonlinear sigmoidal function S is given by:

$$S(M) = 1/\{1 + \exp(- \nu \, (M - \theta))\} - 1/\{1 + \exp(\nu\theta)\} \qquad (3)$$

where θ and ν determine the slope and intercept of the sigmoidal function. In general, interactions like those in the first two equations promote the formulation of stable coalitions between similarly tuned elements within the network (Feldman & Ballard, 1982). These neural coalitions can then produce various cooperative properties such as hysteresis.

The model's parameters were constrained so that the model behaved in what is defined by Wilson and Cowan (1973) as the active transient mode. The system shows hysteresis in this mode, switching back and forth between different states of activity. The perception of local random motion in the model is represented by a steady state of uniform activity across all mechanisms. Global upward flow is represented by a steady state in which all activity is localized about the mechanism selective for upward movement. A transition point is defined by the proportion of signal at which the network switches between these two states of activity. Figure 2 shows the results from this model. The dashed lines represent the transition points from the model calculated using a single parameter set. The model captures both the leftward shift and the narrowing of the hysteresis profile with decreasing signal range. The behavior of the model is straightforward. As signal range decreases, more activity is concentrated in fewer motion selective elements arrayed in the upward direction. As a result, a smaller proportion of signal dots is needed to indicate upward motion. In addition, fewer active elements reduce the opportunity for cooperative interactions in the network, narrowing the hysteresis profile.

The finding that the perception of motion direction is dependent on a cooperative network is useful in many circumstances. For example, observers must extract a mean direction vector from a scene containing a large number of different local vectors in many naturally occurring situations. We are able to determine the average direction in which the ocean's surf moves, despite the fact that individual waves travel among somewhat different paths. We can judge the average direction from which the wind is blowing the leaves on a tree, despite the many random variations in that movement from one leaf to the next, or in any one leaf from one instant to the next. Faced with many multivectoral stimuli in the real world, a cooperative network such as the one described in this model would enhance an observer's signal-to-noise ratio, improving the perception of the mean direction of motion.

Although there is evidence of cooperativity in the generation of coherent motion, these results provide no information regarding the spatial extent of the cooperative interactions. This is not unexpected since in all local regions of the display, the directions of motion of the dots are defined by the same distribution. Nawrot and Sekuler (1990) investigated the issue of spatial interactions by exploring how motions within one region of the visual field influence motion perceived elsewhere. Their stimuli were random dot kinematograms comprised of alternating strips. In one set of interdigitated strips, called biasing bands, half of the dots moved rightward and the

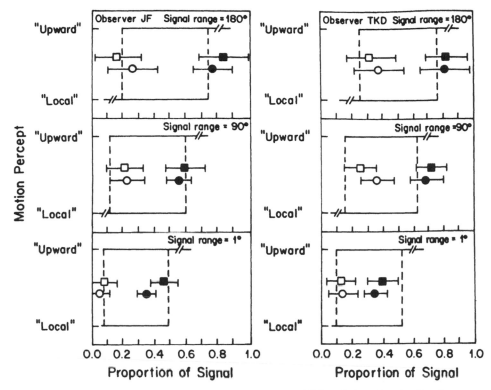

Figure 2 Hysteresis profiles as in Figure 1 but with additional data for the 0.1 degree step size. For a step size of 0.9 degree, data points denoted by closed circles show the proportion of signal dots required for the perceptual transition from local random motion to global upward flow; and data points represented by an open circle represent perceptual transitions from global upward flow to local random motion. For a step size of 0.1 degree, perceptual transitions from local random motion to global upward flow are represented by a closed square, while transitions from global upward flow to local random motion are represented by an open square. The dashed lines mark the transition points calculated from a cooperative model incorporating cooperative interactions among direction-selective motion elements (see text for details). The same parameters set for the model were used to fit the data in all panels. Note that for both step sizes, the model captures both the leftward shift and the narrowing of the hysteresis profile with decreasing signal range. Reproduced with permission from Williams and Phillips (1987).

remaining dots moved at random, choosing directions of motion from a uniform distribution of directions with a range of 360 degrees. In the intervening strips, called probe bands, dots moved at random. When the alternating strips were narrow, motion in one direction in the biasing bands induced a similar perception of movement in the probe strips which contain only random noise; i.e. the entire display appeared to move in the same direction. Nawrot and Sekuler termed this phenomenon 'assimilation'. When the alternating strips were wide, motion in one direction in the biasing bands induced the perception of movement in an opposite direction in the probe strips which contain only random noise (termed 'contrast').

Using a technique similar to that developed by Williams *et al.* (1986), Nawrot and Sekuler demonstrated the presence of hysteresis in the percept of coherent motion in the probe bands. The authors suggested that the hysteresis exhibited by this illusory motion resulted from spatially distributed, cooperative processes. At narrow bandwidths, the effect was facilitatory; at wider bandwidths, inhibitory. The shift from assimilation to contrast with increasing strip size indicates that these facilitatory and inhibitory effects of the cooperative network extend over different distances. Nawrot and Sekuler extended the cooperative model of Williams *et al.* (1986) by incorporating a center–surround configuration for spatial interactions between direction-selective mechanisms. A direction-selective mechanism excited similarly tuned direction-selective mechanisms which are nearby but inhibited similarly tuned mechanisms which are located further away.

5 INDUCED MOTION

> If two or more Objects move with the same velocity, and a third remain at rest, the Moveables will appear fixed, and the Quiescent in Motion the contrary Way. Thus, Clouds moving very swiftly, their Parts seem to preserve their Situation, and the Moon to remove the contrary Way. (Porterfield, 1759, page 424.)

The preceding statement is one of the earliest references to the phenomenon of induced motion. Without the presence of the moving clouds, the moon would appear to be stationary. The issue of how the moon is seemingly set in motion by its surrounding clouds has been the subject of numerous investigations.

Many explanations of induced motion start with the basic concept that induced motion results from image displacements over the retina which are not fully compensated for by ocular motor responses (Kaufman, 1974). For example, if pursuit eye movements follow the inducing stimulus motion (e.g. clouds), then there will be retinal displacements of the stationary object (e.g. the moon) in the opposite direction. This retinal motion would be equivalent to the motion of the stationary object in the opposite direction (Mack, 1986). This retinocentric theory of induced motion has been elaborated to take into account eye position, head position, and the object's position in space (Gogel, 1982).

There are, however, several experimental results which suggest that pursuit eye movements or retinal displacement are not sufficient to account fully for the phenomenon of induced motion. It has been found that induced motion depends on the proximity of the inducing field (Gogel, 1984; Gogel & Koslow, 1971). Also, stimuli have been constructed in which induced motion can occur in opposite directions for two simultaneously presented targets. For example, Gogel (1977) showed that two dots, one on each side of a fixation point, both appear to move toward the fixation point if rectangles surrounding the dots move outwards. The reverse was also true; the dots appear to move away from fixation when the rectangles move inward. A stimulus devised by Nakayama and Tyler (1978) consisted of two moving horizontal lines, one positioned above two stationary horizontal lines and the other positioned below. They found that when the outer lines moved apart, the inner ones appeared to move closer together, and vice versa. A single mechanism which would influence all parts of the visual field would have difficulty accounting for these results.

Recent findings support the idea of an interaction between the region which induces motion and the stationary region. For example, in a stimulus for which a single stationary dot is induced to move by another dot which is moving in a straight line, the threshold for detecting motion is much lower than if only the single moving dot is present (Rock, 1984). For a stimulus pattern in which the stationary center of the pattern appears to move in the opposite direction in a moving surround, researchers have found that the perceived velocity of a moving center is influenced by the velocity and direction of the surrounding motion (Tynan & Sekuler, 1975). Without stationary contours surrounding both the moving adapting pattern and stationary test pattern, the motion aftereffect diminishes or disappears (Day & Strelow, 1971). This strongly suggests the need for relational signals.

The relational motion process can account for induced motion under conditions which have been difficult to explain in terms of single-process models; for example, the simultaneously induced motions in different directions which were discussed above. Although there is not at present a full understanding of the spatial interactions which underlie this relational motion process, the technique used by Nawrot and Sekuler (1990) would seem to be a promising approach. The spatially long-range inhibitory interactions detected by Nawrot and Sekuler (1990) are suitable candidates to account for the relational motion process required for induced motion.

6 CONCLUSIONS

In this chapter we have discussed the spatial integration of local motion signals to generate a global coherent motion percept. Experimental evidence of hysteresis supports the notion that cooperativity underlies the integration process. The results from Williams and Phillips (1987) suggest that cooperative interactions among direction-selective motion mechanisms are required for the global coherent motion percept. These interactions consist of excitatory interactions between mechanisms sensitive to similar directions of motion and inhibitory interactions between mechanisms tuned to different directions of motion. The work of Nawrot and Sekuler (1990) extends these results by demonstrating that cooperative interactions project across space. There appear to be excitatory interactions between similarly tuned direction-selective mechanisms which are proximal in space, and inhibitory interactions between mechanisms tuned to similar directions of motion which are further apart in space.

As evidenced by numerous studies, cooperativity appears to underlie much of motion processing. These include the solution of the correspondence problem as well as spatial integration of local motion signals. Although we have not discussed it in this chapter, there exists a phenomenon called 'motion capture' in which randomly moving dots are captured and move coherently with a superimposed grating or a surrounding contour. Since Chang and Julesz (1984) found that as few as 4% of dots can pull the remaining random dots in a single direction, it is possible that in motion capture it is the peripheral unambiguous motion that is pulling the central ambiguous motion. Therefore, cooperativity could account for this phenomenon as well. As further research is carried out, it may become even more clear that cooperative processes are fundamental to human motion perception.

REFERENCES

Adelson, E. H. & Movshon, J. A. (1982). Phenomenal coherence of moving gratings. *Nature*, **300**, 523–525.

Anderson, S. J. & Burr, D. C. (1987). Receptive field size of human motion detection units. *Vision Res.*, **27**, 621–635.

Baker, C. & Braddick, O. (1982). Does segregation of differently moving areas depend on relative or absolute displacement? *Vision Res.*, **22**, 851–856.

Chang, J. J. & Julesz, B. (1984). Cooperative phenomena in apparent movement perception. *Vision Res.*, **24**, 1781–1788.

Chubb, C. & Sperling, G. (1988). Drift-balanced random stimuli: A general basis for studying non-Fourier motion perception. *J. Opt. Soc. Am.*, **A, 5**, 1986–2007.

Day, R. H. & Strelow, E. R. (1971). Reduction or disappearance of visual after effect of movement in the absence of a patterned surround. *Nature*, **230**, 55–56.

Derrington, A. M., Badcock, D. R. & Holroyd, S. A. (1992). Analysis of the motion of 2-dimensional patterns: Evidence for a second-order process. *Vision Res.*, **32**, 699–707.

Gogel, W. C. (1977). Independent motion induction in separate portions of the visual field. *Bull. Psychonom. Soc.*, **10**, 408–10.

Gogel, W. C. (1982). Analysis of the perception of motion concomitant with a lateral motion of the head. *Percept. Psychophys.*, **32**, 241–250.

Gogel, W. C. (1984). The role of perceptual interrelations in figural synthesis. In P. Dodwell & T. Caelli (Eds.) *Figural Synthesis*, pp. 31–82. Lawrence Erlbaum Associates, Hillsdale, NJ.

Gogel, W. C. & Koslow, M. (1971). The adjacency principle and induced movement. *Percept. Psychophys.*, **11**, 309–314.

Feldman, J. A. & Ballard, D. H. (1982). Connectionist models and their properties. *Cog. Sci.*, **6**, 205–254.

Ferrera, V. C. & Wilson, H. C. (1987). Direction specific masking and the analysis of motion in two dimensions. *Vision Res.*, **27**, 1783–1796.

Kaufman, L. (1974). *Sight and Mind*. Oxford University Press, New York.

Koffka, K. (1935). *Principles of Gestalt Psychology*. Harcourt, Brace, & World, New York.

Levinson, E., Coyne, A. & Gross, J. (1982). Synthesis of visually perceived movement. *Invest. Ophthalmol. Vis. Sci.* (Suppl.), 105.

Mack, A. (1986). Perceptual aspects of motion in the frontal plane. In K. R. Boff, L. Kaufman & J. P. Thomas (Eds.) *Handbook of Perception and Human Performance*, pp. 1–38. John Wiley, New York.

Movshon, J. A., Adelson, E. H., Gizzi, M. S. & Newsome, W. T. (1986). The analysis of moving visual patterns. In C. Chagas, R. Gattas & C. Gross (Eds.) *Pattern Recognition Mechanisms*, pp. 117–180. Springer-Verlag, New York.

Nakayama, K. & Tyler, C. (1978). Relative motion induced between stationary lines. *Vision Res.*, **18**, 1663–1668.

Nawrot, M. & Sekuler, R. (1990). Assimilation and contrast in motion perception: explorations in cooperativity. *Vision Res.*, **30**, 1439–1451.

Porterfield, W. (1759). *A Treatise on the Eye, the Manner, and Phaenomena of Vision.* Hamilton & Balfour, Edinburgh.

Rock, I. (1984). *Perception.* Scientific American Library, New York.

Tynan, P. & Sekuler, R. (1975). Simultaneous motion contrast: velocity sensitivity and depth response. *Vision Res.*, **15**, 1231–1238.

Ullman, S. (1979). *The Interpretation of Visual Motion.* MIT Press, Cambridge.

Van Essen, D. C. (1985). Functional organization of primate visual cortex. In A. Peters, & E. G. Jones (Eds.) *Cerebral Cortex*, pp. 259–329. Plenum Press, New York.

Williams, D. W. & Phillips, G. C. (1987). Cooperative phenomena in the perception of motion direction. *J. Opt. Soc. Am.*, **4**, 878–885.

Williams, D. W. & Sekuler, R. (1984). Coherent global motion percepts from stochastic local motions. *Vision Res.*, **24**, 55–62.

Williams, D. W., Phillips, G. C. & Sekuler, R. (1986). Hysteresis in the perception of motion direction: evidence for neural cooperativity. *Nature*, **324**, 253–255.

Wilson, H. R. & Cowan, J. D. (1973). A mathematical theory of the functional dynamics of cortical and thalamic nervous tissue. *Kybernetik*, **13**, 55–80.

Wilson, H. R., Ferrera, V. P. & Yo, C. (1992). A psychophysically motivated model for two-dimensional motion perception. *Vis. Neurosci.*, **9**, 79–97.

Yo, C. & Wilson, H. R. (1992). Perceived direction of moving two-dimensional patterns depends on duration, contrast, and eccentricity. *Vision Res.*, **32**, 135–147.

Part 5

Higher-order Interpretation of Motion

11

Optic and Retinal Flow

M. G. Harris
The University of Birmingham, UK

This chapter begins with a summary of the main theoretical approaches to optic flow from geometrical and mathematical viewpoints, emphasizing the types of information that are available to a moving observer. It then briefly reviews the neurophysiological and psychophysical evidence for flow-specific mechanisms within the visual system and for the importance of flow in such tasks as locomotory guidance and the recovery of three-dimensional (3D) surface layout.

1 INTRODUCTION

1.1 Optic Flow

As we move about and scan the visual world, the images of things move about, changing their relationships and shapes in a complex dance. J. J. Gibson (e.g. 1950, 1979) was the first to understand this dance, and the first to explain why it was worth learning, by pointing out how it could provide information about our own movements and about the three-dimensional structure of the world. Probably Gibson's greatest contribution was to redefine the dance-floor, emphasizing the amount of information potentially available to an observer in the transforming optic array rather than the instantaneous fragments provided by a pair of retinal images.

The optic array is the three-dimensional bundle of light rays that impinges from all directions upon each point in an illuminated world. Objects in the world can be thought of as labelling specific rays, so producing a global pattern of light intensities. A retinal image provides access to only part of the optic array at any one time, but a stationary observer can sample different parts by eye movements and head rotations. By changing position, the observer can sample the different optic arrays impinging upon neighbouring points in space. However, sampling in this case should not be thought of as a discrete process. Rather, as the observer gradually moves, so each ray gradually moves, thus producing the smooth transformation in the optic array that Gibson called optic flow.

Optic flow is best envisaged as a unit sphere centred upon the viewpoint (i.e. the origin of the optic array). Each ray cuts the sphere at a point and the set of light intensities at these points makes up the optic array. As the observer changes position, each ray traces out a path on the unit sphere and the set of paths make up the pattern of optic flow. More formally, optic flow is a vector field conveniently represented in a

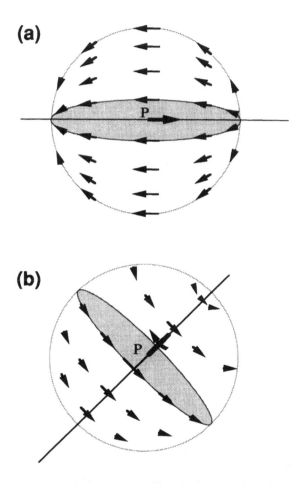

Figure 1   The two global directional components of optic flow. The viewpoint (P) is placed at the centre of a sphere. Optic flow consists of the set of paths traced out on the surface of this sphere as the viewpoint changes. (a) Linear translation of the viewpoint (in this case to the right) produces a characteristic pattern of flow. In the forward half of the array, all motion expands from a single point (the focus of expansion) coincident with the direction of transla-tion. In the backward half of the array all motion contracts to a single point (the focus of contraction). (b) Rotation about the viewpoint produces a different characteristic pattern of flow.

spherical coordinate system. To illustrate this, the optic flow produced by movement of the observer in a straight line is shown in Figure 1a.

The information provided by optic flow is made clearer in Figure 2, which shows a fragment of the pattern produced by movement in a straight line towards a flat surface. In Figure 2a, the surface is at right angles to the direction of approach, i.e. in the frontoparallel plane. In Figure 2b, it is slanted about the vertical axis so that the left side is closer than the right. Comparison of these two figures demonstrates that the *direction* of the flow is unaffected by the 3D layout of the world: all rays move away

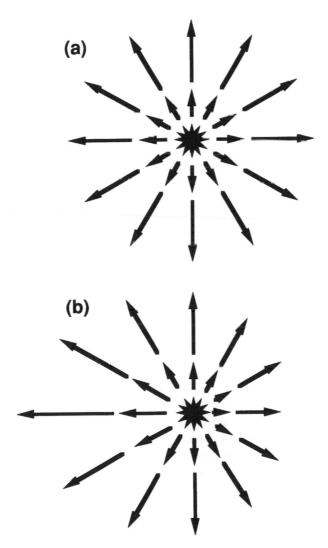

Figure 2 Fragments of the flow pattern in the direction of locomotory heading. (a) Linear translation towards a surface in the frontoparallel plane. (b) Linear translation towards a surface slanted about a vertical axis in the frontoparallel plane so that the left side is closer than the right.

from a common 'focus of expansion' (FOE), or 'focus of radial outflow', the position of which in the optic array corresponds to the direction of observer movement. The *speed* of flow will obviously depend upon the speed of observer movement and, from Figure 1a, upon position in the optic array relative to the direction of movement: for example, in the forward half of the array, speed increases with distance from the FOE. Comparison of Figures 2a and 2b demonstrates that speed of flow also depends upon the 3D layout of the world. In Figure 2b, the surface is closer on the left hand side of the figure than the right and closer things produce faster flow.

The pattern produced by linear translation contains almost all the useful information available for optic flow. In a structured 3D world, the flow pattern is like a dynamic, spherical mosaic in which the different tiles correspond to different surfaces. The overall directions in which the tiles move provide information about the direction of observer movement, the edges of the tiles are delineated by discontinuities in speed, but not direction, that correspond to discontinuities in depth, and, within each tile, smooth spatial speed gradients provide information about the 3D layout of the corresponding surface.

When the observer moves in a curved path, the flow pattern becomes a little more complicated. Curvilinear movement can be broken down into a translation and a rotation (Whittaker, 1944) and the resulting flow pattern can, in turn, be broken down into a translational and a rotational component (Longuet-Higgins & Prazdny, 1980). The additional rotational component is shown in Figure 1b. Like the translational component (Figure 1a) it consists of a global directional pattern, with a single axis of rotation and with speed of flow determined partly by the rate of observer movement and partly by the position in the array relative to this axis. However, unlike translation, the rotational component contains no information about the 3D structure of the world because the speed of flow is not affected by the distance of objects. Two important effects of the addition of the two components, however, are, firstly, that depth discontinuities in the world may now produce discontinuities in the direction of the flow as well as the speed and, secondly, that the FOE will no longer coincide with the direction of observer movement (e.g. Georgeson & Harris, 1978; Farber & McConkie, 1979; Prazdny, 1981).

During general curvilinear movement, then, the observer is faced by a complex flow pattern that must be decomposed into its translational and rotational components. The overall form of each of these can be expressed in terms of the direction of a single axis and an overall speed, and these four terms are potentially very useful for the guidance of locomotion. An analysis of speed and directional discontinuities in the overall flow, or of speed discontinuities in the translational component alone, may be useful in segmenting the pattern into parts corresponding to distinct surfaces. An analysis of local speed gradients, based only upon the translational component, may be useful in recovering the 3D layout of the world.

1.2   Retinal Flow

For an observer who moves about without making eye movements or head rotations, the retinal flow pattern is, essentially, just the visible portion of the optic flow pattern described above. However, when confronted with a moving image, an observer will generally make tracking eye movements, so that retinal flow consists of optic flow plus an additional rotational component. Like the rotational component of optic flow, rotations due to eye movements carry no information about the structure of the world and the whole pattern can be specified by an axis and an overall speed of rotation.[1] In general, then, retinal flow consists of a translational component plus a rotational component, the rotation being the sum of a component due to a curved trajectory and a component due to eye or head movements. These two rotational components will sum and cannot be decomposed instantaneously but, in general, they will follow independent time courses and are, in principle, separable over time.

1.3 Summary

All of the information about surface layout and some information about locomotory heading is provided by the translation component of optic flow (Figure 1a). Especially when dealing with *retinal* flow, this translational component must be disentangled from a rotational component (Figure 1b) that contains no information about surface layout, some information about locomotory heading, and all of the information about eye and head rotations.

In general then, any model of the analysis of retinal flow must address at least the following issues (although the analysis need not necessarily proceed in any particular sequence):

a   The decomposition of flow into its translational and rotational components.
b   The segmentation of the flow into regions corresponding to individual surfaces.
c   The recognition of those regions corresponding to objects moving relative to the rest of the world.
d   The recovery of information about the observer's movements and eye rotations. The important directional aspects of this information, such as locomotory heading, can in principle be recovered from the directional structure of the flow.
e   The recovery of information about surface layout. This requires an analysis of the speed gradients in the flow.

Computational accounts, dealing with some or all of these topics, can be found in, for example, Nakayama and Loomis (1973), Clocksin (1980), Prazdny (1983), Rieger and Lawton (1985), and Waxman and Wohn (1988). In this chapter I shall deal only with those topics that have attracted the most empirical resarch. However, before reviewing the relevant work, it is important to introduce an alternative way to describe optic and retinal flow.

2   DIFFERENTIAL INVARIANTS AND AFFINE TRANSFORMS

Some accounts of the analysis of retinal flow deal directly with the motions of a few individual points and can be understood in terms of the basic geometry described

---

[1] This description assumes that the eyes rotate about the origin of the optic array, which is not strictly true (see, for example, Bingham, 1993).

above. Other accounts, however, use an alternative description that treats the flow as a continuous vector field and computes its first (and potentially higher) order spatial derivatives. This treatment is based on the demonstration (Koenderink & van Doorn, 1975, 1976) that the flow in any small region of the field can be completely described as the sum of a translation (*trans*) and the three differential invariants *div*, *curl* and *def*. These differential invariants are illustrated in Figure 3.

Div is an isotropic expansion, *curl* a rigid rotation, and *def* (deformation) an expansion along one axis accompanied by a complementary compression along the orthogonal axis resulting in a change in shape without a change in area. These operations are termed 'invariant' because, like luminance or wavelength for example, their values do not depend upon the coordinate system in which they are measured. They are 'differential' because they compute the spatial derivative of

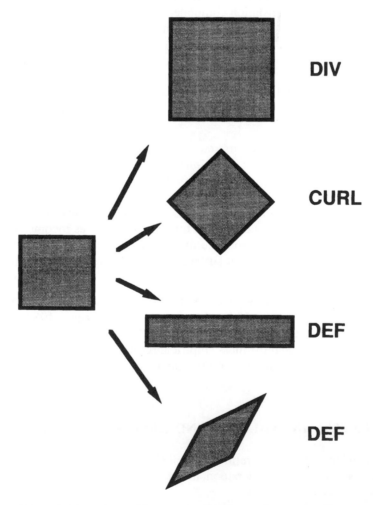

Figure 3 Differential invariants: Changes in shape may be expressed as a combination of expansion (*div*), rotation (*curl*) and deformation about some axis of symmetry (*def*).

velocity in different directions (in fact they are frequently expressed mathematically in terms of partial derivatives). This property is illustrated in Figure 6, which shows possible mechanisms for measuring each component by combining appropriate local velocity estimates. Each diametrically-arranged pair of local elements simply computes the change in speed at a given orientation. Sometimes the change is computed in the direction of the comparison (as in the case of *div*) and sometimes at right angles to the comparison (as in the case of *curl*), reflecting the fact that the speed of flow varies both along and at right angles to the direction of flow (Figure 1).

The basic idea of the differential invariant analysis is that each point in the flow field can be described by just four terms. Two of these (*trans* and *def*) have direction as well as magnitude and are thus vectors. The other two (*div* and *curl*) have only (signed) magnitude and are thus scalars. Each of these terms can be computed throughout the flow to produce four separate two-dimensional fields. One of the most obvious attractions of this description is that neither *div* nor *def* is affected by rotation, so that these terms automatically isolate the component of the flow caused by linear translation of the observer and thus conveniently capture the three-dimensional layout of the world. In fact, mathematical analyses (e.g. Koenderink & van Doorn, 1975, 1976; Longuet Higgins & Prazdny, 1980; Koenderink, 1985, 1986) show that useful information is conveyed both by the set of local terms at a single point in the field and by the global variation of each of the terms across the field.

2.1 Local Analysis

Following a simplified version of Koenderink (1986) the values of the invariants in each direction, **d**, of the field are best understood by expressing the movement of the observer relative to this direction, as shown in Figure 4.

As already stated, observer movement can be decomposed into a translation, **t**, and a rotation, **r**. Each of these terms can in turn be decomposed into a radial component in the direction **d** and a transverse component at right angles to **d**. Thus, translation can be thought of as a combination of a direct approach, t_a (along **d**), and a transverse shift, t_s (at right angles to **d**). Similarly, rotation can be thought of as a rotation, r_a, about **d** (equivalent to a rotation of the head about the line of sight), and a rotation, r_s, about an axis at right angles to **d** (equivalent to an eye rotation: for simplicity in Figure 4 the directions of t_s and r_s are shown to be the same although, of course, they can in general be different). Since, by definition, the radial components are in the direction **d**, only their amplitudes t_a and r_a are important.

When a surface patch is introduced, it is convenient to scale the amplitude of the translation, **t**, by the distance to the surface, *d*, thus eliminating the mutual dependence between speed and distance by expressing everything in units of time. The layout of this surface can also be expressed relative to the direction, **d**, by the tangent vector **n** (at right angles to **d**), with direction giving the surface tilt (direction of maximum increasing nearness) and amplitude the slant (degree of rotation about the axis of tilt). Figure 4b shows a simplified representation of the important terms in which the direction **d** is into the page so that t_a disappears and r_a is a rotation in the image plane. Using this representation, the terms in the direction **d** are given by:

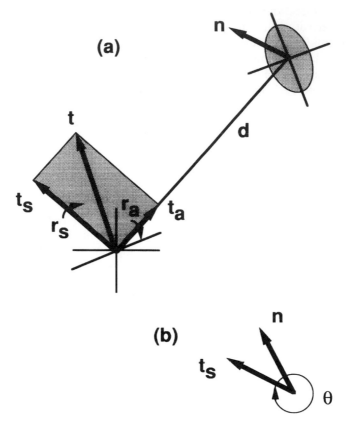

Figure 4 Vector decomposition of movement relative to a surface. See text for details and relevant equations.

$$trans = -\,(\mathbf{t_s} + \mathbf{r_s}) \tag{1}$$

$$def = nt_s \tag{2}$$

(axis of contraction bisects θ)

$$div = -\,\mathbf{n} \cdot \mathbf{t_s} + 2t_a \tag{3}$$

$$(\mathbf{n} \cdot \mathbf{t_s} = nt_s \cos\,\theta)$$

$$curl = -\,\mathbf{n} \times \mathbf{t_s} - 2r_a \tag{4}$$

$$(\mathbf{n} \times \mathbf{t_s} = nt_s \sin\,\theta)$$

It is immediately obvious from equation 1 that a pursuit eye movement can be used to cancel the local translation in any given direction in the flow field and that a knowledge of the direction and amplitude of the eye movement needed to do this gives the transverse component of the observer's movement (Note that this only works locally: an eye movement cannot cancel the translation everywhere.) Moreover, such an eye movement has no effect upon any of the other terms. If the transverse component of movement, $\mathbf{t_s}$, is known, *def* gives immediate access to surface layout. Both *div* and *curl* also provide information about surface layout and about the speed of approach and rate of rotation about the line of sight, respectively.

One interesting implication of the fact that surface layout can be recovered on a local basis is that this stage could occur *before* segmentation of the flow into regions corresponding to surfaces. For example, local estimates of surface layout could form the basis of a three-dimensional segmentation that could deal with curved surfaces more easily than processes based on, for example, discontinuities in the two-dimensional flow (Harris *et al.*, 1992b).

2.2 Global Analysis

The pattern of variation in the differential invariants across the flow field also provides useful information, as exemplified by the simple situation depicted in Figure 5 (Koenderink & van Doorn, 1976).

Figure 5 shows linear movement of the observer in a horizontal plane relative to a flat surface and defines both observer movement, **v**, and direction in the field, **d**, relative to the surface normal, **z**. The two-dimensional functions for the differential invariants are then as follows:

$$div(\mathbf{d}) = (k/2)[3\sin\alpha\sin2\theta\,\cos\phi + \cos\alpha(3\cos2\theta + 1)] \tag{5}$$

$$curl(\mathbf{d}) = -\,k\sin\alpha\sin\theta\sin\phi \tag{6}$$

$$def(\mathbf{d}) = k\sin\theta[(\sin\alpha\cos\theta\cos\phi - \cos\alpha\sin\theta)^2 + \sin^2\alpha\sin^2\phi]^{1/2} \tag{7}$$

where $k = v/z$

Although the *curl* function will also be affected by rotations of the observer, it is particularly attractive because of its simplicity. Equation 6 defines a two-dimensional sinusoid with amplitude determined by the direction of observer movement, and origin

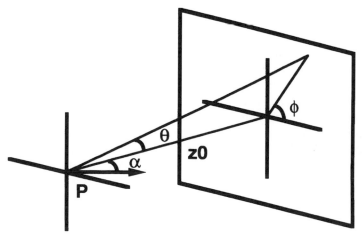

Figure 5 Horizontal movement relative to a flat surface: the optic array is represented in polar coordinates. See text for details and relevant equations.

($\theta = \phi = 0$) in the direction of the surface normal. When moving through a world consisting of several surfaces, the overall *curl* function is like a dynamic mosaic, each tile corresponding to a surface, and each providing a moving window onto a two-dimensional sinusoid. By finding the origin of each of these sinusoids, and thus the directions of the surface normals, surface layout can be recovered. Moreover, because the function is so simple, the origin need not be present in a tile because its location can easily be estimated by extrapolation (Harris *et al.*, 1992b).

The other expressions are rather more complex but, as a further illustration, *div* is potentially useful in the recovery of information about locomotory heading (e.g. Priest & Cutting, 1985; Regan, 1985; Torrey, 1985). The maximum of *div* does not generally lie in the direction of locomotion and, consequently, does not generally coincide with the focus of expansion described in section 1. Instead it occurs halfway between the direction of locomotion and the direction of the surface normal ($\alpha/2$ in Figure 5). For the important case of movement over a horizontal surface, $\alpha = 90°$ and the maximum of *div* occurs in the direction of locomotion at a distance equal to the height of observer's eye. In this case, the ability to locate this maximum is potentially as useful for locomotory guidance as the ability to locate the focus of expansion. More generally, in a world consisting of several surfaces and a mosaic-like *div* function, the location of the maximum and minimum in any tile together define the direction of locomotion relative to the corresponding surface. If the extreme values are not actually present in the function, however, their positions will be difficult to estimate by extrapolation.

2.3 Encoding Differential Invariants

Recovery of the retinal flow field from the time-varying pattern of light intensities on the retina is difficult because local motion sensors can only signal the component of motion at right angles to extended edges (the 'aperture problem'), because the field may be sparse where the world is untextured (e.g. the sky), and because the data will be noisy. These issues are dealt with elsewhere in this volume but, in general, most approaches attempt to solve these problems by integrating the outputs of local motion sensors using a 'smoothness' constraint, attempting to find the best compromise between the local motion data and the requirement that discontinuities in the flow are relatively rare because depth generally varies smoothly in the world (e.g. Horn & Schunk, 1981). Once this difficult step has been achieved, the required operators for *div*, *curl* and *def* could readily be approximated by mechanisms that pool local velocity information in appropriate ways, as illustrated in Figure 6.

Although this two-stage process is intuitively appealing and is no doubt an important reason for the popularity of differential invariants, it is by no means the only approach. Werkhoven and Koenderink (1990a, 1990b) have shown that differential invariants may alternatively be extractable directly from time-varying retinal light intensities, neatly sidestepping the aperture problem and the smoothness constraint, by combining the outputs of conventional luminance filters that take smoothed directional spatial derivatives and the temporal derivatives of their outputs. To emphasize the variety of possible approaches, Koenderink (1986) has also pointed out that the encoding of differential invariants need not depend upon conventional velocity information at all.

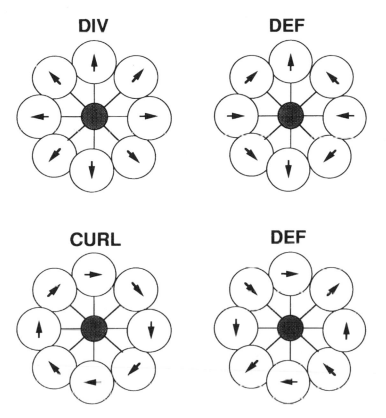

Figure 6 Simple neural mechanisms for the estimation of differential invariants by the combination of directionally-selective units.

As is clear from Figure 3, for example, *curl* and *def* produce characteristic and different changes in orientation so that these components could, in principle, be encoded by mechanisms selective for orientation rather than velocity.

2.4 Affine Transforms

Trans, *div*, *def* and *curl* together define a general *affine* transformation, i.e. a transformation in which parallel lines remain parallel and colinear distance ratios remain unchanged. Perspective transformations are clearly not affine and, since the optic array is a polar (i.e. non parallel) projection, optic flow clearly does involve perspective information. This does not present a theoretical problem for the account described above because differential invariants are calculated over infinitesimal regions of the flow, in which the effects of perspective can be ignored, and perspective is captured by the variation in the values of the invariants across the flow. But it does impose a practical limit on the scale at which invariants can be estimated: large-scale mechanisms, operating over regions of the flow in which the variation in depth is large in comparison with the viewing distance, will not provide a complete description of the flow because they cannot capture the variations due to perspective.

3   EVIDENCE FOR FLOW-SPECIFIC MECHANISMS

As pointed out above, to make use of the information provided by retinal flow, the visual system must solve a number of basic problems. It must first be able to extract the informative aspects of flow, produced by translation of the observer and characterized by expanding motion (Figures 1a and 2), from the largely uninformative distortions introduced by rotational movements about the viewpoint (Figure 1b). Having done this, information about locomotory heading could be extracted by mechanisms that are sensitive only to the directional pattern of optic flow and that are capable of integrating information over large regions of the flow pattern. On the other hand, information about three-dimensional surface layout requires mechanisms that are sensitive to speed as well as direction and that operate only upon local regions of the flow.

3.1   Neurophysiological Evidence for Flow-specific Mechanisms

Although there is at least one report of units in Area 18 of the cat which respond selectively to the expansion or contraction of a small 2° slit of light (Regan & Cynader, 1979), most of neurophysiological evidence for flow-specific mechanisms comes from later stages of the pathway, particularly from two regions of the superior temporal sulcus (STS) of monkeys – the middle temporal (MT) area and the medial superior temporal (MST) area to which it projects. Both regions contain units sensitive to large field unidirectional motion (D cells), and MST also contains units that respond preferentially to rotations (R cells) or expansions (S cells), irrespective of whether the stimulus is a large oriented bar, a large square or a random dot pattern. The receptive field size of MST cells is typically very large, ranging from 20 to 100° (Bruce et al., 1981; Sakata et al., 1985), but is rather smaller (less than 20°) in MT (Saito et al., 1986). In fact, receptive field size in MT varies with retinal eccentricity and, in the fovea, may be less than 1° (Tanaka et al., 1986). These findings have led to the suggestion (e.g. Tanaka et al., 1989) that the units in MST combine the outputs of several MT cells with receptive fields in different retinal positions. Some 'field type' S and R cells in MST require uniform expansion or rotation throughout the receptive field. Other 'figure type' cells respond to patches of rotation or expansion anywhere within the receptive field (Saito et al., 1986), although Tanaka and Saito (1989) point out that these patches must be at least 20° in diameter in order to elicit any response. S and R units respond consistently over a wide range of stimulus speeds and, importantly, Tanaka et al. (1989) report that both types continue to respond strongly to appropriate stimuli in which all the elements move at the same speed (i.e. to expansions and rotations from which the normal speed gradients have been removed). Thus, these mechanisms seem better equipped for the analysis of large-scale directional structure, and for the decomposition of the flow field into its expanding and rotating components, than for the analysis of local speed gradients and the recovery of information about surface layout. However, Duffy and Wurtz (1991a) have recently reported units that respond best to particular *combinations* of expansion, rotation or translation and they suggest, particularly interestingly, that, rather than combining inputs from simple D, S and R cells, these complex units might derive their properties by combining overlapping zones, each sensitive to a differently directed speed gradient (Duffy & Wurtz, 1991b).

3.2 Psychophysical Evidence for Flow-specific Mechanisms

A series of psychophysical studies by Regan and Beverley first suggested the existence of specialized expansion-selective mechanisms within the human visual system. Adaptation to a small, expanding and contracting square depresses sensitivity to a similar test stimulus much more than does adaptation to a square oscillating from side to side with the same amount of linear movement (Regan & Beverley, 1978a). Moreover, expanding squares produce an after effect of movement in depth that has a different time course from the conventional motion after effect produced by adaptation to simple linear motion (Regan & Beverley, 1978b). The underlying mechanisms appear to have fairly localized receptive fields, because the effects of adaptation are confined to the region of the retina that has been directly adapted (Regan & Beverley, 1979), but they do not seem to be concerned with the analysis of speed gradients, because adaptation to a flow pattern containing a uniform rate of radial expansion does produce the expected reduction in sensitivity to a small expanding square (Beverley & Regan, 1982).

Equivalent evidence for rotation-selective mechanisms is less complete, though adaptation to a small rotating field of random dots does selectively impair sensitivity to rotation (Regan & Beverley, 1985). Less direct evidence is based on 'shearing gratings', introduced by Nakayama and Tyler (1981) and consisting of random dot patterns in which all the dots drift in the same direction, with dot speed varying sinusoidally at right angles to the drift direction. Although the shearing is one-dimensional, it should provide a powerful stimulus for localized rotation (and defor-mation) selective mechanisms. Williams (1989) showed that the detection of such a grating is masked by the simultaneous presentation of a second grating only when the two stimuli are of similar spatial frequency, raising the intriguing possibility of rotation (or deformation) mechanisms selective for different scales of spatial speed gradient. Werkhoven and Koenderink's (1990c) report that speed discrimination of rotating patterns is affected by rotating masks only when the radii of the mask and target are very similar is also compatible with this possibility.

Other psychophysical evidence for the existence of specialized expansion and rotation selective mechanisms is provided by a simple comparison of overall sensit-ivity to different types of directionally-structured motion. Freeman and Harris (1992) showed that sensitivity, defined as the slowest detectable speed of motion of a random dot flow pattern, is significantly (and up to three of four times) greater for rotating and expanding patterns than for linearly translating or directionally random stimuli. The most obvious account of these specialized mechanisms is that they combine the outputs of several local elements with appropriate directional selectivities, and simple linear pooling would be the most obvious way to produce mechanisms capable of estimating differential invariants along the lines illustrated in Figure 6. Estimates of suprathreshold speed discrimination of expanding (Sekuler, 1992) and rotating (Werkhoven & Koenderink, 1991) dot patterns suggest that performance can, in fact, be adequately explained by such linear pooling. However, comparison of results across these studies and with those based on unidirectionally translating stimuli (McKee, 1981; McKee & Nakayama, 1984; De Bruyn & Orban, 1988) suggests that suprathreshold discrimination of these different directional patterns is roughly the same: there is no advantage for rotation or expansion as found at threshold.

One obvious explanation of this apparent discrepancy between threshold and suprathreshold performance is that the two tasks reveal different types of mechanism. In Freeman and Harris's stimuli, all the dots moved at the same speed and the task required the observer only to *classify* a particular directional pattern. This task could readily be accomplished by the neural mechanisms discussed above, which are not sensitive to speed gradients and which may serve to decompose the flow pattern into different directional components. Speed *discrimination*, on the other hand, clearly requires mechanisms that can signal speed as well as directional information and Sekuler's task, at least, also required the analysis of expanding speed gradients. This task may thus depend upon different neural processes that are concerned with the analysis of flow speed and thus the recovery of surface layout.

Other psychophysical work also supports the notion that specialized expansion- and rotation-sensitive mechanisms are concerned with an initial analysis of retinal flow into its directional components. The motion after effect produced by a rotating spiral is often explained in terms of such an analysis (e.g. Cavanagh & Favreau, 1980; Hershenson, 1984, 1987) and there is more direct evidence that the visual system can perform the necessary decomposition. Using random dot flow patterns in which the motion of each dot was a combination of a rotational and a expanding component, Freeman and Harris (1992) showed that the detection of expansion was unaffected by the presence of rotation, while the detection of rotation was unaffected by the presence of expansion. However, in this study the focus of expansion always coincided with the focus of rotation and the stimuli contained the normal speed gradients associated with rigid expansion so it is not yet clear under what circumstances the decomposition can be accomplished.

Much of this psychophysical evidence suggests that complex patterns are decomposed into independent expanding and rotating components, and is thus in keeping with the neurophysiological properties of S and R cells. However, other evidence suggests that there may be specialized mechanisms for particular combinations of expansion and rotation, and is more in keeping with the types of combination unit reported by Duffy and Wurtz (1991a). In a suprathreshold task in which obervers adjusted the proportion of dots moving in random directions to those moving according to an appropriate directional pattern, Milne and Snowden (1993) found that adapting to a particular combination of expansion and rotation reduced sensitivity to that particular combination, rather than to pure expansion and rotation.

4   LOCOMOTORY GUIDANCE

One of the first uses proposed for retinal flow was in guiding locomotion (e.g. Gibson, 1950). As pointed out in section 1, when the observer moves in a straight line with a fixed direction of gaze, the location of the focus of expansion (FOE) coincides with the direction of locomotory heading. Alternatively, as pointed out in section 2, heading can also be recovered from differential invariants. Whatever the source, if the information is to be useful in guiding locomotion, heading must be estimated with an accuracy of about 1 degree (Cutting, 1986). Although early studies simulating movement towards a frontoparallel surface were not especially promising (e.g. Carel, 1961; Llewellyn, 1971; Johnston *et al.*, 1973) a more recent series of studies by Warren and his colleagues confirm that human observers can achieve the required degree of accuracy in estimating locomotory heading from random dot flow patterns (e.g. Warren & Hannon, 1988). Warren's stimuli typically simulate movement across a horizontal plane and the observer's task is to indicate whether the depicted heading is to the left or right of a stationary target line. Under these circumstances, estimates remain accurate with very short presentations (two or three frames) and when the directions of individual dot trajectories are randomly perturbed (Warren *et al.*, 1991a). Importantly, performance also remained good with sparse flow patterns (Warren *et al.*, 1988) and when the pattern simulated movement toward a sparse three-dimensional cloud of dots (Warren & Hannon, 1990). These last two findings suggest that differential invariants are not involved in the recovery of locomotory heading because sparse or three-dimensionally irregular stimuli do not provide the smooth velocity fields necessary for their direct computation. It therefore seems more likely that locomotory guidance relies on the ability to locate the FOE.

The disadvantage of this simple strategy is that the introduction of a rotational component, by curvilinear movement or eye rotations, shifts the FOE in the retinal flow pattern so that it no longer coincides with locomotory heading. Under these circumstances, the retinal flow pattern must be decomposed into its rotational and translational components so that the true location of the FOE (in the translational component alone) can be recovered. Nonetheless, estimation of linear locomotory heading remains accurate in the presence of eye rotations (Warren & Hannon, 1990) and curvilinear heading can be accurately estimated from flow patterns simulating circular motion of the observer (Warren *et al.*, 1991b). Again, performance is not affected by sparse patterns or by three-dimensionally irregular stimuli suggesting that differential invariants are not involved – even though *div*, for example, has the apparently important advantage of being unaffected by rotations.

The required decomposition into global translational and rotational components implied by these results might be accomplished by the simple mechanisms, selective for appropriate types of directional structure, described in section 3. Indeed, initial attempts to simulate performance with a neural network suggest that this is a promising approach (Hatsopoulos & Warren, 1991). However, other psychophysical results suggest a more sophisticated strategy may also be required. Warren and Hannon (1990), found that, for simulated motion over a horizontal plane, the introduction of a rotational eye movement had no effect upon the ability to estimate locomotory heading, irrespective of whether the eye rotation was produced actively, by the observer tracking a moving dot in the display, or simulated passively by adding rotation directly to the display. However, for simulated motion toward a fronto-parallel plane, performance was unaffected during active eye rotation but badly disrupted by passive rotation. This suggests, crucially, that when the display simulates a flat, depthless stimulus, decomposition can only be achieved if information about the rotational component is available from eye movement commands, and that direct decomposition of the retinal flow pattern can only be achieved when the stimulus is genuinely three-dimensional. Rieger and Toet (1985) have also reported that accurate performance requires depthy flow patterns.

This reliance upon depth fits neatly with a strategy first proposed by Longuet-Higgins and Prazdny (1980) based on the analysis of directional discontinuities in the flow. Such discontinuities occur at depth edges when rotational and translational components are combined, because the speed of motion changes across the edge in the translational component but not in the rotational component. This can be exploited by subtracting the velocity vectors on each side of the discontinuity, so cancelling the rotational component and leaving a difference vector pointing away from the true (translational) FOE. The intersection of several of these difference vectors gives the location of the translational FOE and thus locomotory heading. This simple technique has been refined and implemented in several computer simulations (e.g. Rieger & Lawton, 1985; Hildreth, 1992). Once the translational FOE is located, the directional structure of the translational component is known and this, in turn, can allow the complete recovery of the rotational component simply by analysing any regions of the flow where motion is orthogonal to the predicted translational component. The recovery of the translational FOE can also be regarded as an important step in the next logical stage of flow processing – the analysis of temporal proximity.

5 TEMPORAL PROXIMITY

Following Lee (1976), temporal proximity is often called time-to-contact or time-to-collision (TTC), in deference to its obvious usefulness in locomotory guidance, and is generally given the symbol τ. τ is simply the distance to contact divided by the speed of approach and so is measured in units of time: it can tell you, for example, that two points will reach you at the same time but it cannot tell you that one is a long way off and approaching quickly while the other is close and approaching slowly. Lee (1976, 1980), Todd (1981) and Tresilian (1990) have provided general mathematical accounts of the relationships between the rate of expansion of the flow pattern and the rate of translation towards a target (term t_a in equation 3). The simple case of linear motion at a constant speed is shown in Figure 7.
By definition

$$\tau = z/-z'$$

where z' is the forward velocity of the observer (or, equivalently, $-z'$ is the approach speed of a looming object).
By similar triangles

$$rz = sf$$

Differentiating with respect to time gives

$$rz' + r'z = 0 \tag{8}$$

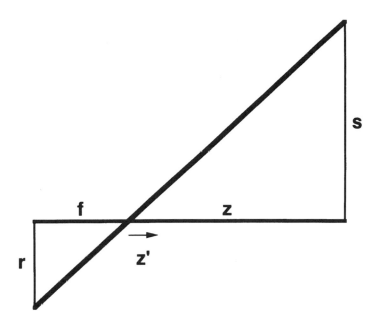

Figure 7   The geometry of temporal proximity. See text for details and relevant equations.

so that

$$r/r' = z/-z' - \tau \qquad (9)$$

Thus, τ can be estimated by measuring the radial speed, r', of any point in the flow pattern and its distance, r, from the FOE.

Differentiating equation 8 again with respect to time gives

$$rz'' + 2r'z' + r''z = 0$$

and, for movement at a constant speed ($z'' = 0$)

$$2r'/r'' = z/-z' = \tau \qquad (10)$$

Thus τ can also be estimated by measuring the radial speed, r', and acceleration, r'', at any point in the image.

Because these equations depend upon the radial (expanding) motion in the flow pattern, estimates of τ rely upon knowledge about the true (translational) FOE, although, from equation 10, only its direction and not its position is strictly required.

There is good evidence, chiefly from the correlation of behaviour with critical values of τ, that animals such as gannets (Lee & Reddish, 1981) and flies (Wagner, 1982) use τ in timing their approach behaviour, and that human beings use it both in timing their own actions (Lee et al., 1982; Macleod & Ross, 1983; Warren et al., 1986; Sidaway et al., 1989; Lee et al., 1992) and the approach of looming objects (von Hofsten, 1980; Lee et al., 1983; Lacquaniti & Maiolo, 1989; Savelsbergh et al., 1991). However, it is clear from Figure 7 that τ provides a point-by-point, time-based map of three-dimensional surface layout and thus has a greater potential use than the timing of locomotory or interceptive action. The more general term temporal proximity is therefore preferable to the functional role implied by time-to-contact.

Experimental studies based, for example, on discrimination between two approaching stimuli, suggest that human observers make direct use of τ, rather than an indirect estimation of distance and speed of approach (Schiff & Detwiler, 1979; Todd, 1981). However, it remains unclear exactly how τ is extracted from the retinal flow. Equation 3 suggests that *div* might be used but that its interpretation is complicated, particularly when the point of interest lies on a surface that is slanted with respect to the frontoparallel plane, because of the $\mathbf{n} \cdot \mathbf{t}_s$ term (see also Tresilian, 1990). Direct measurements of radial flow thus seem more suitable and might be based either on the spatial gradients (equation 9) or the temporal gradients (equation 10) in the radial flow associated with approaching movement. Freeman *et al.* (1994) investigated the role of spatial and temporal gradients by asking observers to estimate the temporal proximity of random dot patterns depicting approaching frontoparallel surfaces. Performance in this task was generally good over a wide range of dot densities. When spatial or temporal gradients were removed, performance deteriorated – as would be expected since these stimuli correspond respectively to distorting or (three-dimensionally) decelerating surfaces, making the task inherently more difficult. Nonetheless, estimates remained surprisingly good when only spatial or only temporal gradients were provided and, indeed, remained above chance even when neither gradient was provided, so that judgements could only be made on the basis of the average speed of the stimuli. Estimates were, however, consistently based on the radial (expanding) component of the pattern because the addition of a tangential (rotating) component, adjusted so that the overall speed of each dot remained constant throughout a trial, did not affect performance. These findings suggest that the visual system can adopt a very flexible strategy, first isolating the informative (expanding) aspect of the flow and then using whatever information it provides, rather than relying rigidly upon one particular aspect such as local speed or acceleration.

6   SURFACE LAYOUT

There can be no doubt that retinal flow can by itself provide a compelling impression of three-dimensional surface layout and that the depth perceived in a random dot flow pattern is direct, like that perceived in a random dot stereogram, rather than indirect, like that inferred from the cues available in a static picture. One example of this is given by temporal proximity, discussed in the previous section, where expanding flow patterns are immediately and compellingly perceived as looming surfaces moving at a constant speed in three dimensions even though the two-dimensional pattern of image speeds is accelerating. A second example stems from demonstrations by Wallach and O'Connell (1953), originally called the kinetic depth effect but now more commonly referred to as 'structure from motion' (SFM). The SFM paradigm can be regarded as a subset of flow in which the stimuli are usually restricted to surfaces rotating in depth under parallel (orthographic) projection. Thus, SFM stimuli tend to involve only the effects of changes in *viewing angle* in the absence of perspective transformations, whereas flow may potentially involve general changes in the *position* of the viewpoint under polar (perspective) projection. SFM is discussed in detail elsewhere in this book, but two observations are particularly relevant.

Firstly, SFM demonstrates that surface layout can be perceived in the absence of perspective and, moreover, that in many tasks the addition of perspective does not greatly improve performance (e.g. Braunstein, 1962, 1968). Secondly, performance is good even when only two frames are presented and, again, does not improve greatly as the number of frames is increased (Braunstein *et al.*, 1987; Todd & Bressan, 1990; Todd & Norman, 1991). Whereas the full 3D structure of the world can theoretically be recovered from two frames under polar projection (Longuet-Higgins, 1981) or from three frames under parallel projection (Ullman, 1979), it cannot be recovered under the restricted conditions described above. The type of 3D structure obtainable in displays restricted to affine transformations (i.e. parallel projection) in which only velocity information is available (i.e. two frames rather than the three required for acceleration) is termed 'affine structure'. General affine structure would provide relative lengths without absolute lengths but SFM displays generally allow rather more than this, specifying 3D shape up to a stretching along the line of sight. Todd and Norman (1991) argue that this type of affine structure is sufficient to account for human performance in a variety of tasks, including the discrimination of surface curvature, the detection of surface rigidity, and the recognition of complex 3D forms.

These arguments suggest that, in recovering surface layout from retinal flow, the human visual system depends, at least partly, on the recovery of affine structure from affine transformations – and affine transformations are, of course, precisely the domain of differential invariants (section 2). Thus, the transformation of a given surface in an SFM display can be completely described by single estimates of *trans*, *div*, *def* and *curl* and, in fact, Koenderink and van Doorn (1991) describe a detailed algorithm for the recovery of the affine structure of multi-faceted objects based only upon deformation.

6.1 The Role of Differential Invariants

Although SFM studies are suggestive of the involvement of differential invariants in the recovery of surface layout, attempts to provide more direct evidence have, thus far, proved rather disappointing.

One simple technique is to look for differences in the ability to estimate surface slant under conditions in which only some differential invariants can be estimated. For example, horizontal motion parallel to a vertically tilted surface produces flow containing *def* and *curl*, but no *div*, while the same motion relative to a horizontally tilted surface produces *div*, the same amount of *def* (though with a different axis), and no *curl*. These facts can readily be confirmed from equations 2–4. Harris *et al.* (1992a) attempted to exploit this asymmetry by comparing judgements of surface slant for the two directions of tilt. Following reports that sensitivity to shearing speed gradients is greater than to compressive speed gradients (Rogers & Graham, 1983; Nakayama *et al.*, 1984), one might expect that the ability to judge surface slant would vary according to the direction of surface tilt. In fact, no such anisotropy was found although of course, as Koenderink and van Doorn (1991) suggest, *def* may play a greater role in the recovery of surface layout than does *div* or *curl*. This possible importance of *def* is further supported by the recent report (Freeman *et al.*, 1993) that the combination of translation and *def* alone is sufficient to produce a compelling and predictable impression of surface slant, and that the perception of slant can be changed in a predicable way by manipulating the amplitude of *def* in otherwise normal random dot patterns.

A second technique makes use of the ability to segment complex patterns into discrete regions. If differential invariants were explicitly represented at an early stage in visual processing one might expect them to provide a powerful segmentation cue and to support the well-known phenomenon of 'pop out' (Treisman, 1988). However, using a dense field of rotating needles, Julesz and Hesse (1970) showed that observers were quite unable to pick out a region in which the needles rotated in the opposite direction, even under free viewing conditions. More recently, Braddick and Holliday (1991) have shown that expansion and deformation are equally ineffectual segmentation cues.

Finally, direct evidence *against* the simple involvement of differential invariants is provided by performance with stimuli consisting of very few dots or which depict transparent surfaces. Models that estimate differential invariants directly from retinal flow, such as that of Werkhoven and Koenderink (1990a, b) or the scheme depicted in Figure 6, require a smooth continuous flow field and should break down under these restricted conditions. In fact, however, estimates of surface slant remain good with sparse patterns containing very few dots distributed randomly over a large field (e.g. Braunstein *et al.*, 1987; Hughes, 1994). As mentioned previously, Warren and his colleagues have demonstrated that the estimation of locomotory heading remains good when the display depicts movement through an incoherent 3D cloud of dots and thus provides no coherent 2D speed gradients. More directly relevant to the recovery of surface layout, it is well known that shape can be recovered from displays depicting two transparent cylinders rotating about a common axis (e.g. Ullman, 1979). Andersen (1989) has shown that observers can simultaneously register up to three overlapping, transparent, moving surfaces, while Eby (1992) and Hughes (1994) have shown that good estimates of 3D structure can be made in such multi-surface displays. It thus seems clear that performance in these tasks does not depend upon estimates of differential invariants taken directly from retinal flow.

7 SUMMARY

Like most areas of relatively low-level vision, the study of optic and retinal flow has been well-served by an interdisciplinary approach. Mathematical and computational theories give precise accounts of what information is available from the retinal flow pattern and how it might best be extracted and represented. These accounts have usefully suggested both psychophysical studies of human performance and neuro-physiological studies of closely-related species. The combination of mathematical and psychophysical approaches has so far been most effective in the study of how human observers direct their locomotion in the presence of eye rotations, and of how they encode temporal proximity. These types of task seem to be accomplished using relatively simple geometric constraints that do not involve differential invariants. The required analysis of the flow pattern into global expansion and rotation components may, instead, involve neural mechanisms selective for large-scale directional structure.

One remaining fundamental problem concerns the possible usage of differential invariants in the recovery of surface layout. According to mathematical accounts, these would provide a compact and convenient way to describe retinal flow that would be particularly suited to this task. However, there is little neurophysiological evidence of the required mechanisms sensitive to small-scale expansion or rotation, and no evidence of mechanisms specifically sensitive to deformation. Similarly, while psychophysical studies suggest that human observers can decompose flow into global rotation and expansion components, there is no evidence that they make use of speed gradients, rather than simple directional structure, in accomplishing this task. Finally, performance with sparse or transparent stimuli rules out the possibility that processing depends upon direct estimation of differential invariants in the initial analysis of retinal flow. While it thus seems sensible to abandon the notion of differential invariants in the sense of specific mechanisms making the required estimates over each and every infinitesimal region of the flow pattern, the more general notion that an affine transform can be conveniently decomposed into a limited set of simpler transforms may yet prove useful. Affine transforms are clearly an adequate and important basis for much of our ability to recover three-dimensional structure from moving displays, and such transforms may be easily and usefully expressed in terms of general measures of expansion, rotation, and deformation. For the moment, it remains to be seen whether such an analysis is useful only as a way for us to understand the workings of the visual system, or whether it is also used by the visual system as a way to understand the world.

ACKNOWLEDGEMENT

The writing of this chapter, and some of the experiments described in it, were supported by grants from the SERC.

REFERENCES

Andersen, G. J. (1989). Perception of three-dimensional structure from optic flow without locally smooth velocity. *J. Exp. Psychol.: Human Percept. Perform.*, **15**, 363–371.

Beverley, K.I. & Regan, D. (1982). Adaptation to incomplete flow patterns: no evidence for 'filling in' the perception of flow patterns. *Perception*, **11**, 275–278.

Bingham, G. P. (1993). Optical flow from eye movements with head immobilized: 'ocular occlusion' beyond the nose. *Vision Res.*, **33**, 777–789.

Braddick, O. J. & Holliday, I. E. (1991). Serial search for targets defined by divergence or deformation of optic flow. *Perception*, **20**, 345–354.

Braunstein, M. L. (1962). Depth perception in rotating dot patterns: effects of numerosity and perspective. *J. Exp. Psychol.*, **64**, 415–420.

Braunstein, M. L. (1968). Motion and texture as sources of slant information. *J. Exp. Psychol.*, **78**, 247–253.

Braunstein, M. L., Hoffman, D. D., Shapiro, L. R., Andersen, G. J. & Bennett, B. M. (1987). Minimum points and views for the recovery of three-dimensional structure. *J. Exp. Psychol.: Human Percept. Perform.*, **13**, 335–343.

Bruce, C., Desimone, R. & Gross, C. G. (1981). Visual properties of neurons in a polysensory area in superior temporal sulcus of the macaque. *J. Neurophysiol.*, **46**, 369–384.

Carel, W. L. (1961). *Visual Factors in the Contact Analog.* General Electric Advanced Electronics Centre Publishers R61 ELC60, Ithaca, NY.

Cavanagh, P. & Favreau, O. E. (1980). Motion aftereffect: a global mechanism for the perception of rotation. *Perception*, **9**, 175–182.

Clocksin, W. H. (1980). Perception of surface slant and edge labels from optical flow: a computational approach. *Perception*, **9**, 253–269.

Cutting, J. E. (1986). *Perception with an Eye for Motion.* MIT Press, Massachusetts.

De Bruyn, C. & Orban, G. A. (1988). Human velocity and direction discrimination measured with random dot patterns. *Vision Res.*, **28**, 1323–1335.

Duffy, C. J. & Wurtz, R. H. (1991a). Sensitivity of MST neurons to optic flow stimuli. I. A continuum of response selectivity to large-field stimuli. *J. Neurophysiol.*, **65**, 1329–1345.

Duffy, C. J. & Wurtz, R. H. (1991b). Sensitivity of MST neurons to optic flow stimuli. II. Mechanisms of response selectivity revealed by small-field stimuli. *J. Neurophysiol.*, **65**, 1346–1359.

Eby, D. W. (1992). The spatial and temporal characteristics of perceiving 3-D structure from motion. *Percept. Psychophys.*, **51**, 163–178.

Farber, J. M. & McConkie, A. B. (1979). Optical motions as information for unsigned depth. *J. Exp. Psychol.: Human Percept. Perform.*, **5**, 494–500.

Freeman, T. C. A. & Harris, M. G. (1992). Human sensitivity to expanding and rotating motion: effects of complementary masking and directional structure. *Vision Res.*, **32**, 81–87.

Freeman, T. C. A., Harris, M. G. & Williams, G. H. (1992). Surface layout from retinal flow. In G. W. Humphreys (Ed.) *Understanding Vision.* Blackwell, Oxford.

Freeman, T. C. A., Harris, M. G. & Meese, T. S. (1993). The effects of masking by and adaptation to relative motion upon human estimation of perceived slant. *Perception*, **22** Supplement, 95.

Georgeson, M. A. & Harris, M. G. (1978). Apparent foveofugal drift of counterphase gratings. *Perception*, **7**, 527–536.

Gibson, J. J. (1950). *The Perception of the Visual World.* Houghton-Mifflin, Boston.

Gibson, J. J. (1979). *The Ecological Approach to Visual Perception.* Houghton-Mifflin, Boston.

Harris, M. G., Freeman, T. C. A. & Hughes, J. A. (1992). Retinal speed gradients and the perception of surface slant. *Vision Res.*, **32**, 587–590.

Harris, M. G., Freeman, T. C. A. & Tyler, P. A. (1994). Human sensitivity to temporal proximity: the role of spatial and temporal speed gradients. *Percept & Psychophys.* **55**, 689–699.

Hatsopoulos, N. G. & Warren, W. H. (1991). Visual navigation with a neural network. *Neural Networks*, **4**, 303–317.

Hershenson, M. (1984). Phantom spiral aftereffect: evidence for global mechanisms in perception. *Bull. Psychonom. Soc.*, **22**, 535–537.

Hershenson, M. (1987). Visual system responds to rotational and size-change components of complex proximal motion patterns. *Percept. Psychophys.*, **42**, 60–64.

Hildreth, E. C. (1992). Recovering heading for visually-guided navigation. *Vision Res.*, **32**, 1177–1192.

von Hofsten, C. (1980). Predictive reaching for moving objects by human infants. *J. Exp. Child Psychol.*, **30**, 369–382.

Horn, B. K. P. & Schunk, B. G. (1981). Determining optical flow. *Artificial Intelligence*, **17**, 185–203.

Hughes, J. (1994). Mechanisms of optic flow analysis. Unpublished PhD Thesis: The University of Birmingham.

Johnston, I. R., White, G. R., & Cumming, R. W. (1973). The role of optical expansion patterns in locomotor control. *Am. J. Psychol.*, **86**, 311–324.

Julesz, B. & Hesse, R. I. (1970). Inability to perceive the direction of rotation movement of line segments. *Nature*, **225**, 243–244.

Koenderink, J. J. (1985). Space, form, and optical deformations. In D. T. Engle, M. Jeanerod & D. N. Lee (Eds.) *Brain Mechanisms and Spatial Vision.* Martinus Nijhoff, Dordrecht.

Koenderink, J. J. (1986). Optic flow. *Vision Res.*, **26**, 161–180.

Koenderink, J. J. & van Doorn, A. J. (1975). Invariant properties of the motion parallax field due to the movement of rigid bodies relative to the observer. *Optica Acta*, **22**, 773–791.

Koenderink, J. J. & van Doorn, A. J. (1976). Local structure of movement parallax of the plane. *J. Opt. Soc. Am.*, **66**, 717–723.

Koenderink, J. J. & van Doorn, A. J. (1991). Affine structure from motion. *J. Opt. Soc. Am. A*, **8**, 377–385.

Lacquaniti, F. & Maiolo, C. (1989). The role of preparation in tuning anticipatory and reflex responses during catching. *J. Neurosci.*, **9**, 134–148.

Lee, D. N. (1976). A theory of visual control of braking based on information about time-to-collision. *Perception*, **5**, 437–459.

Lee, D. N. (1980). The optic flow field: the foundation of vision. (1980). *Phil. Trans. R. Soc. Lond. B*, **290**, 169–179.

Lee, D. N. & Reddish, P. E. (1981). Plummeting gannets: a paradigm of ecological optics. *Nature*, **293**, 293–294.

Lee, D. N., Lishman, J. R. & Thomson, J. A. (1982). Visual regulation of gait in long jumping. *J. Exp. Psychol.: Human Percept. Perform.*, **8**, 448–459.

Lee, D. N., Young, D. S., Reddish, P. E., Lough, S. & Clayton, T. M. H. (1983). Visual timing in hitting an accelerating ball. *Q. J. Exp. Psychol.*, **35A**, 333–346.

Lee, D. N., Young, D. S. & Rewt, D. (1992). How do somersaulters land on their feet? *J. Exp. Psychol.: Human Percept. Perform.*, **18**, 1195–1202.

Llewellyn, K. R. (1971). Visual guidance of locomotion. *J. Exp. Psychol.*, **91**, 245–261.

Longuet-Higgins, H. C. (1981). A computer algorithm for reconstructing a scene from two projections. *Nature*, **293**, 133–135.

Longuet-Higgins, H. C. & Prazdny, K. (1980). The interpretation of a moving retinal image. *Proc. Roy. Soc. Lond. B*, **208**, 385–397.

McKee, S. P. (1981). A local mechanism for differential velocity detection. *Vision Res.*, **21**, 491–500.

McKee, S. P. & Nakayama, K. (1984). The detection of motion in the peripheral visual field. *Vision Res.*, **24**, 25–32.

McLeod, R. W. & Ross, H. E. (1983). Optic flow and cognitive factors in time-to-collision. *Perception*, **12**, 417–423.

Milne, A. B. & Snowden, R. J. (1993). Is there anything special about expansion and rotational flow fields? *Perception*, **22** Supplement, 95.

Nakayama, K. & Loomis, J. J. (1973). Optical velocity patterns, velocity sensitive neurons, a space perception: a hypothesis. *Perception*, **3**, 63–80.

Nakayama, T. & Tyler, C. W. (1981). Psychophysical isolation of movement sensitivity by removal of familiar pattern cues. *Vision Res.*, **21**, 427–433.

Nakayama, K., Silverman, G., MacLeod, D. I. A. & Mulligan, J. (1984). Sensitivity to shearing and compressive motion in random dots. *Perception*, **13**, 229–243.

Prazdny, K. (1981). A note on 'Optical motions as information for unsigned depth'. *J. Exp. Psychol.: Human Percept. Perform.*, **7**, 286–289.

Prazdny, K. (1983). On the information in optic flows. *Computer Vision, Graphics and Image Processing*, **22**, 239–259.

Priest, H. F. & Cutting, J. E. (1985). Visual flow and direction of locomotion. *Science*, **227**, 1063–1065.

Regan, D. (1985). Visual flow and direction of locomotion. *Science*, **227**, 1063–1065.

Regan, D. & Beverley, K. I. (1978a). Looming detectors in the human visual pathway. *Vision Res.*, **18**, 415–421.

Regan, D. & Beverley, K. I. (1978b). Illusory motion in depth: aftereffect of adaptation to changing size. *Vision Res.*, **18**, 415–421.

Regan, D. & Beverley, K. I. (1979). Visually guided locomotion: psychophysical evidence for a neural mechanism sensitive to flow patterns. *Science*, **205**, 311–313.

Regan, D. & Beverley, K. I. (1985). Visual responses to vorticity and the neural analysis of optic flow. *J. Opt. Soc. Am. A*, **2**, 280–283.

Regan, D. & Cynader, M. (1979). Neurons in area 18 of cat visual cortex selectivity sensitive to changing size: nonlinear interactions between responses to two edges. *Vision Res.*, **19**, 699–711.

Rieger, J. H. & Lawton, D. T. (1985). Processing differential image motion. *J. Opt. Soc. Am. A*, **2**, 354–360.

Rieger, J. H. & Toet, L. (1985). Human visual navigation in the presence of 3D rotations. *Biol. Cybernet.*, **52**, 377–381.

Rogers, B. & Graham, M. (1983). Anisotropies in the perception of three-dimensional surfaces. *Science*, **221**, 1409–1411.

Saito, H., Yukie, M., Tanaka, K., Hikosaka, K., Fukada, Y. & Iwai, E. (1986). Integration of direction signals of image motion in the superior temporal sulcus of the macaque monkey. *J. Neurosci.*, **6**, 145–157.

Sakata, H., Shibutani, H., Kawano, K. & Harrington, T. L. (1985). Neural mechanisms of space vision in the parietal association cortex of the monkey. *Vision Res.*, **25**, 453–463.

Savelsbergh, G. J. P., Whiting, H. T. A. & Bootsma, R. J. (1991). Grasping 'tau' *J. Exp. Psychol.: Human Percept. Perform.*, **17**, 315–322.

Schiff, W. & Detwiler, M. L. (1979). Information used in judging impending collision. *Perception*, **8**, 647–658.

Sekuler, A. B. (1992). Simple-pooling of unidirectional motion predicts speed discrimination for looming stimuli. *Vision Res.*, **32**, 2277–2288.

Sidaway, B., McNitt-Gray, J. & Davis, G. (1989). Visual timing of muscle preactivation in preparation for landing. *Ecol. Psychol.*, **1**, 253–264.

Tanaka, H. & Saito, H. (1989). Analysis of motion of the visual field by direction, expansion/contraction, and rotation cells clustered in the dorsal part of the medial superior temporal area of the macaque monkey. *J. Neurophysiol.*, **62**, 626–641.

Tanaka, K., Hikosaka, K., Saito, H., Yukie, M., Fukada, Y. & Iwai, E. (1986) Analysis of local and wide-field movements in the Superior Temporal visual areas of the Macaque monkey. *J. Neurosci.*, **6**, 133–144.

Tanaka, H., Fukada, Y. & Saito, H. (1989). Underlying mechanisms of expansion/contraction and rotation cells in the dorsal part of the medial superior temporal area of the macaque monkey. *J. Neurophysiol.*, **62**, 642–656.

Todd, J. T. (1981). Visual information about moving objects. *J. Exp. Psychol.: Human Percept. Perform.*, **7**, 795–810.

Todd, J. T. & Bressan, P. (1990). The perception of 3-dimensional affine structure from minimal apparent motion sequences. *Percept. Psychophys.*, **48**, 419–430.

Todd, J. T. & Norman, J. F. (1991). The visual perception of smoothly curved surfaces from minimal apparent motion sequences. *Percept. Psychophys.*, **50**, 509–523.

Torrey, C. (1985). Visual flow and direction of locomotion. *Science*, **227**, 1063–1065.

Treisman, A. (1988). Features and objects. *Q. J. Exp. Psychol.*, **40A**, 201–238.

Tresilian, J. R. (1990). Perceptual information for the timing of interceptive action. *Perception*, **19**, 223–239.

Ullman, S. (1979). *The Interpretation of Visual Motion*. MIT Press, Cambridge, MA.

Wallach, H. & O'Connell, D. N. (1953). The kinetic depth effect. *J. Exp. Psychol.*, **45**, 205–217.

Wagner, H. (1982). Flow-field variables trigger landing in flies. *Nature*, **297**, 147–148.

Warren, W. H. & Hannon, D. J. (1988). Direction of self-motion is perceived from optical flow. *Nature*, **336**, 162–163.

Warren, W. H. & Hannon, D. J. (1990). Eye movements and optical flow. *J. Opt. Soc. Am. A*, **7**, 160–169.

Warren, W. H., Young, D. S. & Lee, D. N. (1986). Visual control of step length during running over level terrain. *J. Exp. Psychol.: Human Percept. Perform.*, **12**, 259–266.

Warren, W. H., Morris, M. W. & Kalish, M. (1988). Perception of translational heading from optical flow. *J. Exp. Psychol.: Human Percept. Perform.*, **14**, 646–660.

Warren, W. H., Blackwell, A. W., Kurtz, K. J., Hatsopoulos, N. G. & Kalish, M. L. (1991a). On the sufficiency of the velocity field for perception of heading. *Biol. Cybernet.*, **65**, 311–320.

Warren, W. H., Mestre, D. R., Blackwell, A. W. & Morris, M. W. (1991b). Perception of circular heading from optical flow. *J. Exp. Psychol.: Human Percept. Perform.*, **17**, 28–43.

Waxman, A. M. & Wohn, K. (1988). Image flow theory: a framework for 3-D inference from time-varying imagery. In C. Brown (Ed.) *Advances in Computer Vision, Volume 1*. Lawrence Erlbaum Associates, Hillsdale, NJ.

Werkhoven, P. & Koenderink, J. J. (1990a). Extraction of motion parallax structure in the visual system I. *Biol. Cybernet.*, **63**, 185–191.

Werkhoven, P. & Koenderink, J. J. (1990b). Extraction of motion parallax structure in the visual system II. *Biol. Cybernet.*, **63**, 193–199.

Werkhoven, P. & Koenderink, J. J. (1990c). Interference in rotary motion. *J. Opt. Soc. Am. A*, **7**, 1627–1631.

Werkhoven, P. & Koenderink, J. J. (1991). Visual processing of rotary motion. *Percept. Psychophys.*, **49**, 73–82.

Whittaker, E. T. (1944). *A Treatise on the Analytical Dynamics of Particles and Rigid Bodies*. Dover, New York.

Williams, G. H. (1989). Psychophysical measurement of shear detectors in human vision. Unpublished PhD Thesis: The University of Birmingham.

12

Motion-in-depth

Bruce Cumming
University of Oxford, UK

1 INTRODUCTION

Most of this book considers the problem of how changes in the retinal image over time are used to detect motion. Once appropriate local motion signals are extracted, there remain substantial difficulties in interpreting their significance. One of the central problems concerns how two-dimensional retinal image motion is used to reconstruct three-dimensional (3D) movement. There are two distinct sources of information that make this possible:

1 In animals with substantial binocular overlap, differences between the images of the two eyes can be used to reconstruct three-dimensional properties.
2 Monocular information alone, when combined with simplifying assumptions, can be used to detect motion-in-depth. When the image of an object grows larger on the retina, this produces a strong sensation that the object is moving towards the observer. This reflects an assumption that the object's physical size is not changing.

Sections 2 and 3 discuss binocular and monocular mechanisms respectively. Section 4 discusses how these separate sources of information might be combined. The discussion will concentrate on mechanisms that are useful for detecting the motion of small objects towards or away from an observer. The visual stimulation produced by such motion shares many properties with that produced by an observer moving through a stationary environment. This latter type of optic flow has other special properties and is discussed in a separate chapter by M. G. Harris (Chapter 11). In the interests of brevity, the discussion of many issues is not extensive, and inevitably reflects some personal interpretations. For a fuller discussion of some issues, and an alternative point of view, there is a recent review (Regan, 1991).

VISUAL DETECTION OF MOTION
ISBN 0–12–651660–X

2 BINOCULAR MECHANISMS

Wheatstone (1838) demonstrated that the differences between the images on the two retinae could be used to perceive the three-dimensional layout of a scene. It was over 130 years before this observation was extended to include changes in binocular differences over time (Julesz, 1971; Tyler, 1971; Richards, 1972). Tyler (1971) showed that stereomotion thresholds were higher than monocular motion thresholds, so that although motion was visible with either eye alone, when both eyes were open the target appeared stationary. Tyler concluded that stereomotion was not derived from monocular motion detection. Some results of a more extensive study of stereomotion detection (Regan & Beverley, 1973) are shown in Figure 1. The stimulus consisted of a vertical bar moving in opposite directions in the two eyes, seen against a stationary random dot background. Both Tyler (1971) and Regan and Beverley (1973) found that stereomotion had relatively poor temporal resolution. Thresholds reached a minimum at 1–2 Hz (see Figure 1), and at frequencies greater than 5 Hz no motion-in-

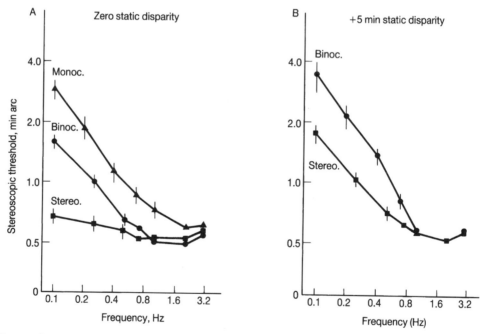

Figure 1 Thresholds for detecting stereomotion and lateral motion, as a function of temporal frequency. Triangles ('Monoc.') plot monocular motion detection thresholds. Circles ('Binoc.') plot binocular thresholds for detecting lateral motion. Squares ('Stereo') plot binocular thresholds for detecting motion towards and away from the observer. In the last two conditions, the monocular stimulation is identical, but the phase relationship of stimulation between the eyes changes: the binocular condition consists of stimulus movement that is in phase for the two eyes; in the stereo condition the two eyes' stimuli are in antiphase. Reprinted from Regan and Beverley, 'Some dynamic features of depth perception', Copyright 1973, pp. 2369–2378, with permission from Pergamon Press Ltd, Headington Hill Hall, Oxford OX3 0BW, UK.

depth was seen. Since this is poorer than the temporal resolution of monocular motion detection (van Nes *et al.*, 1967), it also suggests that stereomotion is detected separately from monocular motion.

At first sight the poor temporal resolution of stereomotion suggests that it will be of limited use for dealing with objects that move rapidly towards an observer. However, extrapolating from the responses to sinusoidal stimuli to other types of motion is only valid for linear systems. Beverley and Regan (1974) measured thresholds for brief disparity pulses. Subjects were able to detect movement in depth for stimuli lasting as little as 10 ms, and at stimulus durations of about 50 ms thresholds were only twice as large as those for pulses lasting 1 second. For a linear system, with a high-frequency cutoff at 5 Hz, the response to a 50 ms pulse would be much smaller than that to a 1 s pulse. This suggests that stereomotion detection is a nonlinear system, which is sensitive to sudden changes in disparity.

2.1 Discriminating Different 3D Trajectories

In order for stereomotion to be useful in everyday life, information about stereomotion must be combined with other motion signals to allow discrimination of different trajectories in three dimensions. In order to understand how this could be done, it is necessary to understand how visual information scales with viewing distance. The angular subtense of an object is approximately proportional to its width (or height) divided by the viewing distance. The relative disparity between the front and back surfaces of the object is approximately proportional to its depth divided by the *square* of the viewing distance. The same relationships hold when discussing the scaling of velocity signals in these two directions (Figure 2). So, two objects moving on parallel trajectories at different viewing distances will result in different ratios of disparity modulation and lateral motion. A large body of work using fixed disparities has shown that human subjects have a limited ability to scale disparity values for viewing distance (see Foley, 1991 for a review). This question has not been investigated in terms of perceived trajectories of moving objects, but the existing evidence on binocular distance perception suggests that subjects would be poor at making absolute judgements of trajectory from stereomotion signals.

This geometrical effect also raises difficulties for expressing experimental results. If trajectories are expressed in terms of the angle they make with respect to the line of sight, then the relationship between the retinal image properties and the trajectory will depend on viewing distance. To avoid this problem, Beverley and Regan (1973, 1975) expressed the direction of motion in terms of the ratio of velocities presented to the eyes: V_l/V_r. This is a useful way of describing stimulus properties in a way which does not depend upon the viewing distance. Figure 3 shows a polar plot of this space. This representation emphasizes the fact that useful information is available even without knowing the absolute direction of motion. For example, if $V_l/V_r \lesssim 0$, then the object is on a trajectory that will pass between the eyes (i.e. it will hit the head). Note that the relationship between this representation and real 3D trajectories is quite distorted for natural viewing distances – 90° of the plot is dedicated to trajectories passing between the eyes, a space which only corresponds to 3.5° at a viewing distance of 1 m.

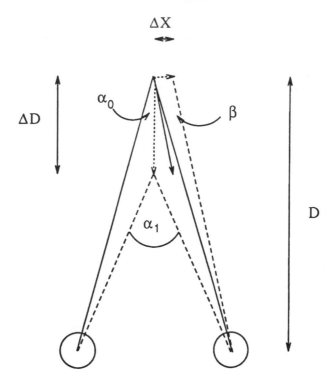

Figure 2 The effect of viewing distance on trajectories defined by stereomotion. The trajectory of an object moving-in-depth (solid arrow) is resolved into a component towards the observer and a component in the frontoparallel plane (dotted arrows). The rate of change of disparity, $\Delta\eta/\Delta t$, depends on $1/D^2$: $(\alpha_{t0} - \alpha_{t1})/\Delta t \approx (I/D - I/(D + \Delta D))/\Delta t \approx I\frac{\Delta D}{\Delta t}/D^2$, while the change of horizontal angular position (β) depends on $1/D$. Consequently, two trajectories which are parallel but at different viewing distances, although they have the same $\Delta D/\Delta X$, have different ratios of disparity change to change in eccentricity.

Beverley and Regan (1973) used adaptation to demonstrate the existence of psychophysical channels sensitive to particular trajectories (particular V_l/V_r ratios). Subjects adapted to a target moving back and forth along one trajectory, and thresholds were then measured for a range of other trajectories. A wide range of trajectories was used for adaptation, and yet the postadaptation threshold elevation always followed one of five patterns (Figure 4). Beverley and Regan accounted for these data with a model consisting of four channels (Figure 5).

The two channels in the centre of Figure 5 are most sensitive to trajectories that pass between the eyes. These could represent a mechanism which is driven by changes in disparity, and which is split into two channels which are selective for the direction of sideways motion. This concurs with subsequent physiological experiments showing that many disparity tuned cells in visual cortex are direction selective (Poggio & Fisher, 1977; Poggio & Talbot, 1981).

The other two channels are most sensitive to velocity ratios close to 1 : 1. An important point to note is that the ends of the abscissae in Figures 4 and 5 could be

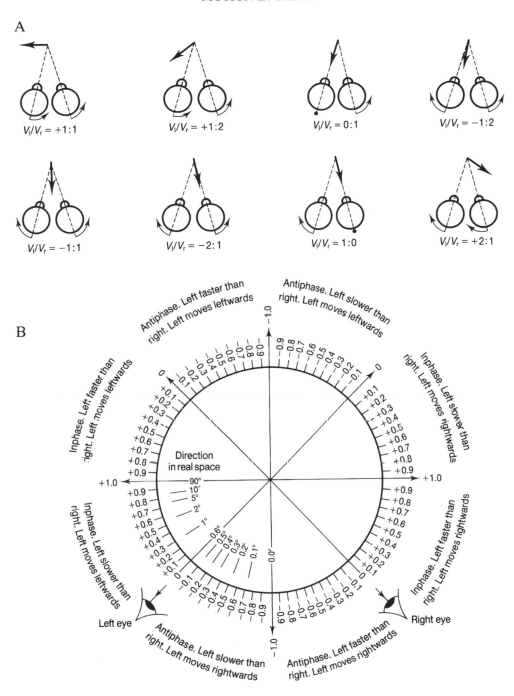

Figure 3 Schematic diagram showing that the ratio of the velocities seen by the two eyes uniquely determines the direction of motion, for a given viewing distance. (A). A variety of trajectories in plan view. (B). A polar plot of V_l/V_r ratios and their relationship to real-world trajectories. The real-world trajectories are calculated for a viewing distance of 3 m. Modified from Beverley and Regan (1973), with permission.

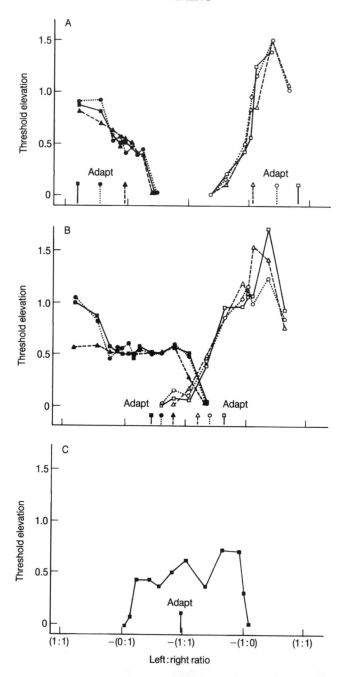

Figure 4 Patterns of threshold elevation for binocular detection of motion along different trajectories. The results for different trajectories of adapting stimulus are shown with different symbols. Thirteen different adapting trajectories gave rise to only five patterns of threshold elevation, suggesting that there are directionally-selective binocular motion filters. Each panel of the figure shows groups of adaptation trajectories that stimulated similar channels. Reproduced with permission from Beverley and Regan (1973).

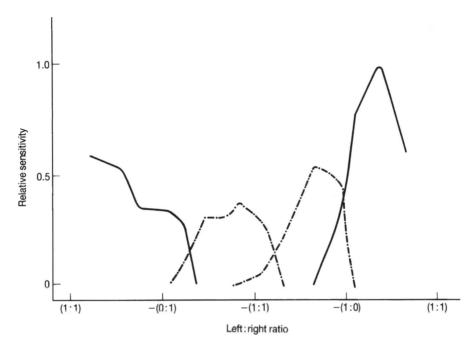

Figure 5 Four binocular motion-in-depth channels proposed to explain the results of adaptation (Figure 4). Note that the leftmost and rightmost channels have the same response at the ratios closest to 1 : 1. These could be considered a single channel, with a response which is symmetrical about its peak sensitivity. Reproduced with permission from Beverley and Regan (1973).

joined into a circle: both ends refer to a velocity ratio of 1, which is motion in the frontoparallel plane. This suggests that these two channels may be just one channel. (The fact that the peak of this channel is not at a ratio of 1 : 1 may reflect an imbalance in the inputs from the two eyes for that subject.) So although the adaptation data suggest that there are psychophysical channels tuned to certain 3D trajectories, these may be no more than one channel for motion close to the frontoparallel plane, and two channels (one for each direction of sideways motion) for motion towards the observer (detecting disparity changes).

Beverley and Regan (1975) reported thresholds for binocular discrimination of different trajectories of 3D motion. They found three peaks in the sensitivity, corresponding to trajectories where the channels postulated by Beverley and Regan (1973) changed the sensitivity most rapidly. However, the trajectories used in this study were chosen such that they all had the same amplitude of disparity oscillation, so stimuli could be discriminated on the basis of their velocity in the frontoparallel plane. Different values of the V_l/V_r ratio resulted in different velocities of lateral motion, so the velocity discrimination would have been performed at different mean velocities depending upon V_l/V_r. Consequently the results of this study are more difficult to interpret than those of the adaptation study (Beverley & Regan, 1973).

2.2 Relative vs. Absolute Motion

An important property of the stimuli used in all of the experiments discussed above is that the object moving in depth was seen against a stationary background. Thus the stimulus motion produced changes in *relative* disparity between the target and the background. Since detection thresholds are substantially larger for absolute disparities than for relative disparities (Westheimer, 1979), it is important to ask whether the same property is true of the stereomotion system. Regan *et al.* (1986a) measured stereomotion thresholds for a single dot stimulus, with and without reference marks. Removing the reference marks increased detection thresholds by a factor of about 5, similar in magnitude to the effect reported by Westheimer. Regan *et al.* also showed that removing reference marks had a similar effect on thresholds for detecting lateral motion. A related phenomenon was reported by Erkelens and Collewijn (1985a), who used a very large random dot stereogram, subtending 93×82 degrees. When the whole visual field was moved in opposite directions for the two eyes, no motion-in-depth was seen. Subsequent investigations (Erkelens & Collewijn, 1985b; Regan *et al.*, 1986a) recorded the positions of both eyes, and studied the effect under circumstances where vergence tracking was poor. Comparing the position of the eyes with the position of the target they demonstrated substantial oscillations in the absolute retinal disparity of the stimulus. In spite of this, no motion-in-depth was perceived. Thus it seems that stereomotion detection relies upon changes in *relative* disparities. The insensitivity to absolute disparity means that retinal changes resulting from changes in convergence angle will not produce the sensation of motion-in-depth. In the real world, movement in depth generally will produce changes in relative disparity between objects. Even if an observer moves towards a stationary scene, the relative disparities between objects will change as a result of the relationship between disparity and viewing distance (discussed above).

2.3 Stereomotion is a Specialized Mechanism

The fact that human observers are able to detect changes in disparity over time does not necessarily indicate the existence of a specialized mechanism for extracting stereomotion. All of the above results could be explained by suggesting that because the subject is aware of different disparities at different times, they *infer* that there has been motion in some cognitive way. Thus they would be able to indicate which stimulus contained motion-in-depth, without any specialized visual mechanism for stereomotion detection.

  However, several lines of evidence suggest that there is a specific brain mechanism for detecting stereomotion:

1 Following selective adaptation for motion-in-depth (Beverley & Regan, 1973), there is little threshold elevation for static disparity detection (Regan, 1991).
2 For short ramp movements in depth, thresholds for detecting receding motion are different from those for approaching motion (Beverley & Regan, 1974).
3 In some subjects, parts of the visual field are blind to stereomotion, despite normal stereoacuity (Richards & Regan, 1973; Regan *et al.*, 1986b; Hong & Regan, 1989).

Regan *et al.* (1986b) also showed some examples of stereomotion blindness only for crossed disparities, while stereomotion detection remained intact for uncrossed disparities (and vice versa in some subjects). Hong and Regan (1989) confirmed this, and showed that some subjects were unable to see approaching stimuli, while they were able to see receding stimuli that moved over the same disparity range.

The claim that there is specialized mechanism for stereomotion detection is slightly different from the earlier suggestion that stereomotion and static stereo are performed by two separate parallel populations of cells in the brain (the 'two population hypothesis': Regan *et al.*, 1986b; Hong & Regan, 1989). The latter implies that the disparity detectors which are used for static stereopsis are separate from the detectors that are used for stereomotion. None of the evidence cited above indicates that the two populations are in fact parallel. All of the observations described so far could be accounted for by a scheme in which the output of static disparity detectors (one population only) is differentiated by a special stereomotion mechanism. This scheme is shown diagrammatically in Figure 6B. All of the stereomotion-specific field deficits described above can be explained by suggesting that only the element that performs the differentiation with respect to time is absent. If stereomotion was independent of static disparity, it should be possible to demonstrate the ability to detect stereomotion in parts of the visual field where subjects are blind to static disparities. One such area, in the visual field of Regan himself, was demonstrated by Richards and Regan (1973).

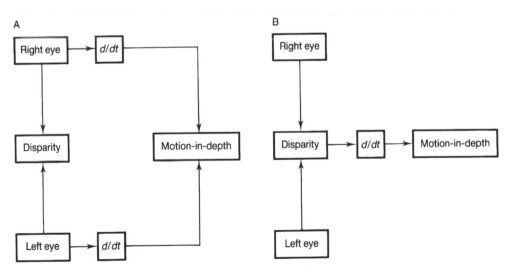

Figure 6 Block diagrams illustrating two possible methods for stereomotion detection. In A (modified from Regan *et al.*, 1979), velocity signals are first calculated separately in the two eyes (indicated by the operator, d/dt). Positional disparities are calculated by a completely independent mechanism. In B, the temporal differentiation is applied to the output of the disparity detection mechanism, providing a mechanism that is independent of the velocity signals measured monocularly. Reprinted from Cumming and Parker, 'Binocular mechanisms for detecting motion in depth', Copyright 1994, with permission from Pergamon Press Ltd, Headington Hill Hall, Oxford OX3 0BW, UK.

However, when stimuli that matched for luminance transients were used, this difference in visual fields was no longer apparent. In subsequent studies that plotted visual fields for stereomotion, no examples of stereomotion without stereopsis were reported.

In summary, although there is good evidence that there is a specialized brain mechanism that detects stereomotion, it is not clear whether this mechanism uses the output of static disparity detectors, or some other input.

2.4 Inter-ocular Velocity Differences vs. Changes in Disparity

One reason why it is thought that stereomotion is independent of static disparities is that it was originally proposed that stereomotion was detected by detecting differences in velocity signals between the two eyes, independent of positional disparity (Regan & Beverley, 1979; Regan et al., 1979). This scheme is illustrated in Figure 6A. For any object moving in the real world, these inter-ocular velocity differences always occur at the same time as changes in disparity (this is a simple piece of geometry). Nonetheless, these two schemes represent quite different brain mechanisms. Unfortunately, because of the geometrical relationship, the idea of inter-ocular velocity differences is often used interchangeably with that of disparity changes (see, for example, Regan et al., 1986a). All of the stimuli for stereomotion discussed so far have contained both inter-ocular velocity differences and disparity changes. They cannot therefore distinguish these two mechanisms.

Random dot techniques can be used to discriminate these possibilities. Julesz (1971, p. 184) used a sequence of random dot stereograms (RDS) in which a new disparity was shown in each stereogram. Each stereogram was also made from a new set of random dots (i.e. a dynamic RDS), so that each eye saw flickering dots, without seeing any consistent velocity signal (because the stimuli are temporally uncorrelated). Nonetheless, subjects reported the sensation of motion-in-depth. A similar effect was reported by Regan (1993). Norcia and Tyler (1984) also showed that disparity oscillations could be seen in dynamic RDS for temporal frequencies up to approximately 7 Hz. However, none of these studies reported thresholds, so it remains possible that stereomotion detection based upon disparity changes is less sensitive than a mechanism based on inter-ocular velocity differences. Thresholds for stereomotion in dynamic RDS were measured by Norcia (1980), and these were somewhat poorer than those reported for temporally correlated stimuli (Tyler, 1971; Regan & Beverley, 1973). However, there were a number of important differences between the stimuli in these studies which could account for the threshold differences. In particular, the stimulus used by Norcia (1980) modulated the absolute disparity of the whole display, not a relative disparity between two planes. The work of Erkelens and Collewijn (1985a, 1985b) and Regan et al. (1986a) on the relevance of relative disparity changes could therefore explain this difference. Cumming and Parker (1994) were the first to compare detection thresholds for stereomotion in dynamic RDS against thresholds in otherwise equivalent temporally correlated stimuli. Thresholds were somewhat lower for dynamic RDS than in the temporally correlated case (Figure 7). Thus sensitivity to changes in relative disparity over time is adequate to account for stereomotion detection in temporally correlated stimuli. For both types of stimulus, thresholds showed a similar dependence upon temporal frequency. Cumming

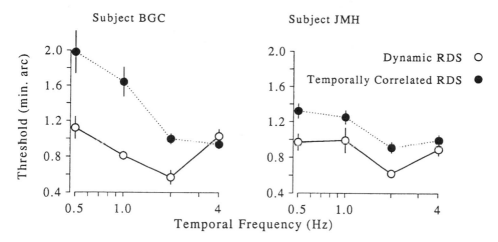

Figure 7 Stereomotion thresholds for two subjects, comparing dynamic RDS (open symbols) with temporally correlated RDS (filled symbols). The temporal correlation gave rise to clear velocity signals in each eye, which were absent from the dynamic RDS. Note that thresholds are generally *higher* for temporally correlated stimuli. Reprinted from Cumming and Parker, 'Binocular mechanisms for detecting motion in depth', Copyright 1994, with permission from Pergamon Press Ltd, Headington Hill Hall, Oxford OX3 0BW, UK.

and Parker also showed that stereomotion shares the same spatial resolution as static stereopsis (Tyler, 1983).

Thus all of the current data concerning stereomotion detection can be accounted for by a scheme (Figure 6B) based upon detecting changes in binocular disparity over time. This does not necessarily exclude the existence of a mechanism using inter-ocular velocity differences (although it is clearly superfluous). Cumming and Parker devised a stimulus to test for the existence of such a mechanism. In this RDS individual dots were shown for 1/15th of a second, during which time their disparity increased continuously. The initial disparities of individual dots were set such that the mean disparity remained constant over time. This, combined with the fact that the movements of the individual dots were beyond the temporal resolution of stereopsis (Norcia & Tyler, 1984), meant that disparity changes could not be used to detect motion-in-depth. Although each eye saw continuous motion in opposite directions, no motion-in-depth was seen. This suggests that inter-ocular velocity differences alone do not support stereomotion. The result cannot be explained by suggesting that stimuli composed of short lifetime dots do not support stereomotion – when a modulation of mean disparity was added to the display, stereomotion was seen.

A second observation that suggests that inter-ocular velocity differences are not used to detect stereomotion comes from Regan and Beverley (1973), who showed that detection thresholds increase with the mean disparity of a display (Figure 8). Although they also showed an effect of mean disparity on binocular detection of lateral motion, this was quantitatively different from the effect on stereomotion. Regan and Beverley did not study the effects of mean disparity on static disparity discrimination. The known dependence of disparity discrimination upon mean disparity (Blakemore, 1970; Westheimer, 1979) might therefore explain the effects of mean disparity upon

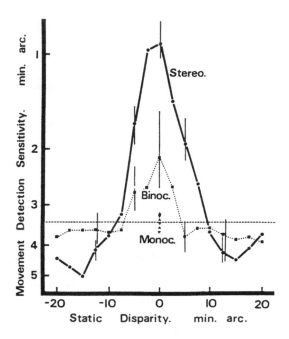

Figure 8 Stereomotion thresholds (solid line) depend upon the mean disparity about which disparity oscillates. The dashed line shows the effects of disparity upon binocular detection of sideways motion. The horizontal dotted line shows the monocular motion detection threshold. Reprinted from Regan and Beverley, 'Some dynamic features of depth perception', Copyright 1973, pp. 2369–2378, with permission from Pergamon Press Ltd, Headington Hill Hall, Oxford OX3 0BW, UK.

stereomotion. Cumming (1994) compared the effects of changes in mean disparity on three thresholds: disparity discrimination in static RDS, stereomotion in dynamic RDS, and stereomotion in temporally correlated RDS. The data, shown in Figure 9, strongly suggest that the effects of disparity upon stereomotion thresholds result from changes in disparity discrimination. This applies both to variations between subjects, and variations across conditions within one subject.

These results indicate that inter-ocular velocity differences are of limited value in *detecting* stereomotion. This does not rule out the possibility that they have some useful role for suprathreshold stimuli. Indeed it has recently been shown that subjects are very poor at stereomotion velocity discrimination in dynamic RDS (Harris & Watamaniuk, 1994). The possibility that inter-ocular velocity differences play a role in judging the velocity of motion-in-depth is currently under investigation.

2.5 Physiological Basis for Stereomotion

Cynader and Regan (1978) showed that there were units in the visual cortex (area 18) of the cat that showed binocular inhibition which was specific for certain 3D trajectories. In the monkey, although Poggio and Talbot (1981) found very few cells

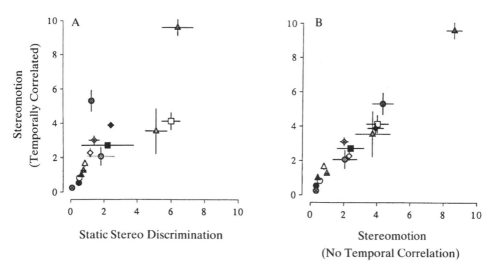

Figure 9 Effects of mean disparity upon static stereo discrimination, and stereomotion thresholds, for both temporally correlated and uncorrelated displays. Different mean disparities are shown with different symbols (circle: 0 min arc; triangle: 5 min arc; square: 20 min arc; diamond: 5 min arc + 1.5 D lenses); different subjects are shown with different fill styles. (A) plots stereomotion thresholds in temporally correlated displays against static disparity discrimination thresholds. (B) plots stereomotion thresholds in temporally correlated displays against stereomotion thresholds in temporally uncorrelated displays (i.e. it compares stimuli with disparity changes only with stimuli containing both disparity changes and inter-ocular velocity differences). Across subjects and across conditions, stereomotion thresholds without inter-ocular velocity differences are good predictors of thresholds measured with inter-ocular velocity differences.

sensitive to stereoscopic motion-in-depth, they found both facilitation and inhibition. One difficulty with these and other early studies (Toyama et al., 1985) is that they explored the effects of trajectory always around a single mean disparity. If one considers the binocular receptive field of a disparity selective cell as a three-dimensional volume, then some trajectories will spend more time in that volume than others. Therefore, the appearance of tuning to three-dimensional trajectories could merely be a result of the tuning to static disparities. If this were the case, one would expect the preferred trajectory to depend upon the mean disparity, which is exactly what was observed by Maunsell and van Essen (1985), recording in the middle temporal (MT) area. They found no cells with a trajectory tuning which was constant across changes in disparity. Regan and Cynader (1982) showed some examples of cells whose trajectory tuning was roughly constant with changes in disparity, but no systematic quantitative analysis of this effect was performed. The most thorough study was performed by Spileers et al. (1990), recording from area 18 in the cat. They found a small proportion of cells which responded to stereomotion (7/42) whose trajectory tuning was unaffected by changes in disparity or in velocity. An example is shown in Figure 10. Interestingly these cells all showed preferred trajectories close to the axes passing through one or other eye. On monocular testing all of the cells were strongly dominated by the inputs from one eye, and showed a

UNIT 9505

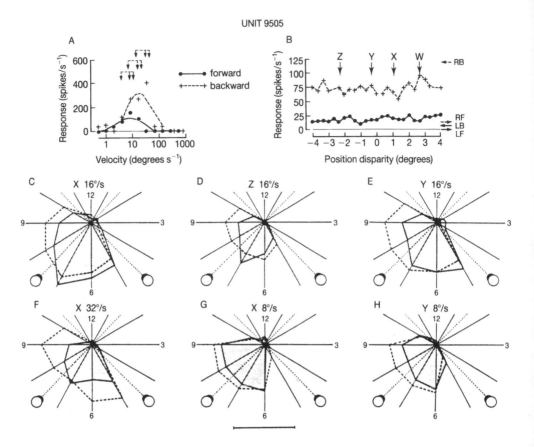

Figure 10 Responses of a cell from area 18 of a cat, selective for stereomotion, independent of disparity and velocity. (A). Monocular velocity selectivity for the preferred direction (dashed line) and the non-preferred direction (solid line). (B). Tuning for static disparities for each direction of motion. Arrows on the right-hand side show monocular responses to each direction of motion. Since the dashed line shows responses always lower than for the right eye monocular motion in the same direction (arrow RB), this cell shows binocular inhibition over a wide range of disparities. C, D and E show the trajectory tuning measured at three different mean disparities. F, G and H show the tuning function at different velocities. For each trajectory tested, the magnitude of the cell response is represented by the radial distance from the centre. The responses predicted on the basis of the monocular velocity sensitivity of the dominant eye alone are shown with dashed lines. Reproduced with permission from Spileers *et al.* (1990).

broad tuning for static disparities. As Spileers *et al.* point out, these properties largely account for the 3D trajectory specificity. Consider the cell shown in Figure 10 which is strongly dominated by the right eye, and shows binocular inhibition (the response is smaller when both eyes are stimulated; see Figure 10B). Trajectories moving towards the left eye produce the highest velocity stimulus to the right eye (strong activation), and zero velocity stimulation to the left eye (minimal inhibition). Thus the preferred trajectory is towards the left eye. In most cases the binocular interactions have relatively little effect on trajectory tuning (compare the dashed lines with the solid lines in parts C–H of Figure 10), and it is not clear how consistent this binocular interaction is across different disparities and velocities. So although these cells may be the physiological substrate of the psychophysical channels that respond to stereomotion (Figure 5), it is not clear that this represents a specialized mechanism, distinct from static disparity processing.

One problem in relating the physiological studies to psychophysics comes from the distinction between absolute and relative disparities. As noted above, changes in absolute disparity alone are insufficient to produce the sensation of motion-in-depth. The physiological studies cited here did not present a stimulus relative to a fixed background. (Although the surrounding laboratory may have been visible, most studies used a dichoptic viewing arrangement which resulted in the backgrounds seen by the two eyes being non-corresponding.) The question of whether cortical units respond to relative or absolute disparities has not been studied explicitly, but the fact that their disparity responses can be explained in terms of their monocular receptive fields (Ferster, 1981; DeAngelis et al., 1991) suggests that they respond to absolute disparities. If this is true, the stimuli which activate these cells are not sufficient to produce the sensation of motion-in-depth in human psychophysical observers.

All of these physiological studies have used stimuli that contained both inter-ocular velocity differences and changes in disparity over time, and so they cannot distinguish the two schemes shown in Figure 6. However, Poggio and Talbot (1981) showed that a very small proportion of cells (7/245) had opposite direction selectivities when tested monocularly in each eye. A very small number of such cells were also found in the work of Spileers et al. (1990). Two of the cells reported by Poggio and Talbot showed opposite direction selectivity for *vertical* motion, which is not a property that is useful for detecting stereomotion. The role of these cells for stereomotion detection is therefore unclear. In monkey MT, Zeki (1974) reported cells that had opposite direction selectivity in the two eyes, suggesting that these could be useful for detecting motion-in-depth. However, Maunsell and van Essen (1985) found no cells in MT that were truly selective for 3D trajectories. Thus the demonstration that a binocular cell has opposite direction selectivity when tested with each eye monocularly cannot be taken by itself as evidence for selectivity to motion-in-depth.

2.6 Summary

The human visual system seems to have a specialized mechanism for detecting stereomotion. This conclusion is supported by adaptation data and observations on stereomotion blindness. This mechanism probably depends upon changes in binocular disparity over time, rather than registering inter-ocular velocity differences. These disparity changes must be changes in *relative* disparity between two points. Although it is possible that the disparity detectors used in stereomotion are different from those used for static stereopsis, there is no conclusive evidence to support such an idea. Stereomotion shares the same dependency upon spatial frequency, temporal frequency, and mean disparity as static stereopsis. Physiological studies have revealed some cells that are selectively activated by certain trajectories in 3D space. Although these may form the basis for the psychophysical stereomotion channels, some properties of the psychophysical channels have yet to be demonstrated in single neurons.

Human stereomotion has rather poor temporal resolution when tested with sinusoidal oscillations. Results using brief pulses suggest that it may perform somewhat better under these circumstances, which could be important for real-world stimuli. An important unanswered question is: how is information from stereomotion converted into a geometrical representation about 3D trajectories? Objects at different viewing distances that give rise to the same patterns of disparity change and lateral retinal motion correspond to very different trajectories. Therefore binocular visual information alone is not sufficient to decide where to place a hand in order, for example, to catch a ball.

3   MONOCULAR DETECTION OF MOTION-IN-DEPTH

A number of properties of monocular images are useful indicators of 3D surface layout. Examples include the effects of texture variation, changes in shading (including specularities) and the shape of occluding contours. In spite of this wide range of available cues, most work on monocular motion-in-depth has concentrated on one cue: the effect of changing size. When an object moves towards an observer its retinal image size increases. Of course this image change could result from changes in the object's physical size, but such image changes give rise to a powerful sensation of motion-in-depth (Wheatstone, 1852).

Several lines of evidence suggest that the visual system contains a specialized mechanism for detecting motion-in-depth from image size changes. Regan and Beverley (1978b) showed that thresholds for detecting looming squares were considerably elevated by periods of adaptation. Following such adaptation there is also a motion-in-depth aftereffect (Regan & Beverley, 1978a). They controlled carefully for the possibility that this was merely a result of adaptation to the lateral motion of the edges of the square, by measuring thresholds for diagonal motion of the squares (Figure 11). The two test stimuli contain the same number of edges moving in each direction, so that the only difference is the phase relationship between the motion of different edges, yet the thresholds for lateral motion were much less elevated than those for motion-in-depth. Because oscillating motion was used, the suggestion that the adaptation is merely the result of local sideways motion adaptation (Simpson, 1993) does not explain the result. When the adapting stimulus moved sideways similar (small) threshold elevations were reported for both test stimuli. This suggests that lateral motion and motion-in-depth share some early processing.

3.1   Looming Objects vs. Size Changes

Although changing image size can give rise to a sensation of motion-in-depth, there is probably a separate channel that detects changing size alone. Two pieces of evidence suggest this.

1   At temporal frequencies between 4 and 16 Hz a changing size stimulus is seen to change in size, but does not appear to move in depth (Regan & Beverley, 1979).

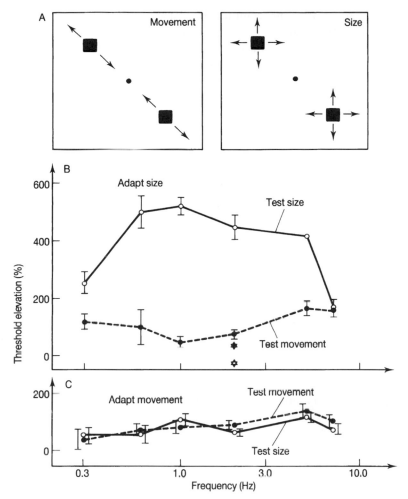

Figure 11   Adaptation to monocular motion-in-depth. A. The stimuli used for lateral motion and expanding motion. B. The effect of adapting to expanding motion on thresholds for motion-in-depth ('test size') and lateral motion ('test movement'). C. The effects of adaptation to lateral movement. Reprinted from Regan and Beverley, 'Looming detectors in the human visual pathway', Copyright 1978, pp. 415–421, with permission from Pergamon Press Ltd, Headington Hill Hall, Oxford OX3 0BW, UK.

2   The time course of recovery following adaptation is different for the sensation of changing size than for motion-in-depth (Beverley & Regan, 1979). The adapting stimulus in this experiment only changed its width. Immediately after the adapting period a stationary square appeared to be changing size, but after a few seconds it appeared to be of fixed size, moving in depth.

The fact that the same stimulus can cause adaptation of both changing size sensation and motion-in-depth suggests that these mechanisms must also share some processing at earlier stages. Given that changes in object size or motion-in-depth can produce similar image changes, it is important for the visual system to try and distinguish these possibilities. One useful approach is to consider whether a pattern of optic flow is compatible with a rigid object moving in depth. For a rigid object approaching an observer without undergoing any rotation, a simple rule (Beverley & Regan, 1980) describes the relative motion of image features: the relative velocity of any two features divided by their separation is a constant. (This rule assumes that the size of the object in depth is small relative to the viewing distance.) Beverley and Regan (1980) suggested that human looming detectors exploited this property to make their responses specific to looming of rigid objects. They showed that the effects of adaptation to oscillating square and rectangular patches were much more specific for stimulus size when this simple geometrical rule was obeyed. For example, when the adapting stimulus was a rectangle which was half as high as it was wide, the most specific adaptation was observed when the velocity of the horizontal and vertical edges were in a ratio of 1 : 2. The simple rule of Beverley and Regan (1980) does not hold for rotating objects. Nonetheless it is well established that it is possible to determine whether or not a set of moving points correspond to a rigid object (Ullman, 1979; Longuet-Higgins & Pradzny, 1981). Even with complex patterns of optic flow, it is straightforward to calculate an expansion component (Koenderink & van Doorn, 1975). How effectively human subjects extract motion-in-depth from such complex patterns of optic flow has not been investigated experimentally. It is possible that catching an American football may require different visual skills than catching a baseball!

3.2 Monocular Judgement of Trajectory

Two pieces of evidence suggest that the mechanism for detecting looming is not subdivided into different channels sensitive to the magnitude of lateral motion (unlike the channels proposed by Beverley and Regan (1973) for stereomotion).

1 Regan and Beverley (1980) showed that adaptation to a range of different trajectories produced similar threshold elevations for detecting looming stimuli without lateral motion. The adapting trajectories were all chosen so that they contained the same amplitude of looming. In order to demonstrate this result, they used stimuli containing a small, high-frequency lateral motion. This was to eliminate the effects of a non-linear interaction seen when the velocity of any single edge approached zero.
2 Regan and Kaushal (1994) measured subjects' ability to discriminate different trajectories of motion specified by combinations of changing size and lateral motion. The simulated velocities in depth were varied at random so that neither sideways velocity nor rate of expansion could be used alone to discriminate trajectories. The discrimination performance showed little dependence upon the mean trajectory about which it was performed. If there were channels tuned for particular trajectories, then one would expect peaks in the discrimination performance at the places where the sensitivity of the channels change most rapidly.

For an object moving in depth, it is possible to calculate the angle between the trajectory and the line of sight, by combining the image expansion rate and the translation rate (Peper *et al.*, 1994; Regan & Kaushal, 1994). This relationship is independent of viewing distance or translation speed. This is therefore quite different from the information provided by binocular mechanisms about 3D trajectories, which does depend upon the viewing distance (see section 2.1 above).

3.3   Using Monocular Expansion

The preceding section demonstrates that the human visual system has developed special mechanisms to deal with detecting images that expand on the retina. This allows the movement of objects in depth to be detected. But in order to interact successfully with the environment (to avoid an oncoming missile, or catch a ball), it would be useful to know more than the fact that an object is moving. It is not possible to calculate the *velocity* of an object moving in depth from the monocular optic flow alone: any flow field produced by one object could be replicated exactly by an object of twice the size, at twice the distance, moving with twice the velocity.

Fortunately, it is possible to extract useful information about object motion without knowing its velocity and distance. Hoyle (1957), in a science fiction novel, pointed out that for an object moving along the line of sight

$$T_o = \frac{\theta}{\dot{\theta}} \qquad (1)$$

where T_0 is the time it will take the object to reach the observer, θ is its angular subtense, and $\dot{\theta}$ is the rate of change of θ. This assumes that the object is rigid and not rotating.

Lee (1974) was the first to point out the potential usefulness of this information in human vision. He refers to the ratio given in equation 1 as τ. He showed that this gives a reliable estimate of time-to-contact both when approaching a frontoparallel surface and when approaching an inclined surface at constant velocity. Lee (1976) extended this theory to show that monitoring the rate of change of τ can be useful for determining braking strategy. If $\dot{\tau}$ is kept constant at a value < -0.5 then velocity smoothly decreases to zero before a collision occurs (deceleration is constant). Lee (1976) showed that the braking behaviour of a human subject while driving corresponded to a nearly constant value of $\dot{\tau}$ of -0.425. This was recently investigated using computer-generated stimuli by Yilmaz and Warren (1994) who randomized target size, and manipulated some of the available distance cues. Across all of their conditions, braking behaviour was well described as an attempt to keep $\dot{\tau}$ below a value of approximately -0.45.

A number of studies have shown that the timing of certain motor actions can be accounted for by suggesting that the response is initiated when τ reaches some critical parameter. Lee *et al.* (1982) showed that long jumpers control their approach by changing the timing of their strides. Warren *et al.* (1986) showed that a similar strategy was used to control walking on a treadmill with an uneven surface. Other studies have explored the usefulness of τ in driving (Macleod & Ross, 1983; Cavello & Laurent, 1988), and table tennis (Bootsma & van Wieringen, 1992). Wagner (1982) showed that, when landing, flies initiated their deceleration when the relative rate of image expansion reached a critical value.

The difficulty with all of these studies using natural stimulation is that they do not exclude a number of other explanations for how the timing is achieved. Most studies used binocular viewing, so any claim that responses are based upon monocular optic flow must be treated with caution. In studies where the size of the object is known to the subject (which again covers all of the studies above) it is particularly straight-forward to estimate time-to-contact from absolute values of image expansion. One recent study examined catching of a luminous ball whose size changed as it approached the subject (Savelsbergh *et al.*, 1991). The time-to-contact specified by optic flow was then roughly double that of fixed size balls used in the study. As shown in Figure 12, the trajectory of the finger grasping movement was intermediate between that seen for a small ball and a large ball. This movement is clearly not merely a time-delayed equivalent of the responses to either rigid ball, although the time at which the grasping velocity was maximal was significantly delayed. This indicates that factors other than τ must have contributed to the control of the response. Unfortunately, the balls always started from the same location, and moved with the same velocity so that the responses may not only depend upon the optic flow stimulation. Nonetheless this result is important: the logic of other studies has been to show that behaviour is compatible with the use of τ. This is not a strong test unless all other cues are removed (which they have not been). The study of Savelsbergh *et al.* shows that when τ is misleading, other cues still provide adequate information.

Two studies using natural stimulation have examined circumstances where the target is not moving at a constant velocity. This is a useful test, since if subjects assume a constant velocity, and initiate actions at some critical value of τ, the timing

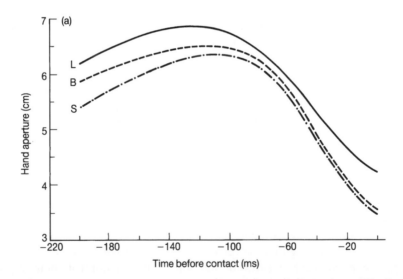

Figure 12   The separation of the fingers as a function of time when subjects attempt to catch a ball with a hand whose wrist is fixed. Experiments used luminous balls in complete darkness. The two solid lines show the responses to a large (L) ball and a small (S) ball. The dotted line shows the response to a ball which deflated as it approached, starting the same size as the large ball, and finishing the same size as the small ball. This deflation approximately doubled the time-to-contact specified by τ. Reproduced with permission from Savelsbergh *et al.*, 1991.

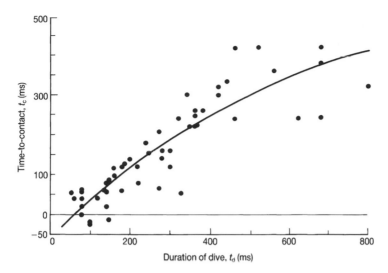

Figure 13   The time at which gannets streamline their wings when diving into the sea, relative to the moment of contact. This is plotted as a function of dive duration. For longer dives, birds approach at higher velocity, and close the wings earlier. The solid line shows predictions based upon using a particular value of τ (820 ms). τ and time-to-contact are not linearly related because the animal accelerates continuously. Reprinted with permission from Lee and Reddish (1981) *Nature*, **293**, 293–294. Copyright Macmillan Magazines Limited.

of the actions will not occur at any fixed value of distance, image size, or image expansion rate. Human subjects, asked to hit a ball which fell from different heights, appear to initiate action at a critical value of τ (Lee *et al.*, 1983). Similarly, Lee and Reddish (1981) showed that when diving into the sea, gannets streamline their wings when τ reaches a critical value (τ_m), independent of the height at which the dive starts (Figure 13). They fitted the relationship between τ_m and the actual time-to-contact with the following equation:

$$t_c = \tau_m + t_d - \sqrt{\tau m^2 + t_d^2} - t_i \tag{2}$$

where t_d is the actual duration of the dive, and t_i is a fixed delay after the critical value of τ is reached before action is initiated. When $\tau_m > t_d$, t_c is more dependent upon t_d than τm, so under these circumstances the data do not provide a good test of the hypothesis that τm is the critical parameter. In fact the value of τm used (820 ms) was larger than the duration of any of the dives recorded. For the majority of dives (those lasting less than 500 ms) the data are well described by a straight line. This allows a simple alternative description of these data: for dives lasting less than 500 ms, the birds streamline their wings 100 ms after the dive begins. For dives lasting longer than 500 ms, the birds close their wings 300–400 ms before time-to-contact (which they estimate taking acceleration into account).

In summary, the optical variable τ is a useful source of information about time-to-contact, and it is sufficient to explain a numer of observations on the timing of motor action. It has the attraction of offering a very simple explanation of these phenomena, but none of the experimental studies demonstrates that other strategies are *not* used.

How information specified by τ is combined with other information relating to distance and velocity requires further investigation.

3.4 Estimates of Sensitivity to τ

A number of psychophysical studies have attempted to show that humans are at least sensitive to τ independent of covarying parameters such as stimulus size. Schiff and Detweiler (1979) asked subjects to judge the time-to-contact when viewing films of approaching black squares. They found that the judged time-to-contact did not vary with image size, but they also found that subjects did underestimate time-to-contact. This underestimate was greater for larger approach velocities. Similar underestimates were reported by Macleod and Ross (1983) and Cavello and Laurent (1988).

  More compelling evidence that humans are sensitive to τ was provided by Todd (1981) who used an interval forced choice task, in which subjects reported which of two stimuli would reach them first. Subjects were able to do this despite large changes in image size between the intervals of each trial. The data suggest a discrimination threshold of 1–2%. The most thorough study was performed by Regan and Hamstra (1993). They used carefully selected pairings of image size and expansion with eight different values of time-to-contact, and eight different values of initial expansion. All 64 combinations of these two stimuli were presented in two different experiments. In the first experiment subjects were asked to judge whether stimuli would hit them earlier or later than the mean of the set. Auditory feedback was given on the basis of time-to-contact. When all the data stimuli specifying the same time-to-contact are pooled (Figure 14) a psychometric function is obtained. When the data are pooled

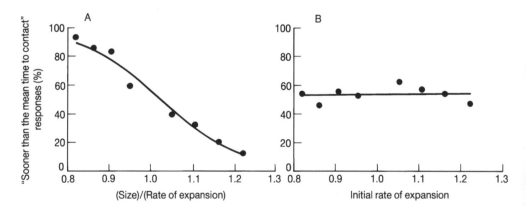

Figure 14 Effects of relative and absolute rates of expansion on judgements of time-to-contact. Eight different values of relative expansion rate were combined with eight values of absolute expansion rate, giving 64 stimuli. When stimuli with the same relative expansion rate are pooled (A), a smooth psychometric function is obtained. When stimuli with the same absolute expansion rate are pooled (B), little effect is seen. Reprinted from Regan and Hamstra, 'Dissociation of discrimination thresholds for time to contact and for rate of angular expansion', Copyright 1993, pp. 447–462, with kind permission from Pergamon Press Ltd, Headington Hill Hall, Oxford OX3 0BW, UK.

according to initial expansion rate, there is little correlation with judged time-to-contact. This indicates that subjects rely on relative expansion rates ($\dot{\theta}/\theta$) rather than absolute expansion rates ($\dot{\theta}$) when judging time-to-contact. The data on time-to-contact suggest a discrimiantion threshold of approximately 10%. One reason that this is larger than the figure suggested by the data of Todd (1981) is that subjects had to make an absolute judgement of time-to-contact in each trial, rather than judging which of two stimuli would arrive sooner.

3.5   Limitations on the Use of τ

Given that the visual system is clearly able to detect relative rates of image expansion, and the wide range of tasks where humans and other animals use visual information to control timing of a response, it is natural to suggest that these are done by calculating τ to estimate time-to-contact. However, as pointed out by Tresilian (1991) none of the experimental observations distinguish this scheme from one in which subjects estimate distance and velocity explicitly. Even if distance were mis-estimated, when velocity is divided by distance the errors would cancel so that time-to-contact would be calculated correctly (this is simply a result of the same geometry which makes equation 1 true). For this reason it is very difficult to determine experimentally whether human subjects use τ to estimate time-to-contact directly. Furthermore, none of the experiments cited above excludes the possibility that other strategies are used for timing the actions studied: they all demonstrate only that the responses are compatible with the use of τ.

Certain actions are in principle difficult to control on the basis of τ alone, because the visual variable specifies time-to-contact *with the eye*. Many human actions involve using the hands at some distance from the eye, so that timing based upon τ could be quite inappropriate. Consider a baseball moving a 8 m/s towards a fielder whose hands are 50 cm in front of the eyes (Tresilian, 1991). The estimate of time-to-contact at the eye is 62 ms longer than the real time-to-contact to the hands. This is too large an error to permit skilled catching. It is not possible to deal with this problem simply by subtracting a constant value from τ. Consider two objects whose size, distance and approach velocity are in some fixed ratio. They therefore have the same value of τ throughout their approach, and yet are always at different distances from the observer. The value of τ at the moment the ball reaches the hands will therefore be different for the two objects. There is a similar problem in using τ to control braking in a car to avoid hitting an obstacle. This is a strategy that brings the car to a halt just as the *eye* reaches the obstacle, by which time the front of the car will be damaged.

A related problem arises when trying to intercept an object which is not moving directly towards the head. Tresilian (1990) extended the original analysis to show that it is possible to calculate the time at which the object reaches any point, p, on a trajectory which is not on a collision course with the eye. Thus is it possible to know the time at which a ball will pass any chosen plane, and time catching movements appropriately. However, it is not possible to know *how far* from the head (within the plane) the ball will pass, thus it is still necessary to use some information about distance in order to catch it. Although it is possible to derive an expression for the lateral distance between the object and the observer in this circumstance (Bootsma,

1991; Bootsma & Peper, 1992; Regan & Kaushal, 1994), the distance is expressed in terms of multiples of the object's physical width. In order to put the arm in an appropriate location an observer must know the size of the approaching object. The effect of known size was explored by Peper *et al.* (1994), when subjects judged whether or not a ball passed within arm's reach. For balls whose size was different from expected, subjects systematically misestimated the lateral passing distance. This suggests that subjects used knowledge of ball size to make trajectory judgements. Once the size of an object is known, it is possible to calculate the viewing distance and hence the rest of the viewing geometry. Given knowledge of the size of a moving object, many different strategies are possible to coordinate interceptive action.

Both of these important limitations in the use of τ arise from the fact that human hands are some distance from the eyes. Any monocular pattern of optic flow produced by a moving object can be reproduced exactly by another object whose distance, size and velocity are appropriately scaled. Since the length of the arm does not scale in the same way, it is necessary to make a judgement about absolute distance to intercept objects with the hands. It may be that for animals who catch things in their mouths, τ is more widely applicable (e.g. gannets – Lee & Reddish, 1981).

3.6 Physiology of Looming Detection

Regan and Cynader (1979) looked for units sensitive to changes in the size of retinal stimuli in area 18 of the cat. Although they identified many units which responded to such stimuli, selective responses were uncommon. Many of the responses they found were explained by changes in total light flux in the receptive field. Another large group changed their properties depending upon the location of the stimulus in the receptive field, suggesting that these were really direction-selective responses to individual edges of the stimulus. Only one cell in the entire sample could not be accounted for in either of these ways.

Units selectively responsive to image expansion have not been reported in the primary visual cortex of monkeys (V1). In the dorsomedial part of the medial superior temporal area (MSTd), however, a number of studies have shown responses that are selective for optic expanding flow fields (Tanaka & Saito, 1989; Tanaka *et al.*, 1989; Duffy & Wurtz, 1991a, 1991b; Orban *et al.*, 1992). Other cells were responsive to rotating or translating fields. These studies all used random dot patterns which avoid some of the problems discussed by Cynader and Regan concerned with changes in light flux. Because they systematically quantified responses to all three stimulus types, Duffy and Wurtz (1991a) and Orban *et al.* (1992) were able to show that many cells responded to more than one of these flow patterns. These two groups both demonstrated that many of these cells responded to the same flow pattern irrespective of retinal position. This establishes that they are really responding to the pattern of flow, not merely responding to a single component in one location. The importance of performing this control is illustrated by the fact that cells in MT that seem to respond to expansion when tested in one location, respond to other flow patterns when tested at other locations (Orban *et al.*, 1992). These studies are discussed in more detail in the chapter by Snowden (Chapter 3).

A number of the reported properties of cells in MSTd suggest that they are primarily concerned with responding to self-motion rather than to motion of objects towards the animal.

1   They have large receptive fields, and respond less well to smaller stimuli (subtending less than 20°). Since psychophysical channels are sensitive to much smaller stimuli, these units are unlikey to be the substrate for such channels.

2   Their responses are determined by the relative *direction* of the flow field components, and are relatively insensitive to the *amplitude* of these components (Tanaka *et al.*, 1989). This is appropriate for sensing egomotion, which is determined by the directions of motion vectors. Variations in the amplitudes of these vectors reflect

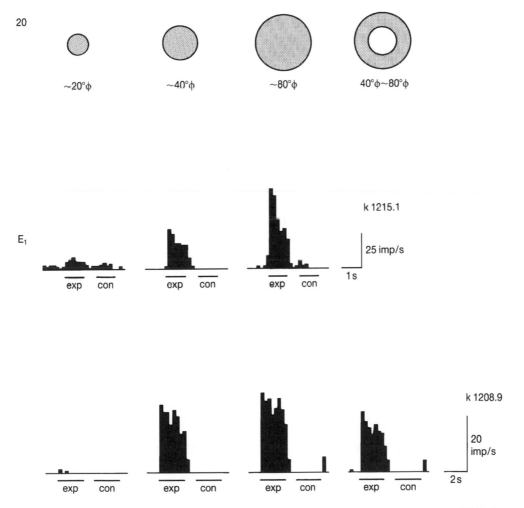

Figure 15   Effects of varying field size on responses of two cells in MSTd (E1 and E2) that responded to expansion flow fields (Tanaka *et al.*, 1989). The top row shows the stimulus configurations, with the unit responses shown beneath. Reproduced with permission from Tanaka *et al.* (1989).

the 3D structure of the environment through which the subject moves (see Chapter 11 by M. G. Harris). Conversely, when a single rigid object moves towards an observer, both the amplitudes and directions of the motion vectors have a specific pattern. Psychophysical channels are most selective for stimuli that conform to this pattern (Beverley & Regan, 1980).

Cells in monkey MSTd that respond to expansion/contraction do not seem to be responding to particular values of τ. Figure 15 shows that such cells are quite broadly tuned for both stimulus size and velocity (Tanaka & Saito, 1989; Duffy & Wurtz, 1991a, 1991b). Tanaka and Saito (1989) showed explicitly that it was not possible to trade-off velocity and stimulus size in defining an optimum stimulus for these cells. In the pigeon, it has recently been reported (Wang & Frost, 1992) that cells in the nucleus rotundus fire vigorously when an expanding stimulus reaches a critical value of τ, independent of image size or velocity alone. The stimuli used were checker patterns, not random dot patterns, so changes in illumination could have contributed to some of these effects. As noted above, τ may be more valuable to birds than primates, and cells reported in the primate that respond to expanding fields appear not to be sensitive to τ. It remains to be seen whether other parts of the primate visual system contain cells responding to τ.

3.7 Summary

Psychophysical evidence strongly suggests that the human visual system has specialized mechanisms that interpret expanding retinal images as motion-in-depth. This mechanism is separate from the system that detects lateral motion, although it may share some early processing. The visual system is sensitive to the relative rate of image expansion, independent of the absolute expansion rate. This parameter can be used to estimate the time-to-contact of an approaching object. Although a number of features of human behaviour can be explained assuming that this information is used without an explicit representation of velocity and distance, there has been no clear demonstration that other strategies are not being employed in these tasks. No neurophysiological correlates of τ have been identified in primates, but neither have any responses that would be specific for the approach of a small object. Performance at a number of tasks (catching a ball in front or to the side of the eyes) cannot be explained simply in this way. Further experimental work is required in order to understand how these feats are achieved.

4 COMBINING MONOCULAR AND BINOCULAR INFORMATION

Most of the experimental work on the perception of motion-in-depth has concentrated on monocular *or* binocular mechanisms. Regan and Beverley (1979) pointed out that, in principle, binocular and monocular information could be combined to calculate the physical size of an object moving in depth. This is illustrated in Figure 16: the rate of change of angular subtense, $\dot{\theta}_s$, divided by the rate of change of disparity, $\dot{\theta}_d$, equals the ratio of the object's size, S, to the subjects interocular separation, I:

$$\dot{\theta}_s / \dot{\theta}_s = S/I. \tag{3}$$

Once the size is known, the rest of the viewing geometry can be calculated, so both the viewing distance and the object's real world trajectory can be calculated.

An idea which is similar in principle was proposed by Richards (1985) who showed theoretically that structure-from-motion (see Chapter 13 by Braunstein) could be combined with stereo to compute metric shape. This idea has recently been tested psychophysically to Johnston *et al.* (1994), who found that motion and stereo did combine to give veridical shape perception. There have been few experimental tests of whether stereo and motion-in-depth combine psychophysically to yield veridical velocity and trajectory percepts. Peper *et al.* (1994) found that subjects misestimated the location of a moving ball, in a way that depended systematically upon the trajectory of motion. These misestimates were observed even during binocular viewing, suggesting that at least the combination of stereo and motion is imperfect.

Few studies have explored the effects of disrupting the normal relationship between monocular and binocular cues on motion-in-depth perception. The use of deflating balls discussed above (Savelsbergh *et al.*, 1991) certainly confirms that disparity changes play an important role even in the presence of monocular optic flow: the mistiming of catching movements was much smaller when viewing was binocular. Judge and Bradford (1988) studied the effect of telestereoscopic viewing on ball catching. Telestereoscopes effectively increase the inter-ocular separation, increasing all of the disparities in a scene. The effect was to make subjects close their hands too early, before the ball reached the hand. This is the effect that would be predicted by equation 3: if $\dot{\theta}_d$ is doubled, the calculated size (and therefore viewing distance) is halved. However, Judge and Bradford offer several other schemes that could explain the data, such as one assuming that the world appeared 'scaled down' as a result of increased convergence. Both of these explanations predict the same misreaching, although the observed effect was slightly smaller than this prediction. More experimental work is required to understand how binocular and monocular cues are combined.

Regan and Beverley (1979) showed that if the stereomotion and changing size signalled opposite directions the sensation of motion-in-depth could be nulled. This suggests that both cues feed into a common motion-in-depth stage. A second piece of evidence that both mechanisms share some common stage is the fact that both changing disparity and changing size share the same limit on temporal frequency (Regan & Beverley, 1979) even though changing size (without the sensation of motion-in-depth) can be perceived at higher frequencies.

Regan and Beverley (1979) used the nulling technique to examine the relative strengths of monocular and binocular cues. Using ramp movements of different durations, they found that the relative effectiveness of disparity changes was greater for stimuli of shorter duration or higher velocity. Whether binocular or monocular cues are more important for everyday tasks will therefore depend upon circumstances. The viewing distance is another important parameter (in addition to stimulus size, velocity, and duration). For objects at far viewing distances, binocular mechanisms are very little use (as pointed out above, disparity scales inversely with the square of the viewing distance). Stereomotion thresholds are rarely much less than 1 min arc (see Figures 1 and 7), which for a typical human observer corresponds to the relative

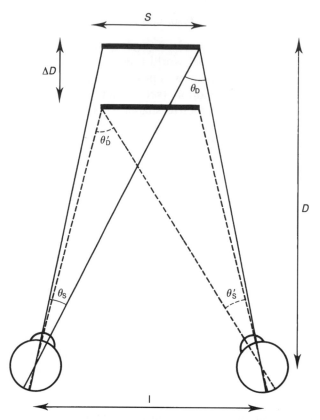

Figure 16 Illustrating how binocular information and monocular information can be combined to calculate object size. $\theta_D \approx I/D$, and $\theta_S \approx S/D$. It follows that $\dot\theta s/\theta_D = S/I$. Thus the physical size of an object can be calculated by combining disparity changes with monocular expansion cues, if the inter-ocular separation is known. Reprinted from Regan and Beverley, 'Binocular and monocular stimuli for motion in depth: changing-disparity and changing-size feed the same motion-in-depth stage', Copyright 1979, pp. 1331–1342, with permission from Pergamon Press Ltd, Headington Hill Hall, Oxford OX3 0BW, UK.

disparity between a point approximately 200 yards away and infinity. This means that stereomotion is blind to all motion-in-depth more than 200 yards away. For tasks such as landing aircraft, stereopsis is therefore unlikely to be useful (which may in part explain the success of the one-eyed pilot, Wiley Post). For closer stimuli, stereomotion is potentially more useful, so the success of the monocular cricket player the Nawab of Pataudi poses more of a challenge to our understanding of motion-in-depth processing.

4.1 Physiology

Very little is known about the combination of monocular and binocular cues physiologically. Many disparity sensitive cells are also direction selective (Poggio & Talbot,

1981). Maunsell and van Essen (1985) showed that direction selective cells in MT are also disparity selective, but no cells were found that were sensitive to the rate of change of disparity. Roy *et al.* (1992) showed that the direction selectivity of some cells in MST reverses when disparities are reversed. Judge (1990) pointed out that such responses could be used to calculate the viewing distance (in principle similar to the methods of Richards (1985) and Regan and Beverley (1979) using stereo and motion to eliminate viewing distance). However, since no responses specific for stereomotion have been reported in MST, it is unclear whether these units are relevant for motion-in-depth. Although it has been reported that some units in the Clare-Bishop area of the cat respond both to stereomotion and to changing size stimulation (Toyama *et al.*, 1985), this study did not control for the possibility that stereomotion responses were a result of static disparity tuning (see section 4.1). Responses of cells in monkey MSTd (where cells selective for expanding stimuli are found) to disparity changes have not been reported, so it remains possible that further study in this area may yield insights into the mechanisms for combining monocular and binocular information about motion-in-depth.

5   CONCLUDING REMARKS

The combination of mathematical analysis with psychophysical and neurophysio-logical experiments has demonstrated the existence of mechanisms in the primate brain specialized for detecting objects moving in depth. There appear to be two separate mechanisms, one based upon binocular disparities, the other based upon the effects of changes in retinal image size. These mechanisms seem to be separate (at least to some extent) from the processing of lateral motion. Since these mechanisms analyse motion along trajectories that are at 90° to one another, the rules for combining them to perceive arbitrary 3D motion may be quite straightforward. However, neither binocular nor monocular cues alone permit a complete reconstruction of 3D traject-ories. Changes in binocular disparity can only be used to calculate the trajectory in space if some information about viewing distance (e.g. from vergence eye position) is used. Although monocular optic flow contains much information about trajectory and time-to-contact, it is not sufficient alone to determine when and where the hand should be placed to intercept an object moving in depth.

Consequently, it remains true that we cannot give a full account of how sportsmen achieve such extraordinary precision when catching or striking a ball. It may be that they use more information than is usually available in an experimental setting (such as the knowledge of familiar size or distance), or that the combination of binocular and monocular cues is important. How these various cues are combined requires further study.

For both monocular and binocular mechanisms, there is considerable uncertainty concerning the relationship between neurophysiological observations on single cells and the performance of humans in psychophysical tasks. In the case of stereomotion, cells that respond to changes in relative disparity, without responding to changes in absolute disparity, have not been demonstrated. Nor have cells that are sensitive to changes in disparity in the absence of monocular motion signals. There is no clear physiological substrate in primates for the human sensitivity to the optical expansion produced by small approaching objects. Despite many successes, there is still a large gap between our understanding of what strategies are available for the visual detection of objects moving in depth, and what physiological mechanisms underlie them.

REFERENCES

Beverley, K. I. & Regan, D. (1973). Evidence for the existence of neural mechanisms selectively sensitive to the direction of movement in space. *J. Physiol. Lond.*, **235**, 17–29.

Beverley, K. I. & Regan, D. (1974). Visual sensitivity to disparity pulses: evidence for direction selectivity. *Vision Res.*, **14**, 357–361.

Beverley, K. I. & Regan, D. (1975). The relation between discrimination and sensitivity in the perception of motion in depth. *J. Physiol. Lond.*, **249**, 387–398.

Beverley, K. I. & Regan, D. (1979). Separable aftereffects of changing-size and motion-in-depth: Different neural mechanisms? *Vision Res.*, **19**, 727–732.

Beverley, K. I. & Regan, D. (1980). Visual sensitivity to the shape and size of a moving object: implications for models of object perception. *Perception*, **9**, 151–160.

Blakemore, C. M. (1970). The range and scope of binocular depth discrimination. *J. Physiol. Lond.*, **211**, 599–622.

Bootsma, R. J. (1991). Predictive information and the control of action. *Int. J. Sport Psychol.*, **22**, 271–278.

Bootsma, R. J. & Peper, C. E. (1992). Predictive visual information sources for the regulation of action with special emphasis on catching and hitting. In L. Proteau & D. Elliot (Eds.) *Vision and Motor Control*, pp. 189–199. North-Holland, Amsterdam.

Bootsma, R. J. and van Wieringen, P. C. W. (1990). Timing an attacking forehand drive in table tennis. *J. Exp. Psychol.: Human Percept. Perform.*, **16**, 21–29.

Cavello, V. and Laurent, M. (1988). Visual information and skill level in time-to-collision estimates. *Perception*, **17**, 623–632.

Cumming, B. G. (1994). The relationship between stereoacuity and stereomotion thresholds. *Perception* (in press).

Cumming, B. G. & Parker, A. J. (1994). Binocular mechanisms for detecting motion-in-depth. *Vision Res.*, **34**, 483–496.

Cynader, M. & Regan, D. (1978). Neurons in cat parastriate cortex sensitive to the direction of motion in three-dimensional space. *J. Physiol. Lond.*, **274**, 549–569.

Cynader, M. & Regan, D. (1982). Neurons in cat visual cortex tuned to the direction of motion in depth: Effect of positional disparity. *Vision Res.*, **22**, 967–982.

DeAngelis, G., Ohzawa, I. & Freeman, R. D. (1991). Depth is encoded in the visual cortex by a specialized receptive field structure. *Nature*, **352**, 156–159.

Duffy, C. J. & Wurtz, R. H. (1991a). Sensitivity of MST neurons to optic flow stimuli. I. A continuum of response selectivity to large-field stimuli. *J. Neurophysiol.*, **65**, 1329–1345.

Duffy, C. J. & Wurtz, R. H. (1991b). Sensitivity of MST neurons to optic flow stimuli. II. Mechanisms of response selectivity revealed by small-field stimuli. *J. Neurophysiol.*, **65**, 1346–1359.

Erkelens, C. J. & Collewijn, H. (1985a). Eye movements and stereopsis during dichoptic viewing of moving random-dot stereograms. *Vision Res.*, **25**, 1689–1700.

Erkelens, C. J. & Collewijn, H. (1985b). Motion perception during dichoptic viewing of moving random-dot stereograms. *Vision Res.*, **25**, 583–588.

Ferster, D. (1981). A comparison of binocular depth mechanisms in areas 17 and 18 of cat visual cortex. *J. Physiol. Lond.*, **11**, 623–655.

Foley, J. (1991). Binocular distance perception. In D. Regan (Ed.) *Binocular Vision*, volume 9 of *Vision and Visual Dysfunction*, pp. 75–92. Macmillan, Basingstoke.

Harris, J. M. & Watamaniuk, S. N. J. (1994). Speed discrimination of binocular motion in depth. *Invest. Ophthalmol. Vis. Sci. ARVO Abstracts*, **35**, 1986.

Hong, X. & Regan, D. (1989). Visual field defects for unidirectional and oscillatory motion. *Vision Res.*, **29**, 809–819.

Hoyle, F. (1957). *The Black Cloud*. Penguin Books, London.

Johnston, E. B., Cumming, B. G. & Landy, M. S. (1994). Integration of stereopsis and motion in specifying 3-D shape. *Vision Res.* (in press).

Judge, S. J. (1990). Vision: knowing where you're going. *Nature*, **348**, 115.

Judge, S. J. & Bradford, C. M. (1988). Adaptation to telestereoscopic viewing measured by one-handed ball-catching performance. *Perception*, **17**, 783–802.

Julesz, B. (1971). *Foundations of Cyclopean Perception*. University of Chicago Press, Chicago.

Koenderink, J. J. & van Doorn, A. J. (1975). Invariant properties of the motion parallax field due to the movement of rigid bodies relative to the observer. *Optica Acta*, **22**, 773–791.

Lee, D. N. (1974). Visual information during locomotion. In *Perception: Essays in Honour of James J. Gibson*, pp. 251–267. Cornell University Press, Ithaca, NY.

Lee, D. N. (1976). A theory of visual control of braking based on information about time to collision. *Perception*, **5**, 437–459.

Lee, D. N. & Reddish, P. E. (1981). Plummeting gannets: a paradigm of ecological optics. *Nature*, **293**, 293–294.

Lee, D. N., Lishman, J. R. & Thompson, J. A. (1982). Visual regulation of gait in long jumping. *J. Exp. Psychol.: Human Percept. Perform.*, **8**, 448–459.

Lee, D. N., Young, D. S., Reddish, P. E., Lough, S. & Clayton, T. (1983). Visual timing in hitting an accelerating ball. *Q. J. Exp. Psychol.*, **35A**, 333–346.

Longuet-Higgins, H. C. & Pradzny, K. (1981). The interpretation of moving retinal images. *Proc. R. Soc. Lond. B*, **208**, 385–397.

Maunsell, J. H. R. & van Essen, D. C. (1985). Functional properties of neurons in middle temporal visual area of the macaque monkey. II. Binocular interactions and sensitivity to binocular disparity. *J. Neurophysiol.*, **9**, 1148–1166.

McLeod, R. W. & Ross, H. E. (1983). Optic flow and cognitive factors in time-to-collision estimates. *Perception*, **12**, 417–423.

Norcia, A. M. (1980). Frequency domain analysis of human stereopsis. PhD thesis, Stanford University, Palo Alto, California.

Norcia, A. M. & Tyler, C. W. (1984). Temporal frequency limits for stereoscopic apparent motion processes. *Vision Res.*, **4**, 395–401.

Orban, G. A., Lagae, L., Verri, A., Raiguel, S., Xiao, D., Maes, H. & Torre, V. (1992). First-order analysis of optical flow in monkey brain. *Proc. Natl. Acad. Sci., NY*, **89**, 2595–2599.

Peper, C. E., Bootsma, R. J., Mestre, D. R. & Bakker, F. C. (1994). Catching balls: How to get the hand to the right place at the right time. *J. Exp. Psychol.: Human Percept. Perform.* (in press).

Poggio, G. F. & Fisher, B. (1977). Binocular interactions and depth sensitivity in striate and prestriate cortex of behaving rhesus monkey. *J. Neurophysiol.*, **40**, 1392–1405.

Poggio, G. F. & Talbot, W. H. (1981). Mechanisms of static and dynamic stereopsis in foveal cortex of the rhesus monkey. *J. Physiol. Lond.*, **315**, 469–492.

Regan, D. (1991). Depth from motion and motion-in-depth. In D. Regan (Ed.) *Binocular Vision*, volume 9 of *Vision and Visual Dysfunction*, pp. 137–169. Macmillan, Basingstoke.

Regan, D. (1993). Binocular correlates of the direction of motion in depth. *Vision Res.*, **33**, 2359–2360.

Regan, D. & Beverley, K. I. (1973). Some dynamic features of depth perception. *Vision Res.*, **13**, 2369–2378.

Regan, D. & Beverley, K. I. (1978a). Illusory motion in depth: aftereffect of adaptation to changing size. *Vision Res.*, **18**, 209–212.

Regan, D. & Beverley, K. I. (1978b). Looming detectors in the human visual pathway. *Vision Res.*, **18**, 415–421.

Regan, D. & Beverley, K. I. (1979). Binocular and monocular stimuli for motion in depth: changing-disparity and changing-size feed the same motion-in-depth stage. *Vision Res.*, **19**, 1331–1342.

Regan, D. & Beverley, K. I. (1980). Visual responses to changing size and to sideways motion for different direction of motion in depth: Linearization of visual responses. *J. Opt. Soc. Am.*, **70**, 1289–1296.

Regan, D. & Cynader, M. (1979). Neurons in area 18 of cat visual cortex selectively sensitive to changing size: nonlinear interactions between responses to two edges. *Vision Res.*, **19**, 699–711.

Regan, D. & Cynader, M. (1982). Neurons in cat visual cortex tuned to the direction of motion in depth: Effect of positional disparity. *Vision Res.*, **22**, 967–982.

Regan, D. & Hamstra, S. (1993). Dissociation of discrimination thresholds for time to contact and for rate of angular expansion. *Vision Res.*, **33**, 447–462.

Regan, D. & Kaushal, S. (1994). Monocular judgement of the direction of motion in depth. *Vision Res.*, **34**, 163–177.

Regan, D., Beverley, K. I. & Cynader, M. (1979). Separate subsystems for position in depth and for motion in depth. *Proc. Royal Soc. Lond. B*, **204**, 485–501.

Regan, D., Erkelens, C. J. & Collewijn, H. (1986a). Necessary conditions for the perception of motion-in-depth. *Invest. Ophthalmol. Vis. Sci.*, **27**, 584–597.

Regan, D., Erkelens, C. J. & Collewijn, H. (1986b). Visual field defects for vergence eye movements and for stereomotion perception. *Invest. Ophthalmol. Vis. Sci.*, **27**, 806–809.

Richards, W. (1972). Response functions for sine- and square-wave modulations of disparity. *J. Physiol. Lond.*, **315**, 469–492.

Richards, W. (1985). Structure from stereo and motion. *J. Opt. Soc. Am.*, **2**, 342–349.

Richards, W. & Regan, D. (1973). A stereo field map with implications for disparity processing. *Invest. Ophthalmol. Vis. Sci.*, **2**, 904–909.

Roy, J. P., Komatsu, H. & Wurtz, R. H. (1992). Disparity sensitivity of neurons in monkey extrastriate area MST. *J. Neurosci.*, **12**, 2478–2492.

Savelsbergh, G. J. P., Whiting, H. T. A. & Bootsma, R. (1991). Grasping tau. *J. Exp. Psychol.: Human Percept. Perform.*, **17**, 315–322.

Schiff, W. & Detweiler, M. L. (1979). Information used in judging impending collision. *Perception*, **8**, 647–658.

Simpson, W. A. (1993). Optic flow and depth perception. *Spatial Vision*, **7**, 35–75.

Spileers, W., Orban, G. A., Gulyas, B. & Maes, H. (1990). Selectivity of cat area 18 neurons for direction and speed in depth. *J. Neurophysiol.*, **63**, 936–954.

Tanaka, K. & Saito, H. (1989). Analysis of motion of the visual field by direction, expansion/contraction, and rotation cells clustered in the dorsal part of the Medial Superior Temporal area of the Macaque monkey. *J. Neurophysiol.*, **62**, 626–641.

Tanaka, K., Fukada, Y. & Saito, H. (1989). Underlying mechanisms of the response specificity of expansion/contraction and rotation cells in the dorsal part of the Medial Superior Temporal area of the Macaque monkey. *J. Neurophysiol.*, **62**, 642–656.

Todd, J. T. (1981). Visual information about moving objects. *J. Exp. Psychol.: Human Percept. Perform.*, **7**, 795–810.

Toyama, K., Komatsu, Y., Kasai, H., Fujij, K. & Umetani, K. (1985). Responses of Clare-Bishop neurons to visual cues associated with motion of a visual stimulus in three-dimensional space. *Vision Res.*, **25**, 407–414.

Tresilian, J. R. (1990). Perceptual information for the timing of interceptive action. *Perception*, **19**, 223–239.

Tresilian, J. R. (1991). Empirical and theoretical issues in the perception of time to contact. *J. Exp. Psychol.: Human Percept. Perform.*, **17**, 865–876.

Tyler, C. W. (1971). Stereoscopic depth movement. Two eyes less sensitive than one. *Science*, **74**, 958–968.

Tyler, C. W. (1983). Sensory processing of binocular disparity. In C. M. Schor & K. J. Ciuffreda (Eds.) *Vergence Eye Movements: Basic and Clinical Aspects*, pp. 199–295. Butterworths, Boston.

Ullman, S. (1979). *The Interpretation of Visual Motion*. MIT Press, Cambridge, Mass.

van Nes, F. L., Koenderink, J. J. & Bouman, M. A. (1967). Spatio-temporal modulation transfer in the human eye. *J. Opt. Soc. Am.*, **7**, 1082–1085.

Wagner, H. (1982). Flow-field variable trigger landing in flies. *Nature*, **297**, 147–148.

Wang, Y. & Frost, B. J. (1992). Time to collision is signalled by neurons in the nucleus rotundus of pigeons. *Nature*, **256**, 236–238.

Warren, W. H., Young, D. S. & Lee, D. N. (1983). Visual control of step length during running over irregular terrain. *J. Exp. Psychol.: Human Percept. Perform.*, **12**, 259–266.

Westheimer, G. (1979). Cooperative neural processes involved in stereoscopic acuity. *Exp. Brain Res.*, **36**, 585–597.

Wheatstone, C. (1838). Contributions to the physiology of vision I.: On some remarkable, and hitherto unobserved, phenomena of vision. *Phil. Trans. R. Soc. B*, **128**, 371–395.

Wheatstone, C. (1852). Contributions to the physiology of vision. II. *Phil. Trans. R. Soc. B*, **142**, 1–18.

Yilmaz, E. H. & Warren, W. H. (1994). Visual control of braking from 2nd-order relative optical expansion $\dot{t}$. *J. Exp. Psychol.: Human Percept. Perform.* Submitted for publication.

Zeki, S. (1974). Cells responding to changing image size and disparity in the cortex of the rhesus monkey. *J. Physiol. London*, **42**, 827–841.

13

Structure from Motion

Myron L. Braunstein

University of California, Irvine, USA

1 INTRODUCTION

Consider two dots moving back and forth in horizontal paths across a cathode ray tube (CRT) (Figure 1a). The dots are moving in phase, but the lower dot is moving faster. What three-dimensional (3D) motions might this display represent? There are three prototypical cases (plus an infinite number of combinations): (1) The dots could be moving at different speeds in 3D (Figure 1b). The motion would be nonrigid and the dots could be moving at an infinite number of speed and distance combinations. (2) The dots could be at different distances from the eye, translating rigidly in 3D, and projected using polar perspective (Figure 1c). This is typically referred to as motion parallax. (3) The dots could be at different distances *from an axis of rotation*, but rotating rigidly in 3D (Figure 1d). What is especially important about this last case is that perspective projection is not necessary to perceive a 3D shape. A parallel projection, unrelated to eyepoint distance, is sufficient. This is structure from motion (SFM). SFM can be defined as the recovery of 3D shape from dynamic two-dimensional (2D) images by processes that do not require polar projection. Because SFM does not require polar projection, it is limited to the study of rotations about axes other than the line of sight, sometimes referred to as rotations in depth. This is not because of some fundamental reason to study rotations and translations separately, but because parallel projections of translations do not provide information for recovery of 3D structure. The same is true of rotations about the line of sight. The lack of information for recovering 3D structure in parallel projections of translations and in any projection of rotation about the line of sight is based on geometric considerations, but has also been demonstrated empirically (Braunstein, 1966). By this definition of SFM, a polar projection of an object rotating in depth would involve both motion parallax and SFM (Braunstein, 1962).

This limited definition of SFM is not based on a belief that SFM is a separate module in the visual system. On the contrary, I will argue below that SFM uses

VISUAL DETECTION OF MOTION
ISBN 0–12–651660–X

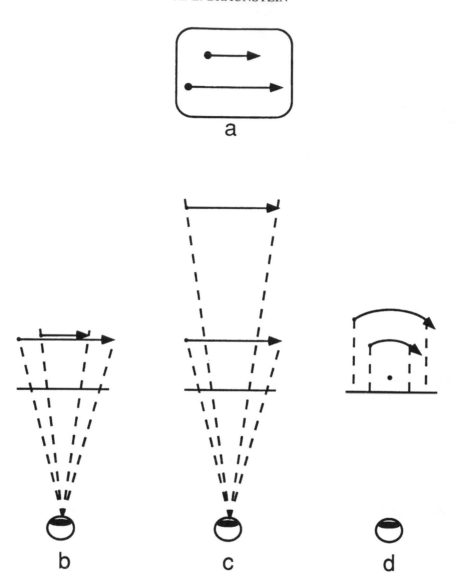

Figure 1 Examples of 3D interpretations compatible with two dots moving at different speeds in a 2D projection (panel a): Nonrigid motion (panel b), rigid translation with polar projection (panel c), and rigid rotation with parallel projection (panel d).

perceptual heuristics that overlap with those used in the perceptual analysis of perspective projections. The critical reason, in my view, for discussing SFM as a separate category of recovering 3D structure from dynamic 2D images is to emphasize that all dynamic information for recovering depth from images does not involve perspective. The inverse relationship between distance from the eye and projected velocity that occurs in direct vision is not the only basis for recovering depth from motion. This is a point raised in early research using computer-generated parallel projections (Braunstein, 1962) and emphasized by Ullman (1979) in his seminal computational analysis of SFM. Although it is useful to distinguish those aspects of recovering depth from dynamic images that are not dependent on perspective projection, a complete understanding of human perception of the 3D environment must consider relationships among all types of dynamic information, as well as relationships between dynamic information and other sources of depth information.

If SFM is not a separate module, one might ask whether it is instructive to study SFM separately at all. After all, we do not experience parallel projections in real life. A parallel projection occurs when the distance from the eye to the observed object is infinite. Parallel projection is approximated at large finite distances, but then the overall projection of the object being observed would be quite small. I would argue that there are two reasons for studying parallel projections. First, it is not unusual for an object of interest to be far enough away so that perspective is of minimal use in the recovery of the depth relationships within the object. Second, we readily perceive depth, and notice little distortion, when presented with artificial parallel or near-parallel projections, for instance, objects photographed through telephoto lenses.

2 OVERVIEW

After briefly reviewing some early observations of SFM (covered in greater detail in Braunstein, 1976), I will discuss the basic information for SFM, first in terms of optical motions and then in terms of underlying constraints. The next two sections will consider relationships of SFM to other sources of depth information – motion parallax and stereo. I will turn next to a critical issue in SFM research: Is metric depth recovered from SFM displays? A related issue – the buildup of SFM over time – is considered in the following section. The next three sections will consider shape perception from SFM, including detection and discrimination of curved surfaces, SFM interpolation, and the effects of boundaries and deforming contours on recovering shape. Since SFM depends on rotation, research on the sensitivity of observers to 3D rotation speed will be reviewed in the next section.

In every experiment covered in this chapter, and indeed in every study of 3D perception, it is necessary to ask whether subjects were performing the assigned task on the basis of a 3D interpretation or on the basis of the 2D image information, without 3D processing. This question is crucial for interpreting the results of previous research and for designing new research, and for this reason I have devoted the remainder of the chapter to a discussion of research methods in the study of SFM.

2.1 Early Observations

Near–far relationships are reversible in parallel projections. This ambiguity is found in an early report by Sinstenden (cf. Boring, 1942, p. 270). A distant windmill, silhouetted against the sky, appeared to reverse direction from time to time. The effect was reported to be so compelling that the owner of the windmill complained about this 'defect' to the windmill's builder. Early laboratory studies of SFM also used silhouettes, with objects rotating between a light source and a translucent screen. The objects included a two-bladed fan, which could be seen as rotating in either direction, clapping (the two blades rotating in opposite directions) or deforming in 2D (Miles, 1931) and vertical rods mounted in a turntable (Metzger, 1934). The systematic study of variables responsible for SFM began with Wallach and O'Connell's classic study in 1953, in which the authors introduced the term 'kinetic depth effect' (KDE). They placed rotating objects between a point light source and a translucent screen, taking care to show only one display to each subject to avoid possible biasing effects of one display on the perception of another. Solid objects in various shapes and arrangements and wire-frame objects were studied. Their conclusion was that 3D structure was perceived when the deforming shadows contained contours, explicit or implicit, that simultaneously changed in length and direction (tilt).

The displays in these early investigations were not actually parallel projections but contained a minimal amount of polar perspective. To quantify the amount of polar perspective in a display, we can compute a 'perspective ratio' (Braunstein, 1962) by dividing the distance from the eye to the most distant point in the display by the distance from the eye to the closest point. In a shadow projection these distances would be measured from the light source and in a computer-generated display they would be measured from the projection point or simulated eyepoint. If the scene contained two points moving perpendicular to the line of sight at the same 3D velocities, one at the near distance and one at the far distance, the ratio of the projected velocities of these two points would be given by the perspective ratio. The ratio is 1.0 in the case of a parallel projection. Ratios as small as 1.03 (Rogers & Graham, 1979) are effective stimuli for motion parallax. With the introduction of computer animation into SFM research (Green, 1959, 1961) it became possible to study the effects of parallel projections precisely, separating SFM from perspective effects.

2.2   Optical Motions that Result in SFM

Wallach and O'Connell (1953) concluded that the necessary condition for SFM was a simultaneous change in the length and direction of an explicit or implicit contour in the projected image. This is a requirement for an especially salient impression of depth, but is not the minimum condition for perceived depth. SFM can be perceived with much simpler stimuli. Johansson (1958) reports that a single spot of light moving back and forth with a sinusoidal (harmonic) velocity will at times appear to be moving in a circular path in depth. The motion of two dots can produce completely compelling perceptions of SFM even if the velocity function of each dot is linear. Börjesson and von Hofsten (1972) showed that rotation in depth is almost always perceived when two dots move towards each other in parallel paths. This is consistent with Wallach and O'Connell's hypothesis, since a line connecting these dots would change in length and direction.

Sinusoidal motion and changes in length and direction are both factors that contribute to a perception of SFM, but neither appears to be necessary. Length changes alone are sufficient for a compelling impression of rotation in depth, but the salience of this perception is increased by sinusoidal motion. Braunstein (1977) showed subjects one to four segments of a horizontal line expanding and contracting on a CRT. The expansion and contraction was either (1) linear and symmetrical, consistent with a constant rate expansion or contraction (or a variable rate rotation); (2) linear, but with the same asymmetry as a polar projection of a rotation; (3) sinusoidal and symmetrical, consistent with a parallel projection of a rotation; or (4) sinusoidal, but with asymmetry consistent with a polar projection. Subjects classified each display as rotating in depth or expanding and contracting in the frontal plane. The proportions of times rotation in depth was selected for the four displays were 0.52, 0.60, 0.69 and 0.77, respectively. These results show that judgments of rotation in depth can be elicited by length changes alone, especially when the rate of change corresponds to a polar projection of a constant angular velocity rotation.

2.3   Constraints Underlying SFM

Before continuing with our discussion of the empirical results from SFM research, let me return to the question of what is SFM. To address this question it is necessary to take a step back and briefly consider the nature of depth perception or, even more broadly, perception in general. Perception, in a formal sense (Bennett *et al.*, 1989b), is a process of inductive inference. This assertion should not be confused with ideas about unconscious inference. Inference does not have to refer to a thought process that has been relegated to an unconscious level. Rather, inference can describe the formal operation of a smart mechanism (Runeson, 1977) or a decoding principle (Johansson, 1970) (see also Braunstein, 1994). Perceptual inference is inductive in that the conclusions (perceptions) do not follow deductively from the premises (physical stimulation). Consider, for example, a typical SFM display – a parallel projection of points on the surface of a sphere rotating about a vertical axis. In the projection, we have dots moving back and forth horizontally at various velocities. With no a priori constraints placed on perceptual inference, this display could be interpreted as consisting of points independently moving back and forth on a 2D frontal plane. It is only because the perceptual system uses constraints such as rigidity and constant angular velocity that a sphere is perceived. These constraints are based on a particular physical environment. Suppose, for example, that a perceptual system evolved in a world in which all objects were white dots and each dot moved with complete independence and randomly varied its speed from moment to moment. It is unlikely that a perceptual system evolved under those circumstances would embody a rigidity constraint or a constant angular velocity constraint. When we study SFM we are studying a set of perceptual processes that are built on constraints – constraints which exploit environmental regularities. The theoretical work is often concerned with constraints that can provide mathematically correct solutions. The empirical work is concerned with determining what constraints are used by human observers.

   The first constraint I will discuss is the one which has received the most attention by far in the perception literature and even more so in the computational vision literature: rigidity. Although rigidity as a constraint has been found in the literature for some time, it was Ullman (1979) who showed precisely how rigidity could be used to recover 3D structure and motion in a parallel projection. In doing so, he also showed how applying a constraint based on an environmental regularity could provide a unique solution (plus reflection) to a perceptual inference. Basically, Ullman showed that the following was possible with displays of three views of four noncoplanar points. By assuming, initially as a hypothesis, that the points moved rigidly through the three views, equations could be formulated in which the 3D distance between each pair of points was equal on each view. This follows from the definition of rigid motion. Ullman showed that, if there is a solution to the system of equations, then there are exactly two solutions (one is the reflection of the other about a plane perpendicular to the line of sight). In the absence of rigid motion, however, the probability is zero of finding a solution to the system of equations. This means that if a solution is found, the hypothesis of rigidity is confirmed and at the same time the depth coordinates and 3D motions of the image points are recovered. Note that in the hypothetical case described above of a world without rigid motion, one might still find a solution to these equations on very rare occasions, but this would have no value in interpreting the 2D projection. To have value, a constraint must exploit an environmental regularity.

Rigidity is clearly a powerful constraint that could, in theory, be used to recover 3D structure from 2D images, even with a very small number of points and views. Is rigidity a sufficient constraint for human subjects? There are both theoretical and empirical difficulties involved in answering this question and it has not as yet been adequately answered. The theoretical problem is the following: Ullman's theorem does not restrict the angular rotation between the views. However, application of the theorem requires solution of the correspondence problem. That is, the visual system must be able to match the points across views. This means that if the theorem is to be compared to human performance it should bc applicd only to a limited subset of interpoint distances and motions – those for which the visual system correctly solves the correspondence problem. Although the theorem assumes correct matching of points across views, it does not incorporate the restrictions on rotation angle between views and interpoint distance that would be required to meet this condition in human vision. For this reason, there is no precise way to compare human performance to predictions of the theorem. Some years ago we (Braunstein *et al.*, 1987) produced displays of three views of four points undergoing rigid motion and found that with the rotation angles between views selected at random, few of the displays could be organized perceptually. To study rigidity, we had to restrict the rotations to angles of around 12° (depending on various other display parameters). (See Mather, 1989, for related results.) With this restriction in place we were no longer studying the rigidity constraint in isolation, for we had added a constraint of limited rotation between views which subjects are very likely to have exploited. Indeed, there is no way to study the rigidity constraint without a restriction of that type. What is needed, therefore, is a theoretical analysis that includes constraints on rotation angle, interpoint distances, and other factors related to solving the correspondence problem, in addition to the rigidity constraint.

The empirical problem in studying use of the rigidity constraint by human observers is that one must move from a theory of competence to a theory of performance. What task should human observers be able to perform if they have recovered the 3D structure? Could they have performed this task as well without recovering the 3D structure? This was a significant problem in our attempts to study the applicability of the rigidity constraint to human vision (Braunstein *et al.*, 1987). We asked subjects to indicate whether two displays represented the same 3D object. We took precautions to avoid direct matches between views in the 2D projections. However, we found that there was a statistical relationship between 3D distances between pairs of points in an object and the 2D projections of these distances that would have allowed subjects to perform at above chance levels. Subjects could have developed a criterion based on similarity of 2D distances and responded 'same' when the similarity of 2D distances for displays in a pair exceeded this criterion and 'different' otherwise. Later in this chapter I will discuss this pervasive problem of controlling 2D information when trying to study 3D perception.

Although it is still unclear whether rigidity is sufficient for perceiving 3D structure in 2D images, it is clear that rigidity is not necessary for perceiving motion in 3D from motion in 2D images. A single dot can be seen as rotating in depth (Johansson, 1958), although a rigidity constraint cannot be applied to the motion of one dot. Nonrigid configurations – dots rotating about the same axis with each dot moving at a different angular velocity – produce an impression of motion in depth at least as salient as that produced by rigid configurations (Braunstein, 1961).

Rigidity appears to be one of several factors involved in SFM. Three of these factors were investigated by Braunstein and Andersen (1984b). Their goal was to find variables that, in combination, could change perceived shape from a sphere to a disc. The displays were dots moving back and forth horizontally or vertically. The first variable was the velocity function, which was either sinusoidal or one of four combinations of a linear and a sinusoidal function. A sinusoidal function is consistent with a parallel projection of a constant angular velocity motion along a circular trajectory in 3D. The linear function was the projection of a constant velocity motion along a linear trajectory perpendicular to the line of sight in 3D; the combination functions were projections of constant speed motions along elliptical trajectories of varying eccentricity. The second variable was the relationship among the point velocities in the image. Either the point velocities varied as a function of vertical position (or horizontal position for horizontal axis rotation) in accordance with rigid rotation of a sphere, were constant across the display, or varied randomly. Only the first case is consistent with rigid rotation in depth. The second case is consistent with rigid translation of dots perpendicular to the line of sight. The third variable was transparency – whether the points moved in one direction or both directions. Subjects used an 11 point scale to judge shape, with a disk marking the 0 position, a 3D ellipsoid marking the middle position and a sphere marking the 10 position. For opaque displays, the ratings varied from less than 2, when the velocities of the individual dots were nearly linear and the image velocities were constant across dots, to over 9, when the individual dot velocities were sinusoidal and the velocity variations across dots were appropriate to a rigid rotation of a sphere. The velocity functions of the individual dots accounted for the most variance, with rigidity next and the influence of transparency relatively small. These results correspond to subjective impressions of these stimuli: A spherical shape is perceived even with nonrigid motion as long as the projected velocity function of each dot is sinusoidal, but the perceived shape changes when the velocity function is altered. These observations suggest that constant angular velocity is a more effective constraint than rigidity in determining perceived shape.

The rigidity constraint, as well as the constant angular velocity constraint and the

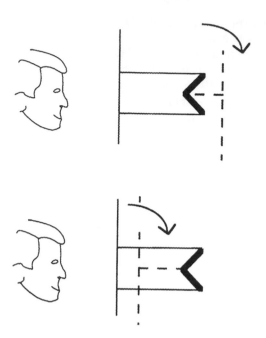

Figure 2 Parallel projections of dihedral angles rotating about a vertical axis (the dashed vertical line). In the top panel the fastest dots in the projection are at the center (at the dihedral edge); in the bottom panel the slowest dots are at the center.

fixed axis constraint (Bennett & Hoffman, 1985), can be applied to both parallel and polar projections. In a polar projection the projected velocities of objects moving rigidly relative to the observer are, in general, inversely proportion to their distances from the eye. This is an approximation rather than a precise rule because it applies only to perspective effects, and not to velocity variations resulting from rotation. It is, in fact, the rule underlying motion parallax. This rule appears to be used by human observers as a perceptual heuristic (see Braunstein, 1976, 1994). Although it may not seem applicable to SFM displays, there is compelling evidence that the rule of inverse proportionality between 3D distance and projected velocity is applied in the perception of parallel projections of rotations. Braunstein *et al.* (1993) presented subjects with parallel projections of dihedral angles, formed by two slanted planes meeting at a horizontal line perpendicular to the line of sight (Figure 2). The dihedral angles oscillated $\pm 19°$ about a vertical axis. Orientation of the angle in depth is indeterminate in a parallel projection, but varying the position of the axis of rotation relative to the angle varies the sign of the velocity gradient in each plane. Placing the axis closer to the dihedral edge (the center of the angle in the vertical dimension) results in the dots at the center moving more slowly in the projection; placing the axis closest to the top and bottom edges makes the dots at the center move fastest. Naive subjects, when asked to judge the orientation of the angle in depth, judged the center to be closest to them more often when the center dots moved fastest. When the ratio of the fastest to slowest speeds was $1.12:1$, the fastest dots were judged closest on 95% of the trials. This percentage dropped to 78% when the ratio was 3 : 1, a drop also found for perspective projections of translations. These results indicate that subjects judge relative depth in parallel projections in accordance with the rules of perspective.

3 SFM IN RELATION TO OTHER SOURCES OF DEPTH INFORMATION

3.1 The Stereokinetic Effect, SFM and Motion Parallax

An interesting relationship between SFM, the stereokinetic effect (SKE) and motion parallax (MP) was proposed recently by Proffitt and his coworkers (Caudek & Proffitt, 1993; Proffitt *et al.*, 1992). The analysis applies to the stereokinetic cone figures (Figure 3b) described by Musatti (1924) and to similar figures, although it may not apply to all stereokinetic figures (see, for example, Braunstein & Andersen, 1984a). For simplicity, consider two views of a stereokinetic disk separated by 180° of rotation (Figure 3a and c). If the change in orientation of the circles is not perceived as the disk

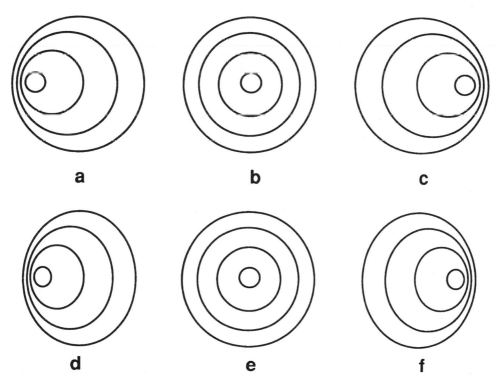

a   b   c

d   e   f

Figure 3 Types of displays studied by Proffitt *et al.* (1992). Panels a–c show translation in the image of the smaller circles without shape change. Panels d–f show shape changes compatible with a parallel projection of a rigid rotation about a vertical axis.

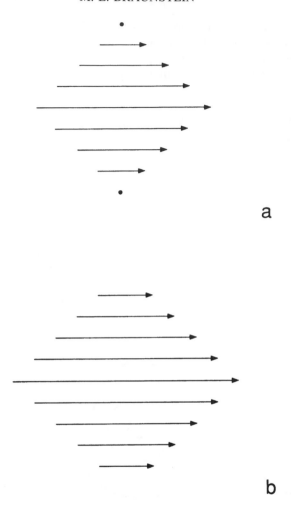

Figure 4 Velocity gradients with relative motion only (panel a) and with common and relative motion (panel b), studied by Caudek & Proffitt (1993).

rotates, the two views shown could be interpreted as an image in which the smaller circles have translated horizontally. Proffitt *et al.* (1992) point out that this image is similar to the projection of a rotation about a vertical axis, except that a projected rotation would show a horizontal compression of the contours – which Proffitt *et al.* refer to as within-contour motion. They argue that SFM (which they refer to as KDE) is a combination of SKE and within-contour shape change. They compared judged depth for stimuli that showed relative motion between contours only, which they labeled SKE displays (Figure 3a–c), and stimuli that also included relative motion within contours, which they labeled KDE displays (Figure 3d–f). Judged depth was significantly greater for the SKE displays than for the KDE displays, in their first two experiments.

In a second series of experiments (Caudek & Proffitt, 1993), displays with relative motion (Figure 4a) only were compared to displays with common motion (Figure 4b). No significant differences were found in judged depth, but depth order was judged more often in accordance with relative velocity when the displays had common motion. Caudek and Proffitt classified the displays with relative motion only as SKE displays (as in Proffitt *et al.*, 1992) and classified the displays with both relative motion and common motion as MP displays. A principal conclusion of their two papers was that SKE determines the amount of perceived depth both in SFM displays, which they regard as combinations of SKE and within contour motion, and in MP displays, which they regard as combinations of SKE and common motion. The specific characteristic of SKE that determines perceived depth, according to their analysis, is relative motion (the difference in velocity between the fastest and slowest features), but because of a compactness heuristic, perceived depth is adjusted according to the projected width of the object.

These are very interesting results, but they can be described somewhat differently. The relative motion that is labeled as SKE in these papers can be regarded as motion parallax (MP). For example, the horizontal shift of the smaller circles in Proffitt *et al.* (1992), shown in Figure 3a–c, is consistent with a perspective projection of a horizontal translation, with the stationary circle at a great (effectively infinite) distance. When the contours are compressed (Figure 3d–f), the images are no longer consistent with MP but are consistent, as Proffitt *et al.* suggest, with SFM. Specifically, the images are consistent with a parallel projection of a rotation about a vertical axis. Proffitt *et al.* were thus comparing MP displays to SFM displays, rather than partial SFM displays to complete SFM displays. The MP displays simulated greater depth, and this is the trend found in subjects' responses in their experiments. According to this revised interpretation, Caudek and Proffitt were comparing two types of MP displays – displays which included an edge at a great (effectively infinite) distance and displays which included only closer distances, rather than SKE and MP displays. To summarize this reinterpretation of the two papers, the basic SKE displays in these studies were MP with a contour at a great distance. Adding compression changed the displays to SFM displays. Adding a common motion to the basic SKE displays changed them to MP displays with all contours at nearer distances.

The situation is somewhat more complex perceptually, because rotation is sometimes perceived when a perspective translation is displayed, even in direct vision. This occurs when the perceived depth indicated by other cues (Ono *et al.*, 1988; Rogers & Collett, 1989), or even by inherent tendencies (Gogel, 1973), is different from the depth indicated by relative motion. So whereas rotation is indicated geometrically in these displays when contours of frontal surfaces are compressed, it is also indicated perceptually when the depth that would be consistent with a motion parallax interpretation is greater than the depth indicated by other cues. For example, if MP indicates that the farthest surface is at a great distance, and accommodation indicates that it is half a meter away, the result may be perceived rotation.

The relationship between SFM and MP was also investigated recently by Braunstein *et al.* (1993). Subjects judged the magnitude of a simulated dihedral angle, formed by two planes slanted in depth and meeting at a horizontal edge, by adjusting a side view of the angle presented on a separate monitor to the subject's side. Two combinations of motion and projection were studied: translation with polar projection and rotation with parallel projection. With the ratio of the fastest to slowest velocity and the average velocity equated, the rotations simulated shallower angles than the translations. Subjects overestimated the depth in the rotation displays and underestimated the depth in the translation displays. Judged depths in the rotation and translation conditions were more similar than the simulated depths but were not the same, indicating that the judgments were affected by image information that distinguished rotations from translations. A subsequent experiment showed that the relevant image information was the compression of the velocity field perpendicular to the axis of rotation. This compression is the same as the within-contour change discussed by Proffitt *et al.* (1992), and their finding of less perceived depth with within-contour changes corresponds to Braunstein *et al.*'s finding of less perceived depth with compression. Both results can be explained by subjects' attributing more of the relative motion to rotation (differences in distance from the axis) and less to perspective effects (differences in distance from the eye) when compression is present. (It should be noted that attributing more of the relative motion to rotation does not necessarily imply less perceived depth, but does so over a wide range of simulated distances from rotation axes, distances from the eye, and rotation speeds, including those studied by Proffitt *et al.* and Braunstein *et al.*) Braunstein *et al.* concluded that the underestimation of depth in translations and the overestimation of depth in rotations was due to subjects' attribution of some of the relative motion in perspective projections of translations to rotation and some of the relative motion in parallel projections of rotation to perspective effects.

3.2   SFM–Stereo Interactions

Two types of interactions have been reported recently between SFM and stereopsis. Pong *et al.* (1990) and Tittle and Braunstein (1993) found that SFM facilitates stereo processing for displays containing transparent surfaces – a condition in which static stereo processing is difficult (Akerstrom & Todd, 1988). This may be related to earlier results showing that a substantial portion of the people who cannot respond according to stereo information in static displays can do so when stereo is combined with SFM (Rouse *et al.*, 1989). Tittle and Braunstein present evidence that SFM facilitates stereo by helping to solve the stereo matching problem.

Nawrot and Blake (1991) report an effect in the opposite direction – stereo influencing SFM processing. Direction of rotation in a parallel projection is ambiguous (unless other information, such as occlusion, is added – Andersen & Braunstein, 1983). Nawrot and Blake had subjects adapt to a parallel projection of a rotating sphere in which the direction of rotation was indicated by binocular disparity. Following adaptation, the direction of rotation of an ambiguous SFM sphere was seen as opposite to the direction of the adapting sphere, indicating adaptation of the SFM perception by the stereo information.

4   METRIC DEPTH FROM SFM

Ullman (1979) showed that for a parallel projection of a rigid rotation of three or more views of four noncoplanar points, it is theoretically possible to recover the 3D distances between these points, up to a scale factor. If the performance of human observers closely matched this competence, the ability to make metric judgments about 3D relationships among points should increase as the number of views exceeded two. Todd and Bressan (1990), on the other hand, have proposed that metric 3D structure is not recovered regardless of the number of views. Instead, they propose that structure is recovered up to an affine scaling in the depth dimension (see also Koenderink & van Doorn, 1991). This proposal has two related implications. First, subjects should be accurate only in judgments that are not affected by rescaling the Z dimension relative to X and Y. Second, accuracy should not increase beyond two views. Todd and Bressan studied two types of judgments – judgments that would require recovering metric structure in Z and judgments that would not be affected by rescaling in Z. For the first type of judgment, which included judging the relative 3D lengths of moving line segments oriented in different directions, accuracy was poor for two-view displays and showed little or no improvement as the number of views was increased. For the second type of judgments, which included discrimination between planar and nonplanar configurations of moving line segments, accuracy was high with two-view displays, but again showed only a small increase with number of views. Similar results were obtained for judgments of depth relative to width (amplitude/period) in corrugated surfaces (Todd & Norman, 1991).

Liter *et al.* (1993) obtained results that support Todd's position. Their initial objective was to determine what 3D structure subjects perceived when presented with a two-view display, since there is an infinite family of mathematically compatible structures for a two-view display, if there is any rigid interpretation (Bennett *et al.*, 1989a; Huang & Lee, 1989; Koenderink & van Doorn, 1991; Todd & Bressan, 1990). Subjects were presented with an oscillating two-view, five-dot SFM display together with a continuously rotating display of one of the mathematically compatible interpretations of the two-view display. The subject could change the continuous display with a joystick, sampling the family of compatible interpretations. The subject was told to select the interpretation that best matched the oscillating display. Although all of the interpretations were mathematically compatible with the two-view display, subjects selected interpretations from a limited range. When additional views were added between the two views, so that there was a single compatible interpretation (plus reflection), subjects did not select the compatible interpretation. Instead, their selections were similar to those made from two views. This is exactly what would be expected from Todd's hypothesis. Liter *et al.* found judged depth in both two-view and multi-view displays to be related to relative motion in the image, defined as the maximum difference in signed velocity between any two points in a display, with curl removed.

The failure of subjects to recover metric structure and the lack of improvement when additional views are added has been studied so far in cases in which rigidity is the principal constraint, or the only constraint, that would be available to a theoretical analysis of the recovery of SFM. It may be that subjects' performance does not match theoretical analyses based on rigidity because human perception does not make direct use of a rigidity constraint. Instead, human perception may use heuristic processes that exploit rigidity as an environmental regularity. An example of such a heuristic process is the use of relative motion between two features in a 2D projection as the basis for perceiving the relative depth of these features in a 3D scene (Liter *et al.*, 1993; Proffitt *et al.*, 1992). The relative motion of two features in an image (the difference in projected velocity with curl removed) is likely to be related to the relative depth of the features in the 3D scene if the features are moving rigidly. In this sense, a relative motion heuristic is based on rigidity. A relative motion heuristic could be applied with only two views and would provide accurate judgments of some 3D relationships, but would not, in general, recover depth veridically in SFM displays. An unanswered question is whether the failure to find accurate recovery of depth, or improvement with more than two distinct views, would occur if heuristics were applied that did not depend on rigidity. In particular, one would expect improvement with increasing numbers of views if a constant angular velocity constraint were applied in interpreting an SFM display. Results from experiments in which the velocity function was manipulated independently of rigidity (Braunstein & Andersen, 1984b) suggest that accurate judgments of 3D shape can be made on the basis of a constant angular velocity constraint, but this study did not include two-view displays. More research is needed on whether it is metric structure in general that is not recovered or only metric structure based on rigidity.

4.1 Temporal Factors

The perception of 3D structure from motion appears to build up over time. Hildreth *et al.* (1990) had subjects judge which of three points was intermediate in depth. They found an increase in accuracy as the extent of angular rotation increased, reaching a plateau between 30° and 45° (660 to 990 ms). They compared the increase to predictions from a version of Ullman's incremental rigidity scheme. A buildup of SFM over time was also reported by Eby (1992) for displays of 256 dots positioned on the surface of half-ellipsoids. Eby showed the same sequence repeatedly, with 1 second inter-stimulus intervals between repetitions, so that the effects of total duration could be separated from any effects due to presenting new views. Judged depth was under-estimated for all durations, but increased with duration, reaching a plateau in 16° (about 270 ms). The shorter duration, compared to Hildreth *et al.*'s result, is probably due to Eby's use of a larger number of dots forming a surface. (It would be interesting to examine buildup for a large number of dots that did not form a surface.)

The perception of a surface in an SFM display does not require that individual dots remain present in the display over extended intervals. Todd (1985) found good correspondence between simulated slant and judged slant with as many as 88% of the dots in a display randomly repositioned at each frame transition. Additional evidence that the perception of 3D shape does not require point lifetimes of more than two views can be found in Dosher *et al.* (1989) and Husain *et al.* (1989).

5 SHAPE PERCEPTION FROM SFM

5.1 Detection and Discrimination of Curved Surfaces

The ability of subjects to discriminate between planar and curved surfaces in motion depends on the relationship between the direction of motion and the direction of curvature. Cornilleau-Pérès and Droulez (1989) found that oscillating surface patches, curved in one direction, were most discriminable from planar patches when the motion direction was perpendicular to the direction of curvature. For example, a section of a cylinder with a vertical axis would have its maximum curvature in the horizontal direction. Rotated about a horizontal axis, the motion direction would be vertical. Under these conditions, the patch was maximally discriminable from a planar patch. If the section was rotated about a vertical axis it was less discriminable from a planar patch. These results were confirmed and extended to additional shape comparisons by Norman and Lappin (1992).

Detection of a smooth surface in an SFM display requires only a few feature points, with the number depending on characteristics of the set of surfaces to be detected. Turner et al. (in press), using methods based on Uttal's (1985) stereo research, asked subjects to determine whether each display in a series of SFM displays consisted of a smooth surface, from a specified set of surfaces, or of dots randomly arranged in a volume (a 'noise' display). The noise displays were produced by randomly rearranging the Z coordinates in the surface displays. In one experiment sinusoidal surfaces of the form $z = A*\sin(\omega*x)*\sin(\omega*y)$ were varied in amplitude (A = 0.25 or 0.90) and frequency ($\omega = \pi$, 2π, or 3π, with x and y varying between -1 and 1). Sensitivity, as measured with d', increased with numerosity of the points, increased with amplitude, and decreased with frequency. These findings are similar to Andersen's (1991) results with motion parallax displays of translating surfaces with sinusoidal corrugations. In another experiment, Hines et al. examined detection of surfaces that varied in shape index and curvedness (Koenderink, 1990). The shape index values were 0.0 (a saddle), 0.5 (a parabolic arch), and 1.0 (a paraboloid of rotation). Sensitivity increased with the shape index and with curvedness. Hines et al. propose a measure of smoothness that can be applied to sparse dot displays and can account for the effects of corrugation frequency and amplitude, or of shape index and curvedness.

5.2   Interpolation of Surfaces between Visible Features

The perception of continuous 3D surfaces involves interpolation between visible feature points. Saidpour et al. (1992) developed an SFM probe to assess interpolation between visible features in SFM displays. The probe moved in phase with the 2D motion of the features in the SFM display, but its amplitude could be varied by the subject (using a joystick). The simulated motion of the probe was always rigid with respect to the surface features, but it appeared to approach or recede in depth as its amplitude was increased or decreased by the subject. The probe dot was located in a region devoid of feature points – the 'gap'. The subject's task was to place the probe dot on the surface. For smoothly curved surfaces, subjects tended to place the probe dot slightly outside the surface, as if they perceived a bulge in the region of the gap. When a discontinuity was simulated in the region of the gap, however, subjects placed the probe dot inside the surface, as if they were smoothing over the discontinuity (Saidpour et al., 1994). Probe dot placement in that case matched the predictions obtained by applying 2D interpolation algorithms proposed by Ullman (1976) and by Kellman and Shipley (1991) to cross-sections perpendicular to the discontinuous edge, and were consistent with Grimson's (1981) 3D interpolation model. Although subjects showed a bias in both studies, they were very precise in their placement of the probe dot, with standard deviations as low as 2% of the simulated depth in some conditions.

5.3   Boundaries and Occluding Contours

Perceived shape in SFM displays can be manipulated by altering the occlusion boundaries of the flow field. Ramachandran *et al.* (1988) demonstrated that dots on the surface of a rotating cylinder, when viewed through a triangular aperture, appeared to be on the surface of a cone. Dots on two adjacent cylinders appeared to be on a single large cylinder. Thompson *et al.* (1992) report a series of experiments in which shape was determined by occlusion at boundaries, even when boundary cues conflicted with the differential motion within the pattern.

The research on aperture shape demonstrates an important effect of stationary occluding contours on perceived shape. Another type of occluding contour is the deforming contour in the projection of a rotating object. Except in special cases, such as parallel projections of a rotating sphere or of a cylinder rotating about its axis, the projected contour deforms as an object rotates in depth. Effects of contour deformations on the perceived shape of objects have been studied by displaying silhouettes of rotating objects, thus eliminating any information available from the motion of texture elements on the surface. Todd (1985) reports a demonstration using displays simulating parallel projections of rotating silhouettes, including a single rotating ellipsoid and two overlapping ellipsoids (Figure 5). The single ellipsoid was seen as a deforming 2D image, whereas the overlapping ellipsoids were seen as rotating rigidly in depth. Beusmans (1990) showed subjects deforming contour displays simulating projections of rotating ellipsoids of varying thickness. With zero thickness the object was a 'wire-frame' ellipse. Subjects were able to discriminate

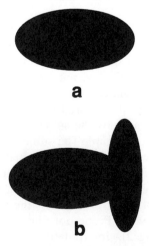

Figure 5   Separate (panel a) and overlapping (panel b) ellipsoids, of the type studied by Todd (1985).

between the rotating ellipse and the ellipsoids, with accuracy increasing as the thickness of the ellipsoid increased. Accuracy was highest for axes of rotation in the image plane and was reduced for slanted axes. Accuracy in recovering the shape of objects from the projections of occluding contours was studied by Cortese and Andersen (1991). Subjects judged the horizontal cross-section of ellipsoids rotating about a vertical axis. Shape recovery was accurate when the major axis was small. More eccentric shapes were reported as the rate of contour deformation decreased. In one experiment subjects judged both shape and amount of rotation. Eccentricity of shape responses was predictable from rotation responses, suggesting that perceived shape in deforming contour displays depends on the perceived extent of rotation.

6 SENSITIVITY TO 3D ROTATION SPEED

Cortese and Andersen's (1991) research demonstrates the importance of perceived rotation in the recovery of SFM, at least for deforming contour displays. This is not surprising, since relative motion between image features in the projection of a rotating object is determined by both the 3D structure of the object and the rate of angular rotation. For this reason it is important to understand the variables that affect the perception of rotation. Kaiser (1990) had subjects judge which of two objects, presented simultaneously, was rotating faster. The displays simulated cubes rotating about the same or different axes (horizontal vs. vertical), rotating in the same or different directions (when the axis was the same), and rotating in phase or 45° out of phase. A staircase method was used to measure the difference threshold (DT) and the point of subjective equality (PSE). The DT was similar across these conditions and the PSE did not differ significantly from the point of objective equality (POE). To examine the influence of edge transitions, a bisphenoid, which has three faces along the motion direction, was compared to a cube. The bisphenoid would have to rotate 33% faster to match the edge transitions of the cube. The PSE was 9% higher than the POE, indicating some influence of edge transitions, but demonstrating that relative rotation speed was not judged primarily in accordance with these transitions. Another experiment compared cubes of different sizes. Both the 2D flow rate and the tangential edge velocities in 3D are affected by size. A cube of half the size had to rotate 18.5% faster to be judged as equal in rotation speed, compared to an increase of 100% that would be required if the 2D flow rate or tangential edge velocities determined the judgments. Kaiser and Calderone (1991) attempted to separate the effects of 2D flow rate from 3D tangential velocities by occluding vertical strips of displays simulating dots arranged regularly or randomly on the surfaces of spheres. Occluders placed over the right and left edges of a sphere rotating about a vertical axis covered the regions (near the equator) with the greatest 3D tangential velocities. An occluder placed over the center of the sphere covered the regions with the greatest 2D projected velocities. The spheres with the center occluded had to rotate 6% (regular) or 9% (random) faster than the spheres with the edges occluded, indicating that rotation speed judgments were influenced by 2D velocities and not by (or not as much by) 3D tangential velocities. The sphere with the center covered would have had to rotate 160% faster to match the 2D velocities. Petersik (1991) obtained results consistent with those of Kaiser and Kaiser and Calderone in experiments in which subjects estimated rotation magnitudes for spheres with dots either located on the surface only or distributed throughout the volume. Judged rotation magnitude was less for the filled spheres, but not as much less as it would be if subjects were judging rotation magnitude on the basis of 2D linear velocity. Overall, these results indicate that there is some influence of 2D motion on judgments of 3D angular velocity, but judgments are primarily in accord with the 3D angular velocities.

7 METHODOLOGICAL ISSUES IN SFM RESEARCH

A pervasive methodological problem faces the researcher designing psychophysical studies of the recovery of SFM. How does one assure that the subject is performing the assigned task on the basis of a 3D analysis of the motion display, and not on the basis of some 2D cue? We cannot assure this with absolute certainty. If the subject is provided only with 2D images, and the subject's task is possible, it must in principle be possible to perform the task using information in the 2D images. The possibility of responding to SFM displays by using simple relationships in the 2D images was addressed recently by Sperling *et al.* (1989). They asserted that all prior KDE research (including both MP and SFM studies) is suspect because of artifactual cues that may have allowed subjects to make supposedly 3D judgments from 2D information without 3D processing. For example, they argue that a subject asked to judge whether a simulated dihedral angle was concave or convex (Braunstein & Andersen, 1981) could have responded simply by noting whether the slowest or the fastest dots in the projection were located at the dihedral edge in the 2D image. They argued that use of such simple 2D strategies could be avoided by using a more difficult psychophysical task that would require a larger number of velocity comparisons for a correct response. Sperling *et al.* introduced a new KDE display simulating a planar region containing up to three bumps or depressions. The possible locations of bumps or depressions in a given display were at the vertices of either an upright or inverted equilateral triangle. Since each position on each triangle could have either a bump, a depression, or neither, there were $3 \times 3 \times 3$ possible arrangements of features on the two triangles, or 54 combinations. (Because the two triangles with no bumps or depressions were identical, there were a total of 53 unique combinations.) Subjects were trained to identify the 53 patterns according to a labeling system based on the locations of bumps and depressions. Effects of amplitude and number of dots representing the surface were investigated.

Performing accurately in this task on the basis of 2D velocity information would require comparing the velocity at six locations – each of the potential feature locations – with the surrounding velocity. To determine whether this task could have been performed on the basis of 2D velocity comparisons, Sperling *et al.* conducted a control experiment in which subjects performed a purely 2D task requiring the same number of velocity comparisons. Dot velocities were varied in six regions of a 2D velocity field corresponding to the potential locations of the six features in the 3D task. They found that subjects could perform the 2D task as accurately as the 3D task, suggesting that it was possible to perform the 3D task using the 2D velocities directly. They introduced the expression 'alternative computations' to describe this possibility, which they wished to distinguish from the 'artifacts' that they had pointed out in their discussion of previous KDE research. According to Sperling *et al.*, an alternative computation 'uses the same stimulus and velocity flow field', whereas an artifactual computation 'uses an incidental property of the display . . . that may be quite unrelated to the KDE computation'.

In a commentary on Sperling *et al.*'s paper, Braunstein and Todd (1990) argued that the distinction between artifacts and alternative computations was a useful one, but was not being uniformly applied by Sperling *et al.* In particular, Braunstein and Todd maintained that labeling a computation based on a single velocity comparison in the image as artifactual and one based on six velocity comparisons as an alternative computation was arbitrary. They argued that velocity variations are an appropriate source of information for perceived depth in SFM displays, and the presence of 2D velocity variations from which the required 3D judgment could be computed was not an artifact. If a subject responded in an SFM experiment on the basis of velocity variations in the image, without 3D processing, this would have constituted an alternative computation. In a reply to the commentary, Sperling *et al.* (1990) maintained that the number of velocities that must be compared is a legitimate basis for distinguishing between artifacts and alternative computations, citing evidence that tasks that could be performed by comparing two velocities are possible with second-order motion, but tasks requiring comparison of a larger number of velocities could not be performed accurately in such displays (with a limited duration).

Braunstein and Todd (1990) discuss possible methods for controlling the effects of artifacts and alternative computations in the study of KDE. Ideally, artifacts should be eliminated. For example, if one is studying KDE, one should eliminate texture effects that could be used to respond to the experimental task (unless the purpose is to study the combined effects of texture and motion). Some potential artifacts are easily avoided. Stereopsis, for example, can be eliminated by using monocular viewing. For other potential artifacts, such as texture density effects, the problem is more difficult. Texture elements are needed to carry the motion in KDE displays (unless second-order motion is used, but second-order motion has not been found effective in SFM – see Landy *et al.*, 1991) and texture density naturally covaries with motion in these displays in most cases. The same is true of the brightness, size, and shape of texture elements.

Texture density on individual frames of a display can be controlled by adding or subtracting feature points to maintain a uniform density (Sperling *et al.*, 1989), but this procedure produces a scintillation over time that could itself provide information about 3D structure (Braunstein & Todd, 1990). One can begin with texture density uniform in 2D, and for some displays (e.g. translations perpendicular to the line of sight and parallel projections of rotations of planar surfaces about an axis in the image plane) the density will remain uniform, although not constant in the case of rotation (Braunstein *et al.*, 1993). Density can also be maintained over views when a surface of revolution is rotated about its axis (Braunstein *et al.*, 1989). In these specific cases, density can be controlled, but in general texture density cannot be eliminated as a source of depth information. In research in which texture density effects cannot be eliminated, static controls are sometimes used to demonstrate that the most extreme texture effects in the dynamic displays would not allow the subject to respond accurately in the task used in the experiment. Also, it may be argued that previous studies (e.g. Braunstein, 1968) have shown relatively little effect of texture density for random textures.

Artifacts can sometimes be controlled by selecting a limited subset of stimuli for which the artifactual computation would not be effective in allowing the subject to perform the required judgments. For example, if the subject's task is to judge which of two lines is longer in 3D, the simulated 3D lengths and slants of the lines can be selected so that the longer line in 3D is the longer line in the 2D projection on half of the trials (Todd & Bressan, 1990). This prevents subjects from effectively using 2D line length to make the judgments.

The possibility of alternative computations usually presents a more difficult problem for the researcher studying SFM. This possibility cannot be eliminated by controlling the stimulus displays because the stimulus displays are 2D images that must contain the information for an SFM analysis. It is not possible to prove with certainty that alternative computations were not used because an alternative computation must exist if the SFM computation is possible for a display. Instead, one can present arguments as to why it was unlikely that the subject responded on the basis of an alternative computation. Four types of arguments may be used to support this assertion. The first argument is that an alternative computation would be too difficult for the subject (or at least more difficult than the SFM computation). In stereo research using contoured stereograms, an alternative computation of depth order might consist of noting the relative horizontal positions of contours in the 2D images. Researchers generally consider this problem to be eliminated when random dot stereograms (Julesz, 1971) are used, because comparing relative displacements of matching pixels in the two images by any method except stereopsis is considered too difficult for human subjects. I use an example from stereopsis because there is no similarly clear-cut case in SFM research.

The second argument is probably the most common, but at the same time the least reliable and most controversial. This is the argument that subjects, usually during debriefing at the close of an experiment, reported either that the displays appeared 3D, or that they were making their judgments on the basis of a 3D perception, or both. Many SFM researchers report such introspective evidence, although some are ambivalent about its use. For example, Sperling *et al.* (1990) use introspection to support their claim that KDE computations were used in their own experiment ('We have demonstrated these KDE shapes to visitors, and to hundreds of observers at numerous lectures and have not yet received even one report of an observer who did not perceive a vivid 3-D shape', p. 446) but decry the use of introspection in general ('An important component of the development of psychology has been the move away from introspection', p. 446). Evidence based on introspection must be used with caution. We know from signal detection research that subjects can respond well above chance accuracy to a signal in a forced choice task even when they are unable to report that they have perceived the signal. Reporting that displays appear 3D, and lack of awareness of an alternative 2D computation, is not conclusive evidence that an alternative computation was not used.

A third argument, somewhat related to the second in that it may rely on responses during debriefing, is that the subject would have no basis for associating the alternative computation with the response required in the SFM task. For example, it is not clear why a naive subject would use the relative magnitude of the linear and sinusoidal components in the velocity function for dots moving back and forth horizontally on a CRT to rate perceived 3D shape on a scale of flat to spherical (Braunstein & Andersen, 1984b). This argument would not apply to knowledgeable subjects or to subjects given feedback.

A fourth argument that can be made only in some cases is that the results of the experiment would have been different if subjects had used a 2D alternative computation. For example, Braunstein *et al.* (1993) found that texture elements with faster velocities in an oscillating SFM display were judged to be closer. One might be concerned that subjects were using relative motion as an alternative 2D computation, except that faster moving texture elements were judged as closer less often as the ratio of the fastest to the slowest velocity was increased. This is the same result found for MP by Braunstein and Andersen (1981), and is exactly the opposite of the result that would be expected if subjects were using the 2D velocities directly.

The possibility of alternative computations is a concern in all SFM research, but the likelihood that one's results have been seriously affected by alternative computations may depend on the research question being addressed and the psychophysical method used to address this question. All SFM research is concerned with relating perceived 3D structure to 2D image variables, but two types of research questions can be distinguished. The first type of question is concerned with identifying relevant image variables and constraints. The objective is to learn about the subject's 3D interpretation, and there may or may not be an objectively correct answer. For example, a subject might be shown two-view displays and asked to select the closest match for each display from a set of compatible 3D interpretations. The second type of research question is concerned with the subject's sensitivity to changes in an SFM display. The objective is to determine how closely the subject's response matches the correct response. For example, a subject might be shown two-view displays that either simulate rigid 3D motion or deviate by different amounts from rigid motion and asked whether or not the 3D motion in each display is rigid. These two types of research questions can be distinguished in other areas of perception as well. For example, a study in which subjects were asked to sort color samples according to similarity, in order to determine how physical dimensions map onto a subjective color space, would be addressing the first type of question. A study in which subjects were asked whether two patches varying along a physical dimension are the same or different, to determine sensitivity to differences on that dimension, would be addressing the second type of question.

These two types of questions, when asked separately, are logically addressed in sequence. We generally want to identify the relevant information before attempting to measure sensitivity. Sometimes, however, a stimulus variable might be of interest on theoretical grounds and empirical research may move immediately to studies of sensitivity. For example, Braunstein *et al.*'s (1990) study of sensitivity to nonrigid motion was motivated by theoretical accounts of the role of rigidity in SFM. A similar distinction between types of questions in perceptual research was made by Bennett *et al.* (in press) as part of a formal analysis of perceptual processing. They note that the first type of question logically requires recovering a 3D interpretation, whereas the second type of question can be addressed entirely within the space of 2D projections.

The methods used to address the first type of question include categorization, ratings and adjustment. Categorization was the method used by Miles (1931) to examine the alternative percepts that occurred when the shadow of a rotating two-bladed fan was displayed. It was the method used by Johansson (1950) to determine the perceptions resulting from various combinations of 2D point motions, by Wallach and O'Connell (1953) to determine the necessary conditions for the KDE, and by Börjesson and von Hofsten (1972) to determine the relationship between relative and common motion and perceived rotation and translation in depth. It is a highly informative method early in the development of a research area, and may still be a useful method at more advanced stages of research. For example, Börjesson and Ahlström (1993) recently studied the dominance of alternative motion percepts by having subjects categorize motions of groups of dots in which one dot participated in two alternative types of motion.

Since studies using categorization do not compare subjects' performance to a 'correct' answer, feedback is not used. The purpose is usually to determine what the subject regards as the appropriate response, not to see how closely the subject can match what the experimenter regards as the appropriate response. The questions asked may be about what 2D or 3D object is seen or which of a set of 2D or 3D motions is perceived in a display. Ratings are similar to categorization except that a quantitative scale is provided. This implies prior knowledge by the researcher of the appropriate dimension for the scale. Ratings of coherence (Braunstein, 1962; Green, 1961) and of perceived depth (Braunstein, 1962) were used in early SFM studies using computer-generated displays. As with studies using categorization, studies using ratings do not usually specify a correct answer and feedback is not provided.

The method of adjustment straddles the boundary between the two types of questions discussed here by providing two measures: a measure of bias and a measure of variability. The first measure is informative about subjective appearance; the second measure is informative about sensitivity to changes in appearance. Feedback is not usually used. It might be used if only variability was of interest, but it would make the bias measure meaningless. Consider, for example, Saidpour et al.'s (1994) interpolation study, described above, in which subjects were asked to adjust a probe point until it appeared to be on a simulated surface. The conclusion that subjects smoothed across the simulated discontinuity was based on an examination of the constant error, relative to the simulated position of the surface. If Saidpour et al. had used feedback, it is possible that the subjects would have learned to place the probe point on the simulated surface by compensating for this constant error. A study using feedback might still be informative about a subject's precision in positioning the probe point, but it would not be informative about perceptual smoothing over discontinuities.

The second type of question – how sensitive is the observer to variations in information from a particular source – is usually studied with discrimination paradigms, although as noted, the method of adjustment can be used to address this type of question as well. The method of constant stimuli and staircase procedures are typical and feedback is often used. A study by Lappin and Fuqua (1983), examining the precision with which subjects could judge whether the middle of three points on a line slanted in depth was exactly centered between the other two points, illustrates the effective use of this paradigm in SFM research. Adjustment methods could be applied to this type of question as well, by having the subjects move the middle dot until it appeared to be centered between the other two dots.

The concerns about alternative computations in these two cases are somewhat different. In a categorization, rating, or adjustment experiment it is possible that a subject will use some 2D information directly without a 3D computation, but one could argue, as indicated above, that since the requested judgment is a 3D judgment, the subject would have no reason to assume that a particular 2D computation was appropriate to the task. Still, alternative computations are possible in this situation. For example, the subjects could learn a particular association between 2D image information and 3D appearance on some trials and use this relationship on other trials to give a 3D judgment on the basis of the 2D image formation. Or the subject, unable to make the 3D judgment, could make a judgment on the basis of whatever 2D information is varying among trials. These are conditions that one might try to discover through debriefing.

In discrimination tasks it seems likely that subjects will use any information that can produce the required discrimination, whether or not a 3D computation is involved. This is especially likely when feedback is given, but it is also possible without feedback. If the difficulty of the discrimination varies across trials, as it usually does in a discrimination paradigm, the subject might learn the relevant 2D information on the easier trials and apply this knowledge on the more difficult trials. Note that this is most likely when the 2D computation is easier than the 3D computation, a circumstance that should be avoided if possible. Debriefing may not reveal whether a 2D or 3D computation was used, especially if a 2D computation is used only on the more difficult trials. The subject may still report a 3D percept and even a 3D strategy.

8   SUMMARY AND CONCLUSIONS

The concept of SFM as a distinct source of information about the 3D environment rests on the observation that a 3D structure is perceived when objects rotate about axes other than the line of sight, even in the absence of polar perspective. This concept has been reinforced by a number of analyses showing that the information in parallel projections is mathematically sufficient, under constraints such as rigidity, to recover the 3D structure. Human observers, however, do not appear to compute metric solutions in the manner suggested by these mathematical analyses, but instead appear to use heuristic processes based on similar constaints. Systematic errors are found in shape judgments, interpolation, and judgments of rotation speed, and these errors provide clues as to the actual processing rules. Research examining these processing rules has moved from simple introspection to sophisticated adjustment and discrimination paradigms, and we are beginning to see precise quantitative measures of what is perceived in an SFM display. With this increased precision, however, have come questions about whether subjects are responding on the basis of 3D perceptions or alternative 2D computations, and this is a major issue in interpreting current research.

ACKNOWLEDGMENTS

This chapter includes research supported by National Science Foundation Grant DBS-9209973. I am grateful to George J. Andersen, Jeffrey C. Liter, Asad Saidpour, and

Jessica Turner, for comments on an earlier draft and to David W. Eby, Donald D. Hoffman, and Scott N. Richman for helpful discussions.

REFERENCES

Akerstrom, R. A. & Todd, J. T. (1988). The perception of stereoscopic transparency. *Percept. Psychophys.*, **44**, 421–432.

Andersen, G. J. (1991). Interpolation of smooth 3-D shape from motion. *Invest. Ophthalmol. Vis. Sci.*, **32** (3, Supplement), 1277.

Andersen, G. J. & Braunstein, M. L. (1983). Dynamic occlusion in the perception of rotation in depth. *Percept. Psychophys.*, **34**, 356–362.

Bennett, B. & Hoffman, D. (1985). The computation of structure from fixed axis motion: nonrigid structures. *Biol. Cybernet.*, **51**, 293–300.

Bennett, B., Hoffman, D., Nicola, J. & Prakash, C. (1989a). Structure from two orthographic views of rigid motion. *J. Op. Soc. Am. A*, **6**, 1052–1069.

Bennett, B., Hoffman, D. & Prakash, C. (1989b). *Observer Mechanics: A Formal Theory of Perception*. Academic Press, New York.

Bennett, B. M., Hoffman, D. D., Prakash, C. & Richman, S. N. (in press). Observer theory, Bayes theory, and psychophysics. In *Proceedings of the January 1993 Cape Cod Conference on Visual Perception and Psychophysics*.

Beusmans, M. H. (1990). *Visual Perception of Solid Shape from Occluding Contours* (Tech. Rep. No. 90–40). Irvine: University of California, Dept. of Information and Computer Science.

Boring, E. G. (1942). *Sensation and Perception in the History of Experimental Psychology*. Appleton-Century-Crofts, New York.

Börjesson, E. & Ahlström U. (1993). Motion structure in 5-dot patterns as a determinant of perceptual grouping. *Percept. Psychophys.*, **53**, 2–12.

Börjesson, E. & von Hofsten, C. (1972). Spatial determinants of depth perception in two-dot motion patterns. *Percept. Psychophys.*, **11**, 263–268.

Braunstein, M. L. (1961). Rotation of dot patterns as stimuli for the perception of motion in three dimensions: The effects of numerosity and perspective. *Dissertation Abstracts*, **21**, 3860. (University Microfilms No. 61–1720).

Braunstein, M. L. (1962). Depth perception in rotating dot patterns: Effects of numerosity and perspective. *J. Exp. Psychol.*, **64**, 415–420.

Braunstein, M. L. (1966). Sensitivity of the observer to transformations of the visual field. *J. Exp. Psychol.*, **72**, 683–687.

Braunstein, M. L. (1968). Motion and texture as sources of slant information. *J. Exp. Psychol.*, **78**, 247–253.

Braunstein, M. L. (1976). *Depth Perception through Motion*. Academic Press, New York.

Braunstein, M. L. (1977). Minimal conditions for the perception of rotary motion. *Scand. J. Psychol.*, **18**, 216–223.

Braunstein, M. L. (1994). Decoding principles, heuristics and inference in visual perception. In G. Jansson, S. S. Bergstrom & W. Epstein (Eds.), *Perceiving Events and Objects*. Erlbaum, Hillsdale, NJ.

Braunstein, M. L. & Andersen, G. J. (1981). Velocity gradients and relative depth perception. *Percept. Psychophys.*, **29**, 145–155.

Braunstein, M. L. & Andersen, G. J. (1984a). A counterexample to the rigidity assumption in the perception of structure from motion. *Perception*, **13**, 213–217.

Braunstein, M. L. & Andersen, G. J. (1984b). Shape and depth perception from parallel projections of three-dimensional motion. *J. Exp. Psychol.: Human Percept. Perform.*, **10**, 749–760.

Braunstein, M. L. & Todd, J. T. (1990). On the distinction between artifacts and information. *J. Exp. Psychol.: Human Percept. Perform.*, **16**, 211–216.

Braunstein, M. L., Hoffman, D. D., Shapiro, L. R., Andersen, G. J. & Bennett, B. M. (1987). Minimum points and views for the recovery of three-dimensional structure. *J. Exp. Psychol.: Human Percept. Perform.*, **13**, 335–343.

Braunstein, M. L., Hoffman, D. D. & Saidpour, A. (1989). Parts of visual objects: An experimental test of the minima rule. *Perception*, **18**, 817–826.

Braunstein, M. L., Hoffman, D. D. & Pollick, F. E. (1990). Discriminating rigid from nonrigid motion: Minimum points and views. *Percept. Psychophys.*, **47**, 205–214.

Braunstein, M. L., Liter, J. C. & Tittle, J. S. (1993). Recovering 3-D shape from perspective translations and orthographic rotations. *J. Exp. Psychol.: Human Percept. Perform.*, **19**, 598–614.

Caudek, C. & Proffitt, D. R. (1993). Depth perception in motion parallax and stereo-kinesis. *J. Exp. Psychol.: Human Percept. Perform.*, **19**, 32–47.

Cornilleau-Pérès, V. & Droulez, J. (1989). Visual perception of surface curvature: Psychophysics of curvature detection induced by motion parallax. *Percept. Psychophys.*, **46**, 351–364.

Cortese, J. M. & Andersen, G. J. (1991). Recovery of 3-D shape from deforming contours. *Percept. Psychophys.*, **49**, 315–327.

Dosher, B. A., Landy, M. S. & Sperling, G. (1989). Ratings of kinetic depth in multidot displays. *J. Exp. Psychol.: Human Percept. Perform.*, **15**(4), 816–825.

Eby, D. W. (1992). The spatial and temporal characteristics of perceiving 3-D structure from motion. *Percept. Psychophys.*, **51**, 163–178.

Gogel, W. C. (1973). The organization of perceived space: I. Perceptual interactions. *Psychol. Forsch.*, **36**, 195–221.

Green, B. F., Jr. (1959). *Mathematical notes on 3-D rotations, 2-D perspective transformations, and dot configurations* (Group Report No. 58–5). Massachusetts Institute of Technology, Lincoln Laboratory, Lexington, MA.

Green, B. F., Jr. (1961). Figure coherence in the kinetic depth effect. *J. Exp. Psychol.*, **62**, 272–282.

Grimson, W. E. L. (1981). *From images to surfaces: a computational study of the human early visual system.* MIT Press, Cambridge, MA.

Hildreth, E. C., Grzywacz, N. M., Adelson, E. H. & Inada, V. K. (1990). The perceptual buildup of three-dimensional structure from motion. *Percept. Psychophys.*, **48**, 19–36.

Huang, T. & Lee, C., (1989). Motion and structure from orthographic projections. *IEEE Transactions on Pattern Analysis and Machine Intelligence*, **11**, 536–540.

Husain, M., Treue, S. & Andersen, R. A. (1989). Surface interpolation in three-dimensional structure-from-motion perception. *Neural Computation*, **1**, 324–333.

Johansson, G. (1950). *Configurations in Event Perception.* Almqvist & Wiksell, Stockholm, Sweden.

Johansson, G. (1958). Rigidity, stability, and motion in perceptual space. *Acta Psychol.*, **14**, 359–370.

Johansson, G. (1970). On theories for visual space perception: A letter to Gibson. *Scand. J. Psychol.*, **11**, 67–74.

Julesz, B. (1971). *Foundations of Cyclopean Perception.* University of Chicago Press, Chicago.

Kaiser, M. K. (1990). Angular velocity discrimination. *Percept. Psychophys.*, **47**, 149–156.

Kaiser, M. K. & Calderone, J. B. (1991). Factors influencing perceived angular velocity. *Percept. Psychophys.*, **50**, 428–434.

Kellman, P. J. & Shipley, T. F. (1991). A theory of visual interpolation in object perception. *Cogn. Psychol.*, **23**, 141–221.

Koenderink, J. J. (1990). *Solid Shape.* MIT Press, Cambridge, MA.

Koenderink, J. J. & van Doorn, A. J. (1991). Affine structure from motion. *J. Opt. Soc. Am. A*, **8**, 377–385.

Landy, M. S., Dosher, B. A., Sperling, G. & Perkins, M. E. (1991). The kinetic depth effect and optic flow: II. First- and second-order motion. *Vision Res.*, **31**, 859–876.

Lappin, J. S. & Fuqua, M. A. (1983). Accurate visual measurement of three-dimensional moving patterns. *Science*, **221**, 480–481.

Liter, J. C., Braunstein, M. L. & Hoffman, D. D. (1993). Inferring structure from motion in two-view and multi-view displays. *Perception,* **22**, 1441–1465.

Mather, G. (1989). Early motion processes and the kinetic depth effect. *Q. J. Exp. Psychol.: Human Exp. Psychol.*, **41**, 183–198.

Metzger, W. (1934). Tiefenerscheinungen in optischen Bewungsfeldern. *Psychol. Forsch.*, **20**, 195–160.

Miles, W. R. (1931). Movement interpretations of the silhouette of a revolving fan. *Am. J. Psychol.*, **43**, 392–405.

Musatti, C. L. (1924). Sui fenomeni stereocinetici. *Archivio italiano di psicologia*, **3**, 105–120.

Nawrot, M. & Blake, R. (1991). The interplay between stereopsis and structure from motion. *Percept. Psychophys.*, **49**, 230–244.

Norman, J. F. & Lappin, J. S. (1992). The detection of surface curvatures defined by optical motion. *Percept. Psychophys.*, **51**, 386–396.

Ono, H., Rogers, B. J., Ohmi, M. & Ono, M. E. (1988). Dynamic occlusion and motion parallax in depth perception. *Perception*, **17**, 255–266.

Petersik, J. T. (1991). Perception of three-dimensional angular rotation. *Percept. Psychophys.*, **50**, 465–474.

Pong, T., Kenner, M. A. & Otis, J. (1990). Stereo and motion cues in preattentive vision processing: Some experiments with random-dot stereographic image sequences. *Perception*, **19**, 161–170.

Proffitt, D. R., Rock, I., Hecht, H. & Schubert, J. (1992). The stereokinetic effect and its relation to the kinetic depth effect. *J. Exp. Psychol.: Human Percept. Perform.*, **18**, 3–21.

Ramachandran, V. S., Cobb, S. & Rogers-Ramachandran, D. (1988). Perception of

3-D structure from motion: The role of velocity gradients and segmentation boundaries. *Percept. Psychophys.*, **44**, 390–393.

Rogers, B. J. & Collett, T. S. (1989). The appearance of surfaces specified by motion parallax and binocular disparity. *Q. J. Exp. Psychol. Section A – Human Exp. Psychol.*, **41**, 697–717.

Rogers, B. & Graham, M. (1979). Motion parallax as an independent cue for depth perception. *Perception*, **8**, 125–134.

Rouse, M. W., Tittle, J. S. & Braunstein, M. L. (1989). Stereoscopic depth perception by static stereo-deficient observers in dynamic displays with constant and changing disparity. *Optometry Vision Sci.*, **66**, 355–362.

Runeson, S. (1977). On the possibility of smart perceptual mechanisms. *Scand. J. Psychol.*, **18**, 172–179.

Saidpour, A., Braunstein, M. L. & Hoffman, D. D. (1992). Interpolation in structure from motion. *Percept. Psychophys.*, **51**, 105–117.

Saidpour, A., Braunstein, M. L. & Hoffman, D. D. (1994). Interpolation across surface discontinuities in structure-from-motion. *Percept. Psychophys.*, **55**, 611–622.

Sperling, G., Landy, M. S., Dosher, B. A. & Perkins, M. E. (1989). Kinetic depth effect and identification of shape. *J. Exp. Psychol.: Human Percept. Perform.*, **15**, 826–840.

Sperling, G., Dosher, B. A. & Landy, M. S. (1990). How to study the kinetic depth effect experimentally. *J. Exp. Psychol.: Human Percept. Perform.*, **16**, 445–450.

Thompson, W. B., Kersten, D. & Knecht, W. R. (1992). Structure-from-motion based on information at surface boundaries. *Biol. Cybern.*, **66**, 327–333.

Tittle, J. S. & Braunstein, M. L. (1993). Recovery of 3-D shape from binocular disparity and structure from motion. *Percept. Psychophys.*, **54**, 157–169.

Todd, J. (1985). The perception of structure from motion: Is projective correspondence of moving elements a necessary condition? *J. Exp. Psychol.: Human Percept. Perform.*, **11**, 689–710.

Todd, J. T. & Bressan, P. (1990). The perception of 3-dimensional affine structure from minimal apparent motion sequences. *Percept. Psychophys.*, **48**, 419–430.

Todd, J. T. & Norman, J. F. (1991). The visual perception of smoothly curved surfaces from minimal apparent motion sequences. *Percept. Psychophys.*, **50**, 509–523.

Turner, J., Braunstein, M. L., Andersen, G. J. (in press). Detection of surfaces in structure from motion. *J. Exp. Psychol.: Human Percept. Perform.*

Ullman, S. (1976). Filling-in the gaps: The shape of subjective contours and a model for their generation. *Biol. Cybernet.*, **25**, 1–6.

Ullman, S. (1979). *The Interpretation of Visual Motion*. MIT Press, Cambridge, MA.

Uttal, W. R. (1985). *The Detection of Nonplanar Surfaces in Visual Space*. Erlbaum, Hillsdale, NJ.

Wallach, H. & O'Connell, D. N. (1953). The kinetic depth effect. *J. Exp. Psychol.*, **45**, 205–217.

Part 6
Motion Detection and Eye Movements

14

Visual Motion Caused by Movements of the Eye, Head and Body

Laurence R. Harris

Department of Psychology, York University, Toronto, Canada

1 VISUAL MOTION ON THE RETINA RESULTING FROM SELF MOTION

1.1 Uses and Problems

Most of the chapters in this book up to this point have considered the processing of visual movement seen by an eye that is stable in space. Of course this never happens. The eyes are always in motion and most retinal image motion is caused by movement of the observer. Compensatory eye movements can often reduce the retinal movement but can never remove it completely. This chapter discusses the visual consequences of self motion including all motions of the eyes in space. The eyes can move both under the influence of the extraocular muscles and as a consequence of head movement. The head in turn can move both under the influence of the neck musculature and as a consequence of body movement. Normally the eyes move as a consequence of a combination of these causes.

It is misleading to think of the retinal image as primary when considering visual perception. This notion is easily dispelled by comparing the small, inverted, optically-distorted and blood-vessel-interrupted retinal image with the pristine quality of visual experience. Self motion is particularly removed from its retinal origins since it is derived from many sensory sources and can even arise independently of visual information. The traditional view that self motion contributes a distortion to the retinal image which must be hidden from perception is misleading.

Observers must distinguish between self- and externally-generated retinal motion to

VISUAL DETECTION OF MOTION
ISBN 0–12–651660–X

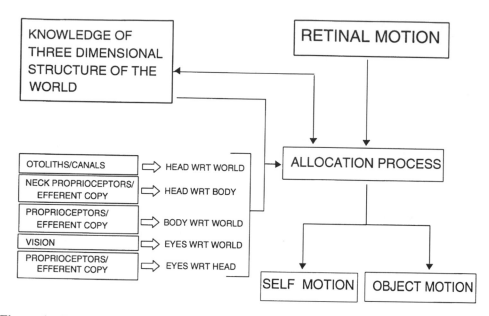

Figure 1 The sources of information about self motion which can help in the allocation of retinal motion to self motion and object motion components. The various sources are shown as giving information with respect to (WRT) different reference frames. Information about the three-dimensional structure of the world is also required. There is a reciprocal relationship in which self motion helps in obtaining three-dimensional structure information and three-dimensional structure helps in obtaining self motion information.

use and perceive retinal motion appropriately. Three basic sources of information about self motion are available during natural movement: the visual consequences, the motor commands and signals from proprioceptors (Figure 1). Care must be taken to distinguish information that is *present* and information that is *used*.

Once retinal motion is attributed to self motion, it can provide important information about that self motion. This is *proprioceptive vision* (Simpson *et al.*, 1988a, 1988b; Nakayama, 1985; Harris & Jenkin, 1993). The same retinal motion is also important for determining the three-dimensional structure of the world: a topic which is beyond the scope of this chapter (see Regan, 1991, for a recent review). These concepts are interactive since a knowledge of the three-dimensional structure of the world is needed to interpret self motion information and vice versa.

1.2   Coordinate Systems, Units and Vectors

Understanding the consequences of self motion requires using appropriate coordinate systems and units. The literature is often confusing on this point with often one system used to describe visual movement but another to describe the self movement that created it. A coordinate system requires a reference frame and a geometry within that frame. Possible reference frames include the retina, the head, the body or some part of external space. Possible geometries include the familiar Euclidean or Cartesian system of three orthogonal axes. This geometry is an arbitrary and biologically-implausible choice (Simpson et al., 1981, 1988a) and biological axes are much more likely to be non-orthogonal and non-linear. There is no 'correct' solution and representations in one system can always be converted into another. Often additional information is required for the conversion, however. For example, converting information relative to the retina into a head-referenced system requires information about the position of the eyes in the head. Schemes that claim to be 'independent of coordinate systems' (e.g. Viirre et al., 1986, p. 446) are misleading, merely referring to convenient mathematical techniques for moving between systems.

The choice of units is equally important. The motion of each eye in space is a combination of rotation and translation. The speed of rotation is the rate of change of angle (e.g. degrees/second) whereas the speed of a translation is the rate of change of distance (e.g. metres/second). Notice that neither of these units (or the units of angle, distance or time that make them up) is likely to have linear biological counterparts. Translational velocity can only be calculated using extraretinal information.

When talking about rotations, the terms 'horizontal', 'vertical' and 'torsion' are inherently inappropriate and reflect historical attempts to represent three-dimensional rotations on two-dimensional paper. A convenient method that avoids some of the pitfalls of these terms (see Tweed & Vilis, 1987; Carpenter, 1988) is to express the rotation as being about an axis of specified orientation. This is called a vectorial representation. Vectors can be usefully used not only to describe the rotation of the eye or head, but also when describing the retinal consequences (see Figure 2).

1.3 Allowing for the Contribution of Eye Movements

Dealing with self-generated sensory information is not a problem unique to vision; it is a general problem of all sensory systems. Distinguishing self-produced from external sounds is a similar problem. Once self-generated sensory stimuli have been identified, a possible way to handle them is to remove or 'cancel' them from the percept leaving any activity of external origin uncontaminated (Von Holst & Mittelstaedt, 1950).

When self- and externally-generated components are independent, as they are for sounds, removing the contribution of internally-generated stimuli is indeed often appropriate. For vision, however, the self-generated and external aspects of the retinal image interact: which external object features are present on which parts of which retina depends on the self-generated movements. The interaction between the internal and external components contains important information for interpreting both the internal and external generators. Although retinal image motion as a consequence of self motion presents a problem for clear vision and a challenge to interpretation of the image, simply 'cancelling' it and pretending that nothing has happened is not a valid solution – things that it is important to know about have happened. For example there might have been changes in the direction of heading. New areas of the visual scene might have come into view and so on.

A simple subtraction process of the motion of the eye from the retinal motion, as was implied by Von Holst and Mittelstaedt (1950), cannot in any case be adequate. Consider the coordinate and the unit systems involved. It is the eye movements in three-dimensional space including both angular and linear components that need to be taken into account in the interpolation of the complex visual movement arising simultaneously on two retinae.

1.4 How Well Does the System Need to Work? Tolerance of Retinal Image Slip

Carpenter (1988), using Green and Campbell's (1965) data, calculates that visual acuity should be degraded by the equivalent of 2 dioptres of myopia for a 1 d/s movement. Potential visual degradation is often cited as a *raison d'être* for compensatory eye movements: to maintain a stable retinal image. When the head is kept still, image motion can be kept down to about 0.5 d/s by compensatory eye movements (Skavenski *et al.*, 1979). However, during normal head movement, image slippage, even at the fixation point, is often in excess of 5 d/s (Steinman & Collewijn, 1980; Steinman & Levinson, 1990; Sperling, 1990). Omnipresent retinal slip generates two obvious questions: Why is vision not degraded? And why is the image movement unnoticed?

Self-generated retinal slip does not seems to degrade vision as much as it should (Murphy, 1978; Westheimer & McKee, 1975, 1978; see Steinman & Levinson, 1990, for a comprehensive review). Movement of the image due to self motion alters the contrast sensitivity function (CSF): high spatial frequencies require more contrast to be detected and lower spatial frequencies are easier to see (Steinman *et al.*, 1985). The changes may actually be beneficial to vision (Tulunay-Keesey & VerHoeve, 1987). Compensatory eye movements may keep or introduce desirable movement, especially enhancing the visibility of large objects (Skavenski *et al.*, 1979; see Carpenter, 1991). The changes to the CSF during active motion are less pronounced than when the same movement is externally generated (cf. Steinman *et al.*, 1985; with Kelly, 1979).

There seems to be a general depression of motion sensitivity that accompanies self motion. Reduction in sensitivity is a well-known correlate of saccadic eye movement (where it is called saccadic suppression: section 2.5.2.1) but the reduction actually occurs during all types of self motion (see, for example, section 4.4). Physical factors such as blur and relative motion have an effect, but the extent of the difficulty of seeing object motion indicates a central suppression mechanism. It seems that detection of one kind of motion (self motion) affects the ability to detect other kinds of motion (object motion). Such an interaction suggests a common mechanism for *all* motion detection arising from both visual and non-visual sources.

2 HANDLING RETINAL MOTION DUE TO EYE MOVEMENT ALONE

The centre of rotation of the eye is essentially fixed in the head (but see Harris *et al.*, 1993a), and therefore, if the head does not move, the only movements of the eye with respect to the head, body or space are pure rotations. The sources of information which might be used to determine whether image motion was caused by eye rotation in the head are vision, efference copy of eye movement commands and proprioception from the extraocular muscles.

Different perceptual tasks have different uses for eye movement information. The

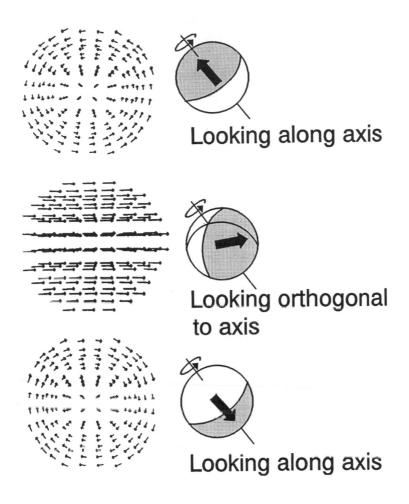

Looking along axis

Looking orthogonal to axis

Looking along axis

Figure 2   The optic flow associated with pure rotation of the eye. On the left side of the figure, the small arrows trace out the paths of dots equally spaced in the visual field. The three circles are planar projections of the hemispherical view looking in different directions with respect to the axis of rotation. The views are shown shaded on the right-hand side of the figure. Describing the rotation needs only an instantaneous representation of the orientation of the axis and the velocity of rotation: a vectorial description. When looking along the axis (top and bottom sections of figure), the movement in the centre of the field is pure rotation, there is no translation of these features across the retina. When looking orthogonal to the axis of rotation (centre section), the motion in the centre of the field is pure translation. These planar projections of three dimensional events need careful interpretation. Remember that *the observer is at the centre of the circles* (in the plane of the paper). The arrows at the top and bottom of the centre circle represent the movement of dots pointed to by the axis of rotation: the same movement that is represented in the centre of the upper and lower circles. Similarly the arrows around the edges of the top and bottom circles represent the movement orthogonal to the axis: the same movement that is represented by the arrows in a horizontal line through the centre of the centre circle (left-hand side redrawn from Andersen, 1986, p. 54).

perception of a stable world needs to take into account the resulting position change. The perception of object movement during pursuit needs to take into account the eyes' velocity. The nature of the signals that drive smooth pursuit is considered in the chapter by Krauzlis. Eye-in-orbit information is important when the head *is* free to move too, in order to distinguish eye rotation from that of the head or body. Rotations of each of these have different perceptual and practical problems associated with them.

Optokinetic responses (OKN) to full-field motion normally represent the response to the visual consequences not of an eye movement but of a head movement. OKN is therefore considered below under head movements (section 3).

2.1   Visual Cues Concerning Eye Movement

The obvious visual cue that suggests that retinal motion was caused by eye rotation is that the entire field moves together at the same speed.[1] All parts of the field have a constant angular velocity added on to any existing movement. Because the retina is hemispherical, images from different parts of the visual field have different retinal motions depending on their visual direction with respect to the axis of rotation. The part of the image directly orthogonal to the axis translates across the retina whereas those parts of the image at the ends of the axes rotate and do not change their retinal locus (Figure 2). Full-field rotation above 0.5 deg/s is perceived by a stationary observer and the orientation of the axis can be judged to within 5 deg (Harris & Lott, 1993).

2.2   Efference Copy Describing Eye Movements

Evidence for the use of an efference copy of the motor signals controlling eye movements (Helmholtz, 1866; Sperry, 1950) is largely circumstantial. Other sources of information might be too slow. There is some evidence that compensation for saccadic eye movement commences before the movement itself (Duhamel *et al.*, 1992). However, the visual world is not very stable actually during eye movements: vision is suppressed during fast saccades and the world often seems to move around all over the place during slower pursuit movements.

A theoretical reason for proposing an efference copy is for engineering stability in the control of eye movements. Making an internal comparisons between the intended movement and the movement-carried-out-so-far allows an eye control system to function at high gain and therefore helps make fast and accurate eye movements

---

[1] Actually, since the nodal point of the eye lies some 6.2 mm in front of the centre of rotation of the eye (Alpern, 1962), the consequences of even a pure rotation of the eye will be affected by translation of the nodal point. The effect of this translation depends on the distance of the object. For the image of an object at 25 cm, the displacement of the nodal point will alter the maximum velocity by 2.5% (cf. section 3.1. and Appendix A).

(Robinson, 1975; Guthrie *et al.*, 1983; Van Gisbergen *et al.*, 1981 but cf. Sparks, 1986).

For an efference copy to be useful, either for eye movement control or for perceptual cancellation, it has to describe the relevant eye movement fully in three dimensions and each eye has to be dealt with separately since they might be doing different things. Using a vectorial representation of eye-in-head information to interpret visual (eye-in-space) information needs considerable sophistication and some extraretinal information. Furthermore, a record of position deduced by keeping track of continuous movements is extremely vulnerable to cumulative error.

2.3   Proprioception Describing Eye Movements

When skeletal muscles are pulled, various associated proprioceptors are activated including joint receptors and muscle spindles (see Matthews, 1982). These signals are important for skeletal muscle control. Some component of the signal probably represents limb position and may underlie the sense of where limbs are in space (Roll *et al.*, 1991a,b). Using the proprioceptor system of skeletal musculature as an analogy, oculomotor proprioceptors have been proposed as a source of eye position information (Sherrington, 1918). The eye's motion is very different from the essentially one-dimensional movement of a hinged skeletal muscle joint, however, and the problem of defining reference points from which to describe the eye's position is formidable. The distribution and variation in type of proprioceptors in the oculomotor muscles is rich and unusual (Spencer & Porter, 1988) and there is a massive representation of oculomotor proprioception throughout the brain (Abrahams & Rose, 1975; Donaldson & Long, 1980; Batini *et al.*, 1974; Kimura *et al.*, 1991). Circumstances suggest that proprioception might carry useful information about the activity of extraocular muscles (Steinbach, 1987). Whether extraocular proprioceptive information under normal conditions is related to eye position, dynamic eye movement control or both, is unknown.

2.4   Comparison of Proprioception and Efference Copy Contributions

A stable world is perceived during fixation periods between eye movements when visual, efference copy and proprioceptive information are all available (Howard, 1993). The relative contribution of these three cues during and between eye movements has traditionally been assessed by looking at each one alone. There has been little work on the use of vision in this context, perhaps because it has been seen as the problem rather than the solution.

During attempted eye movements under paralysis, an efference copy should be present but without visual or proprioceptive signals about eye movement. If efference copy were used then, under these circumstances, visual motion should be perceived, as the presumed motion of the eye is subtracted from the unexpectedly stable image. Early experiments reported just such a phenomenon (Mach, 1886; Brindley & Merton, 1960; Kornmüller, 1931). But careful repetition, ensuring *complete* paralysis, found no illusory visual movement during attempted eye movements (Brindley *et al.*, 1976;

Stevens *et al.*, 1976). The expected brisk illusory motion of a stabilized image during normal saccades is also not experienced (Grüsser *et al.*, 1987). It seems that some other relevant signals are required to make use of an efference copy.

Pulling the eye around can be sensed without vision, presumably from proprioceptive information (Skavenski, 1972). By carefully manipulating an occluded eye with a finger while the unobstructed eye viewed a small fixation point in a dark room, Gauthier *et al.* (1990a, 1990b) and Bridgeman and Stark (1991) claimed, rather improbably, to be able to evoke the same proprioceptive inflow as might accompany a natural movement. Poked eye movements were perceived as 16% to 25% of their actual sizes. Pushing on the uncovered eye evoked both efferent (resisting the push) and afferent activity from which the efferent-only contribution could be calculated as indicating movement of 61% of the actual movement (Stark & Bridgeman, 1983, Bridgeman & Stark 1991). Bridgeman and Stark (1991) suggested that proprioceptive and efferent information might be additive although it is unclear what advantages such a combination might impart.

A dim spot fixated in front of a dark background appears to drift around ('*autokinesis*', Aubert, 1887; see Howard, 1982). Although a comprehensive explanation of autokinesis is lacking, the phenomenon suggests that, in the absence of both efference copy of movement commands and visual information, proprioceptive knowledge about eye position is inadequate to achieve visual stability.

2.5 Does it Work? How Good is Vision During Eye Movement?

There are two consequences of the successful division of the visual signal into internally- and externally-generated components. Firstly, the perceptual stability of the world is not disrupted by eye movements. Secondly, external object movement can be discerned during eye movements.

2.5.1 Judging eye position and world stability

The retina only has good resolution in the fovea. Knowledge of where the eyes are pointing compared with where they were pointing during previous fixations is required to relate sequentially foveated views in space. Knowledge of change-of-position, however, is not the same as knowledge about motion. Although position can be derived from motion, eye position is probably determined largely from information available when the eye is at rest (Howard, 1993). The eyes can be repositioned with an accuracy of 2–4 deg after a gaze change (Hansen & Skavenski, 1977; Lemij & Collewijn, 1989). This level of repeatability seems crude compared with our perception of features as being in precisely the same place every time we fixate them. But in fact target shifts of 2–4 deg during saccades are not detectable (Bridgeman, 1983).

Knowledge of eye position, even using visual, proprioceptive and efferent copy information, would be a very unreliable way to confirm that the world indeed did not move during saccades. Such a system would be fraught with missed movements and false positives.

2.5.2 *Judging object motion perception during eye movements*

2.5.2.1 *During saccadic eye movements*
One of the most obvious things about visual perception during saccades is that it is suppressed (Volkmann *et al.*, 1978; Volkmann, 1986; Matin, 1974). Whilst a lot of the reduction in visual sensitivity can be attributed to the optical consequences of high-speed image movement (Carpenter, 1991), the effect is larger that can be accounted for by physical blur of the image, especially at low saccadic velocities (Burr *et al.*, 1982) and some central suppression mechanism is required although its contribution during large saccades is 'feeble' (Carpenter, 1988). Here we are concerned with just one aspect: the ability to see motion during a saccade: that is, the success with which retinal and object motions can be disentangled.

The threshold for motion detection during saccadic eye movements increases linearly with both saccadic amplitude and velocity but is more closely correlated to amplitude (Bridgeman *et al.*, 1975). Interestingly it does not matter which direction the displacement is in (Bridgeman & Stark, 1979; Ilg & Hoffmann, 1993) suggesting a generalized loss of sensitivity comparable to that seen during other self motions. The phenomenon may be explainable entirely by the high retinal speeds (and thus difficult discriminations) involved. Identical retinal stimuli were not compared in the eye moving and in the eye stationary controls. Brooks *et al.* (1980) found that when saccadic retinal velocities were accurately simulated on a stationary eye, motion discriminations were comparable to those during actual saccadic movements. More experiments and detailed reporting of the stimuli used is required before firm conclusions can be made about what happens during and following a saccade.

2.5.2.2 *During smooth pursuit eye movements*
How well is motion of the visual image that results from smooth pursuit eye movements compensated for? The question has two aspects: how well can we perceive the external motion of a pursued target and how well can we ignore (that is, assign to the consequence of the tracking eye movement) the retinal movement of everything else?

Two sources of information contribute to identifying the motion of a pursued object: (i) the visual movement of the object relative to its background and (ii) knowledge about the pursuit eye movement. The perceived speed of a pursued target moving in the fronto-parallel plane is only about 70% of the perceived speed of the same target when it is not pursued. The reduction of perceived velocity is independent of the presence of a background and is known as the *Aubert–Fleischl phenomenon* (Fleischl, 1882; Aubert, 1886, 1887; Gibson *et al.*, 1957; Dichgans *et al.*, 1969, 1975; Mack & Herman, 1972, 1973, 1978). The phenomenon suggests that efference copy and proprioceptive information about smooth pursuit together underestimate the eye velocity. Alternatively, estimates of the speed of targets that are not pursued may be too fast.

Judgements of *external* target movement should be in *linear* terms (m/s). The conversions from either the angular retinal velocity of targets that are not pursued or the angular eye velocity associated with targets that are pursued, to external linear movement requires distance information (see Gogel, 1982 and Appendix B). Perhaps the two assessments are equally accurate in angular terms but differ in their access to distance information. Targets presented during smooth pursuit can be hit accurately (Hansen, 1979) suggesting that at least some levels of perception can do the job.

And what of the other side of the coin, the appearance of the background?

> If you move your finger back and forth at arms length, and follow it with your eyes, the appearance of the background is actually rather hard to describe. It appears to move, yet we know it isn't really moving; whereas during a saccade we have no sense at all that anything has moved except our gaze.
>
> Carpenter (1988) p. 333

The illusory movement of a stationary background as a pursued target passes in front of it is known as the *Filehne effect* (Filehne, 1922). The phenomenon is compatible with the velocity of pursuit eye movements being underestimated (Mack & Herman, 1973, 1978; Wertheim & Bles, 1984; Wertheim, 1985). If the eyes pursue a target accurately but are registered as moving at only 70% of their true velocity then the perceived target movement cannot account for all the relative motion present. The remaining 30% must be due to external movement.

Wertheim (1981), measured thresholds for imposed movement of the background during smooth pursuit. He found that background drift up to 10–15% of the eye velocity could not be distinguished from stationary backgrounds. Even suprathreshold movements of the background in either direction were underestimated (Wertheim & Van Gelder, 1990).

Taken together these observations suggest that knowledge of eye movements is used in the perceptual process. Vision, efference copy and proprioception contribute but the process is imprecise to say the least. The retinal motion that is actually due to eye movement but that is not attributed to this cause should be interpreted as indicating alarming movement of the outside world. Normally the problem is solved by suppression.

3 HANDLING RETINAL MOTION DUE TO HEAD ROTATION

The head's axis of rotation cannot pass through both eyes[2] and so head rotation, even with the rest of the body still, causes both rotation and translation of the eyes. Rotation and translation have their own distinctive retinal consequences. *Both* have to be correctly attributed to the head rotation that caused them in order to reveal any remaining motion of external origin.

During head rotation, some sources of information in addition to vision and oculomotor afferents and efferents are potentially available. These include (i) activity of the vestibular system, (ii) activity of neck proprioceptors, (iii) an efference copy of instructions to move the head and possibly (iv) activity of an efference copy of compensatory eye movement control signals. The possible existence of the latter signal does not necessarily follow from the existence of an efferent copy of pursuit and saccadic eye movements (Bedell, 1990).

---

[2] The only exception, rotation about a horizontal axis through the centres of rotation of both eyes, actually does not occur naturally. The natural axis for pitch rotations of the head lies well behind the eyes and so even pitch of the head is normally associated with eye translation.

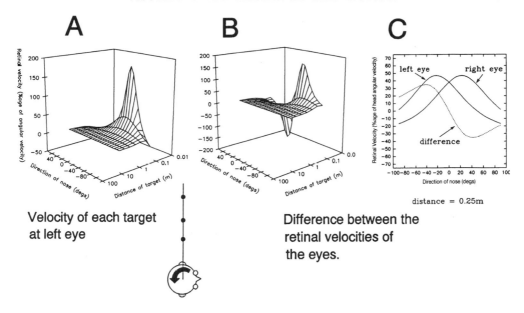

Velocity of each target
at left eye

Difference between the
retinal velocities of
the eyes.

distance = 0.25m

Figure 3 The angular retinal velocity introduced by the linear component of an angular head movement (see Appendix A for derivation). These retinal velocities are *in addition* to those due to the angular velocity. The velocity depends on the distance and eccentricity of each point and is plotted here as a percentage of the angular head velocity for objects from 0.125 to 8 m distant as the head sweeps through 180° as shown in the insert. *Panel A* shows a 3D plot of the effect of eccentricity and distance for objects seen with the left eye. Notice that at 12.5 cm (the closest object distance shown), the *additional* velocity reaches some 150% of the angular velocity when the object is in the direction pointed to by the line joining the eye and the centre of the head's rotation. Thus, during a head velocity with a peak of 200 deg/s, an object at that distance and eccentricity would have an angular velocity of 500 deg/s at the eye. *Panel B* plots the difference between the two eyes for the same X/Y values as panel A. There are dramatic differences (up to around 150% of the angular velocity or up to 300 deg/s for a 200 deg/s head movement) at quite modest eccentricities for close (0.125 m) targets. *Panel C* plots the retinal velocity for each eye and the difference between them for one distance (25 cm). At this distance the retinal velocities reach almost 50% of the head velocity. Thus, if the head had an angular velocity of 200 deg/s, corresponding to taking 900 ms to traverse the horizontal axis (from + to − 90 deg), the peak angular velocity in each eye (ignoring any compensatory eye movement) would be close to 300 deg/s and the difference would be about 70 deg/s alternating from fastest in the left eye as the object lined up with that eye to fastest in the right eye when the object lined up with that eye some 44 deg further on in the head movement. The difference between the two eyes would be zero when the object lined up with the subject's nose. Notice also that the angular speed of the object is actually reduced by the linear component at eccentricities greater than about 45 deg. (Interocular distance: 6 cm; eyes to centre of head rotation: 8 cm.)

3.1   Visual Cues Concerning Head Rotation

During a head rotation, the motion of each point on each retina is the vector sum of motion due to rotation and the motion due to translation of the eyes. The rotation component is equivalent to rotation around an axis passing through the eye, parallel to the head's rotation axis (see section 2.1 and Figure 2).

The direction of the translation component is tangential to the circle described by the movement of the eye in space and it follows that the direction is therefore *different for each eye* (Figure 3). The translation speed depends on the distance between the axis of rotation and the eye and on the rate of rotation.

The retinal consequences of translation depend on target distance. Since distance is not available from the retinal image, even from both retinal images, it follows that the direction and speed of eye translation cannot be deduced from retinal image movement alone. Objects at different distances have different retinal velocities generated by the translation component of head rotation which can be conveniently expressed as a percentage of the head rotation velocity (Figure 3). When the distance between the object and eye is large relative to the distance between the eye and the axis of rotation, this component of image movement during head rotation becomes negligible. The retinal velocity of distant points is determined exclusively by the angular rotation (see Figure 2). At close viewing distances the effects of the translation of the eye associated with head rotation can be dramatic (Figure 3, Appendix A).

Since the distances from a given point to each eye are usually different, the retinal velocities, as well as the translation direction, are usually different in the two eyes for each point (Figure 3C).

3.1.1   Assignment of visual motion to self motion (circularvection)

Visual motion compatible with head rotation often evokes the sensation of physical rotation even when the observer is stationary (*circularvection*: Mach, 1875; Fischer & Kornmüller, 1930; Howard, 1982). Circularvection can be used as an indicator of when visual movement is attributed to head rotation. Contrary to intuition, it is not necessary for a large part of the field to move (Brandt *et al.*, 1973) provided that the moving area is perceived as the background (Ohmi & Howard, 1987; Howard & Heckmann, 1989).

The speed of perceived circularvection, for a given angular velocity, depends on the perceived distance of the evoking stimulus (Wist *et al.*, 1975). The rate of head rotation during circularvection is derived from a retinal image whose movement, were it genuinely generated by a head rotation, is partly due to the rotation component and partly to the consequent translation of the eyes. Misperception of the relative contribution of translation arises from a misperception of depth. More systematic studies of the roles of the rotational and translational components of head rotation are needed.

3.2   Vestibular Cues Concerning Head Rotation

The vestibular system is an important proprioceptive sense for monitoring head movement. The system comprises two parts, the semicircular canals which transduce angular accelerations, and the otolith organs which transduce linear accelerations (see Benson, 1980; Wilson & Melvill Jones, 1979 for reviews)

3.2.1 Semicircular canals and head rotation

The semicircular canals provide important information about head rotation. Although the canals are activated by angular *acceleration*, they perform a mechanical integration so that they signal the head's angular *velocity* over the normal physiological range (Fernandez & Goldberg, 1971). There are three canals roughly orthogonal to each other and so the axis of rotation is mechanically broken down into three vectors. These three vectors are to some extent kept separate within the brain (e.g. vestibular nucleus; Curthoys & Markham, 1971) although there is convergence between signals coming from different canals (Baker *et al.*, 1984). Thus head rotation velocity and axis of rotation information are provided by the semicircular canal system during head rotation.

3.2.2 Otoliths and head rotation

The otolith system operates independently of the semicircular canals. Otoliths are sensitive to linear acceleration (Benson, 1980; Wilson & Melvill Jones, 1979). The system consists of arrays of hair cells that respond maximally when bent in a certain direction (Loe *et al.*, 1973). Thus, *which* hair cells respond provides the direction of an acceleration and *how much* they respond is related to the magnitude of the acceleration.

During a head rotation those otolithic hair cells tuned to directions radiating out from the rotation axis are activated by the centrifugal force. If the distance of the otoliths from the rotation axis were known, their activation under these circumstances could provide information about the rate, but not the direction, of the rotation. It is not known if this information is used when the axis is within the head, but when it is outside the head, as in eccentric rotation or as a component of curvilinear translation, the concurrent otolith activation does play a role (see section 4.4.1).

The otoliths can provide useful information about the rate and plane of head rotation when the axis of rotation is within the head but not orthogonal to gravity. Exactly which hair cells are most active at any one time depends on the head's orientation with respect to gravity. If the axis of rotation is tilted, orientation varies systematically as rotation proceeds. The associated ripple of otolith stimulation evokes strong compensatory eye movements in some species (monkeys: Raphan *et al.*, 1981; cats: Harris, 1987; Darlot & Denise, 1988) but not in man (Guedry, 1965; Harris & Barnes, 1985; Darlot *et al.*, 1988; Benson & Bodin, 1966).

3.3 Neck Muscle Proprioception and Head Rotation

Neck muscle proprioceptors might potentially provide head-on-body information although this system shares many of the difficulties of the extraocular proprioceptive system. The freedom of the head functionally to move in a ball-and-socket-like arrangement makes reference points difficult to define or discover. There is no monosynaptic reflex in neck muscles (Abrahams *et al.*, 1975a,b) suggesting an unusual innervation. Circumstantial evidence suggests that neck proprioceptors are important since they are extensively represented in the brain (Boyle & Pompeiano, 1981a,b; Anastasopoulos & Mergner, 1982; Mergner *et al.*, 1985).

The contribution of neck proprioception to the sense of head movement and position has been investigated by holding the subject's head still in space and twisting the body beneath it (e.g. Mergner *et al.*, 1991) thus achieving the same relative displacement of the head-on-body as during a natural head movement. Visual, vestibular and efference copy sources are silent during these body-alone rotations. Eye movements and sensations of rotation are evoked which appear roughly in accord with the supposed head movement (*the cervico-ocular reflex*: Bles & Dejong, 1982). The difference between neck proprioceptor activity during these imposed neck twists and their activity during natural head movements, however, makes these experiments hard to interpret. Both vibratory stimulation of neck muscles (Biguer *et al.*, 1988; Roll *et al.*, 1991a,b) and rotation of the body beneath a stationary head (Mergner *et al.*, 1992a,b) causes illusory movements of a head-stationary visual target suggesting illusions of head movement (cf. section 3.5.2).

3.4 Efference Copy Concerning Head Rotation

Obviously, when the head is voluntarily rotated on the shoulders, motor command signals must be present in the brain. It might be useful to take an efference copy of these signals to act as a reference during movement control or to aid in assessing the direction of gaze in space. The use of efference copy for head movement control and monitoring has all the same problems as it does for eye movement control (difficulty of defining a frame of reference, cumulative error, etc.; cf. section 2.2). In addition, the head presents a potentially variable load to its muscles and so head position derived from the efferent commands is likely often to be inaccurate. Since the vestibular proprioceptive sense is so fast and effective (10 ms; Lisberger, 1984; Snyder & King, 1992) there can be little need for other sources of fast information about head position. One function of efferent copy of head motor commands might paradoxically be to keep track of the position of the *body* relative to the head since the primary spatial orientation detectors are in the head.

Predictable head movements evoke eye movements that are more efficient in their phase and gain relations to the visual stimulus than those evoked by unpredictable head movements (Barnes, 1991, 1992; Barnes & Lawson, 1992). This suggests that, at least under some circumstances, the oculomotor system can access a motor command in anticipation of the head movement.

3.5 Does it Work? How Good is Vision during Head Rotation?

The correct assignment of retinal movement to an ongoing head rotation allows the same two perceptual processes to occur that were potentially disrupted by movements of the eyes in the head. It allows the maintenance of perceptual stability of the world and the perception of external object motion despite the retinal movement that almost inevitably accompanies head movement.

Considerable extraretinal information is available concerning the head rotation.

Since a sensation of motion can be induced by vestibular information alone (see Benson, 1980), the canal signal might contribute centrally to the interpretation of visual movement. But it is important not to count the signal twice: if compensatory eye movements are effective at removing the angular component of the visual signal at source, then a central representation of the canal signal is only useful in providing a context for the remaining linear movements. Extraretinal distance information is essential since retinal movement of objects during translation of the eye depend on their distance from the eye. The dependence on distance has two consequences. Firstly, the accuracy with which motion can be interpreted as resulting from self motion depends on the accuracy of distance estimation as well as the accuracy of self movement estimation. And secondly, external object motion needs to be detected in the presence of many different retinal velocities associated with objects at different distances. These problems get worse in section 4 when unconstrained translations are considered.

3.5.1   The effect of eye movements compensatory for head rotation on retinal image motion

Head rotation causes retinal motion due to both the rotation and the associated translation of the eyes in space. Any contribution of rotation can theoretically be cancelled by ocular counter-rotation of appropriate speed and about the appropriate axis (with the caveat described in footnote 1). Complete removal of the angular component occurs when the vestibulo-ocular reflex (VOR) has unity gain (eye velocity output as a ratio of head velocity input). Although some animals routinely shown high gains to vestibular stimulation alone (e.g. cats: Harris, 1987; Blakemore & Donaghy, 1980), humans usually exhibit a much lower gain when measured in the dark (e.g. Barr et al., 1976). The contribution of the VOR when other systems (e.g. vision) are present is more difficult to assess (Collewijn, 1989a, 1989b), but there is no doubt that, during natural active head rotation, the compensatory eye movements for the angular component are highly effective (light: Steinman et al., 1985; dark: Tomlinson et al., 1980).

Figure 3 shows that, due to the translation of the eyes during head rotation, the retinal images of objects at different distances have different angular velocities. Thus eye movements cannot completely stabilize the whole field at once. The background could be stabilized by counter-rotating the eyes at the rate of head rotation and any contribution of translation ignored. Alternatively the image of a particular target could be stabilized. Fixating a particular target results in retinal motion of the images of all features at other distances from the eye.

Even when fixating on distant targets, compensatory eye velocity does not perfectly compensate for retinal movement (Steinman et al., 1985). But the distance of the target is taken into account and the speed of compensatory eye movements increases when closer targets are fixated (cats: Blakemore & Donaghy, 1980; monkeys: Viirre et al., 1986; Snyder & King, 1992; humans: Biguer & Prablanc, 1981; Gresty & Bronstein, 1986; Hine & Thorn, 1987). Adjustment of eye velocity for target distance occurs in the dark and with too short a latency and at too high a speed for vision to play a role (Snyder & King, 1992; Snyder et al., 1992). It is necessary to postulate a central representation of target distance that can influence eye movements of vestibular origin.

The role of vision in generating eye movements that help maintain a stable image during head rotation may, under natural conditions, be restricted to long-duration, constant-velocity movements. Optokinetic eye movements experimentally evoked by full-field visual motion alone have two components, one which builds up quickly and another which builds up more slowly (Cohen *et al.*, 1977). The swift and efficient vestibularly-evoked response to head rotation does not usually require visual support (Miles, 1993) and leaves little role for vision. On the other hand the slow-build-up of a central representation of head velocity during long duration head rotation ('velocity store': Raphan *et al.*, 1979; 'delayed optokinetic nystagmus (OKN)': Miles, 1993) is ideally suited to smoothing over the inadequacies of the canals' response to long-duration stimuli.

To summarize: during natural head rotation, the motion due to the angular component can be effectively removed by the vestibulo-ocular reflex. The additional, distance-dependent translation component is also effectively removed for targets at the distance of regard but retinal movement of objects at other distances is inevitable and made worse by fixation of close targets.

3.5.2   Judging object motion during head rotation

Anecdotally, a swinging tree branch seems to move less if viewed while nodding or shaking one's head. Laboratory studies confirm this insensitivity: object movement as fast as 35% of the speed of self rotation is still perceived as earth-stationary (Wertheim & Bles, 1984; Wallach, 1985). The threshold for object motion is increased by up to three times during passive or active head movements (0.5–1.5 Hz: Probst & Wist, 1982; Probst *et al.*, 1986) and is also increased by circularvection or even just neck muscle stimulation (Probst *et al.*, 1986). Probst and his colleagues assumed that reflex eye movements kept the image completely stable during these measurements and therefore image motion on the retina could not contribute to the degradation. The retinal image is unlikely to be stable during head movements (see sections 1.4, 4.4.1) but the likely slippage seems unlikely to be enough to explain the considerable degradation reported: some kind of central suppression mechanism is required. There have been no supra-threshold measures of perceived velocity during head rotation.

Although motion sensitivity is reduced during head movement, an object that is actually observer-stationary seems to move slightly faster than the observer and in the same direction. The effect occurs with or without a background (*oculogyral effect*: Graybiel & Hupp, 1946; Elsner, 1971; Ross, 1974; Howard, 1982). During optokinetic stimulation a similar illusory movement of an observer-stationary target is seen (*induced motion*: Duncker, 1929). These illusory motions could reflect mis-estimates in the internal representation of reflex vestibular or optokinetic eye movements and their suppression by fixation (Raymond *et al.*, 1984; Whiteside *et al.*, 1965; Post *et al.*, 1984; Post & Leibowitz, 1985; Post, 1986 cf. section 2.5.2.2), but it is likely other factors are involved.

The targets associated with these illusory movements are observer-stationary, that is they are seen to be in orbit around the rotation axis and to be translating through space.

Their perceived linear velocity must be deduced from their angular velocity and perceived distance with respect to the centre of their orbit (not the eye). Mis-estimates of either of these parameters or errors in the deduction process will result in a mis-estimate of the target's linear velocity and might contribute to illusory movements.

Which of the various cues to head rotation is responsible for altering sensitivity to external motion? Is it that object motion detection is centrally attenuated (as during saccadic suppression) in response to the detection of head rotation? Or is all retinal motion channelled to the head movement processing system once a head movement has been detected and anything left over discarded (rather than being detected as object motion)? Or is it that the stimulus conditions are just too complex: object motion cannot be distinguished from or is masked by the other motion cues introduced by head rotation? The answers to these important questions await appropriate research programmes.

4   HANDLING RETINAL MOTION DUE TO BODY MOTION

Vision is an integrated part of behaviour. So the visual consequences of self motion normally include components due to movements of the eye, head and body together. Analysing retinal events during unrestrained body movement involves extracting the independent contributions of rotation and translation. Extracting each in the presence of the other, especially since some of each eye's translation is due to head rotation (section 3.1), is not trivial. The decomposition is important, however, because the components denote different things and need to be used differently. Rotation changes the visual direction of all objects except those at the end of the axis of rotation but does not change their distance from the centre of rotation. Translation changes the visual direction of all objects except those in the direction of travel and changes the distance to all points. Translation is also associated with various challenges to balance and locomotion guidance as well as the perceptual consequences of moving through the world.

Movements of the body generate extraretinal information that includes: (i) activity of the vestibular system, (ii) activity of proprioceptors from throughout the neck and body, and (iii) an efference copy of instructions to move the body.

Can these factors help interpret the movement of the retinal image due to body movement? How do they interact with information about eye and head rotations? Knowledge of head position in space contributes to the full description of the direction of the gaze in space needed to relate the location of a foveated point to other points in space.

4.1   Visual Cues Concerning Head Translation

The angular velocity of the image of each part of each object due to translation of the observer through space depends on its distance from the eye and the observer's speed and direction of heading with respect to that point. Since these parameters cannot be the same for any two points, translation introduces relative movement between the images of different objects on the retina (motion parallax). Since the parameters cannot be the same for the two eyes' images of any single point in space, the angular velocity in each eye is different. The distribution of relative velocities within and between retinae contains useful information concerning the three-dimensional structure of space and the observer's movements within it (Regan & Beverley, 1979; Regan, 1992; see Chapter 12 by Cumming).

As an observer translates through space, objects move relative to the head and eye through an equal distance in the opposite direction. The movement produces a pattern of instantaneous velocities on the retina in directions that radiate out from the direction of travel. An example is given in Figure 6A and the velocities are plotted in detail in Figure 4 for points on a horizontal plane that transects the eyes during

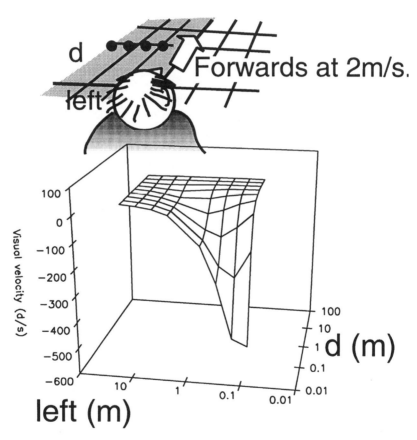

Figure 4 The angular velocity at the eye for points on a plane level with the head (inset) during forward motion at 2 m/s (fast walking speed, around 4 mph). Instantaneous angular velocities have been calculated for a logarithmically-spaced grid of points stretching from 0.1 m to the left of the subject out to 25.6 m and up to 25.6 m in front using the formulae described in Appendix B. The distance ahead ('d') and sideways ('left') have been plotted in metres on logarithmic scales. Although retinal angular velocities for distant objects are small, closer objects can reach high velocities. For example, an object 1.6 m forwards and off to the side by 1.6 m has a retinal velocity of 36 deg/s. An object at 0.1 m has a velocity in excess of 500 deg/s.

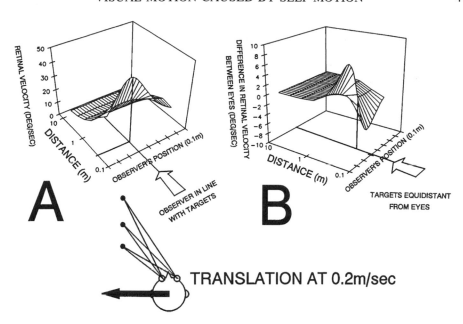

Figure 5 The effect of object distance and eccentricity during lateral translation. This is the effect of looking sideways while walking or out of the side window of a vehicle while travelling but has been calculated for the very slow speed of 0.2 m/s. The 'distance' axis is logarithmic. The 'observer's position' axis starts 0.2 m to the right of being lined up with the targets and ends 0.4 m on the other side (see inset). *Panel A* shows the retinal velocities for targets from 0.25 to 8 m distant. The maximum velocity for each target occurs when it is lined up with the observer. An object at 0.25 m has a retinal velocity in excess of 30 d/s. At 0.5 m, the velocity still exceeds 10 d/s. These retinal speeds are proportional to the translation velocity. *Panel B* shows the difference in retinal velocity between the two eyes for these same targets. The maximum differences occur approximately as the targets line up with one eye and then the other. (Interocular distance: 6 cm; eyes to centre of head rotation: 8 cm, as for Figure 3.)

forwards motion. The retinal speeds depend on the distance of each object from the eye. An object level with the eyes, 1.6 m forwards and 1.6 m to the left, for example, has a retinal velocity of 36 deg/s to the left when walking at 2 m/s. To fixate this object requires an eye rotation at 36 deg/s around an axis orthogonal to the plane defined by the direction of translation, the centre of rotation of the eye and the point in question. The retinal consequences of the eye rotation (section 2, Figures 2 and 6B) add to the existing retinal motion.

Translation in any direction produces a similar pattern of retinal movement with the motion of each point radiating out from the direction of travel. The retinal speed is zero for a point in the direction of travel and reaches a maximum when its visual direction is orthogonal to the direction of travel (Figure 5). Figure 5A plots the retinal velocities of a number of points arranged in a straight line that would correspond to those highlighted in the insert to Figure 4, but with the observer looking sideways rather than straight ahead and moving more slowly. The direction in which the eye is looking does not alter the distribution of retinal velocities although it will, of course,

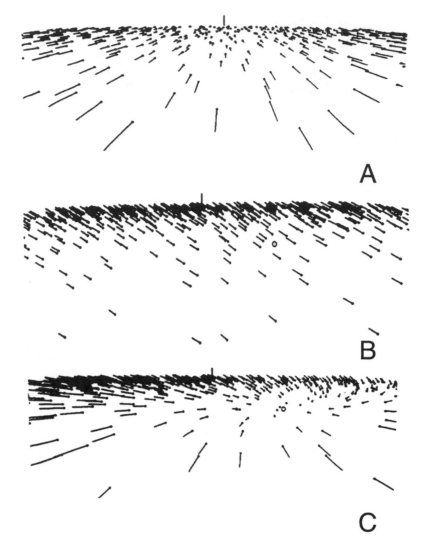

Figure 6 *Panel A* The pattern of optic flow on a stable retina produced by forwards translation across a ground plane. The point of heading is shown by the little flag pole in the centre. *Panel B* shows the effect of eye rotation alone on such a display. Compare the central region of the centre circle of Figure 2. *Panel C* shows the effect of having both rotation and translation present at the same time. This simulates the effect of fixating an earth-fixed point (shown by the circle) during translation towards the flag pole. It could also simulate translation along a curved path. The point fixated is quite distant so the rotation velocity is modest. Redrawn from Warren and Hannon (1990, Figure 1, p. 161).

determine which velocities fall on which part of the retina. The *difference* between the velocities in two eyes depends on the orientation of the head relative to the direction of travel (Figure 5B and Appendix B).

Translation of the eye due to *translation* of the head and due to *rotation* of the head generates identical instantaneous retinal image velocities, of course. But the two situations can be distinguished because, in the latter case, the direction of translation of each eye is tangential to the orbit and therefore is different (cf. section 3.1). The presence of rotation will also be independently identified both by its (non-translational) contribution to image movement and extraretinal cues (section 3).

Figure 6C shows what happens to the retinal image when angular (Figure 6B) and translation (Figure 6A) components are present at the same time. Although it may be mathematically possible to extract Figures 6A and 6B from 6C (Gibson, 1950, 1954; Koenderink & van Doorn, 1975; see Warren & Hannon, 1990 and Warren *et al.*, 1991 for useful reviews), it would still be necessary to know the combination of eye, head and body rotation that caused the rotation contribution of Figure 6B. It requires extraretinal information to find that out.

4.1.1   Assignment of visual motion to self motion (linearvection)

The assignment of visual motion to linear self motion generates a *sensation* of self motion in a particular direction and with a particular velocity (*linearvection*: Fischer & Kornmüller, 1930; Berthoz *et al.*, 1975; see Howard, 1982 and Miles & Wallman, 1993 for reviews). The pattern of retinal motion consistent with the linearvection might subsequently be calculated for each point using its perceived distance and deviations attributed to external movement.

Research into the extraction of information from optic flow caused by translation has concentrated on deriving the direction of heading of the observer. But optic flow provides information about rotation and translation in space and, as such, can be used to update the positions of *all* points in space, at least with respect to the eye (e.g. Rieser, 1983; Rieser *et al.*, 1986). There is nothing special about the point that represents the direction of heading. In fact direction of heading seems particularly artificial information to extract from a small display which presents the same image to each eye[3] and which is often not associated with the sensation of self motion. Amazingly, however, the direction of heading can be extracted to within 1 deg from such displays simulating translation alone (Figure 6A: Warren, 1976; Royden *et al.*, 1992a, 1992b; Warren & Hannon, 1990; Warren *et al.*, 1991). In the presence of a rotation component above 2 deg/s, however, extraretinal information about the rotation is required for accurate heading judgement (Figure 6C: Royden *et al.*, 1992a). Further experiments will undoubtedly show a role of vestibular and other sensory input in the interpretation of optic flow with many components.

4.2   Vestibular Cues Concerning Head Translation

An important cue that normally helps the rapid assignment of visual flow to linear translation is concurrent activity of the otoliths which transduce linear accelerations of

---

[3] The images to the two eyes will actually be slightly different during stimulus presentations but the difference arises from the distance of the subject to the viewing screen rather than, as should be the case, the simulated distance of each dot in the display.

the head (see section 3.2.2). If the otoliths do not report head acceleration, as when only visual cues are provided to a stationary observer, linearvection is a very sluggish phenomenon (Howard, 1982). The canal system is not sensitive to linear translation (Goldberg & Fernandez, 1975) and so its levels of activity are not relevant here.

On earth, the otoliths are subjected to the constant acceleration of gravity. In addition, from time to time, the otoliths are subjected to accelerations generated by linear movement of the head. Applying two simultaneous accelerations to a body is exactly the same as applying a single acceleration of appropriate magnitude in their resultant direction. So the otolith system at any one time can only signal a single acceleration in a single direction from which the components of externally-generated gravity and self-generated motion must be deduced. This case is akin to the problem of identifying self-generated sounds (section 1.3) – unlike in the visual version of this problem, the two components do not interact. The components *can* be separated if at least one direction and one acceleration value are known. The use of extra-otolithic information such as tilt signalled by the canals (Guedry, 1974) or knowledge of the otolith's dynamics (Mayne, 1974) might help.

The eye movements evoked by linear acceleration of lower animals such as rabbits indicates that they attribute all otolith activity to the direction of gravity (Baarsma & Collewijn, 1975). But the eye movements of higher animals such as cats (Fukushima & Fukushima, 1991), monkeys (Paige & Tomko, 1991a,b) and humans (Paige, 1989; Baloh *et al.,* 1988; Berthoz *et al.,* 1987; Buizza *et al.,* 1979; Israel & Berthoz, 1989), during linear translation suggests that linear acceleration information can be extracted in those species. The decomposition is not perfect, however, and especially during high accelerations, perceptual confusion often arises between interpreting the otoliths as signalling an added linear acceleration or tilt (Guedry, 1974).

Interactions can occur between vision and otoliths where neither are able to operate unambiguously alone: the optic flow due to translation is difficult to disentangle from simultaneous rotation and otolithic information due to linear acceleration is difficult to disentangle from gravity. Together each helps reduce the other's ambiguities and rapid and appropriate sensations of movement through the environment result.

4.3 Other Cues Concerning Head Translation

There are many other potential sources of information in addition to vestibular and visual cues concerning body movements associated with head translation. Proprioceptors all over the body, for example in the legs and ankle joints, could contribute. Copies of relevant motor commands, such as to walk or run in a particular direction, might also be useful. How such high-level information might be represented in the brain and how it might be used in interpreting optic flow is unknown. Cognitive factors also play an important role (Howard, 1982) and expectancies can, for example, strongly influence whether the sensation of vection is experienced.

4.4 Does it Work? How Good is Vision during Head Translation?

The successful assignment of the appropriate components of retinal movement to head translation, when coupled with knowledge of eye and head rotations, means that perceptual stability of the world and the perception of external object motion can continue during unrestrained movement in the real world.

Extraretinal distance information is particularly significant when dealing with the retinal consequences of free translation. The accuracy of motion assignment depends on the accuracy of distance estimation, and external object motion needs to be detected within a retinal image containing many different retinal velocities resulting from objects at different distances.

4.4.1   The effect of eye movements compensatory for head translation on retinal image motion

Compensatory eye movements tend to oppose retinal image movement that would otherwise result from head movement even if it is not appropriate to achieve stabilization (sections 1.4, 4.4.1). During head *rotation*, the dominant eye movement required is counter-rotation of both eyes at the same speed as the head rotation around axes parallel to the head rotation axis. During head *translation*, the compensatory eye movements required for retinal movement depend critically on the distance of the target being fixated and on its position with respect to the direction of travel. The eye movements required for fixation of close targets during translation are more like pursuit than reflex compensatory eye movements. The velocity for each eye needs to be updated continuously because the stimulus velocity varies with direction and distance which are themselves continuously changed by translation. Each eye needs to rotate around an axis orthogonal to the plane defined by the direction of translation, the centre of rotation of the eye and the point in question. For each point, the orientation of the required axis is different for each eye: the movements of each eye are geometrically required to be independent.

Only a part of each eye's image can possibly be stabilized at one time. The images of objects further away than a couple of metres usually require no stabilizing during translation at normal human speeds since they move on the retina at less than the speeds normally left by compensatory eye movements (Steinman & Collewijn, 1980; Sperling, 1990; Appendix B). The velocity of an eye movement that stabilizes one point (Figure 2) is added vectorially to the entire retinal image, speeding some points, slowing others and changing the direction of motion on the retina of most.

Compensatory eye movements evoked by linear motion are certainly sensitive to distance information (Baloh *et al.*, 1988; Israel & Berthoz, 1989; Post & Leibowitz, 1982; Schwarz & Miles, 1991; Paige, 1989; Schwarz *et al.*, 1989). Target distance might be obtained from accommodation or vergence angle (Demer, 1992; Schwarz *et al.*, 1989; Schwarz & Miles, 1991). In the absence of a new target the old distance estimate is retained for at least 250 ms as the modulator of eye velocity (Schwarz & Miles, 1991). Divergence of the eyes was occasionally seen in the dark during fixation of a remembered, earth-fixed target during lateral translation of the observer (Schwarz & Miles, 1991). Divergence is consistent with each eye following the target under separate control. The possibility of independent control should be investigated systematically by varying the position of the target and hence the movements of each eye needed to maintain fixation during translation movements.

Translation of the eyes can be along parallel straight lines or along curved paths. A special example of curvilinear translation is the path of the eyes as they transcribe an orbit around the centre of rotation of the head (section 3). For each eye, the result is equivalent to translation in the direction tangential to the orbit and rotation about its own axis. The feature that distinguishes curvilinear from linear translation is that the simultaneous direction of translation of each eye is different. The difference reaches a maximum when rotation is about a point in between the two eyes and falls off as the distance from that point increases. Theoretically, if the direction and velocity of the translation of each eye could be deduced it would, by triangulation, provide full information about the rate of rotation and the location of the axis. Altering the position of the axis of rotation while keeping fixation distance constant (using increasingly eccentric rotations: Viirre et al., 1986; Gresty et al., 1987; Sargent & Paige, 1991; Snyder & King, 1992; Snyder et al., 1992) causes appropriate adjustments in eye velocity. The adjustment occurs in the dark, suggesting an otolithic contribution (see section 4.2).

Are the eye movements evoked by translation due to vestibular, visual or other drives? Miles (1993) suggested that the fast-build-up component of optokinetic nystagmus may be particularly useful in maintaining vision in the presence of many different retinal velocities because of the dominant role of the centre of the field in the generation of the eye movements (Brandt et al., 1973). But visually-driven eye movements can be easily confused by differences between the central and peripheral fields (Abadi & Pascal, 1991). The most accurate eye movements evoked by translation during body movements are in the light when multiple cues are available (Wall et al., 1992; Harris et al., 1993b; Solomon & Cohen, 1992a,b; Bles & Kotaka, 1986).

In conclusion, eye movements can never remove the retinal consequences of translation. They can be extremely accurate and resourceful in stabilizing one object's image but usually at the expense of the rest of the field. Their contribution to the optic flow over the rest of the image is additive. Their presence is thus potentially confusable with various other situations. Some extraretinal cues are certainly used to identify their contribution and others might theoretically do so but have yet to be systematically investigated.

4.4.2   Judging object motion during head translation

Perceptual thresholds for object motion are raised and object velocity is underestimated during head translation (Pavard & Berthoz, 1977; Berthoz & Droulez, 1982; Probst et al., 1984, 1986). The perceived direction of motion of objects is also distorted and the *accuracy* of direction judgements gets worse when the judgements are made during head translation (standard errors increased from $\pm$ 1.8° to $\pm$ 4.5°: Swanston & Wade, 1988; Swanston et al., 1992).

The uncorrectable retinal image motions associated with translational head movement probably contribute to the perceptual degradation under some conditions. But visual simulations of translation are not always associated with changes in motion thresholds suggesting that extraretinal factors are important and possibly the only contributors (Brenner, 1991a,b, 1993).

In addition to difficulties and misperceptions of external motion during head translation, illusory movement is often perceived in objects that are in fact stationary during the translation, for example the moon (*apparent concomitant motion*: Gogel,

1982; Gogel & Tietz, 1973; Post & Leibowitz, 1982; cf. section 3.5.2). Apparent concomitant motion presumably arises from the deduction of linear motion from angular retinal events using inaccurate distance information. The general reduced sensitivity to motion found during head movements probably reduces the confusion that might otherwise arise from such misassignments (Probst *et al.*, 1986).

5 CONCLUSIONS

The argument that considering a moving eye in a fixed head is the first step on the path towards looking at retinal image motion in the real world has introduced many misleading ideas which need to be revised – as when learning to touch-type after typing with one finger. Examples are the concepts of 'cancellation' (section 1.3), 'efference copy' (section 2.2) and describing eye movements as 'horizontal' or 'vertical' (section 1.2). These and related concepts become all but useless when applied to the unconstrained movements of normal behaviour.

This chapter has described the retinal consequences of self movement elaborating the effect of a nested chain of contributors including eye-in-orbit, head-on-shoulders, etc. Each leaves retinal trademarks such that they can, theoretically, be recovered from the conglomerate, especially in association with the appropriate extraretinal information. Little is known about whether some sources of movement information can be used for human performance. It is probable that information, although technically available perhaps in a single sense (e.g. the optic flow) is normally available through an interaction of the senses (e.g. otoliths and vision).

The extraction of different contributors to visual motion (e.g. eye rotation during translation) allows different tasks to be carried out simultaneously. Think of the sub-tasks involved in fixating a jogger while cycling around a corner. Features of the observer's self motion and relationship to space are available and are not irreversibly disrupted by the additional optic flow introduced by fixating a moving target. The multiple assignment of visual image movement must introduce some cumulative noise. Some movement is assigned to eye rotation, some to head rotation, some to translation of the eye resulting from head rotation, some to head translation and any remaining movement is available to be assigned to external causes. It is perhaps not surprising that object movement perception is poor under such conditions!

The retinal movements generated by self motion can be thought of as relative to the eye itself or relative to the images of other points. They can be thought of as instantaneous velocities or as patterns that change over time. This gives us four ways of describing the retinal events. It is likely that the same retinal information might be processed in more than one of these ways simultaneously. For example, relative motion between neighbouring retinal points is not sensitive to rotation since neighbouring points are affected similarly by rotation (Figure 2). Local differences therefore represent a technique for identifying translation relatively uncontaminated by the effects of rotation. Other processing systems that look for common features across the retina would be more suited to identify the rotation.

How well the brain uses self-motion information and its tolerances and fussiness are important both in understanding and in exploiting the visual system. The dependence of the retinal consequences of translation on the distance of each point from each eye introduces a special challenge for proponents of virtual reality (VR). To experience virtual reality, the angular and linear accelerations of an observer's head are monitored and miniature television screens, mounted in goggles, provide a visual input. The screens simulate what an observer would see were they to move, with the patterns of head movement actually detected, around a computer-simulated landscape. As the head rotates to the left, the images are shifted by the program to the right. To simulate the relationships described in this chapter in which the movement of each point in each eye has motions attributable to several sources which depend on their perceived depth presents a formidable programming challenge.

Perhaps if VR participants could be kept mobile enough, the consequent reduction in sensitivity to motion would keep players tolerant of program shortcomings in the same way that the brain is forgiving of the often unwanted and potentially confusing retinal motion present during natural self motion.

ACKNOWLEDGEMENTS

LRH is supported by NSERC of Canada and the Institute for Space and Terrestrial Science. I would like to thank John Findlay, Ian Howard, Lori Lott, Mike Swanston, and Nick Wade for their useful comments on an early version of this chapter.

APPENDIX A   The Geometry of Retinal Velocities due to the Linear Displacement of the Eyes by an Angular Head Movement

Given:

$$\text{angular velocity} = \frac{\text{linear velocity} \times \cos^2 \text{(eccentricity)}}{\text{perpendicular distance}} \; \text{[rads/s]}$$

In this situation, therefore:

$$\dot{\theta}_E = \frac{\dot{M}.\cos^2(\theta_E)}{d} \; \text{[rads/s]}$$

where:

$\dot{\theta}_E$ = angular velocity of eye [°/s]
$\dot{M}$ = linear velocity of eye [m/s]
  = head rotation rate × h [rads/s × m]
θ_E = eye position [°]
d = orthogonal distance to target (see Figure 7)

d and θ_E can be expressed in terms of dist$_H$ (the distance from the head to the target), h (the distance between the axes of rotation of the head and eye) and θ_H (head position) (Figure 7).

$$d = \text{dist}_H.\cos(\theta_H) - h$$

$$\theta_E = \tan^{-1}\left(\frac{\text{dist}_H.\sin(\theta_H)}{d}\right)$$

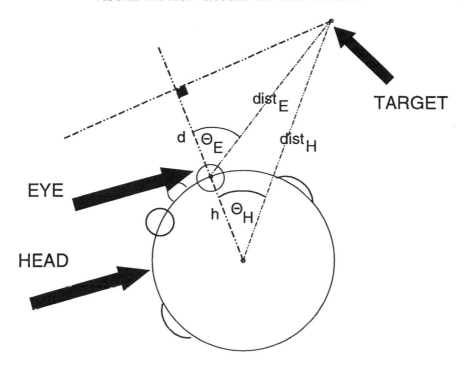

TARGET

EYE

HEAD

Figure 7   The variables needed to analyse retinal motion of a target during head rotation. The target has a bearing θ_E and dist$_E$ from the centre of the eye and bearing θ_H and dist$_H$ from the centre of the head.

For a head rotation about an axis of fixed distance from the eyes (h) and of fixed distance from a target ($dist_H$), the only things that vary during rotation are the head velocity and the head position (θ_H).

The difference between the eyes (see Figure 8) is calculated by a simple increment to θ_H given by:

$$\Delta\theta_H = \pm\, \sin^{-1}\left(\frac{\text{interocular distance}}{2h}\right)$$

APPENDIX B   The Geometry of Retinal Velocities due to Linear Head Translations

Given:

$$\text{angular velocity} = \frac{\text{linear velocity} \times \cos^2\,(\text{eccentricity})}{\text{closest distance}}\quad\text{[rads/s]}$$

$$\dot{\theta}_E = \frac{\dot{M}.\cos^2(\theta_E)}{cd}$$

interocular distance

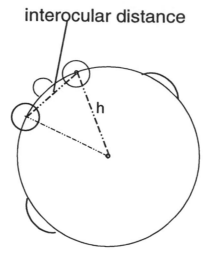

Figure 8 Interocular distance is the distance between the eyes. The eyes are 'h' from the centre of the head.

$$= \frac{\dot{M}.cd^2}{cd.\text{dist}^2_\text{E}}$$

$$= \frac{\dot{M}.cd}{dist^2_\text{E}}$$

Where

$\dot{\theta}_\text{E}$ = eye velocity [°/s]
cd = orthogonal distance (see Figure 9)
$\dot{M}$ = linear velocity of the eye [m/s]

For a given direction and speed of translation and for a given point of closest distance cd, the only thing that varies is distance from the eye (dist$_\text{E}$) (Figure 9).

$$\text{dist}_\text{E} = \sqrt{cd^2 + dx^2}$$

and dx varies linearly with the head translation

$$\Delta dx = \dot{M}.t$$

The maximum speed, θ_E occurs when $dx = 0$. For each eye, cd and the initial value of dx will be different depending on the interocular distance and the direction of travel, the orientation of the head relative to the direction of travel and the bearing of the target. But M and Δdx are the same resulting in different speeds at the two eyes and different times when $dx = 0$.

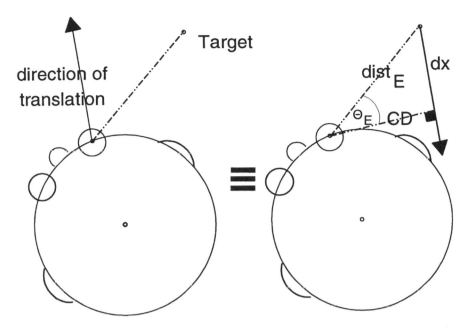

Figure 9 The variables needed to analyse the retinal motion of a target during translation. The target is at $dist_E$ from the eye. Translation of the eye past a stationary point (left side) is exactly equivalent to translation of a point past a stationary head (right side). *cd* indicates the closest approach that the target will make to that eye.

REFERENCES

Abadi, R. V. & Pascal, E. (1991). The effects of simultaneous central and peripheral field motion on the optokinetic response. *Vision Res.*, **31**, 2219–2225.

Abrahams, V. C. & Rose, P. K. (1975). Projections of extraocular, neck muscle, and retinal afferents to superior colliculus in the cat: their connections to cells of origin of tectospinal tract. *J. Neurophysiol.*, **38**, 10–18.

Abrahams, V. C., Richmond, F. & Rose, P. K. (1975a). Absence of monosynaptic reflex in dorsal neck muscles of the cat. *Brain Res.*, **92**, 130–131.

Abrahams, V. C., Richmond, F. & Rose, P. K. (1975b). Basic physiology of the head–eye movement system. In G. Lennerstrand & P. Bach-y-Rita (Eds.) *Basic Mechanisms of Ocular Motility and their Clinical Implications*, pp. 473–476. Pergamon, Oxford.

Alpern, M. (1962). Kinematics of the eye. In H. Davson (Ed.) *The Eye*, pp. 15–27. Academic Press, New York.

Anastasopoulos, D. & Mergner, T. (1982). Canal–neck interaction in vestibular nuclear neurons of the cat. *Exp. Brain Res.*, **46**, 269–280.

Andersen, G. J. (1986). The perception of self-motion. *Psychol. Bull.*, **99**, 52–65.

Aubert, H. (1886). Die Bewegungsempfindung. *Pflügers Arch.*, **39**, 347–370.

Aubert, H. (1887). Die Bewegungsempfindung. Zweiter Mitteilung. *Pflügers Arch.*, **40**, 459–480.

Baarsma, E. A. & Collewijn, H. (1975). Eye movements due to linear accelerations in the rabbit. *J. Physiol. Lond.*, **245**, 227–247.

Baker, J., Goldberg, J., Hermann, G. & Peterson, B. (1984). Optimal response planes and canal convergence in secondary neurons in vestibular nuclei of alert cats. *Brain Res.*, **294**, 133–137.

Baloh, R. W., Beykirch, K., Honrubia, V. & Yee, R. D. (1988). Eye-movements induced by linear acceleration on a parallel swing. *J. Neurophysiol.*, **60**, 2000–2013.

Barnes, G. R. (1991). Predictive control of head and eye-movements during head-free pursuit in humans. *J. Physiol. Lond.*, **438**, P215.

Barnes, G. R. (1992). Mathematical modeling of visual and nonvisual mechanisms of head–eye coordination. In A. Berthoz, W. Graf & P. P. Vidal (Eds.) *The Head–Neck Sensory Motor System*, pp. 449–454. Oxford University Press, Oxford.

Barnes, G. R. & Lawson, J. F. (1992). Visual and vestibular contributions to head-eye coordination during head-free pursuit. In A. Berthoz, W. Graf & P. P. Vidal (Eds.) *The Head–Neck Sensory Motor System*, pp. 443–448, Oxford University Press, Oxford.

Barr, C. C., Schultheis, L. W. & Robinson, D. A. (1976). Voluntary, non-visual control of the human vestibulo-ocular reflex. *Acta Otolaryngol*, **81**, 365–375.

Batini, C. & Buisseret, P. (1974). Sensory peripheral pathway from extrinsic eye muscles. *Arch. Italiennes Biol.*, **112**, 18–32.

Batini, C., Buisseret, P. & Kado, R. T. (1974). Extraocular proprioceptive and trigeminal projections to the purkinje cells of the cerebellar cortex. *Arch. Italiennes Biol.*, **112**, 1–17.

Bedell, II. E. (1990). Directionalization of visual targets during involuntary eye movement. *Optometry Vis. Sci.*, **67**, 583–589.

Benson, A. J. (1980). The vestibular system. In H. B. Barlow & J. Mollon (Eds.) *The Senses*. Cambridge University Press, Cambridge.

Benson, A. J. & Bodin, M. A. (1966). Interaction of linear and angular accelerations on vestibular receptors in man. *Aerospace Med.*, **37**, 144–154.

Berthoz, A. & Droulez, J. (1982). Linear self motion perception. In A. H. Wertheim, W. A. Wagenaar & H. W. Leibowitz (Eds.) *Tutorials on Motion Perception*. Plenum, New York.

Berthoz, A., Pavard, B. & Young, L. (1975). Perception linear horizontal self motion induced by peripheral vision. *Exp. Brain Res.*, **23**, 471–489.

Berthoz, A., Israel, I., Vieville, T. & Zee, D. (1987). Linear head displacement measurement by the otoliths can be reproduced through the saccadic system. *Neurosci. Lett.*, **82**, 285–290.

Biguer, B. & Prablanc, C. (1981). Modulation of the vestibulo-ocular reflex in eye-head orientation as a function of target distance in man. In A. F. Fuchs & W. Becker (Eds.) *Progress in Oculomotor Research*, pp. 525–530. Elsevier, North Holland, Amsterdam.

Biguer, B., Jeannerod, M., Hein, A. & Donaldson, I. M. L. (1988). Neck muscle vibration modifies the representation of visual-motion and direction in man. *Brain*, **111**, 1405–1424.

Blakemore, C. & Donaghy, M. (1980). Coordination of head and eyes in the gaze changing behavior of cats. *J. Physiol. Lond.*, **300**, 317–335.

Bles, W. & Dejong, J. M. B. V. (1982). Cervico-vestibular and visuo-vestibular interaction – self-motion perception, nystagmus, and gaze shift. *Acta Otolaryngol.*, **94**, 61–72.

Bles, W. & Kotaka, S. (1986). Stepping around: nystagmus, self-motion perception and Coriolis effects. In E. L. Keller & D. S. Zee (Eds.) *Adaptive Processes in Visual and Oculomotor Systems*, pp. 465–471. Pergamon, Oxford.

Boyle, R. & Pompeiano, O. (1981a). Cerebellar influences on the response characteristics of vestibulospinal neurons to sinusoidal tilt. *Arch. Italiennes Biol.*, **119**, 208–225.

Boyle, R. & Pompeiano, O. (1981b). Convergence and interaction of neck and macular vestibular inputs on vestibulospinal neurons. *J. Neurophysiol.*, **45**, 852–868.

Brandt, T., Dichgans, J. & Koenig, W. (1973). Differential central and peripheral contributions to self-motion perception. *Exp. Brain Res.*, **16**, 476–491.

Brenner, E. (1991a). Judging object motion velocity during smooth pursuit – effects of moving backgrounds. *Invest. Ophthalmol. Vis. Sci.*, **32**, 826–826.

Brenner, E. (1991b). Judging object motion during smooth pursuit eye-movements – the role of optic flow. *Vision Res.*, **31**, 1893–1902.

Brenner, E. (1993). Judging an object's velocity when its distance changes due to ego-motion. *Vision Res.*, **33**, 487–504.

Bridgeman, B. (1983). Phasic eye-movement control appears before tonic control in human-fetal development. *Invest. Ophthalmol. Vis. Sci.*, **24**, 658–659.

Bridgeman, B. & Stark, L. (1979). Omnidirectional increase in threshold for image shifts during saccadic eye movements. *Percept. Psychophys.*, **25**, 241–243.

Bridgeman, B. & Stark, L. (1991). Ocular proprioception and efference copy in registering visual direction. *Vision Res.*, **31**, 1903–1913.

Bridgeman, B., Hendry, D. & Stark, L. (1975). Failure to detect displacement of the visual world during saccadic eye movements. *Vision Res.*, **15**, 719–722.

Brindley, G. S. & Merton, P. A. (1960). The absence of position sense in the human eye. *J. Physiol. Lond.*, **153**, 127–130.

Brindley, G. S., Goodwin, G. M., Kulikowski, J. J. & Leighton, D. (1976). Stability of vision with a paralysed eye. *J. Physiol.*, 65P–66P.

Brooks, B. A., Yates, J. T. & Coleman, R. D. (1980). Perception of images moving at saccadic velocities during saccades and during fixation. *Exp. Brain Res.*, **40**, 71–78.

Buizza, A., Leger, A., Berthoz, A. & Schmid, R. (1979). Otolithic–acoustic interaction in the control of eye movement. *Exp. Brain Res.*, **36**, 509–522.

Burr, D. C., Holt, J., Johnstone, J. R. & Ross, J. (1982). Selective depression of motion sensitivity during saccades. *J. Physiol. Lond.*, **333**, 1–15.

Carpenter, R. H. S. (1988). *Movements of the Eyes*. Pion, London.

Carpenter, R. H. S. (1991). The visual origins of ocular motility. In R. H. S. Carpenter (Ed.) *Eye Movements*, pp. 1–10. Macmillan Press, London.

Cohen, B., Matsuo, V. & Raphan, T. (1977). Quantitative analysis of the velocity characteristics of optokinetic nystagmus and optokinetic after-nystagmus. *J. Physiol. Lond.*, **270**, 321–344.

Collewijn, H. (1989a). The vestibulo-ocular reflex – is it an independent subsystem. *Rev. Neurol.*, **145**, 502–512.

Collewijn, H. (1989b). The vestibuloocular reflex – an outdated concept. *Progr. Brain Res.*, **80**, 197–209.

Curthoys, I. S. & Markham, C. H. (1971). Convergence of labyrinthine influences on units in the vestibular nuclei of the cat. I. Natural stimulation. *Brain Res.*, **35**, 469–490.

Darlot, C. & Denise, P. (1988). Nystagmus induced by off-vertical rotation axis in the cat. *Exp. Brain Res.*, **73**, 78–90.

Darlot, C., Denise, P., Cohen, B., Droulez, J. & Berthoz, A. (1988). Eye-movements induced by off-vertical axis rotation (Ovar) at small angles of tilt. *Exp. Brain Res.*, **73**, 91–105.

Demer, J. L. (1992). Mechanisms of human vertical visual–vestibular interaction. *J. Neurophysiol.*, **68**, 2128–2146.

Dichgans, J., Korner, F. & Vogt, K. (1969). Vergleichende Skalierung des afferenten und efferenten Bewegungssehens beim Menschen: Linearen Funktionen mit verschiedenen Ausstiegssteilkeit. *Psychol. Forsch.*, **32**, 277–295.

Dichgans, J., Wist, E., Diener, H. C. & Brandt, T. (1975). The Aubert Fleischl phenomenon: A temporal frequency effect on perceived velocity in afferent motion perception. *Exp. Brain Res.*, **23**, 529–533.

Donaldson, I. M. L. & Long, A. C. (1980). Interactions between extraocular proprioceptive and visual signals in the superior colliculus of the cat. *J. Physiol.*, **298**, 85–110.

Duhamel, J. R., Colby, C. L. & Goldberg, M. E. (1992). The updating of the representation of visual space in parietal cortex by intended eye-movements. *Science*, **255**, 90–92.

Duncker, K. (1929). Ueber induzierte Bewegung. *Psychol. Forsch.*, **12**, 180–259.

Elsner, W. (1971). Power laws for the perception of rotation and the oculogyral illusion. *Percept. Psychophys.*, **9**, 418–420.

Fernandez, C. & Goldberg, J. M. (1971). Physiology of peripheral neurons innervating semicircular canals of the squirrel monkey. II. Response to sinusoidal stimulation and dynamics of peripheral vestibular system. *J. Neurophysiol.*, **34**, 661–675.

Filehne, W. (1922). Über das optische Wahrnehmen von Bewegungen. *Z. Sinnephysiol.*, **53**, 134–145.

Fischer, M. H. & Kornmüller, A. E. (1930). Optokinetisch ausgelöste Bewegungs-wahrnehmung und optokinetischer Nystagmus. *J. Psychol. Neurol. (Leipzig)*, **41**, 273–308.

Fleischl, E. V. (1882). Physiologisch optische Notizen. *Sitzungsbez. Akad. Wissensch.*, **3**, 7–25.

Fukushima, K. & Fukushima, J. (1991). Otolith-visual interaction in the control of eye-movement produced by sinusoidal vertical linear acceleration in alert cats. *Exp. Brain Res.*, **85**, 36–44.

Gauthier, G. M., Nommay, D. & Vercher, J. L. (1990a). Ocular muscle proprioception and visual localization of targets in man. *Brain*, **113**, 1857–1871.

Gauthier, G. M., Nommay, D. & Vercher, J. L. (1990b). The role of ocular muscle proprioception in visual localization of targets. *Science*, **249**, 58–61.

Gibson, J. J. (1950). *The Perception of the Visual World*. Houton Mifflin, Boston.

Gibson, J. J. (1954). The visual perception of objective motion and subjective movement. *Psychol. Rev.*, **61**, 304–314.

Gibson, J. J., Smit, O. W., Steinschneider, A. & Johnson, C. W. (1957). The relative accuracy of visual perception of motion during fixation and pursuit. *Am. J. Psychol.*, **70**, 64–68.

Gogel, W. C. (1982). Analysis of the perception of motion concomitant with a lateral motion of the head. *Percept. Psychophys.*, **32**, 241–250.

Gogel, W. C. & Tietz, J. D. (1973). Absolute motion parallax and the specific motion tendency. *Percept. Psychophys.*, **13**, 284–314.

Goldberg, J. M. & Fernandez, C. (1975). Responses to peripheral vestibular neurones to angular and linear accelerations in the squirrel monkey. *Acta Otolaryngol.*, **80**, 101–110.

Graybiel, A. & Hupp, E. D. (1946). The oculogyral illusion: A form of apparent motion which may be observed following stimulation of the semicircular canals. *J. Aviation Med.*, **17**, 3–27.

Green, D. G. & Campbell, F. W. (1965). Effect of focus on the visual response to a sinusoidally modulated spatial stimulus. *J. Opt. Soc. Am.*, **55**, 1154–1157.

Gresty, M. A. & Bronstein, A. M. (1986). Otolith stimulation evokes compensatory reflex eye-movements of high-velocity when linear motion of the head is combined with concurrent angular motion. *Neurosci. Lett.*, **65**, 149–154.

Gresty, M. A., Bronstein, A. M. & Barratt, H. (1987). Eye-movement responses to combined linear and angular head movement. *Exp. Brain Res.*, **65**, 377–384.

Grüsser, O. J., Krizic, A. & Weiss, L. R. (1987). Afterimage movement during saccades in the dark. *Vision Res.*, **27**, 217–226.

Guedry, F. E. (1965). Orientation of the rotation axis relative to gravity: its influence on nystagmus and the sensation of rotation. *Acta Otolaryngol.*, **60**, 30–48.

Guedry, F. E. (1974). Psychophysics of vestibular sensation. In H. H. Kornhuber (Ed.) *Handbook of Sensory Physiology*, pp. 2–154. Springer-Verlag, New York.

Guthrie, B. L., Porter, J. D. & Sparks, D. L. (1983). Corollary discharge provides accurate eye position information to the oculomotor system. *Science*, **221**, 1193–1195.

Hansen, R. M. (1979). Spatial localization during pursuit eye movements. *Vision Res.*, **19**, 1213–1221.

Hansen, R. M. & Skavenski, A. A. (1977). Accuracy of eye position information for motor control. *Vision Res.*, **17**, 919–926.

Harris, L. R. (1987). Vestibular and optokinetic eye-movements evoked in the cat by rotation about a tilted axis. *Exp. Brain Res.*, **66**, 522–532.

Harris, L. R. & Barnes, G. B. (1985). The orientation of vestibular nystagmus is modified by head tilt. In M. D. Graham and J. L. Keiminck (Eds.) *The Vestibular System*, pp. 571–581. Raven Press, New York.

Harris, L. R. & Jenkin, M. (1993). Spatial vision in humans and robots. In L. R. Harris & M. Jenkin (Eds.) *Spatial Vision in Humans and Robots*, pp. 1–7. Cambridge University Press, Cambridge.

Harris, L. R. & Lott, L. A. (1993). Thresholds for full-field visual motion indicate an axis-based coding system similar to that of the vestibular system. *Neurosci. Abstr.*, **19**, 316.21.

Harris, L. R., Goltz, H. & Steinbach, M. J. (1993a). The effect of gravity on the resting position of the cat's eye. *Exp. Brain Res.*, **96**, 107–116.

Harris, L. R., Lathan, C. E. & Wall, C. (1993b). The effect of z-axis linear acceleration on the range of optokinetic performance in humans. *Invest. Ophthalmol. Vis. Sci.*, **34**, 3964.

Helmholtz, H. von. (1866). *Handbuch der physiologischen Optik (Handbook of Physiological Optics)*. Voss, Leipzig.

Hine, T. & Thorn, F. (1987). Compensatory eye-movements during active head rotation for near targets – effects of imagination, rapid head oscillation and vergence. *Vision Res.*, **27**, 1639–1657.

Howard, I. P. (1982). *Human Visual Orientation*. John Wiley, New York.

Howard, I. P. (1993). The stability of the visual world. In F. A. Miles & J. Wallman (Eds.) *Visual Motion and its Role in the Stabilization of Gaze*, pp. 103–118. Elsevier, North Holland, Amsterdam.

Howard, I. P. & Heckmann, T. (1989). Circularvection as a function of the relative sizes, distances, and positions of two competing visual displays. *Perception*, **18**, 657–665.

Ilg, U. J. & Hoffmann, K. P. (1993). Motion perception during saccades. *Vision Res.*, **33**, 211–220.

Israel, I. & Berthoz, A. (1989). Contribution of the otoliths to the calculation of linear displacement. *J. Neurophysiol.*, **62**, 247–263.

Kelly, D. H. (1979). Motion and vision. II. Stabilized spatio-temporal threshold. *J. Opt. Soc. Am. A – Opt. Image Sci.*, **69**, 1340–1349.

Kimura, M., Takeda, T. & Maekawa, K. (1991). Contribution of eye muscle proprioception to velocity-response characteristics of eye-movements – involvement of the cerebellar flocculus. *Neurosci. Res.*, **12**, 160–168.

Koenderink, J. J. & van Doorn, A. J. (1975). Invariant properties of the motion parallax due to movement of rigid bodies relative to an observer. *Opt. Acta.*, **22**, 773–791.

Kornmüller, A. E. (1931). Eine experimentelle Anästhesie der äusseren Augenmuskeln am Menschen und ihre Auswirkungen. *J. Psychol. Neuro.*, **41**, 354–366.

Lemij, H. G. & Collewijn, H. (1989). Differences in accuracy of human saccades between stationary and jumping targets. *Vision Res.*, **29**, 1737–1748.

Lisberger, S. G. (1984). The latency of pathways containing the site of motor learning in the monkey vestibulo-ocular reflex. *Science*, **225**, 74–76.

Loe, P. R., Tomko, D. L. & Werner, G. (1973). The neural signal of angular head position in primary afferent vestibular nerve axons. *J. Physiol. Lond.*, **230**, 29–50.

Mach, E. (1875). *Grundlinien der Lehre von den Bewegungsempfindungen*. Leipzig.

Mach, E. (1886). *Beitrage zur Analyse der Empfindungen (The Analysis of Sensations)*. Gustav Fischer, Jena.

Mack, A. & Herman, E. (1972). A new illusion: The underestimation of distance during smooth pursuit eye movements. *Percept. Psychophys.*, **12**, 471–473.

Mack, A. & Herman, E. (1973). Position constancy during pursuit eye movement: An investigation of the Filehne illusion. *Q. J. Exp. Psychol.*, **25**, 71–84.

Mack, A. & Herman, E. (1978). The loss of position constancy during pursuit eye movements. *Vision Res.*, **18**, 55–62.

Matin, E. (1974). Saccadic suppression: a review and analysis. *Psychol. Bull.*, **81**, 899–917.

Matthews, P. B. C. (1982). Where does Sherrington's muscular sense originate? Muscles, joints, corollary discharges? *Ann. Rev. Neurosci.*, **5**, 189–218.

Mayne, R. (1974). A systems concept of the vestibular organs. In H. H. Kornhuber

(Ed.) *Handbook of Sensory Physiology. Vestibular System*, pp. 493–580. Springer-Verlag, New York.

Mergner, T., Becker, W. & Deecke, L. (1985). Canal–neck interaction in vestibular neurons of the cat's cerebral cortex. *Exp. Brain Res.*, **61**, 94–108.

Mergner, T., Siebold, C., Schweigart, G. & Becker, W. (1991). Human perception of horizontal trunk and head rotation in space during vestibular and neck stimulation. *Exp. Brain Res.*, **85**, 389–404.

Mergner, T., Rottler, G., Kimmig, H. & Becker, W. (1992a). Role of vestibular and neck inputs for the perception of object motion in space. *Exp. Brain Res.*, **89**, 655–668.

Mergner, T., Schweigart, G. & Hlavacka, F. (1992b). Human self-motion perception during vestibular-proprioceptive interaction. *Eur. J. Neurosci.*, **55**, 304–304.

Miles, F. A. (1993). The sensing of rotational and translational optic flow by the primate optokinetic system. In F. A. Miles & J. Wallman (Eds.) *Visual Motion and its Role in the Stabilization of Gaze*, pp. 393–403. Elsevier, North Holland, Amsterdam.

Miles, F. A. & Wallman, J. (1993). *Visual Motion and its Role in the Stabilization of Gaze*. Elsevier, North Holland, Amsterdam.

Murphy, B. J. (1978). Pattern thresholds for moving and stationary gratings during smooth pursuit eye movement. *Vision Res.*, **18**, 521–530.

Nakayama, K. (1985). Biological image motion processing: a review. *Vision Res.*, **25**, 625–660.

Ohmi, M. & Howard, I. P. (1987). Circularvection as a function of foreground–background relationships. *Perception*, **16**, 17–22.

Paige, G. D. (1989). The influence of target distance on eye-movement responses during vertical linear motion. *Exp. Brain Res.*, **77**, 585–593.

Paige, G. D. & Tomko, D. L. (1991a). Eye-movement responses to linear head motion in the squirrel-monkey. 1. Basic Characteristics. *J. Neurophysiol.*, **65**, 1170–1182.

Paige, G. D. & Tomko, D. L. (1991b). Eye-movement responses to linear head motion in the squirrel-monkey. 2. Visual–vestibular interactions and kinematic considerations. *J. Neurophysiol.*, **65**, 1183–1196.

Pavard, B. & Berthoz, A. (1977). Linear acceleration modifies the perception of a moving visual scene. *Perception*, **6**, 529–540.

Post, R. B. (1986). Induced motion considered as a visually-induced oculogyral illusion. *Perception*, **15**, 131–138.

Post, R. B. & Leibowitz, H. W. (1982). The effect of convergence on the vestibulo-ocular reflex and implications for perceived movement. *Vision Res.*, **22**, 461–465.

Post, R. B. & Leibowitz, H. W. (1985). A revised analysis of the role of efference in motion perception. *Perception*, **14**, 631–643.

Post, R. B., Shupert, C. L. & Leibowitz, H. W. (1984). Implications of OKN suppression by smooth pursuit for induced motion. *Percept. Psychophys.*, **36**, 493–498.

Probst, T. & Wist, E. R. (1982). Impairment of object-motion perception during head movements. *Perception*, **11**, A33–A34.

Probst, T., Krafczyk, S., Brandt, T. & Wist. E. (1984). Interaction between perceived self motion and object motion impairs vehicle guidance. *Science*, **225**, 536–538.

Probst, T., Brandt, Th. & Degner, D. (1986). Object motion detection affected by concurrent self-motion perception: psychophysics of a new phenomenon. *Behav. Brain Res.*, **22**, 1–11.

Raphan, T., Matsuo, V. & Cohen, B. (1979). Velocity storage in the vestibulo-ocular reflex arc (VOR). *Exp. Brain Res.*, **35**, 229–248.

Raphan, T., Cohen, B. & Henn, V. (1981). Effects of gravity on rotary nystagmus in monkeys. *Ann. NY Acad. Sci.*, **374**, 44–55.

Raymond, J. E., Shapiro, K. L. & Rose, D. J. (1984). Optokinetic backgrounds affect perceived velocity during ocular tracking. *Percept. Psychophys.*, **36**, 221–224.

Regan, D. M. (1991). Depth from motion and motion-in-depth. In D. M. Regan (Ed.) *Binocular Vision*, pp. 137–169. Macmillan Press, London.

Regan, D. M. (1992). Visual judgements and misjudgements in cricket and the art of flight. *Perception*, **21**, 91–115.

Regan, D. M. & Beverley, K. I. (1979). Binocular and monocular stimuli for motion in depth: changing-disparity and changing-size feed the same motion-in-depth stage. *Vision Res.*, **19**, 1331–1342.

Rieser, J. J. (1983). The generation and early development of spatial inferences. In H. L. Pick, Jr. & L. P. Acredolo (Eds.) *Spatial Orientation: Theory, Research and Application*, pp. 39–71. Plenum, New York.

Rieser, J. J., Guth, D. A. & Hill, E. W. (1986). Sensitivity to perspective structure while walking without vision. *Perception*, **15**, 173–188.

Robinson, D. A. (1975). Adaptive gain control of vestibuloocular reflex by the cerebellum. *J. Neurophysiol.*, **39**, 954–969.

Roll, J. P., Roll, R. & Velay, J. L. (1991a). Proprioception as a link between body space and extra-personal space. In J. Paillard (Ed.) *Brain and Space*, pp. 112–132. Oxford University Press, Oxford.

Roll, R., Velay, J. L. & Roll, J. P. (1991b). Eye and neck proprioceptive messages contribute to the spatial coding of retinal input in visually oriented activities. *Exp. Brain Res.*, **85**, 423–431.

Ross, H. (1974). *Behavior and Perception in Strange Environments*. George Allen and Unwin, London.

Royden, C. S., Banks, M. S. & Crowell, J. A. (1992a). The perception of heading during eye-movements. *Nature*, **360**, 583–587.

Royden, C. S., Laudeman, I. V., Crowell, J. A. & Banks, M. S. (1992b). The influence of eye-movements on heading judgments. *Invest. Ophthalmol. Vis. Sci.*, **33**, 1051.

Sargent, E. W. & Paige, G. D. (1991). The primate vestibuloocular reflex during combined linear and angular head motion. *Exp. Brain Res.*, **87**, 75–84.

Schwarz, U. & Miles, F. A. (1991). Ocular responses to translation and their dependence on viewing distance. 1 Motion of the observer. *J. Neurophysiol.*, **66**, 851–864.

Schwarz, U., Busettini, C. & Miles, F. (1989). Occular responses to linear motion are inversely proportional to viewing distance. *Science*, **245**, 1394–1396.

Sherrington, C. S. (1918). Observations on the sensual role of the proprioceptive nerve supply of the extrinsic ocular muscles. *Brain*, **41**, 332–343.

Simpson, J. I., Graf, W. & Leonard, C. (1981). The coordinate system of visual climbing fibers to the flocculus. In A. F. Fuchs & W. Becker (Eds.) *Progress in Oculomotor Research*, pp. 475–484. Elsevier, North Holland, Amsterdam.

Simpson, J. I., Leonard, C. S. & Soodak, R. E. (1988a). The accessory optic-system – analyzer of self-motion. *Ann. NY Acad. Sci.*, **545**, 170–179.

Simpson, J. I., Leonard, C. S. & Soodak, R. E. (1988b). The accessory optic-system of rabbit. 2. Spatial-organization of direction selectivity. *J. Neurophysiol.*, **60**, 2055–2072.

Skavenski, A. A. (1972). Inflow as a source of extraretinal eye position information. *Vision Res.*, **12**, 221–229.

Skavenski, A. A., Hansen, R. M., Steinman, R. M. & Winterson, B. J. (1979). Quality of retinal image stabilization during small natural and artificial body rotations in man. *Vision Res.*, **19**, 675–683.

Snyder, L. H. & King, W. M. (1992). Effect of viewing distance and location of the axis of head rotation on the monkey's vestibulo-ocular reflex. I. Eye movement responses. *J. Neurophysiol.*, **67**, 861–874.

Snyder, L. H., Lawrence, D. M. & King, W. M. (1992). Changes in vestibulo-ocular reflex (VOR) anticipate changes in vergence angle in monkey. *Vision Res.*, **32**, 569–575.

Solomon, D. & Cohen, B. (1992a). Stabilization of gaze during circular locomotion in light. 1. Compensatory head and eye nystagmus in the running monkey. *J. Neurophysiol.*, **67**, 1146–1157.

Solomon, D. & Cohen, B. (1992b). Stabilization of gaze during circular locomotion in darkness. 2. Contribution of velocity storage to compensatory eye and head nystagmus in the running monkey. *J. Neurophysiol.*, **67**, 1158–1170.

Solomon, D. & Cohen, B. (1992c). Visual, vestibular, and somatosensory control of compensatory gaze nystagmus during circular locomotion. In A. Berthoz, W. Graf & P. P. Vidal (Eds.) *The Head–Neck Sensory Motor System*, pp. 576–581. Oxford University Press, Oxford.

Sparks, D. L. (1986). Translation of sensory signals into commands for control of saccadic eye-movements – role of the primate superior colliculus. *Physiol. Rev.*, **66**, 118–171.

Spencer, R. F. & Porter, J. D. (1988). Structural organization of the extraocular muscles. In J. Büttner-Ennever (Ed.) *Neuroanatomy of the Oculomotor System*, pp. 33–79. Elsevier Science Publishers, North Holland.

Sperling, G. A. (1990). Comparison of perception in the moving and stationary eye. In E. Kowler (Ed.) *Eye Movements and their Role in Visual and Cognitive Processes*, pp. 307–351. Elsevier, North Holland, Amsterdam.

Sperry, R. W. (1950). Neural basis of the spontaneous optokinetic response produced by visual inversion. *J. Comp. Physiol. Psychol.*, **43**, 482–489.

Stark, L. & Bridgeman, B. (1983). Role of corollary discharge in space constancy. *Percept. Psychophys.*, **34**, 371–380.

Steinbach, M. J. (1987). Proprioceptive knowledge of eye position. *Vision Res.*, **27**, 1737–1744.

Steinman, R. M. & Collewijn, H. (1980). Binocular retinal image during natural active head rotation. *Vision Res.*, **20**, 415–429.

Steinman, R. M. & Levinson, J. Z. (1990). The role of eye movement in the detection of contrast and spatial detail. In E. Kowler (Ed.) *Eye Movements and their Role in Visual and Cognitive Processes*, pp. 115–211. Elsevier, North Holland, Amsterdam.

Steinman, R. M., Levinson, J. Z., Collewijn, H. & Van der Steen, J. (1985). Vision in the presence of known natural retinal image motion. *J. Opt. Soc. Am. A*, **2**, 226–233.

Stevens, J. K., Emerson, R. C., Gerstein, G. L., Kallos, T., Neufeld, G. R., Nicholas, C. W. & Rosenquist, A. C. (1976). Paralysis of the awake human: visual perceptions. *Vision Res.*, **16**, 93–98.

Swanston, M. T. & Wade, N. J. (1988). The perception of visual motion during movements of the eyes and of the head. *Percept. Psychophys.*, **43**, 559–566.

Swanston, M. T. & Wade, N. J. (1992). Motion over the retina and the motion aftereffect. *Perception*, **21**, 569–582.

Swanson, M. T., Wade, N. J., Ono, H. & Shibuta, K. (1992). The interaction of perceived distance with the perceived direction of visual-motion during movements of the eyes and the head. *Percept. Psychophys.*, **52**, 705–713.

Tomlinson, R. D., Saunders, G. E. & Schwarz, D. W. F. (1980). Analysis of human vestibulo-ocular reflex during active head movements. *Acta Otolaryngol.*, **90**, 184–190.

Tulunay-Keesey, U. & Verhoeve, J. N. (1987). The role of eye-movements in motion detection. *Vision Res.*, **27**, 747–754.

Tweed, D. & Vilis, T. (1987). Implications of rotational kinematics for the oculomotor system in three dimensions. *J. Neurophysiol.*, **58**, 832–849.

Van Gisbergen, J. A. M., Robinson, D. A. & Gielen, S. (1981). A quantitative analysis of generation of saccadic eye movements by burst neurons. *J. Neurophysiol.*, **45**, 417–442.

Viirre, E., Milner, K., Tweed, D. & Vilis, T. (1986). A reexamination of the gain of the vestibuloocular reflex. *J. Neurophysiol.*, **56**, 439–450.

Volkmann, F. C. (1986). Human visual suppression. *Vision Res.*, **26**, 1401–1416.

Volkmann, F. C., Riggs, L. A., White, K. D. & Moore, R. K. (1978). Contrast sensitivity during saccadic eye movements. *Vision Res.*, **8**, 1193–1199.

Von Holst, E. & Mittelstaedt, H. (1950). Das Reafferenzprinzip. *Naturwissenschaften*, **37**, 464–476.

Wall, C., Lathan, C. E. & Harris, L. R. (1992). Otolith input enhances vertical optokinetic nystagmus in Z-axis. *Invest. Ophthalmol. Vis. Sci.*, **33**, 1152.

Wallach, H. (1985). Perceiving a stable environment. *Sci. Am.*, **252** (4), 92–98.

Warren, R. (1976). The perception of ego motion. *J. Exp. Psychol.: Human Percept. Perform.*, **2**, 448–456.

Warren, W. H. & Hannon, D. J. (1990). Eye-movements and optical-flow. *J. Opt. Soc. Am. A – Opt. Image Sci.*, **7**, 160–169.

Warren, W. H., Blackwell, A. W., Kurtz, K. J., Hatsopoulos, N. G. & Kalish, M. L. (1991). On the sufficiency of the velocity field for perception of heading. *Biol. Cybernet.*, **63**, 311–320.

Wertheim, A. H. (1981). On the relativity of perceived motion. *Acta Psychol.*, **48**, 97–110.

Wertheim, A. H. (1985). How extraretinal is extraretinal? *Perception*, **14**, 1 A8.

Wertheim, A. H. & Bles, W. (1984). *A Reevaluation of Cancellation Theory: Visual, Vestibular and Oculomotor Contributions to Perceived Object Motion*. Institute for Perception Technical Report IZF-1984–8. TNO Institute for Perception, Soesterberg, The Netherlands.

Wertheim, A. H. & Van Gelder, P. (1990). An acceleration illusion caused by underestimation of stimulus velocity during pursuit eye movements: the Aubert Fleischl phenomenon revisited. *Perception*, **19**, 471–482.

Westheimer, G. & McKee, S. P. (1975). Visual acuity in the presence of retinal-image motion. *J. Opt. Soc. Am. A – Opt. Image Sci.*, **65**, 847–850.

Westheimer, G. & McKee, S. P. (1978). Stereoscopic acuity in the presence of retinal-image motion. *J. Opt. Soc. Am. A – Opt. Image Sci.*, **68**, 450–455.

Whiteside, T. C. D., Graybiel, A. & Niven, J. I. (1965). Visual illusions of movement. *Brain*, **88**, 193–210.

Wilson, V. J. & Melvill Jones, G. (1979). *Mammalian Vestibular Physiology*. Plenum, New York.

Wist, E. R., Diener, H. C., Dichgans, J. & Brandt, T. (1975). Perceived distance and the perceived speed of self-motion: linear vs angular velocity? *Percept. Psychophys.*, **17**, 549–554.

15

The Visual Drive for Smooth Eye Movements

Richard J. Krauzlis

National Institutes of Health, Bethesda, USA

1 INTRODUCTION

The complete array of visual motion presented to a moving observer is determined by the geometric relationship between the viewpoint of the observer and the environment in which the observer moves. However, an observer typically uses only a subset of this array in guiding his/her behavior. This selective use of visual motion reflects the influence of a biological strategy. In this review, I hope to illustrate this point by providing an overview of what is known about how three animals – the rabbit, the cat, and the monkey – use visual motion to generate smooth eye movements.

1.1 The Retinal Representation of the Visual World

It is informative to first consider two basic features of the visual system: the field of view projected onto the retina and the inhomogeneous distribution of projections from the retina. For the rabbit, the eyes provide a panoramic view of the world (Figure 1). The eyes are directed laterally with respect to the head and the field of view for each eye extends 190°. This provides the rabbit with monocular vision over almost the entire visual surround and an additional portion of binocular vision in front. In the retina, there is a horizontally elongated region in which the density of ganglion cells is highest, the visual streak. For the cat, the eyes are directed forward. The field of view provided by the optics of the eye extends 181°, but only 143° of this extent is served by the retina. The cat has a larger region of binocular vision (98°) than the rabbit, but the total field of view is only 187°. In the retina, a local region in which the density of ganglion cells is highest, the area centralis, is surrounded by a lower density halo that stretches toward the nasal retina. Similarly, the frontal-eyed monkey has a large

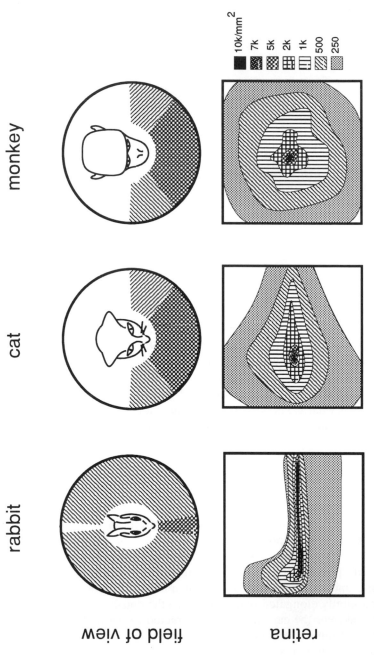

Figure 1 Schematic diagram showing how the visual world is represented by the retina of the rabbit, cat, and monkey. In the top row, the visual surround for each animal is depicted as a circle and the portions of the visual surround served by the retinae of the two eyes are indicated by different shadings: single hatching represents the angular extent of the visual field served by one eye, cross-hatching represents the extent served by both eyes, and the absence of shading indicates a blind spot. In the bottom row, the squares depict flattened representations of the retina for each animal and the different shadings indicate different densities of ganglion cells, as indicated by the scale to the right. These schematic depictions are based on data published for the rabbit (Hughes, 1971, 1972), cat (Hughes, 1975, 1976), and monkey (Webb & Kaas, 1976).

binocular field of view and a central retinal region densely packed with ganglion cells. These differences in the construction of the eyes serve to emphasize different portions of the visual world and reflect differences in behavior. The panoramic vision of the rabbit aids in the detection of predators, while the frontal vision of the cat and monkey aids in the search for food.

Because the effect of eye movements is to re-map the projection of the visual world onto the retina, there is a strong linkage between the control of eye movements and the processing of visual inputs. This linkage is reciprocal, in the sense that eye movements affect visual inputs and, in turn, visual inputs provide an important drive for eye movements. The range of eye movements made by a species can therefore reveal the goals of vision. Research over the last three decades has provided a large body of information about the relationship between smooth eye movements and their primary visual drive, visual motion. This review will focus on the range of smooth eye movements made by the rabbit, cat, and monkey, and on the neural pathways that mediate the appropriate visual motion signals. In describing the data, I will rely on a basic conceptual model, namely, that visual motion represents a command signal used by motor pathways to move the eyes. I hope to show that the strategy used by the rabbit to construct this command is very different from the strategy used by the cat or monkey. While the rabbit uses a global analysis of visual motion, the cat and monkey emphasize the central portion of the visual field. While the analysis in the rabbit is mediated by subcortical structures, the cat and monkey rely heavily on inputs from the cerebral cortex.

2 THE RABBIT

2.1 The Optokinetic Response

When the visual world is rotated around a rabbit, the rabbit's eyes rotate smoothly to follow the motion of the visual scene, minimizing the motion of the images projected onto the retina. These eye movements are called the optokinetic response (OKR). In natural settings, such motion of the visual world occurs only when the rabbit turns its head. During head turns, the eyes are initially stabilized by the vestibulo-ocular reflex (VOR) which is driven by head motion signals derived from the semicircular canals of the inner ear. The VOR has a very short latency (around 15 ms) and brisk dynamics, but it operates open-loop – head motion signals do not indicate whether the eyes are actually stabilized. The optokinetic response compensates for this limitation, serving as a visual backup to the VOR. If the performance of the VOR is inadequate, motion of the visual world on the retina will ensue and this visual motion will produce an optokinetic response.

The properties of the visual drive for optokinetic eye movements can be studied in the laboratory by presenting full-field visual stimuli to a subject whose head is immobilized and by recording the eye movements that are evoked (Judge *et al.*, 1980). Continuous motion of the visual surround is usually achieved by rotating a patterned drum around the subject or by projecting a full-field stimulus with a slowly spinning planetarium. As the visual field is rotated, the eyes smoothly follow, generating 'slow phases' which reduce the motion of the projected image on the retina. When the following movements have led the eyes far from the center of the orbit, saccadic eye movements, or 'quick phases', reset the position of the eye (Figure 2A). This pattern of slow following movements and quick resetting saccades is called optokinetic nystagmus (OKN).

2.1.1   Sensitivity to low speeds

One distinctive feature of the optokinetic response of the rabbit is its sensitivity to very low speeds. If the visual field is rotated more slowly than 1°/s, the slow phase eye movements start with a latency of about 100 ms and match the speed of the stimulus in less than one second. In fact, the rabbit can match its eye speed to motion as slow as 0.002°/s. For stimulus speeds faster than a few °/s, the eyes start to move at a slightly shorter latency, but do not immediately match the speed of the stimulus (Figure 2B). Instead, eye speed increases slowly over time. For example, for a stimulus speed of 30°/s, eye speed reaches 1°/s in the first second and reaches a steady speed of about 10°/s after one minute (Collewijn, 1969, 1972).

2.1.2   Directional asymmetries

A second feature of the rabbit's optokinetic response is the presence of pronounced directional asymmetries. If the rabbit views a scene monocularly, a vigorous optokinetic response can be produced by horizontal motion in the temporal to nasal direction (i.e. flowing toward the front of the rabbit), but horizontal motion in the nasal to temporal direction is ineffective (Collewijn, 1969). The functional consequence of this asymmetry is that rabbits respond well to rotational disturbances of the visual scene, such as occur during head turns, but do not respond to translational disturbances that occur when the rabbit moves forward. For example, if we set aside the effects of the VOR, when a rabbit rotates its head to the right, the leftward motion of the surround will provide a strong visual drive through the right eye (which sees temporal to nasal motion), but a weak input through the left eye (which sees nasal to temporal motion). The right eye input will drive both eyes to the left, stabilizing their orientation with respect to the visual surround. However, if the rabbit moves forward, both eyes will see motion in the nasal to temporal direction and the eyes will not turn. The absence of eye movements during forward movement may allow the rabbit to use motion parallax to interpret the structure of the visual surround. For vertical motion of the visual surround, the optokinetic response of the rabbit is very nearly symmetric (Dubois & Collewijn, 1979; Erickson & Barmack, 1980).

2.1.3   Optokinetic after-nystagmus

If a rabbit has been viewing an optokinetic stimulus for some time and the stimulus is then removed, leaving the rabbit in total darkness, the slow eye movements persist for many seconds. This persistence of the optokinetic response is called optokinetic after-nystagmus (OKAN). This after-nystagmus is the complement of the gradual rise in eye velocity that occurs when the stimulus is first presented. Both of these effects reflect the fact that the visual drive is not used in isolation, but is accompanied by a 'velocity storage' mechanism that improves the response of optokinetic eye movements. In a

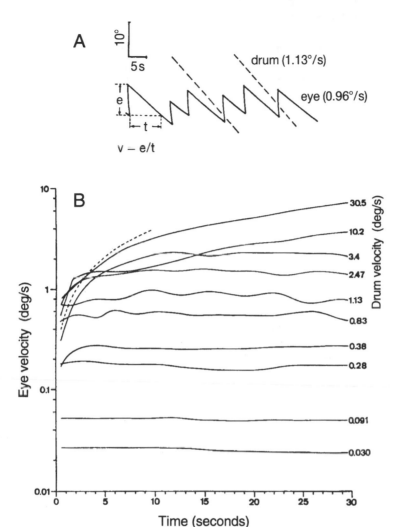

Figure 2 The optokinetic response in the rabbit. A. Example of optokinetic nystagmus (solid trace) elicited by a drum (dashed line) rotated at 1.13°/s. The amplitude of the optokinetic response is measured from the smooth movements which follow the motion of the drum. B. Amplitude of the optokinetic response during the first 30 seconds after the onset of drum motion. The responses were measured during viewing of ten different drum speeds, as indicated by the numbers to the right of the plotted eye velocities. Modified from Collewijn (1969).

laboratory setting, velocity storage serves to integrate visual motion inputs over time and makes it possible for the rabbit to be sensitive to motion of the visual surround over a wide range of speeds. In more natural settings, the after-nystagmus produced by velocity storage helps to cancel the vestibular nystagmus that occurs in the wake of a prolonged head movement.

2.2 Neural Pathways

The heart of the circuitry underlying the optokinetic response in the rabbit is contained in a set of three nuclei located in the midbrain which comprise the accessory optic system (AOS) and a fourth, closely related nucleus, the nucleus of the optic tract (NOT). The analysis of visual motion accomplished by these nuclei is built up in a straightforward way from inputs provided by retinal ganglion cells. The outputs provided by these nuclei reflect the close functional relationship of the rabbit's optokinetic response to the vestibular system.

2.2.1 Retinal inputs

In the retina of the rabbit, about one-quarter of the retinal ganglion cells are direction-selective (Barlow & Hill, 1963). These retinal ganglion cells can be subdivided into 'on-off' ganglion cells, which respond best to speeds over 1°/s, and 'on' ganglion cells, which respond best to speeds around 0.5°/s (Barlow et al., 1964; Oyster et al., 1972). Based on the differences in speed selectivity, the on-direction-selective ganglion cells are considered to be the probable source of the visual drive for the rabbit's optokinetic response (Oyster et al., 1972). The population of on-direction-selective ganglion cells can be subdivided into three groups, based upon their preferred directions of motion. These preferences are aligned with the directions of eye motion produced by contraction of the extraocular muscles (Oyster & Barlow, 1967; Oyster, 1968). The three groups of on-direction-selective ganglion cells therefore encode visual motion along spatial axes that can be used directly by the output motor pathways of the optokinetic system.

2.2.2 The accessory optic nuclei

Direction-selective ganglion cells project directly to three nuclei in the anterior midbrain of the rabbit: the medial (MTN), lateral (LTN), and dorsal (DTN) terminal nuclei (Hayhow, 1959; Oyster et al., 1980). The projection to nuclei on each side is almost exclusively from the contralateral retina (Erickson & Cotter, 1983). Neurons in the three terminal nuclei are characterized by a selectivity to speed and direction that closely matches the properties of the on-direction selective ganglion cells (Simpson et al., 1979). However, units in the accessory optic nuclei have very large receptive fields and, consequently, are driven best by large textured patterns. The receptive fields are typically about 50° across, are located in the contralateral visual field, and almost always include the visual streak (Soodak & Simpson, 1988).

The preferred directions of motion differ in the three nuclei. In the DTN, units prefer horizontal motion in the temporal to nasal direction, while in the LTN, units prefer downward motion. In the MTN, approximately two-thirds of the units prefer upward motion and the remaining third prefer downward motion like units in the LTN (Simpson et al., 1979; Soodak & Simpson, 1988; Simpson et al., 1988). However, the details of their direction selectivity reveal that these units are selective for global visual motion that would result from self-movement (Simpson et al., 1979, 1988). In particular, the direction selectivity in portions of the large receptive fields is arranged so that units are best activated by visual motion that would result from rotating the head about one spatial axis. Furthermore, the axes of rotation preferred by these units match the axes of head rotation detected by the three pairs of semicircular canals. For example, as shown in Figure 3, some units in the MTN have distinctly different preferred directions in different parts of the receptive field (Simpson *et al.*, 1988). In the anterior quarter of the receptive field (from the nose to 45°), the preferred direction might be upward. In the posterior half of the receptive field, the preferred direction would then be downward. This combination of preferred *linear* directions of motion corresponds to a single preferred axis of *rotational* motion – in this case, an axis pointing 45° back from straight ahead. This also corresponds to the rotational axis that most strongly modulates vestibular afferents from one pair of semicircular canals, in this case, the anterior vertical canal on the ipsilateral side and the posterior vertical canal on the contralateral side. Following this logic, units in the accessory optic nuclei can each be functionally related to one set of canals, based upon which rotational axis provides the strongest visual modulation.

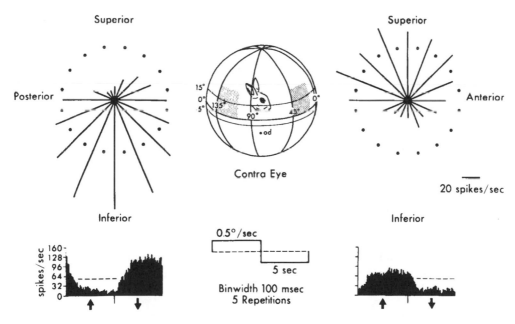

Figure 3 Regional differences in direction selectivity within the receptive field of an MTN neuron. The direction-selective tuning curves for each of the two shaded regions of the visual world of the contralateral eye (center panel) were obtained using a 30° rear-projected stimulus moving at 0.5°/s. In the left and right panels, the spontaneous discharge rate is indicated by the circle of dots, and the length of each line in a particular direction indicates the steady-state modulation for movement in that direction. The histograms below the two tuning curves illustrate the respective responses for vertical movement. For this neuron, the change in the spatial organization of the directional preferences occurred at about 50° contralateral azimuth. From Simpson *et al.* (1988).

2.2.3   The nucleus of the optic tract

The nucleus of the optic tract (NOT) in the pretectum is not a component of the accessory optic nuclei, but it is an essential member of the optokinetic pathways and is closely related to the DTN in the response properties of its neurons. Furthermore, the accessory optic and pretectal nuclei in the midbrain are anatomically linked. For example, the MTN has ipsilateral projections to both the DTN and the NOT (Giolli et al., 1984, 1985). As in the DTN, units in the NOT respond best to horizontal motion in the temporal to nasal direction (Collewijn, 1975a; Simpson et al., 1979). Some NOT units show a selectivity for low stimulus speeds, also like DTN units, while others show a preference for slightly higher speeds. Electrical stimulation of the NOT produces horizontal OKN with slow phases directed to the ipsilateral side (Collewijn, 1975b). Conversely, small unilateral lesions of the NOT abolish horizontal OKN to the ipsilateral side and bilateral lesions abolish OKN in both directions.

2.2.4   Absence of a cortical projection

The visual pathways underlying the optokinetic response in the rabbit are entirely subcortical. There is no evidence of a projection from cortical visual areas I and II of the rabbit to the nuclei of the accessory optic system or to the NOT (Giolli & Guthrie, 1971; Giolli et al., 1978). Consistent with the absence of a cortical projection, decortication has little effect on the optokinetic response of the rabbit (Hobbelen & Collewijn, 1971).

2.2.5   The nucleus reticularis tegmenti pontis

The link between the accessory optic and pretectal nuclei that process visual motion and the oculomotor nuclei that control movements of the eyes appears to be provided by the nucleus reticularis tegmenti pontis (NRTP). The NRTP receives projections from the NOT and the MTN, as well as projections from other visual nuclei, including the superior colliculus and the ventral lateral geniculate nucleus (Torigoe et al., 1986; Maekawa et al., 1984). The NRTP also receives projections from several eye movement related nuclei, such as the deep cerebellar nuclei, the superior, lateral, and medial vestibular nuclei, and the paramedian pontine reticular formation (Torigoe et al., 1986), and has return projections to many of these same nuclei (Balaban, 1983; Maekawa et al., 1981).

2.2.6   The inferior olive and cerebellum

One of the best studied projections from the optokinetic visual nuclei is to the dorsal cap of the inferior olive (Takeda & Maekawa, 1976; Maekawa & Takeda, 1979; Giolli et al., 1985; Mizuno et al., 1974). As in the accessory optic and pretectal nuclei, neurons in the dorsal cap show strong directional selectivity and prefer directions of visual motion that match the principal axes of the semicircular canals (Simpson et al., 1981; Leonard et al., 1988).

The dorsal cap, in turn, provides climbing fiber inputs to the floccular lobe of the cerebellum (Alley *et al.*, 1975). This projection is of interest, because the flocculus has been shown to play a critical role in producing adaptive changes in the vestibuloocular reflex. The visual motion signals carried by climbing fibers indicate how successfully the VOR has stabilized gaze during head turns. It has been suggested that this information could be used to modify the efficacy of synapses either in the flocculus or on the targets of the flocculus which would then act to change the amplitude of the VOR. Despite its role in adaptive mechanisms, the cerebellum does not appear to be directly involved in producing optokinetic responses in the rabbit. Removal of the cerebellum does not change the smooth component of OKN (Collewijn, 1970), although such lesions do cause disruptions in the occurrence of resetting saccades during optokinetic nystagmus.

2.3   Visual Motion as a Proprioceptive Sense

The optokinetic response of the rabbit is an example of how visual motion can be used for proprioception. As suggested originally by Simpson *et al.* (1979) the accessory optic system of the rabbit serves to signal self-motion as indicated by the global slip of the visual world on the retina. The results of this global analysis are cast into a rotational coordinate system which is shared by both the semicircular canals (Simpson *et al.*, 1979) and the extraocular muscles (Ezure & Graf, 1984). This choice of coordinate system facilitates the union of the visual *reafferent* information provided by the accessory optic system and the vestibular *afferent* information provided by the labyrinths of the inner ear.

3   THE CAT

3.1   The Optokinetic Response

Considering the types of smooth eye movements that it can make, the cat is rather more like a rabbit than a monkey. Like the rabbit, the cat is virtually incapable of smoothly tracking a small spot with its eyes. Even after extensive training, the cat can produce maximum eye velocities of only 0.5 to 1.0°/s when pursuing a stimulus 0.5° in diameter (Figure 4). Instead, the cat tracks the target by making a series of saccadic eye movements (Evinger & Fuchs, 1978).

  If a textured background is presented along with the spot, however, the cat is able to produce faster smooth eye movements which have a latency of about 90 ms (Figure 4; Evinger & Fuchs, 1978). If the stimulus speed is less than 10°/s, these optokinetic responses match target speed within several hundred milliseconds. For speeds greater than 10°/s, the cat is unable to match the speed of the stimulus and instead uses saccades to follow the target.

3.1.1   Comparison with the rabbit

As with the rabbit, the cat can gradually generate faster smooth eye velocities if a large moving stimulus is presented for several seconds. With an optokinetic drum, slow eye movements up to 28°/s can be achieved after exposures of 10–20 s (Evinger & Fuchs, 1978; Benson & Guedry, 1971; Grasse & Cynader, 1988). The optokinetic response of the cat also shows pronounced directional asymmetries, but the biases are somewhat different from those observed in the rabbit. Similar to the rabbit, the cat shows a horizontal bias for motion in the temporal to nasal direction when the stimulus is viewed monocularly (Cynader & Harris, 1980). When the stimulus is viewed with both eyes, the amplitude of the response is symmetric horizontally. Unlike the rabbit, which has a symmetric vertical optokinetic response, the cat displays a bias for upward motion (Grasse & Cynader, 1988). For both the rabbit and the cat, the directional asymmetries have the effect of minimizing responses to motion that might occur during forward locomotion. The rabbit, with its laterally placed eyes, experiences primarily horizontal motion in the nasal to temporal direction during locomotion. The cat, with its more

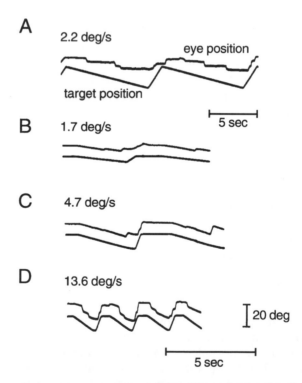

Figure 4 Smooth eye movements made by a cat in response to movement of a target spot or the combined movement of a target spot and a background. A. When tracking a spot (0.5° diameter) that moves at 2.2°/s, the cat's smooth eye movements are interrupted by frequent catch-up saccades. B–D. With combined motion of a spot and background, the cat makes fewer saccades and achieves higher eye speeds. Adapted from Evinger and Fuchs (1978).

forward placed eyes, experiences primarily downward vertical motion. In both cases, these directions of motion are least effective in driving the optokinetic response.

3.2   Neural Pathways

Several of the key structures mediating smooth eye movements in the cat are homologous to the nuclei already described for the rabbit. However, there are at least two major differences between the organization of the optokinetic pathways in the cat and the rabbit. First, the anatomy of the accessory optic and pretectal nuclei is somewhat different. For example, in the rabbit the MTN is the largest of the three terminal nuclei; in the cat, the LTN and MTN are approximately the same size (Hayhow, 1959; Marcotte & Updyke, 1982). Second, in the rabbit the cortex exerts no influence over smooth eye movements. In the cat, the cerebral cortex provides extensive inputs to the accessory optic and pretectal nuclei that change the response properties of neurons in the visual nuclei and, consequently, the range of smooth eye movements that the cat can make.

3.2.1   The accessory optic and pretectal nuclei

In general respects, the accessory optic and pretectal nuclei of the cat are similar to those of the rabbit. In the cat, neurons in these nuclei have large receptive fields, typically with diameters of 60–100°. The receptive fields are usually located near the center of the contralateral visual field and always include the area centralis. The best stimuli for evoking responses from units in these nuclei are large textured patterns swept across the receptive field. As in the rabbit, neurons in the MTN and LTN respond best to vertical motion, while neurons in the MTN and NOT prefer horizontal motion (Grasse & Cynader, 1982, 1984; Hoffman, 1981). Also as in the rabbit, small unilateral lesions in the region of the NOT and DTN produce a deficit in the horizontal optokinetic response to that side (Hoffman & Distler, 1986).

However, there are several features of the cat's accessory optic and pretectal nuclei that differ from the rabbit. First, the distribution of preferred directions is not the same. In the cat's MTN, more units prefer downward than prefer upward motion, the opposite of the bias in the rabbit. In the cat's LTN, equal numbers of units prefer upward and downward motion, while in rabbit, most prefer downward. Second, the range of preferred speeds is different. In the LTN, DTN and NOT, individual neurons are driven best by speeds of 1–100°/s, although the most effective speeds in driving units are still rather slow, between 1 and 10°/s (Grasse & Cynader, 1982; Hoffman & Distler, 1986). In the LTN, there is also a correlation between the preferred direction and the preferred speed: units with upward directional selectivity tend to prefer higher speeds (Grasse et al., 1984). In the MTN, units prefer low speeds, around 1°/s, as they do in the rabbit (Grasse & Cynader, 1982).

Finally, unlike in the rabbit, the accessory optic and pretectal nuclei of the cat receive inputs from both eyes. In the LTN, DTN and NOT, units are always driven by the contralateral eye, but some fraction of the units can also be driven by monocular stimulation of the ipsilateral eye (Grasse & Cynader, 1984; Hoffman & Schoppmann, 1981). The exact fraction is different in the LTN (78%), DTN (93%), and NOT (46%). In the MTN, as in all of the visual nuclei of the rabbit, there is only a minor contribution from the ipsilateral eye (Grasse & Cynader, 1982). The input from the ipsilateral eye is not mediated by a direct retinal projection, since almost all of the retinal inputs to the midbrain visual nuclei come from the contralateral eye (Farmer & Rodieck, 1982; Hayhow, 1959; Marcotte & Updyke, 1982). Instead, the binocular responses are mediated by an additional input from the visual cortex.

3.2.2   Cortical projections to the accessory optic and pretectal nuclei

There is a widespread projection from visual areas in the cerebral cortex to the visual nuclei in the midbrain of the cat (Berson & Graybiel, 1980; Garey et al., 1968; Marcotte & Updyke, 1982). When the cortex is removed, the horizontal optokinetic response becomes asymmetric. With monocular viewing, only motion in the temporal to nasal direction can elicit an optokinetic response, as is the case in the normal rabbit (Wood et al., 1973).

Consistent with the effect on smooth eye movements, lesions of the cerebral cortex cause distinct changes in the visual properties of the accessory optic nuclei. After decortication, units in the LTN and DTN can no longer be driven through the ipsilateral eye. In addition, although units remain directionally-selective, there are changes in the distribution of preferred directions across the population. In the LTN and DTN, decortication results in the loss of units that prefer upward motion. Concomitant with this loss, there are fewer units that prefer higher speeds (Grasse et al., 1984). Likewise, in the NOT, decortication removes the ipsilateral drive to units and reduces the sensitivity to stimuli moving faster than 10°/s (Hoffman, 1981). Thus, the major properties which distinguish the accessory optic and pretectal nuclei of the cat from those of the rabbit are abolished when the cortex is removed.

3.2.3   Cortical projections to the basilar pontine nuclei

In parallel with the projections from cortex to the accessory optic and pretectal nuclei, there are projections from the same cortical areas to neurons in the basilar pontine nuclei. Neurons in the pontine nuclei, in turn, project to several regions of the cerebellum. The visual cortico-pontine projections in the cat arise primarily from extrastriate regions areas 18, 19, 21, from area 7 in parietal cortex, and from the peripheral field representation of area 17 (Brodal, 1972a, 1972b; Fries & Albus, 1980). This projection terminates in the ventromedial portion of the rostral pons and may involve the convergence of inputs from multiple cortical areas onto single pontine neurons (Fries & Albus, 1980). Visual neurons in the pontine nuclei have properties similar to those observed in the accessory optic and pretectal nuclei. They display a strong selectivity for direction of motion and have large receptive fields (Glickstein et al., 1972).

3.2.4   The cerebellum

Visual neurons in the pontine nuclei project heavily to the flocculo-nodular lobe of the cerebellum (Robinson et al., 1984). The flocculus also receives projections from the vestibular nuclei, the NRTP, and other eye movement related nuclei in the brainstem

(Blanks, 1990). The flocculus can influence eye movements through return projections to the vestibular nuclei. The cerebellum has the distinctive anatomical feature of being organized into longitudinal zones which differ in their patterns of connectivity. In the case of the flocculus, these zones are defined by the subregions of the vestibular nuclei with which they share the strongest connections (Sato et al., 1982, 1988). Each subdivision can be interpreted as corresponding to a functional unit that exerts control over a particular set of extraocular muscles (Sato & Kawasaki, 1984, 1990).

The flocculus also receives a climbing fiber input from the inferior olive (Gerrits & Voogd, 1982). In the cat, as in the rabbit, the inferior olive receives a strong projection from the accessory optic and pretectal nuclei. However, in the cat, this link is made by the NOT and the MTN, but not by the LTN or DTN (Walberg et al., 1981). Neurons in the inferior olive display visual properties like those in the accessory optic nuclei. The projection made by climbing fibers onto the flocculus also conforms with the zonal organization exhibited by the floccular efferents to the vestibular nucleus (Sato et al., 1983).

Unlike in the rabbit, lesions of the cerebellum in the cat do affect the optokinetic response. The characteristic of the induced deficit is a narrowing of the range of attainable eye speeds. Whereas the intact cat is able to generate slow phases up to about 40°/s, the cerebellectomized cat cannot generate smooth eye movements faster than about 20°/s (Keller & Precht, 1979; Kato et al., 1982). The deficit is also directionally selective – only smooth movements toward the side of the lesion are affected. The effects of cerebellar lesions suggest that some of the behavioral consequences of cortical lesions may be due to loss of cortical inputs to the cerebellum, in addition to the loss of cortical inputs to the accessory optic and pretectal nuclei.

3.2.5 Nucleus reticularis tegmenti pontis

The persistence of the optokinetic response after cerebellar lesions indicates that, as in the rabbit, there is a pathway conveying visual inputs that does not go through the cerebellum. Several pieces of evidence indicate that this additional pathway involves the NRTP. First, the NRTP of the cat receives projections from widespread regions of cerebral cortex, similar to those received by the visual pontine nuclei (Brodal & Brodal, 1971). Second, lesions of the NRTP have a large effect on the ability to generate slow eye movements, and the magnitude of this deficit is somewhat larger than that observed after cerebellar lesions (Keller & Precht, 1979). After such lesions, cats cannot generate smooth eye velocities faster than 10°/s. Unilateral lesions affect eye movements directed to the same side as the lesion and bilateral lesions affect eye movements in both directions (Precht & Strata, 1980; Kato et al., 1982). Finally, lesions of the NRTP have the additional effect of reducing the number of neurons in the vestibular nuclei that show optokinetic visual responses (Precht & Strata, 1980). In contrast, lesions of the cerebellum have only a slight effect on the optokinetic responses recorded in the vestibular nuclei (Keller & Precht, 1979). This indicates that the NRTP is normally able to transmit visual inputs to neurons in the vestibular nuclei along a pathway that does not involve the cerebellum. The identity of this pathway is not clear, since there is no anatomical evidence for either direct projections from the pretectum to the NRTP (Berman, 1977) or from the NRTP to the caudal brainstem (Gerrits & Voogd, 1986).

3.3   Cortical Colonization of the Brainstem

The influence of the cortex on the optokinetic response of the cat represents a fundamental difference from the rabbit. However, the colonization of the accessory optic and pretectal nuclei by the cortex is a developmentally labile event. If the normal development of the visual cortex is disrupted, the influence of the cortex is largely aborted. For example, raising a cat in darkness has effects similar to those observed after cortical lesions in adult animals. Behaviorally, the optokinetic response shows a strong bias for motion in the temporal to nasal direction. In the accessory optic nuclei, units lose their responses to higher speeds and to inputs from the ipsilateral eye (Grasse & Cynader, 1986). If the cat is deprived of vision through only one eye, nuclei on *both* sides are dominated by inputs from the contralateral eye. In nuclei contralateral to the deprived eye, units are more strongly influenced by the direct retinal inputs from the deprived eye and only weakly influenced by the cortical inputs arising from the non-deprived eye (Grasse & Cynader, 1987).

The effects of dark-rearing and monocular deprivation may be more accurately regarded as arrested development, rather than abnormal development. Kittens under the age of 5 weeks have an asymmetric optokinetic response with a horizontal bias for temporal to nasal motion (van Hof-van Duin, 1976, 1978). It is believed that this temporal to nasal response is mediated by crossed projections directly from the nasal retina to the accessory optic and pretectal nuclei (Figure 5). This projection, conveying information about the temporal visual field, is present at birth. In contrast, the responses to motion in the nasal to temporal direction are mediated by projections through visual cortex. This projection conveys information about the nasal visual field and develops post-natally. The delay in the maturation of the cortical projection makes it more sensitive to disruptions in early visual experience. These developmental factors appear to be shared across cats and primates, since both monkeys and humans show directional asymmetries early in infancy which persist into adulthood if they are visually deprived (Atkinson, 1979; Naegele & Held, 1982; Schor & Levi, 1980).

4   THE MONKEY

4.1   Smooth Eye Movements

Unlike the rabbit and the cat, the monkey is able to smoothly track visual targets of all sizes. In addition to generating optokinetic responses to full-field stimuli, two other types of eye movements have been described for the monkey. *Ocular following* eye movements are brief involuntary movements in response to quick movements of the central portion of the visual field. *Pursuit* eye movements are sustained voluntary movements that keep the eyes pointed at small moving targets, even against texture backgrounds. In considering each of these eye movements, it should be kept in mind that while there are differences between them, they do not represent completely independent eye movement systems. As shall become clearer in the discussion of

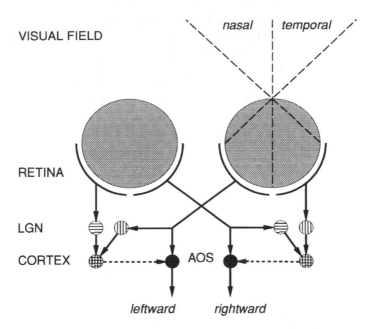

Figure 5 Schematic diagram of the interaction between cortical and subcortical pathways for smooth eye movements in the cat. Accessory optic nuclei (AOS, black circles) on each side receive direct inputs from the contralateral retina and mediate eye movements toward the ipsilateral side. The inputs to the AOS arise from the nasal retina, which is excited by an optical projection of the temporal visual field. Visual units in cortex (cross-hatched circles) combine inputs from both eyes, conveyed by the lateral geniculate nucleus (LGN, single-hatched circles). The inputs to the cortex combine information from the contralateral nasal retina and the ipsilateral temporal retina. As described in the text, visual deprivation reduces the efficacy of the projection from the cortex to the AOS (dashed-line arrow).

the neural pathways, they share a dependence on cortical projections to the cerebellum and brainstem.

4.1.1   The optokinetic response

In the monkey, the optokinetic response consists of two distinct components, an 'early' component which is responsible for a quick rise in smooth eye velocity, and a 'delayed' component with an amplitude that increases slowly over the course of many seconds. When an optokinetic stimulus is first presented, the eyes accelerate at rates of 200–450°/s^2, much higher than those observed in the rabbit or cat. For speeds less than about 30°/s, eye velocity matches stimulus velocity within 200 ms. For speeds greater than about 60°/s, eye velocity reaches about 60% of stimulus speed in the first 200 ms, and then rises more slowly, reaching a maximum constant speed after 10–15 seconds (Cohen *et al.*, 1977; Lisberger *et al.*, 1981a). This delayed component of the monkey's optokinetic response is qualitatively similar to the slow rise in eye velocity that is observed in rabbits and cats for stimulus speeds over 1°/s. However, the monkey is able to match speeds up to about 100°/s (Cohen *et al.*, 1977;

Zee *et al.*, 1981) and can reach peak velocities of up to 180°/s. The optokinetic response of adult monkeys also differs in that it does not show directional asymmetries in the horizontal plane, even with monocular stimulation (Koerner & Schiller, 1972). However, the monkey does show a vertical asymmetry – the upward optokinetic response is larger than the downward (Takashi & Igarashi, 1977). Like the rabbit and cat, the monkey shows an OKAN, and this also consists of two components. When the lights are put out, smooth eye velocity initially drops by about 10–20%, then decreases more slowly for 20–60 seconds (Kreiger & Bender, 1956; Cohen *et al.*, 1977; Zee *et al.*, 1981).

The delayed component of the optokinetic response displays a linkage to the VOR that is reminiscent of the rabbit's optokinetic response. The strongest evidence of this linkage is the common adaptation displayed by the delayed component and the VOR. Adaptation of the VOR can be achieved by having a monkey wear magnifying or minifying spectacles. Such spectacles change the compensatory eye movements required to stabilize gaze during head turns. After wearing the spectacles for a few days, the VOR adapts and the amplitude of the eye movements made during head movements become larger or smaller, so that retinal images remain stabilized. Concomitant with such adaptive modification of the VOR, changes are also observed in the amplitude of the delayed component of the optokinetic response. In contrast, no changes are observed in the early component of the optokinetic response (Lisberger *et al.*, 1981a).

4.1.2   Ocular following

Ocular following eye movements are driven best by fast movements of textured stimuli about 40° across. The latency of ocular following is very short, 50–55 ms, with the shortest latencies and largest eye accelerations occuring for stimulus speeds of 50–100°/s. These eye movements have several interesting properties (Miles *et al.*, 1986). First, the latency of ocular following is a function of contrast and spatial frequency. This indicates that the onset of ocular following might be triggered simply by local changes in luminance. Second, the amplitude of ocular following is larger for stimuli 20–60° across than for stimuli that occupy the entire visual field. In fact, the response to motion in the center of the visual field is enhanced if there is motion in the peripheral visual field in the opposite direction. It has been suggested that this peripheral modulation might assist in the tracking of objects. Third, the responses are largest if the stimuli are presented immediately in the wake of a saccadic eye movement (Kawano & Miles, 1986). This effect is related to the retinal stimulation associated with the saccade, rather than the saccade itself, because visual stimulation at saccadic velocities produces similar results. Finally, the amplitude of ocular following also depends upon viewing distance (Busettini *et al.*, 1991). When the stimulus is close to the subject, the responses are larger than when the stimulus is further away, even when care is taken to ensure that the retinal stimulation is identical in the two situations.

It has been argued that ocular following is concerned with compensating for disturbance of gaze that result from *translation*, analogous to how the delayed component of the optokinetic response is concerned with *rotational* disturbances of gaze (Miles *et al.*, 1992; Miles, 1993). This idea is supported by the fact that the best stimulus for ocular following is local motion such as would occur during translation through an enviroment, rather than full-field motion that occurs during rotation. The idea is also supported by the observation that ocular following shares some properties with the translational vestibulo-ocular reflex (TVOR). When a subject is translated in the dark, compensatory eye movements are made that stabilize gaze. These reflexive eye movements are driven by the otolith organs of the inner ear, in analogous fashion to how the semicicular canals drive the rotational VOR. It has been observed that eye movements mediated by the TVOR show a dependence on viewing distance like that found for ocular following (Schwarz et al., 1989; Schwarz & Miles, 1991).

4.1.3 Pursuit eye movements

When tracking small visual targets, monkeys typically make a combination of pursuit and saccadic eye movements. Pursuit eye movements act to match eye speed to the speed of the target, minimizing the motion of the target's retinal image. Saccadic eye movements act to reduce position offsets, placing the image of the target on the high acuity region of the retina (Rashbass, 1961). Monkeys are able to accurately pursue small targets moving at speeds up to 30°/s (Fuchs, 1967). Higher pursuit speeds are possible (Lisberger et al., 1981b), but at higher speeds pursuit does not match target speed and is therefore less useful, since visual acuity is degraded for retinal speeds in excess of a few °/s (Westheimer & McKee, 1975).

Several distinct phases can be indentifed as a monkey pursues a moving visual target (Figure 6). After the target has been selected, it usually takes 80–130 ms for pursuit to start, although the latency can be as short as 65 ms under some conditions (Lisberger & Westbrook, 1985). During this *latent* phase, the speed of the retinal image is equal to the speed of the target. At the *initiation* of pursuit, the eye accelerates at a nearly constant rate that is roughly proportional to target speed. This initiation phase is generally considered to have a duration equal to one latent period. During the *transition* phase, eye velocity continues to increase and often overshoots target velocity slightly before settling to a constant value, during *sustained* pursuit, that approximates target velocity. During sustained pursuit, a constant visual drive is not needed to maintain eye velocity. In fact, if retinal image motion is eliminated by briefly turning the target off or by electronically stabilizing the image of the target, pursuit velocity continues undiminished (Morris & Lisberger, 1987). This persistence of pursuit indicates that the neural equivalent of a mathematical integrator converts the visual drive for pursuit into an eye velocity signal that can be maintained in the absence of further visual inputs.

4.2 Neural Pathways

The visual drive for smooth eye movements in the monkey is mediated primarily by pathways from areas in the cerebral cortex that project to the cerebellum via the basilar pontine nuclei. The analysis of motion by visual cortex has been viewed as a largely serial process, starting in striate cortex in the occipital lobe and continuing in adjacent extrastriate areas, in particular, the middle temporal (MT) and the medial superior temporal (MST) areas. Removal of the occipital lobe has a devastating effect

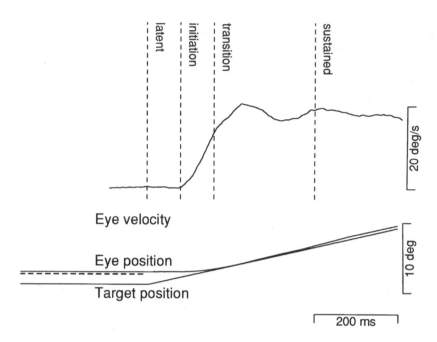

latent

initiation

transition

sustained

20 deg/s

Eye velocity

Eye position

Target position

10 deg

200 ms

Figure 6   Pursuit eye movement made by a monkey tracking a 0.25° spot moving at a constant velocity of 15°/s. Bottom traces show position of the target and the eye as a function of time; dashed trace indicates when a fixation LED was illuminated. Top trace shows average eye velocity from 22 presentations of the target motion. Dashed vertical lines mark the onset of different phases of pursuit, as described in the text.

on many aspects of vision, including visually-driven eye movements (Zee *et al.*, 1987). The effects are not specific for eye movements driven by visual motion, however, because saccadic eye movements are also affected.

4.2.1   The middle temporal area (MT)

The middle temporal area (MT) is specialized for the processing of visual motion. Inputs to MT arise primarily from the direction-selective cells in layers IVB and VI of primary visual cortex (Lund *et al.*, 1975). Neurons in MT are sharply tuned for the direction and speed of stimulus motion. Across the population of neurons, all directions of motion are represented. The preferred speeds of individual neurons range from 2 to over 200°/s; most have a preferred speed near 30°/s (Maunsell & Van Essen, 1983a; Mikami *et al.*, 1986; Albright, 1984). Lesions of MT cause deficits in the generation of pursuit, although the animals usually recover fully within a few weeks (Newsome *et al.*, 1985). If the lesion is small and does not include the representation of the fovea, then the deficit affects only those eye movements made to stimuli located in parts of the visual field formerly represented by the lesioned cortex. This retinotopic deficit only affects the first few hundred milliseconds of pursuit, because the monkey quickly makes a saccadic eye movement that moves the target to a portion of the visual field unaffected by the lesion.

In the portion of MT that represents the fovea (MTf), units discharge during pursuit of small targets, as well as during passive visual stimulation (Komatsu & Wurtz, 1988a). This pursuit-related response appears to be a visual response driven by the residual motion that occurs during imperfect tracking, because it disappears if the target is turned off even briefly. However, lesions of MTf produce a more severe pursuit deficit that differs in two respects from other MT lesions. First, it affects both the initial and sustained aspects of pursuit. Second, it is specific for pursuit directed to the same side as the lesion (Dursteler et al., 1987).

4.2.2 The medial superior temporal area (MST)

Area MT projects directly and heavily to an adjoining area in parietal cortex, the medial superior temporal area (MST). MST consists of subregions containing neurons which differ in their receptive field sizes, preferred stimuli and their responses during pursuit eye movements (Komatsu & Wurtz, 1988a). One subregion, located in dorsal MST (MSTd), contains units that have large receptive fields and respond best to large-field stimuli. These units also tend to respond during pursuit of a small target even when the visual stimulus is removed, indicating the presence of an extraretinal input. For most of these units, the preferred direction of pursuit is opposite to the preferred direction for large-field stimulation. The lateral portion of MST (MSTl) contains a mixture of units like those in MSTd and units that have smaller receptive fields. Like neurons in MTf, many of the units with small receptive fields respond during pursuit only while the visual stimulus is present (Newsome et al., 1988). For these small-field neurons, the preferred direction of pursuit is the same as the preferred direction for small-field stimulation (Komatsu & Wurtz, 1988b).

As with area MT, lesions of MST can cause two types of deficits in pursuit. One type of deficit is similar to the effects of lesioning extrafoveal MT – a retinotopic deficit affecting the initiation of pursuit. However, because the representation of visual space is less refined in MST, lesions tend to affect larger portions of the visual field (Dursteler & Wurtz, 1988). The second type of deficit affects pursuit toward the same side as the lesion. This directional deficit is similar to that observed after lesions of MTf, and is most commonly seen after lesions affecting MSTl rather than MSTd. These lesions also result in deficits in the optokinetic response, again affecting ipsilaterally directed eye movements.

Electrical stimulation of areas MT and MST can also influence pursuit eye movements. The effect is characterized by an acceleration of the eye toward the same side as the stimulation. However, stimulation is only effective when applied during ongoing pursuit; stimulation during fixation has little effect. The effect also depends upon the cortical location of the stimulation. The most effective sites are in MTf and MSTl; stimulation in MSTd is usually ineffective (Komatsu & Wurtz, 1989).

4.2.3 Parietal cortex

There are two other areas in parietal cortex that contain neurons active during pursuit. One of these areas, called the ventral intraparietal area (VIP), is located within the intraparietal sulcus and receives projections from MT. Neurons in this area are activated during pursuit eye, but it is not known whether this activation is due to residual visual motion during imperfect tracking or to a nonvisual signal related to pursuit (Colby *et al.*, 1993). Neurons in the posterolateral part of area 7a also fire during pursuit. These neurons have been called 'visual-tracking' neurons, because they fire during pursuit of a small target in an otherwise dark room, and continue to respond when the stimulus is briefly turned off (Sakata *et al.*, 1983). Some of these neurons also fire during movements made with a combination of eye and head movements (Kawano *et al.*, 1984).

4.2.4   Frontal cortex

The study of cortical inputs for smooth eye movements has largely focused on the processing of visual motion accomplished in occipito-parietal cortex, but recent studies have shown that a region of the frontal cortex is also important. The frontal eye fields (FEF), located along the arcuate sulcus in the frontal lobe, have largely been studied for their role in initiating saccadic eye movements. However, lesions of the FEF cause only a minor deficit in initiating saccades, but cause a large deficit in pursuit (Lynch, 1987; Keating 1991; MacAvoy *et al.*, 1991). The results of making smaller lesions and recording from single neurons show that the critical region for smooth eye movements is immediately adjacent to the region involved with saccadic eye movements (MacAvoy *et al.*, 1991). As shown in Figure 7, neurons in this part of the FEF show directionally selective responses during pursuit, and often begin firing before the onset of pursuit. However, these neurons respond only weakly during passive visual stimulation or during saccadic eye movements (MacAvoy *et al.*, 1991).

Similar to the effects of MTf and MST lesions, lesions of the FEF result in a difficulty in generating ipsilateral pursuit, but do not affect the accuracy of saccades made to moving targets. However, the recovery after FEF lesions appears to be much slower (Lynch, 1987). Also, lesions of the FEF result in deficits in 'predictive' pursuit. Anticipatory responses, which can occur when the motion of the target is predictable, are abolished by FEF lesions (MacAvoy *et al.*, 1991; Keating, 1991, 1993). Sustained tracking during periods when the target is not illuminated is also severely affected. However, the deficit cannot be explained solely by a reduction in predictive ability, because pursuit is also deficient in situations where visual motion, and not prediction, provide the primary drive for pursuit. For example, the initiation of pursuit is deficient even for unpredictable target motions.

Stimulation of the FEF can generate smooth eye movements. As in MST, the effect of stimulation is primarily an acceleration of the eye toward the ipsilateral side, although other directions of movement can also be evoked (Gottlieb *et al.*, 1993). However, unlike MST, stimulation of the FEF can elicit smooth eye movements even during active fixation. Also, rather high rates of eye acceleration (50–200°/s) can be achieved with only modest currents. The average latency of the evoked eye movements is 39 ms (Gottlieb *et al.*, 1993).

4.2.5   The dorsolateral pontine nucleus (DLPN)

Large anatomical projections from areas in the occipital, parietal and frontal cortex converge onto the lateral, dorsolateral and dorsomedial portions of the basal pontine nuclei (Brodal, 1978; Glickstein *et al.*, 1980, 1985; May & Andersen, 1986; Kunzle & Akert, 1977; Leichnetz, 1982; Huerta *et al.*, 1986). Among the pontine nuclei, the

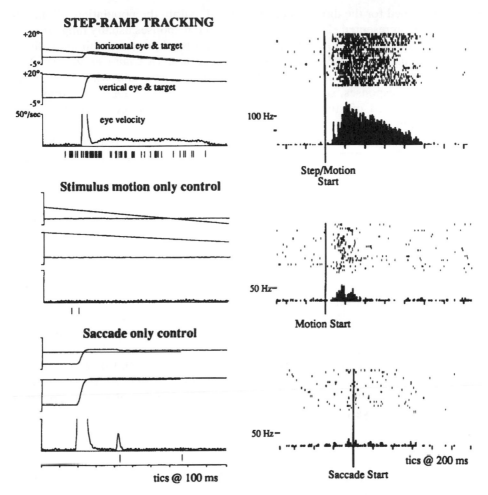

Figure 7 Activity of a single neuron in the smooth-pursuit zone of the frontal eye fields. This unit was recorded in the left hemisphere and responded during pursuit directed to the left and down. The top row shows the results obtained during step-ramp tracking. Comparison of eye position and velocity traces with the cell's activity (tics below the velocity trace) for one representative trial (left top panel) shows that firing preceded onset of the eye movement. Histogram (right top panel) sums the cell's activity over all 56 correctly performed trials and the raster lines above the histogram show the last 25 correct trials. The middle row shows that this same cell responded much less during passive visual stimulation with the same stimulus. The bottom row shows that this cell also had no activity in conjunction with saccades to a stationary target. From MacAvoy *et al.* (1991).

DLPN has been studied most, because of its strong connections with areas MT and MST. The DLPN contains a mixture of units that are driven purely by visual stimulation, units that respond during pursuit with no apparent visual response, and units that both respond to visual stimulation and fire during pursuit (Suzuki & Keller, 1984; Mustari *et al.*, 1988; Their *et al.*, 1988). The *visual* responses of DLPN neurons are broadly tuned for the direction of motion and, within the population, all directions of motion are represented. The latency of the visual responses usually range from 70 to 90 ms after the onset of stimulus motion, but can be shorter than 50 ms with large-field stimuli that elicit an ocular following response (Kawano et al., 1992). The receptive fields are large and do not have clear borders, and while larger stimuli produce larger responses, small stimuli can also be effective. DLPN neurons are also tuned for the speed of stimulus motion and the preferred speed for most units is between 20 and 100°/s.

The *tracking* responses of DLPN neurons are characterized by their selectivity for the direction of smooth eye movement. The latency of the tracking responses during pursuit is much more variable than the latency of the visual responses, ranging from 100 ms before to 100 ms after the eyes start to move in the preferred direction (Suzuki & Keller, 1984; Mustari et al., 1988; Their et al., 1988). For many of these units, the response during pursuit is not visual, since they continue to respond when the target is briefly turned off. For units that respond both during pursuit and to visual stimuli, the preferred direction for tracking can be either the same or the opposite as the preferred direction for visual stimulation. Many of the DLPN units that fire during pursuit also respond during ocular following (Kawano et al., 1992). For most of these units, the onset of the neural response precedes the onset of the eye movement.

Lesions of the DLPN produce deficits similar to those observed after lesions of MTf or MSTl. There is both a deficit in the initiation of pursuit and a directional deficit during sustained pursuit. DLPN lesions affect not only pursuit, but also impair the early component of the optokinetic response (May et al., 1988). Likewise, stimulation of the DLPN produces effects similar to those found for MTf or MSTl. Stimulation during fixation is ineffective, but stimulation during sustained pursuit produces an acceleration of the eye toward the ipsilateral side (May et al., 1985).

4.2.6   The dorsomedial pontine nucleus (DMPN)

The dorsal medial pontine nucleus (DMPN) also contains neurons that have directionally-selective visual responses (Keller & Crandall, 1983). The DMPN receives projections from occipital and parietal cortex, like the DLPN, but also receives a projection from the FEF. Neurons in the DMPN are driven best by large textured stimuli, but they are not sharply tuned for a particular range image speeds. Instead, the response of most units increases for speeds up to about 10°/s and shows only small increments or decrements for speeds up to 100°/s. Some of these units also discharge during pursuit of a small target. At its caudal boundary, the DMPN merges with the NRTP. As described before, the NRTP plays a critical role in the optokinetic response of the rabbit and cat. However, in the monkey, the NRTP contains neurons that discharge during saccades, not during smooth eye movements (Crandall & Keller, 1985).

4.2.7   The cerebellar flocculus and paraflocculus

Visual nuclei of the basilar pons project to several regions of the cerebellum, including the flocculus, paraflocculus and vermis (Brodal 1979, 1982; Langer et al., 1985a). Each of these cerebellar regions can influence smooth eye movements by acting on extraocular nuclei in the brainstem (Langer *et al.*, 1985b; Yamada & Noda, 1987; Noda *et al.*, 1990). Anatomical data provide a basis for distinguishing between the flocculus and the ventral paraflocculus (Gerrits & Voogd, 1989), but most of the available physiological data do not follow these distinctions. For this reason, I will refer to the combined flocculus and ventral paraflocculus of the monkey as the 'floccular region'.

The anatomical connections of the floccular region are well-suited to mediate the conversion of visual motion information provided by the cerebral cortex into the commands for pursuit eye movements required by motoneurons. The major projections to the floccular region relay signals concerning the motion of the eye and head and arise via mossy fiber projections from the vestibular nuclei and the nucleus prepositus (Langer *et al.*, 1985a). The floccular region also receives visual inputs from mossy fibers arising from the DLPN, DMPN and the NRTP (Brodal, 1980, 1982; Langer *et al.*, 1985a). The projections from the floccular region target several portions of the ipsilateral vestibular nucleus and can cause smooth eye movements with 10 ms through inhibition of interneurons in these nuclei (Langer *et al.*, 1985b; Belknap & Noda, 1987; Lisberger & Pavelko, 1988). Furthermore, lesions of the floccular region cause large and lasting deficits in generating pursuit eye movements and the early component of the optokinetic response (Zee *et al.*, 1981).

The activity of floccular Purkinje cells (P-cells), which provide the sole outputs from the cerebellum, also provides strong evidence that the floccular region represents an interface between the sensory and motor systems used for smooth eye movements. Early studies of the floccular region recognized the convergence of eye velocity, head velocity, and visual signals and suggested that the output of the floccular region may provide a neural signal encoding target velocity (Miles & Fuller, 1975). It was subsequently shown that the firing rate of P-cells in the floccular region reflects intrinsic signals related to movements of the eye and head, rather than an extrinsic signal like target velocity. The modulation in simple-spike firing rate during eye and head movements of the same amplitude is approximately equal and the effects of the two signals sum linearly (Lisberger & Fuchs, 1978; Miles *et al.*, 1980; Stone & Lisberger, 1990). Many P-cells in the floccular region have therefore been identified as 'gaze-velocity' cells, since they encode the velocity of changes in gaze accomplished with any combination of eye and head movements.

During continuous smooth movements of the eyes, P-cells in the floccular region show a continuous modulation in their simple-spike firing rate. For example, during sustained pursuit, P-cells show a tonic change in firing rate that is proportional to eye speed (Figure 8). This modulation is related to eye velocity, because it persists when the target is electronically stabilized on the retina (Stone & Lisberger, 1990), and is therefore believed to represent a motor component of the command for smooth tracking. Through its reciprocal connections with the vestibular nuclei, the floccular region may form part of a loop that retains a copy of the gaze velocity command provided to the output motor pathways (Lisberger & Fuchs, 1978; Miles *et al.*, 1980; Stone & Lisberger, 1990). Continuance of activity around this feedback loop could serve as the eye velocity integrator for pursuit.

P-cells in the floccular region also respond to visual motion. A distinctive feature of the visual responses of these P-cells is that they show robust responses if the moving stimulus is the object of a smooth eye movement, but show only modest responses

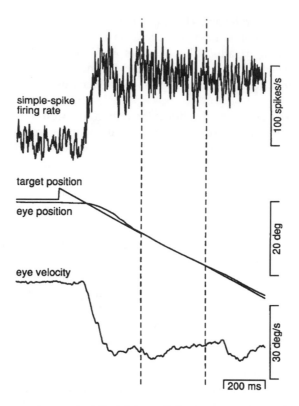

Figure 8 Response of a typical Purkinje cell in the ventral paraflocculus during pursuit with a stabilized target. Top to bottom: traces are averaged simple-spike firing rate, target position, eye position, and eye velocity. Ramp target motion was toward the side of the recording at a speed of 30°/s. Vertical dashed lines delimit a 310 ms interval during which the target was electronically stabilized and moved exactly with the eye. From Stone and Lisberger (1990).

during passive presentation of moving visual stimuli. Only a minority of P-cells can be appreciably modulated by the motion of a large-field visual stimulus or the motion of a small spot presented while the monkey fixates a stationary target (Noda & Warabi, 1987; Stone & Lisberger, 1990). In contrast, large responses are evoked when the monkey is required to pursue a target, initially at rest, that then moves at a constant velocity (Stone & Lisberger, 1990). Similar responses are also observed at the onset of ocular following (Shidara & Kawano, 1993). Although the simple-spike activity recorded in this situation also reflects the sensitivity of the P-cells to eye velocity, the increase in firing rate occurs too early and is too large to be produced by inputs conveying eye velocity information. The large increase in simple-spike firing rate occurs approximately 100 ms after the onset of target motion, consistent with the latency produced by visual pathways through the cerebral cortex. It has therefore been suggested that the transient pulse in simple-spike firing rate exhibited by P-cells reflects visual signals that are used to accelerate the eye at the onset of pursuit and ocular following (Stone & Lisberger, 1990; Shidara & Kawano, 1993).

4.2.8   The cerebellar vermis

While lesions of the floccular region cause large deficits in smooth eye movements, removal of the entire cerebellum results in a nearly complete loss. This indicates that other regions of the cerebellum are also important in generating smooth eye movements. One possibility is the 'oculomotor' vermis, which consists of vermal lobules VI and VII. The cerebellar vermis has historically been viewed as a modulatory influence on the generation of saccadic eye movements, but it has been reported that lesions of the vermis also cause deficits in pursuit (Suzuki & Keller, 1983). Considering its anatomical connections, the oculomotor vermis is well-suited to mediate smooth eye movements. The vermis receives visual projections from the pontine nuclei, including the DLPN and DMPN, as well as eye-movement-related projections from the fastigial nucleus and the vestibular nuclei. The oculomotor vermis can affect smooth eye movements through its return projections to the mediocaudal portion of the fastigial nucleus (Yamada & Noda, 1987). The fastigial nucleus, one of the deep cerebellar nuclei, is an important station in the traffic of visual and eye movement information. It receives the same inputs as the oculomotor vermis and sends large projections to the NRTP and the visual pontine nuclei (Batton *et al.*, 1977; Gonzalo-Ruiz & Leichnetz, 1990; Noda *et al.*, 1990).

The P-cells in the oculomotor vermis respond to a combination of visual, eye velocity and head velocity signals. As was originally proposed for the floccular region, it has been suggested that these signals are used to construct a neural analogue of target velocity (Kase *et al.*, 1979; Suzuki & Keller, 1988a, 1988b). Comparison of activity in the floccular region and vermis indicates that although the types of signals present in the two structures are similar, the form in which they appear is quite different. First, the firing rate of vermal P-cells is modulated during sustained pursuit, but the amplitude of the modulation is lower and more variable than that observed in the floccular region. The strength of the eye velocity signal conveyed by the vermis therefore appears to depend on some, as yet unidentified, factor. Second, vermal P-cells respond to moving visual stimuli presented either during fixation or during pursuit. Unlike in the floccular region, the visual responses are not contingent on the generation of a smooth eye movement. Therefore, while the activity in the floccular region is more closely associated with eye movements that are actually made, the activity in the oculomotor vermis is more closely associated with visual signals that might be used in making an eye movement.

The uvula, lobule IX of the vermis, also receives a strong projection from visual pontine nuclei. Lesions of the uvula produce deficits in contralateral pursuit, but the functional role of the uvula appears to be distinct from that of the floccular region or the vermis. Purkinje cells in the uvula are not reliably activated by brief movements of spots or textured patterns, the stimuli that are effective in driving neurons elsewhere in the pathways mediating smooth eye movements. However, uvular P-cells do show modulation during sustained optokinetic stimulation (Heinen & Keller, 1990). The uvula is therefore unlikely to provide a major drive for smooth eye movements, although it may be important for compensating for the consequences of smooth eye movements.

4.2.9   The accessory optic and pretectal nuclei

Although overshadowed by the large nuclei of the basilar pons, the accessory optic and pretectal nuclei play an important role in the monkey. As in the cat and rabbit, the NOT and DTN are associated with horizontal movements, and the LTN is associated with vertical movement. However, the MTN is essentially non-existent in the primate. Retinal inputs to these nuclei come from contralateral eye (Giolli, 1963; Hendrickson *et al.*, 1970; Itaya & van Hoesen, 1983; Weber, 1985; Weber & Giolli, 1986; Cooper & Magnin, 1986). The accessory optic and pretectal nuclei are connected with each other and with pontine nuclei, including the DLPN and DMPN (Mustari *et al.*, 1990). In addition, the NOT has projections to eye movement related nuclei in the caudal brainstem (Mustari *et al.*, 1990).

As in the cat and rabbit, unilateral lesions of the NOT cause an ipsilateral deficit in the generation of the optokinetic response (Kato *et al.*, 1986, 1988). Electrical stimulation of the NOT produces smooth eye movements toward the ipsilateral side (Mustari & Fuchs, 1990). Neurons in the NOT and adjacent DTN respond well to the motion of large textured stimuli and have binocular receptive fields. They are also broadly tuned for the direction of motion and, across the population of neurons, display a preference for ipsilateral motion (Hoffman & Distler, 1989; Mustari & Fuchs, 1990). However, neurons in the monkey NOT prefer higher speeds (64°/s) than are observed in the cat NOT. In addition, some units prefer small stimuli (less than 10°) and respond well during smooth pursuit of a small visual target. The responses during tracking are visual in origin, because they disappear when the target is transiently turned off even though tracking continues.

Neurons in the LTN also prefer moving stimuli and have large binocular receptive fields (Mustari & Fuchs, 1989). The majority of LTN units (88%) respond best to upward motion. The average preferred speed of LTN units (13°/s) is lower than units in the NOT and DTN. Some LTN units can be driven by either large stimuli or small spots. For these units, the preferred direction for the two types of stimuli can be either the same or the opposite. In addition, about a third of the units continue to discharge during pursuit when the target spot is briefly turned off, indicating the presence of a non-visual input.

4.3   Unresolved Issues

The diverse responses related to smooth eye movements in the monkey have not yet been mapped onto a unifying scheme similar to that proposed for the rabbit. There are several unresolved issues which might hold the keys to understanding how these pathways are organized. First, it is not known how the visual inputs for smooth eye movements are restricted to those portions of the visual field which hold the target. This issue is especially relevant during pursuit of small targets, when only a small portion of the visual field conveys inputs related to target motion, and the remainder of the visual field provides potentially confounding visual inputs. One possibility is that other cues are used to segregate the visual inputs. For example, the disparity of the two retinal images might be used to eliminate motion not contained in the depth plane of the target. Indeed, many MT neurons are selective for disparity, as well as for the direction of motion (Maunsell & Van Essen, 1983b), and some neurons in MST have different preferred directions for stimuli at different disparities (Roy & Wurtz, 1990). However, the use of auxiliary cues still requires that the target has previously been identified and binocularly foveated.

Second, the function of the non-visual signals often observed on visual neurons is unclear. While it is not surprising to find such signals in a pre-motor structure like the cerebellum, their role in cortical processing is enigmatic. It has been suggested that these signals might be used to construct a neural analogue of target velocity, similar to that been proposed for cerebellar P-cells. In some models of the pursuit system, such a signal is used as a command for smooth eye velocity (Robinson et al., 1986). A second suggestion is that the non-visual signals act like a switch, enabling the use of visual motion inputs by the oculomotor system (Grasse & Lisberger, 1992). A related possibility is that the non-visual signals reflect cognitive inputs that are functionally interchangeable with visual motion inputs. Such inputs might be used to guide predictive pursuit or pursuit of imagined targets.

Finally, the significance of the multiple neural pathways for smooth eye movements is not known. For example, do the parallel projections from parieto-occipital cortex and frontal cortex to the pontine nuclei represent a redundancy of function or a specialization of function? Likewise, what distinguishes the roles of the cerebellar floccular region and vermis? These questions touch upon a fundamental issue in sensorimotor research, because similar cortico-ponto-cerebellar pathways control voluntary movements of other parts of the body.

5 EPILOGUE: BIOLOGICAL STRATEGIES IN THE ANALYSIS OF VISUAL MOTION

In the rabbit, smooth eye movements reflect a global analysis of visual motion that is accomplished by visual nuclei in the brainstem. In the cat, both the range of eye movements and the visual properties of visual nuclei in the brainstem are heavily influenced by inputs from the cerebral cortex. In the monkey, the ability to smoothly track targets of all sizes depends critically on visual inputs provided by the cerebral cortex and mediated by the cerebellum. It seems relevant to ask, then, what strategy is reflected in how the primate cortex processes visual inputs for smooth eye movements? The frontal vision and exploratory behavior of the primate make it impossible to stabilize the retinal image of the entire visual world. This situation presents at least two major problems. First, the piece of the visual world containing the target must be selected. Second, local analysis of that piece, perhaps also influenced by motion elsewhere in the visual field, must then provide the appropriate inputs for smooth eye movements. How is this done?

One possibility is that smooth eye movements remain a visual reflex, as they are in the rabbit. With this *visual-based* strategy, the selection of the target might be accomplished by some other system. The appropriate inputs for smooth eye movements could then be obtained by applying a set of simple rules, based upon the expected retinal location of the target. For example, saccadic eye movements might be relied upon to foveate the target with both eyes. The visual drive would then always be obtained primarily from the fovea. In addition, retinal locations that do not correspond in the two eyes (i.e. are not in the same plane of depth) might be excluded.

A second possibility is that smooth eye movements are tied to higher functions, such as prediction and object recognition. With this *object-based* strategy, the issues of target selection and motion analysis would be combined. Visual motion would be an important input for delineating and identifying targets, but would not directly drive smooth eye movements. Instead, the results of local motion analysis would be cast as attributes of objects. The drive for smooth eye movements would be the motion of the object, and visual motion would be an important, but not an exclusive, source of information about what that motion was.

These possibilities are speculative, but hopefully they also indicate why smooth eye movements remain a vital area of research. A full understanding of how visual motion drives smooth eye movements will likely require a deeper understanding of many other issues of brain function.

ACKNOWLEDGEMENTS

I gratefully acknowledge F. A. Miles for his comments on this manuscript. I also thank G. A. Bush, C. J. Duffy, and F. A. Miles for many helpful and entertaining discussions during the preparation of this manuscript.

REFERENCES

Albright, T. D. (1984). Direction and orientation selectivity of neurons in visual area MT of the macaque. *J. Neurophysiol*, **52**, 1106–1130.

Alley, K. E., Baker, R. & Simpson, J. I. (1975). Afferents to the vestibulo-cerebellum and the origin of the visual climbing fibers in the rabbit. *Brain Res.*, **98**, 582–589.

Atkinson, J. (1979). Development of optokinetic nystagmus in the human infant and monkey infant: an analogue to development in kittens. In *Developmental Neurobiology of Vision*, pp. 277–288. Plenum Press, New York.

Balaban, C. D. (1983). A projection from nucleus reticularis tegmenti pontis of Bechterew to the medial vestibular nucleus in rabbits. *Exp. Brain Res.*, **51**, 304–309.

Barlow, H. B. & Hill, R. M. (1963). Selective sensitivity to direction of movement in ganglion cells of the rabbit retina. *Science*, **139**, 412–414.

Barlow, H. B., Hill, R. M. & Levick, W. R. (1964). Retinal ganglion cells responding selectively to direction and speed of image motion in the rabbit. *J. Physiol.*, **173**, 377–407.

Batton, R. R., Jayaraman, A., Ruggiero, D. & Carpenter, M. B. (1977). Fastigial efferent projections in the monkey: an autoradiographic study. *J. Comp. Neurol.*, **174**, 281–306.

Belknap, D. B. & Noda, H. (1987). Eye movements evoked by microstimulation in the flocculus of the alert macaque. *Exp. Brain Res.*, **67**, 352–362.

Benson, A. J. & Guedry, F. E. (1971). Comparison of tracking task performance and nystagmus during sinusoidal oscillation in yaw and pitch. *Aerosp. Med.*, **42**, 593–603.

Berman, N. (1977). Connections of the pretectum in the cat. *J. Comp. Neurol.*, **174**, 227–254.

Berson, D. M. & Graybiel, A. M. (1980). Some cortical and subcortical fiber projections to the accessory optic nuclei in the cat. *Neuroscience*, **5**, 2203–2217.

Blanks, R. H. I. (1990). Afferents to the cerebellar flocculus in cat with special reference to pathways conveying vestibular, visual (optokinetic) and oculomotor signals. *J. Neurocytol.*, **19**, 628–642.

Brodal, P. (1972a). The corticopontine projection from the visual cortex in the cat. I. The total projection and the projection from area 17. *Brain Res.*, **39**, 297–317.

Brodal, P. (1972b). The corticopontine projection from the visual cortex in the cat. II. The projection from areas 18 and 19. *Brain Res.*, **39**, 319–335.

Brodal, P. (1978). The corticopontine projection in the rhesus monkey: origin and principles of organization. *Brain*, **101**, 251–283.

Brodal, P. (1979). The pontocerebellar projection in the rhesus monkey: an experimental study with retrograde axonal transport of horseradish peroxidase. *Neuroscience*, **4**, 193–208.

Brodal, P. (1980). The projection from the nucleus reticularis tegmenti pontis in the rhesus monkey. *Exp. Brain Res.*, **38**, 29–36.

Brodal, P. (1982). Further observations on the cerebellar projections from the pontine nuclei and the nucleus reticularis tegmenti pontis in the rhesus monkey. *J. Comp. Neurol.*, **204**, 44–55.

Brodal, A. & Brodal, P. (1971). The organization of the nucleus reticularis tegmenti pontis in the cat in the light of experimental anatomical studies of its cerebral cortical afferents. *Exp. Brain Res.*, **13**, 90–110.

Busettini, C., Miles, F. A. & Schwarz, U. (1991). Ocular responses to translation and their dependence on viewing distance. II. Motion of the scene. *J. Neurophysiol.*, **66**, 865–878.

Cohen, B., Matsuo, V. & Raphan, T. (1977). Quantitative analysis of the velocity characteristics of optokinetic nystagmus and optokinetic after-nystagmus. *J. Physiol.*, **270**, 321–344.

Colby, C. L., Duhamel, J.-R. & Goldberg, M. E. (1993). Ventral intraparietal area of the macaque: anatomic location and visual response properties. *J. Neurophysiol.*, **69**, 902–914.

Collewijn, H. (1969). Optokinetic eye movements in the rabbit: input–output relations. *Vision Res.*, **9**, 117–132.

Collewijn, H. (1970). The normal range of horizontal eye movements in the rabbit. *Exp. Neurol.*, **28**, 132–143.

Collewijn, H. (1972). Latency and gain of the rabbit's optokinetic reactions to small movements. *Brain Res.*, **36**, 59–70.

Collewijn, H. (1975a). Direction-selective units in the rabbit's nucleus of the optic tract. *Brain Res.*, **100**, 489–508.

Collewijn, H. (1975b). Oculomotor areas in the rabbit's midbrain and pretectum. *J. Neurobiol.*, **6**, 3–22.

Cooper, H. M. & Magnin, M. (1986). A common mammalian plan of accessory optic system organization revealed in all primates. *Nature*, **324**, 457–459.

Crandall, W. F. & Keller, E. L. (1985). Visual and oculomotor signals in nucleus reticularis tegmenti pontis in alert monkey. *J. Nerophysiol.*, **54**, 1326–1345.

Cynader, M. & Harris, L. (1980). Eye movement in strabismic cats. *Nature*, **286**, 64–65.

Dubois, M. F. W. & Collewijn, H. (1979). The optokinetic reactions of the rabbit: relation to the visual streak. *Vision Res.*, **19**, 9–17.

Dursteler, M. R. & Wurtz, R. H. (1988). Pursuit and optokinetic deficits following chemical lesions of cortical areas MT and MST. *J. Neurophysiol.*, **60**, 940–965.

Dursteler, M. R., Wurtz, R. H. & Newsome, W. T. (1987). Directional pursuit deficits following lesions of the foveal representation within the superior temporal sulcus of the macaque monkey. *J. Neurophysiol.*, **57**, 1262–1287.

Erickson, R. G. & Barmack, N. H. (1980). A comparison of horizontal and vertical optokinetic reflexes of the rabbit. *Exp. Brain Res.*, **40**, 448–456.

Erickson, R. G. & Cotter, J. R. (1983). Uncrossed retinal projections to the accessory optic nuclei in rabbits and cats. *Exp. Brain Res.*, **49**, 143–146.

Evinger, L. C. & Fuchs, A. F. (1978). Saccadic, smooth pursuit, and optokinetic eye movements of the trained cat. *J. Physiol.*, **285**, 209–229.

Ezure, K. & Graf, W. (1984). A quantitative analysis of the spatial organization of the vestibuloocular reflexes in lateral- and frontal-eye animals. I. Orientation of semicircular canals and extraocular muscles. *Neuroscience*, **12**, 85–93.

Farmer, S. G. & Rodieck, R. W. (1982). Ganglion cells of the cat accessory optic system: Morphology and retinal topography. *J. Comp. Neurol.*, **205**, 190–198.

Fries, W. & Albus, K. (1980). Responses of pontine nuclei cells to electrical stimulation of the lateral and suprasylvian gyrus in the cat. *Brain Res.*, **188**, 255–260.

Fuchs, A. F. (1967). Saccadic and smooth pursuit eye movements in the monkey. *J. Physiol.*, **191**, 609–631.

Garey, L. J., Jones, E. G. & Powell, T. P. S. (1968). Interrelationships of striate and extrastriate cortex with the primary relay sites of the visual pathway. *J. Neurol. Neurosurg. Psychiatry*, **31**, 135–157.

Gerrits, N. M. & Voogd, J. (1982). The climbing fiber projection to the flocculus and adjacent paraflocculus in the cat. *Neuroscience*, **7**, 2971–2991.

Gerrits, N. M. & Voogd, J. (1986). The nucleus reticularis tegmenti pontis and the adjacent rostral paramedian reticular formation: differential projections to the cerebellum and the caudal brain stem. *Exp. Brain Res.*, **62**, 29–45.

Gerrits, N. M. & Voogd, J. (1989). The topographical organization of climbing and mossy fiber afferents in the flocculus and ventral paraflocculus in rabbit, cat and monkey. *Exp. Brain Res. Suppl.*, **17**, 26–29.

Giolli, R. A. (1963). An experimental study of the accessory optic system in the cynomolgus monkey. *J. Comp. Neurol.*, **121**, 89–108.

Giolli, R. A. & Guthrie, M. D. (1971). Organization of subcortical projections of visual areas I and II in the rabbit. An experimental degeneration study. *J. Comp. Neurol.*, **142**, 351–376.

Giolli, R. A., Towns, L. C., Takahashi, T. T., Karamanlidis, A. N. & Williams, D. D. (1978). An autoradiographic study of the projections of visual cortical area 1 to the thalamus, pretectum and superior colliculus of the rabbit. *J. Comp. Neurol.*, **180**, 743–752.

Giolli, R. A., Blanks, R. H. I. & Torigoe, Y. (1984). Pretectal and brain stem projections of the medial terminal nucleus of the accessory optic system of the rabbit and rat as studied by anterograde and retrograde neuronal tracing methods. *J. Comp. Neurol.*, **227**, 228–251.

Giolli, R. A., Blanks, R. H. I., Torigoe, Y. & Williams, D. D. (1985). Projections of medial terminal accessory optic nucleus, ventral tegmental nuclei, and substantia nigra of rabbit and rat as studied by retrograde axonal transport of horseradish peroxidase. *J. Comp. Neurol.*, **232**, 99–116.

Glickstein, M., Stein, J. & King, R. A. (1972). Visual input to the pontine nuclei. *Science*, **178**, 1110–1111.

Glickstein, M., Cohen, J. L., Dixon, B., Gibson, A., Hollins, M., Labossiere, E. & Robinson, F. (1980). Corticopontine visual projections in macaque monkeys. *J. Comp. Neurol.*, **190**, 209–229.

Glickstein, M., May, J. & Mercer, B. E. (1985). Corticopontine projection in the macaque: the distribution of labelled cortical cells after large injections of horseradish peroxidase in the pontine nuclei. *J. Comp. Neurol.*, **235**, 343–359.

Gonzalo-Ruiz, A. & Leichnetz, G. R. (1990). Afferents of the caudal fastigial nucleus in a New World monkey (*Cebus apella*). *Exp. Brain Res.*, **80**, 600–608.

Gottlieb, J. P., Bruce, C. J. & MacAvoy, M. G. (1993). Smooth eye movements elicited by microstimulation in the primate frontal eye field. *J. Neurophysiol.*, **69**, 786–799.

Grasse, K. L. & Cynader, M. S. (1982). Electrophysiology of medial terminal nucleus of accessory optic system in the cat. *J. Neurophysiol.*, **48**, 490–504.

Grasse, K. L. & Cynader, M. S. (1984). Electrophysiology of lateral and dorsal terminal nuclei of the cat accessory optic system. *J. Neurophysiol.*, **51**, 276–293.

Grasse, K. L. & Cynader, M. S. (1986). Response properties of single units in the accessory optic system of the dark-reared cat. *Dev. Brain Res.* **27**, 199–210.

Grasse, K. L. & Cynader, M. S. (1987). The accessory optic system of the monocularly deprived cat. *Dev. Brain Res.*, **31**, 229–241.

Grasse, K. L. & Cynader, M. S. (1988). The effect of visual cortex lesions on vertical optokinetic nystagmus in the cat. *Brain Res.*, **455**, 385–389.

Grasse, K. L. & Lisberger, S. G. (1992). Analysis of a naturally occurring asymmetry in vertical smooth pursuit eye movements in a monkey. *J. Neurophysiol.*, **67**, 164–179.

Grasse, K. L., Cynader, M. S. & Douglas, R. M. (1984). Alterations in response properties in the lateral and dorsal terminal nuclei of the cat accessory optic system following visual cortex lesions. *Exp. Brain Res.*, **55**, 69–80.

Hayhow, W. R. (1959). An experimental study of the accessory optic fiber system in the cat. *J. Comp. Neurol.*, **113**, 281–313.

Heinen, S. J. & Keller, E. L. (1990). Cerebellar uvula correlates of moving visual stimuli. *Soc. Neurosci. Abstr.*, **16**, 1083.

Hendrickson, A., Wilson, M. E. & Toyne, M. J. (1970). The distribution of optic nerve fibers in *Macaca mulatta*. *Brain Res.*, **23**, 425–427.

Hobbelen, J. F. & Collewijn, H. (1971). Effects of cerebro-cortical and collicular ablations upon the optokinetic reactions in the rabbit. *Doc. Opthalmol.*, **30**, 227–236.

Hoffman, K.-P. (1981). Cortical versus subcortical contributions to the optokinetic reflex in the cat. In B. Lennerstrand, D. S. Zee & E. L. Keller (Eds.) *Functional Basis of Ocular Motility Disorders*, pp. 303–310. Pergamon Press, Oxford.

Hoffman, K.-P. & Distler, C. (1986). The role of direction selective cells in the nucleus of the optic tract of cat and monkey during optokinetic nystagmus. In E. L. Keller & D. S. Zee (Eds.) *Adaptive Processes in Visual and Oculomotor Systems*, pp. 261–266. Pergamon Press, Oxford.

Hoffman, K.-P. & Distler, C. (1989). Quantitative analysis of visual receptive fields of neurons in the nucleus of the optic tract and dorsal terminal nucleus of the accessory optic tract in macaque monkey. *J. Neurophysiol.*, **62**, 416–428.

Hoffman, K.-P. & Schoppman, A. (1981). A quantitative analysis of the direction-specific response of neurons in the cat's nucleus of the optic tract. *Exp. Brain Res.*, **42**, 146–157.

Huerta, M. F., Krubitzer, L. A. & Kaas, J. H. (1986). Frontal eye field as defined by intracortical microstimulation in squirrel monkeys, owl monkeys, and macaque monkeys. I. Subcortical connections. *J. Comp. Neurol.*, **253**, 415–439.

Hughes, A. (1971). Topographical relationships between the anatomy and physiology of the rabbit visual system. *Docum. Opthal. (den Haag)*, **30**, 33–159.

Hughes, A. (1972). A schematic eye for the rabbit. *Vision Res.*, **12**, 123–138.

Hughes, A. (1975). A quantitative analysis of the cat retinal ganglion cell topography. *J. Comp. Neurol.*, **163**, 107–128.

Hughes, A. (1976). A supplement to the cat schematic eye. *Vision Res.*, **16**, 149–154.

Itaya, S. K. & Van Hoesen, G. W. (1983). Retinal axons to the medial terminal nucleus of the accessory optic system in old world monkeys. *Brain Res.*, **269**, 361–364.

Judge, S. J., Richmond, B. J. & Chu, F. C. (1980). Implantation of magnetic search coils for measurement of eye position: an improved method. *Vision Res.*, **20**, 535–538.

Kase, M., Noda, H., Suzuki, D. & Miller, D. C. (1979). Target velocity signals of visual tracking in vermal Purkinje cells of the monkey. *Science*, **205**, 717–720.

Kato, I., Harada, K. Nakamura, T., Sato, Y. & Kawasaki, T. (1982). Role of the nucleus reticularis tegmenti pontis on visually induced eye movements. *Exp. Neurol.*, **78**, 503–516.

Kato, I., Harada, K., Hasegawa, T., Igarashi, T., Koike, Y. & Kawasaki, T. (1986). Role of the nucleus of the optic tract in monkeys in relation to optokinetic nystagmus. *Brain Res.*, **364**, 12–22.

Kato, I., Harada, K., Hasegawa, T. & Ikarashi, T. (1988). Role of the nucleus of the optic tract of monkeys in optokinetic nystagmus and optokinetic after-nystagmus. *Brain Res.*, **474**, 16–26.

Kawano, K. & Miles, F. A. (1986). Short-latency ocular following responses of monkey. II. Dependence on a prior saccadic eye movement. *J. Neurophysiol.*, **56**, 1355–1380.

Kawano, K., Sasaki, M. & Yamashita, M. (1984). Response properties of neurons in posterior parietal cortex of monkey during visual-vestibular stimulation. I. Visual tracking neurons. *J. Neurophysiol.*, **51**, 340–351.

Kawano, K., Shidara, M. & Yamane, S. (1992). Neural activity in dorsolateral pontine nucleus of alert monkey during ocular following responses. *J. Neurophysiol.*, **67**, 680–703.

Keating, E. G. (1991). Frontal eye field lesions impair predictive and visually-guided pursuit eye movements. *Exp. Brain Res.*, **86**, 311–323.

Keating, E. G. (1993). Lesions of the frontal eye field impair pursuit eye movements, but preserve the predictions driving them. *Behav. Brain Res.*, **53**, 91–104.

Keller, E. L. & Crandall, W. F. (1983). Neuronal responses to optokinetic stimuli in pontine nuclei of behaving monkey. *J. Neurophysiol.*, **49**, 169–187.

Keller, E. L. & Precht, W. (1979). Visual-vestibular responses in vestibular nucleus neurons in the intact and cerebellectomized, alert cat. *Neuroscience*, **4**, 1599–1613.

Koerner, F. & Schiller, P. H. (1972). The optokinetic response under open and closed loop conditions in the monkey. *Exp. Brain Res.*, **14**, 318–330.

Komatsu, H. & Wurtz, R. H. (1988a). Relation of cortical areas MT and MST to pursuit eye movements. I. Localization and visual properties of neurons. *J. Neurophysiol.*, **60**, 580–603.

Komatsu, H. & Wurtz, R. H. (1988b). Relation of cortical areas MT and MST to pursuit eye movements. III. Interaction with full-field visual stimulation. *J. Neurophysiol.*, **60**, 621–644.

Komatsu, H. & Wurtz, R. H. (1989). Modulation of pursuit eye movements by stimulation of cortical areas MT and MST. *J. Neurophysiol.*, **62**, 31–47.

Kreiger, H. P. & Bender, M. B. (1956). Optokinetic afternystagmus in the monkey. *EEG Clin. Neurophysiol.*, **8**, 97–106.

Kunzle, H. & Akert, K. (1977). Efferent connections of cortical area 8 (frontal eye field) in *Macaca fascicularis*. *J. Comp. Neurol.*, **173**, 147–164.

Langer, T., Fuchs, A. F., Scudder, C. A. & Chubb, M. C. (1985a). Afferents to the flocculus of the cerebellum in the rhesus macaque as revealed by retrograde transport of horseradish peroxidase. *J. Comp. Neurol.*, **235**, 1–25.

Langer, T., Fuchs, A. F., Chubb, M. C., Scudder, C. A. & Lisberger, S. G. (1985b). Floccular efferents in the rhesus macaque as revealed by autoradiography and horseradish peroxidase. *J. Comp. Neurol.*, **235**, 26–37.

Leichnetz, G. R. (1982). Inferior frontal eye field projections to the pursuit-related dorsolateral pontine nucleus and middle temporal area ('MT') in the monkey. *Vis. Neurosci.*, **207**, 394–402.

Leonard, C. S., Simpson, J. I. & Graf, W. (1988). Spatial organization of visual messages of the rabbit's cerebellar flocculus. I. Typology of inferior olive neurons of the dorsal cap of Kooy. *J. Neurophysiol.*, **60**, 2073–2090.

Lisberger, S. G. & Fuchs, A. F. (1978). Role of primate flocculus during rapid behavioral modification of the vestibulo-ocular reflex. I. Purkinje cell activity during visually guided horizontal smooth pursuit eye movements and passive head rotation. *J. Neurophysiol.*, **41**, 733–763.

Lisberger, S. G. & Pavelko, T. A. (1988). Brain stem neurons in modified pathways for motor learning in the primate vestibulo-ocular reflex. *Science*, **242**, 771–773.

Lisberger, S. G. & Westbrook, L. E. (1985). Properties of visual inputs that initiate horizontal smooth pursuit eye movements in monkeys. *J. Neurosci.*, **5**, 1662–1673.

Lisberger, S. G., Miles, F. A., Optican, L. M. & Eighmy, B. B. (1981a). Optokinetic response in monkey: Underlying mechanisms and their sensitivity to long-term adaptive changes in vestibuloocular reflex. *J. Neurophysiol.*, **45**, 869–890.

Lisberger, S. G., Evinger, C., Johnamson, G. W. & Fuchs, A. F. (1981b) Relationship between eye acceleration and retinal image velocity during foveal smooth pursuit eye movements in man and monkey. *J. Neurophysiol.*, **46**, 229–249.

Lund, J. S., Lund, R. D., Hendrickson, A. E., Bunt, A. H. & Fuchs, A. F. (1975). The origin of efferent pathways from the primary visual cortex, area 17, of the macaque monkey as shown by retrograde transport of horseradish peroxidase. *J. Comp. Neurol.*, **164**, 287–303.

Lynch, J. C. (1987). Frontal eye field lesions in monkey disrupt visual pursuit. *Exp. Brain Res.*, **68**, 437–441.

MacAvoy, M. G., Gottlieb, J. P. & Bruce, C. J. (1991). Smooth-pursuit eye movement representation in the primate frontal eye field. *Cerebral Cortex*, **1**, 95–102.

Maekawa, K. & Takeda, T. (1979). Origin of descending afferents to the rostral part of dorsal cap of inferior olive which transfers contralateral optic activities to the flocculus. An HRP study. *Brain Res.*, **172**, 393–405.

Maekawa, K., Takeda, T. & Kimura, M. (1981). Neural activity of nucleus reticularis tegmenti pontis. The origin of visual mossy fiber afferents to the cerebellar flocculus of rabbits. *Brain Res.*, **210**, 17–30.

Maekawa, K., Takeda, T. & Kimura, M. (1984). Responses of the nucleus of the optic tract neurons projecting to the nucleus reticularis tementi pontis upon optokinetic stimulation in the rabbit. *Neurosci. Res.*, **2**, 1–25.

Marcotte, R. R. & Updyke, B. V. (1982). Cortical visual areas of the cat project differentially onto the nuclei of the accessory optic system. *Brain Res.*, **242**, 205–217.

Maunsell, J. H. R. & Van Essen, D. C. (1983a). Functional properties of neurons in middle temporal visual area of macaque monkey. I. Selectivity for stimulus direction, speed and orientation. *J. Neurophysiol.*, **49**, 1127–1147.

Maunsell, J. H. R. & Van Essen, D. C. (1983b). Functional properties of neurons in middle temporal visual area of macaque monkey. II. Binocular interactions and sensitivity to binocular disparity. *J. Neurophysiol.*, **49**, 1148–1167.

May, J. G. & Andersen, R. A. (1986). Different patterns of corticopontine projections from separate cortical fields within the inferior parietal lobule and dorsal prelunate gyrus of the macaque. *Exp. Brain Res.*, **63**, 265–278.

May, J. G., Keller, E. L. & Crandall, W. F. (1985). Changes in eye velocity during smooth pursuit tracking induced by microstimulation in the dorsolateral pontine nucleus of the macaque. *Soc. Neurosci. Abstr.*, **11**, 79.

May, J. G., Keller, E. L. & Suzuki, D. A. (1988). Smooth-pursuit eye movement deficits with chemical lesions in the dorsolateral pontine nucleus of the monkey. *J. Neurophysiol.*, **59**, 952–977.

Mikami, A., Newsome, W. T. & Wurtz, R. H. (1986). Motion selectivity in macaque visual cortex. I. Mechanisms of direction and speed selectivity in extrastriate area MT. *J. Neurophysiol.*, **55**, 1308–1327.

Miles, F. A. (1993). The sensing of rotational and translational optic flow by the primate optokinetic system. In F. A. Miles & J. Wallman (Eds.) *Visual Motion and its Role in the Stabilization of Gaze*, pp. 393–403. Elsevier, Amsterdam.

Miles, F. A. & Fuller, J. H. (1975). Visual tracking and the primate flocculus. *Science*, **189**, 1000–1002.

Miles, F. A., Fuller, J. H., Braitman, D. J. & Dow, B. M. (1980). Long-term adaptive changes in primate vestibulo-ocular reflex. III. Electrophysiological observations in flocculus of normal monkey. *J. Neurophysiol.*, **43**, 1437–1476.

Miles, F. A., Kawano, K. & Optican, L. M. (1986). Short-latency ocular following responses of monkey. I. Dependence on temporospatial properties of visual input. *J. Neurophysiol.*, **56**, 1321–1354.

Miles, F. A., Schwarz, U. & Busettini, C. (1992). Decoding of optic flow by the primate optokinetic system. In A. Berthoz, P.-P. Vidal & W. Graf (Eds.) *The Head–neck Sensory-motor system*, pp. 471–478. Oxford University Press, New York.

Mizuno, N., Nakamura, Y. & Iwahori, N. (1974). An electron microscope study of the dorsal cap of the inferior olive in the rabbit, with special reference to the pretecto-olivary fibers. *Brain Res.*, **77**, 385–395.

Morris, E. J. & Lisberger, S. G. (1987). Different responses to small visual errors during initiation and maintenance of smooth-pursuit eye movements in monkeys. *J. Neurophysiol.*, **58**, 1351–1369.

Mustari, M. J. & Fuchs, A. F. (1989). Response properties of single units in the lateral terminal nucleus of the accessory optic system in the behaving primate. *J. Neurophysiol.*, **61**, 1207–1220.

Mustari, M. J. & Fuchs, A. F. (1990). Discharge patterns of neurons in the pretectal nucleus of the optic tract (NOT) in the behaving primate. *J. Neurophysiol.*, **64**, 77–90.

Mustari, M. J., Fuchs, A. F. & Wallman, J. (1988). Response properties of dorsolateral pontine units during smooth pursuit in the rhesus monkey. *J. Neurophysiol.*, **60**, 664–686.

Mustari, M. J., Fuchs, A. F & Kaneko, C. R. S. (1990). Descending connections of the macaque nucleus of the optic tract. *Soc. Neurosci. Abstr.*, **16**, 904.

Naegele, J. R. & Held, R. (1982). The postnatal development of monocular optokinetic nystagmus in infants. *Vision Res.*, **22**, 341–346.

Newsome, W. T., Wurtz, R. H., Dursteler, M. R. & Mikami, A. (1985). Deficits in visual motion processing following ibotenic acid lesions of the middle temporal visual area of the macaque monkey. *J. Neurosci.*, **5**, 825–840.

Newsome, W. T., Wurtz, R. H. & Komatsu, H. (1988). Relation of cortical areas MT and MST to pursuit eye movements. II. Differentiation of retinal from extraretinal inputs. *J. Neurophysiol.*, **60**, 604–620.

Noda, H. & Warabi, T. (1987). Responses of Purkinje cells and mossy fibers in the flocculus of the monkey during sinusoidal movements of a visual pattern. *J. Physiol.*, **387**, 611–628.

Noda, H., Sugita, S. & Ikeda, Y. (1990). Afferent and efferent connections of the oculomotor region of the fastigial nucleus in the macaque monkey. *J. Comp. Neurol.*, **302**, 330–348.

Oyster, C. W. (1968). The analysis of image motion by the rabbit retina. *J. Physiol.*, **199**, 613–635.

Oyster, C. W. & Barlow, H. B. (1967). Direction-selective units in rabbit retina: distribution of preferred directions. *Science*, **155**, 841–842.

Oyster, C. W., Takahashi, E. & Collewijn, H. (1972). Direction-selective retinal ganglion cells and control of optokinetic nystagmus in the rabbit. *Vision Res.*, **12**, 183–193.

Oyster, C. W., Simpson, J. I., Takahashi, E. & Soodak, R. E. (1980). Retinal ganglion cells projecting to the rabbit accessory optic system. *J. Comp. Neurol.*, **190**, 49–61.

Precht, W. & Strata, P. (1980). On the pathway mediating optokinetic responses in vestibular nuclear neurons. *Neuroscience*, **5**, 777–787.

Rashbass, C. (1961). The relationship between saccadic and smooth tracking eye movements. *J. Physiol. Lond.*, **159**, 326–338.

Robinson, D. A., Gordon, J. L. & Gordon, S. E. (1986). A model of the smooth pursuit eye movement system. *Biol. Cybernet.*, **55**, 43–57.

Robinson, F. R., Cohen, J. L., May, J., Sestokas, A. K. & Glickstein, M. (1984). Cerebellar targets of visual pontine cells in the cat. *J. Comp. Neurol.*, **223**, 471–482.

Roy, J.-P. & Wurtz, R. H. (1990). The role of disparity-sensitive cortical neurons in signalling the direction of self-motion. *Nature*, **348**, 160–161.

Sakata, H., Shibutani, H. & Kawano, K. (1983). Functional properties of visual tracking neurons in posterior parietal association cortex of the monkey. *J. Neurophysiol.*, **49**, 1364–1380.

Sato, Y. & Kawasaki, T. (1984). Functional localization in the three floccular zones related to eye movement control in the cat. *Brain Res.*, **290**, 25–31.

Sato, Y. & Kawasaki, T. (1990). Operational unit responsible for plane-specific control of eye movement by cerebellar flocculus in cat. *J. Neurophysiol.*, **64**, 551–564.

Sato, Y., Kawasaki, T. & Ikarashi, K. (1982). Zonal organization of the floccular Purkinje cells projecting to the group y of the vestibular nuclear complex and the lateral cerebellar nucleus in cats. *Brain Res.*, **234**, 430–434.

Sato, Y., Kawasaki, T. & Ikarashi, K. (1983). Afferent projections from the brainstem to the three floccular zones in cats. I. Climbing fiber projections. *Brain Res.*, **272**, 27–36.

Sato, Y., Kanda, K. & Kawasaki, T. (1988). Target neurons of floccular middle zone inhibition in medial vestibular nucleus. *Brain Res.*, **446**, 225–235.

Schor, C. M. & Levi, D. M. (1980). Disturbances of small-field horizontal and vertical optokinetic nystagmus. *Invest. Ophthalmol. Vis. Sci.*, **19**, 668–683.

Schwarz, U. & Miles, F. A. (1991). Ocular responses to translation and their dependence on viewing distance. I. Motion of the observer. *J. Neurophysiol.*, **66**, 851–864.

Schwarz, U., Busettini, C. & Miles, F. A (1989). Ocular responses to linear motion are inversely proportional to viewing distance. *Science*, **245**, 1394–1396.

Shidara, M. & Kawano, K. (1993). Role of Purkinje cells in the ventral paraflocculus in short-latency ocular following responses. *Exp. Brain Res.*, **93**, 185–195.

Simpson, J. I., Soodak, R. E. & Hess, R. (1979). The accessory optic system and its relation to the vestibulocerebellum. In R. Granit & O. Pompeiano (Eds.) *Reflex Control of Posture and Movement*, pp. 715–724, Elsevier.

Simpson, J. I., Graf, W. & Leonard, C. (1981). The coordinate system of visual climbing fibers to the flocculus. In A. F. Fuchs & W. Becker (Eds.) *Progress in Oculomotor Research*, pp. 475–484. Elsevier, Amsterdam.

Simpson, J. I., Leonard, C. S. & Soodak, R. E. (1988). The accessory optic system of rabbit. II. Spatial organization of direction selectivity. *J. Neurophysiol.*, **60**, 2055–2072.

Soodak, R. E. & Simpson, J. I. (1988). The accessory optic system of rabbit. I. Basic visual response properties. *J. Neurophysiol.*, **60**, 2037–2054.

Stone, L. S. & Lisberger, S. G. (1990). Visual responses of Purkinje cells in the cerebellar flocculus during smooth pursuit eye movements in monkeys. I. Simple spikes. *J. Neurophysiol.*, **63**, 1241–1261.

Suzuki, D. A. & Keller, E. L. (1983). Sensory-oculomotor interactions in primate cerebellar vermis: a role in smooth pursuit control. *Soc. Neurosci. Abstr.*, **9**, 606.

Suzuki, D. A. & Keller, E. L. (1984). Visual signals in the dorsolateral pontine nucleus of the alert monkey: their relationship to smooth-pursuit eye movements. *Exp. Brain Res.*, **53**, 473–478.

Suzuki, D. A. & Keller, E. L. (1988a). The role of the posterior vermis of monkey cerebellum in smooth-pursuit eye movement control. I. Eye and head movement related activity. *J. Neurophysiol.*, **59**, 1–18.

Suzuki, D. A. & Keller, E. L. (1988b). The role of the posterior vermis of monkey cerebellum in smooth-pursuit eye movement control. II. Target velocity related Purkinje cell activity. *J. Neurophysiol.*, **59**, 19–40.

Takahashi, M. & Igarashi, M. (1977). Comparison of vertical and horizontal optokinetic nystagmus in the squirrel monkey. *Oto-Rhino-Laryngol.*, **39**, 321–329.

Takeda, T. & Maekawa, K. (1976). The origin of the pretecto-olivary tract. A study using the horseradish peroxidase method. *Brain Res.*, **117**, 319–325.

Their, P., Koehler, W. & Buettner, U. W. (1988). Neuronal activity in the dorsolateral pontine nucleus of the alert monkey modulated by visual stimuli and eye movements. *Exp. Brain Res.*, **70**, 496–512.

Torigoe, Y., Blanks, R. H. I. & Precht, W. (1986). Anatomical studies on the nucleus reticularis tegmenti pontis in the pigmented rat. II. Subcortical afferents demonstrated by the retrograde transport of horseradish peroxidase. *J. Comp. Neurol.*, **243**, 88–105.

van Hof-van Duin, J. (1976). Early and permanent effects of monocular deprivation on pattern discrimination and visuo-motor behavior in cats. *Brain Res.*, **111**, 261–276.

van Hof-van Duin, J. (1978). Direction preference of optokinetic responses in monocularly tested normal kittens and light deprived cats. *Arch. Ital. Biol.*, **116**, 472–477.

Walberg, F., Nordby, T., Hoffmann, K.-P. & Hollander, H. (1981). Olivary afferents from the pretectal nuclei in the cat. *Anat. Embryol.*, **161**, 291–304.

Webb, S. V. & Kaas, J. H. (1976). The sizes and distribution of ganglion cells in the retina of the owl monkey, *Aotus trivirgatus*. *Vision Res.*, **16**, 1247–1254.

Weber, J. T. (1985). Pretectal complex and accessory optic system of primates. *Brain Behav. Evol.*, **26**, 117–140.

Weber, J. T. & Giolli, R. A. (1986). The median terminal nucleus of the monkey: evidence for a 'complete' accessory optic system. *Brain Res.*, **365**, 164–168.

Westheimer, G. & McKee, S. P. (1975). Visual acuity in the presence of retinal image motion. *J. Opt. Soc. Am.*, **65**, 847–850.

Wood, L. L., Spear, P. D. & Braun, J. J. (1973). Direction-specific deficits in horizontal optokinetic nystagmus following removal of visual cortex in the cat. *Brain Res.*, **60**, 231–237.

Yamada, J. & Noda, H. (1987). Afferent and efferent connections of the oculomotor cerebellar vermis in macaque monkey. *J. Comp. Neurol.*, **265**, 224–241.

Zee, D. S., Yamazaki, A., Butler, P. H. & Gucer, G. (1981). Effects of ablation of the flocculus and paraflocculus on eye movements in the primate. *J. Neurophysiol.*, **46**, 878–899.

Zee, D. S., Tusa, R. J., Herdman, S. J., Butler, P. H. & Gucer, G. (1987). Effects of occipital lobectomy upon eye movements in primate. *J. Neurophysiol.*, **58**, 883–907.

Index